Modeling and Computing
for
Geotechnical Engineering

An Introduction

M.S. Rahman

North Carolina State University
Department of Civil, Construction and Environmental Engineering
Raleigh, North Carolina, USA

M.B. Can Ülker

Istanbul Technical University
Earthquake Engineering and Disaster Management Institute
Istanbul, Turkey

CRC Press
Taylor & Francis Group
Boca Raton   London   New York

CRC Press is an imprint of the
Taylor & Francis Group, an **informa** business

A SCIENCE PUBLISHERS BOOK

CRC Press
Taylor & Francis Group
6000 Broken Sound Parkway NW, Suite 300
Boca Raton, FL 33487-2742

First issued in paperback 2021

© 2018 by Taylor & Francis Group, LLC
CRC Press is an imprint of Taylor & Francis Group, an Informa business

No claim to original U.S. Government works

Version Date: 20180519

ISBN 13: 978-1-4987-4541-3 (hbk)
ISBN 13: 978-0-367-78090-6 (pbk)

Visit the Taylor & Francis Web site at
http://www.taylorandfrancis.com

and the CRC Press Web site at
http://www.crcpress.com

Dedicated to

My Wife: Nuzhat for her encouragement and support
My Children: Feraz & Asra, Akef & Sadia and Asem & Soha for their love and respect
My Grand Children: Zaki, Noor and Tawfiq for being the hope and meaning of the future!

M.S. Rahman

Fidan: my lovely wife, the love of my life and my best friend
whose constant support and immense patience have made this book possible
Peren Candan: my precious little daughter, my PeCan pie, my everything
whose birth has made my life such a blessing!

M.B. Can Ülker

MATLAB® codes and MAPLE® worksheets are available for those who have bought the book. Please contact the author at mbulker@itu.edu.tr or canulker@gmail.com. Kindly provide the invoice number and date of purchase.

Acknowledgements

This book has evolved from our lecture notes for an introductory graduate course on 'Modeling and Computing for Geotechnical Engineering'. While teaching this course we have made use of various resources, which we would like to acknowledge. Here, we wish to mention some of these. The works of late professor John Booker and his colleagues (Booker and Small, 1985) have inspired the chapter on the Semi Analytical Method. For the chapter on the Finite Difference Method, we have borrowed materials from G.D. Smith (1975), and C.S. Desai and J. Christian (1977). For the chapter on the Finite Element Method, we have referred to J.N. Reddy (1993), and borrowed, modified and adapted a few MATLAB® programs from Y. Hwon and H. Bang (1996).

We would also like to thank our graduate students who helped us in this book, namely Giray Baksı for sketching most of the crucial figures and Esra Tatlıoğlu for checking the overall format. Our sincere thanks go to Vijay Primlani of CRC for all he did to keep us engaged in this adventure of ours.

Preface

Modeling and computing are now frequently used in Geotechnical Engineering. Often readily available computer packages are used for the analysis of real world geotechnical problems, and the results from these, along with traditional methods supported by engineering judgement, form the basis for the safe and economic design of geotechnical engineering systems.

This book has been written primarily to introduce the basic aspects of 'modeling and computing' for geotechnical problems to those who do not have adequate background in this area. It is based on the lecture notes prepared during the teachings of a lower level graduate course with the same title in the Department of Civil, Construction and Environmental Engineering at North Carolina State University and also taught with a slightly modified name in the Institute of Earthquake Engineering and Disaster Management at Istanbul Technical University. We believe this book will be quite useful to both graduate students and practicing geotechnical engineers who want to acquire basic knowledge of modeling and computing and make use of them in solving important problems. It will also facilitate their learning of advanced topics and motivate them in adapting such topics in their routine calculations, either through the use of various computer packages commonly available in practice or by direct implementation of the underlying theories into computer programs.

There are three major sections in the book presenting the necessary elements of modeling and computing for a geotechnical engineering problem. The first section titled "Basic Mechanics" provides a review of the basics of continuum mechanics including states of stresses and strains as well as physical laws yielding the equations governing the static response of geotechnical engineering systems. In the second main section titled "Elemental Response: Constitutive Models", four types of such models are discussed: Chapter 4 (Elasticity), Chapter 5 (Plasticity) and Chapter 6 (Viscoelasticity and Viscoplasticity). In the third section, "System Response: Methods of Analyses", there are four chapters: Chapter 7 (Analytical Methods), Chapter 8 (Semi-Analytical Methods), Chapter 9 (Finite Difference Methods), and Chapter 10 (Finite Element Method). The focus in this last section is on the formulation of mathematical models for some frequently encountered geotechnical engineering problems as well as the analytical and numerical methods utilized for solving associated governing equations. The solution procedures are subsequently implemented into a number of computer programs, which are developed using the software packages; MAPLE® and MATLAB®. Both packages are increasingly used in many engineering institutions and also in practice, thus we believe many readers will find these programs quite useful. Please note that the codes placed in the book may look different and require adjustment of some of the lines due to margins. We hope that this book will benefit the graduate students, the practicing engineers in the field and the geotechnical engineering community at large.

Raleigh, NC, U.S.A.   **M.S. Rahman**
Istanbul, Turkey   **M.B. Can Ülker**
October, 2017

Contents

Introduction

Modeling and computing have become an essential part of the analysis and design of engineering systems. This is also becoming the case for 'Geotechnical systems' such as foundations, earth dams, tunnels, retaining walls and other soil-structure systems. In Figure 1.1, there are a few examples of such systems.

In a very generalized sense, the essential objective of 'modeling and computing' of an engineering system is to understand and predict the behavior of a system subjected to a variety of possible conditions/scenarios (with respect to both external stimuli and system parameters) which provide the basis for a rational design. Such understanding is achieved by developing a model for the actual system and then solving/computing for the response of the system under different set of conditions (i.e. simulation). The essence of this is to predict the response for a set of external stimuli (i.e. forces), which is obtained by mapping the variables (representing both the external forces and the system parameters) in the input space to a set of response variables in the output space. A general representation of all the systems identified in Figure 1.1 is abstracted as shown in Figure 1.2.

The modeling (and computing) of a *physical* system involves the following phases:

1. Idealization of the actual physical problem to be studied by considering the important aspects and ignoring those which may not be as important.
2. Formulation of a mathematical model in terms of a set of equations governing the global response of the idealized system and defining its initial and boundary conditions by: (a) Involving all the relevant physical laws and constraints, and (b) Using the constitutive laws governing the behavior of the system at the *elemental* level.
3. Evaluating the response of the system by: (a) Developing a solution to the governing equations, (b) Computing the key response variables through implementing the solution schemes into a computer program and (c) Graphically presenting and visualizing the results. General steps of mathematical modeling are shown in Figure 1.3. All three phases mentioned above are necessary in developing the numerical models of several problems in geotechnical engineering and geomechanics. Analysis of these problems can be abstracted to the prediction of the response involving coupled flow and deformation of saturated or partially saturated soils or soil-structure systems subjected to external loadings (including failure) as shown in Figure 1.4.

In this book, the focus is on the formulation of mathematical models for some common geotechnical problems. The relevant physical laws and models for the constitutive behavior are introduced. The schemes for solving the governing equations utilized to develop numerical models are also presented. With reference to the method of solution, three general classes of methods are considered: (i) *Analytical methods*, (ii) *Semi-analytical methods* and (iii) *Numerical methods*. Their solution procedures are subsequently implemented into a number of computer programs, which are developed in computing packages of MAPLE® and MATLAB®.

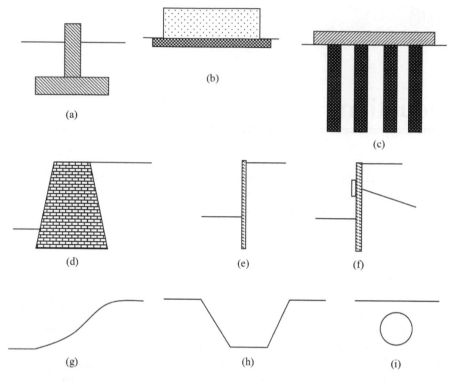

Figure 1.1. Typical geotechnical systems: (a) Shallow foundation, a footing, (b) Shallow foundation, a mat, (c) Pile raft, (d) Gravity retaining wall, (e) Cantilever sheet pile wall, (f) Anchored cantilever sheet pile wall, (g) Man-made or natural slope, (h) Open excavation, (i) Pipeline

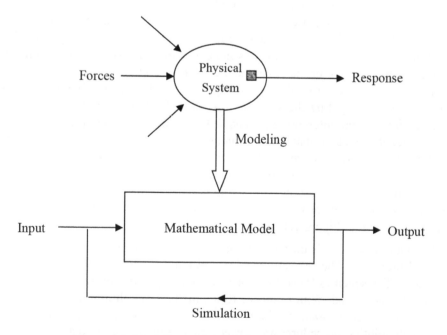

Figure 1.2. Modeling and simulation of a physical system

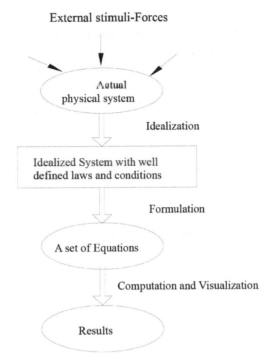

External stimuli-Forces

Actual physical system

Idealization

Idealized System with well defined laws and conditions

Formulation

A set of Equations

Computation and Visualization

Results

Figure 1.3. Steps of mathematical modeling

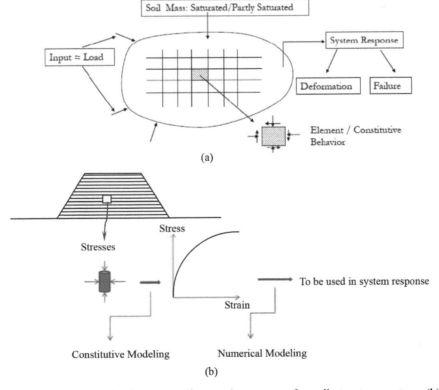

Soil Mass: Saturated/Partly Saturated

Input ≈ Load

System Response

Deformation   Failure

Element / Constitutive Behavior

(a)

Stress

Stresses

Strain

To be used in system response

Constitutive Modeling   Numerical Modeling

(b)

Figure 1.4. (a) Evaluation steps of general response of a soil-structure system, (b) Example of an embankment problem

The book is organized in three major sections presenting the essential elements of modeling and computing for a geotechnical engineering problem. In the first section titled "Basic Mechanics" where there are two chapters, we will start by stating the basics of mechanics in terms of states of stresses and strains as well as physical laws and the equations governing the analysis of a geotechnical engineering system in terms of conservation of momentum and mass. In the second main section called "Elemental Response: Constitutive Models", four types of models are discussed in these chapters: Chapter 4 (Elasticity), Chapter 5 (Plasticity) and Chapter 6 (Viscoelasticity and Viscoplasticity). In the third section titled "System Response: Methods of Analyses", there are four chapters: Chapter 7 (Analytical Methods), Chapter 8 (Semi-Analytical Methods), Chapter 9 (Finite Difference Methods), and Chapter 10 (Finite Element Method).

This introductory book does not cover two very important but relatively advanced topics: (i) Dynamic response analysis and (ii) Nonlinear response analysis of problems in geotechnical engineering. Authors are aiming to cover these two concepts along with other advanced topics within the subjects of a second book which is already in preparation. Surely, such a project will be a natural continuation to this first book.

PART I

Basic Mechanics

2

Stresses and Strains

2.1 Introduction

In this chapter, we first summarize the basic concepts from continuum mechanics followed by their adaptations for the mechanics of fluid-filled particulate (porous) media (soils) which we idealize as continua.

Most of the basic concepts utilized in this book involve treating the soil as a continuum. The continuum assumption, disregarding the discrete nature at small scale, implies that any material attribute is continuously distributed and defined at all points in a material body. For example the mass density, ρ, of a soil can be associated with any point, $P(x, y, z)$ in the body and is defined as the limiting ratio of an elemental mass, ΔM to volume, ΔV as,

$$\rho = \lim_{\Delta V \to 0} \frac{\Delta M}{\Delta V} \qquad (2.1.1)$$

It should be noted that if we reduce the change in volume, ΔV to zero, in a real soil we will find ρ to be varying hugely, depending upon the location of the point whether it is on the solid particle, water or air. Therefore, we simply consider the density defined above as a representative average value over a finite elemental volume, ΔV (Figure 2.1-1).The same continuum notion is also adopted for all other quantities of interest.

Certain other definitions are made at this point to characterize such a material. A *homogeneous material* is one having identical properties at all points in its volume. With respect to its properties, a material is *isotropic* if that property is the same in all directions at a point. A soil mass/layer is *anisotropic* if its physical or engineering properties vary in different directions. The forces, which act on all elements of the volume of a continuum are known as *body forces*. Some examples are gravity and inertia forces. These forces are represented by the

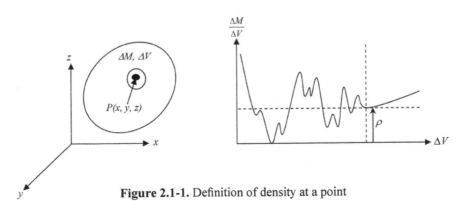

Figure 2.1-1. Definition of density at a point

symbol b_i (force per unit mass), or as γ_i (force per unit volume). They are related through the density by the equation,

$$\rho b_i = \gamma_i \text{ or, } \rho b = \gamma \qquad (2.1.2)$$

Surface forces (tractions) are the forces acting on a surface element, whether it is a portion of the bounding surface of the continuum or an arbitrary internal surface. These are designated by f_i (force per unit area). *Contact forces* between bodies are the other type of surface forces.

2.2 Reference Coordinate System: Notations

Our goal is to analyze the effects of loads on a material body. In order to do this, we must first adopt a coordinate system with reference to which we can define the location of a point, the stretch of a line segment and the deformation of an element of the material body. For simplicity, we will only make use of a standard orthogonal, rectangular and right handed coordinate system as shown below in Figure 2.2-1.

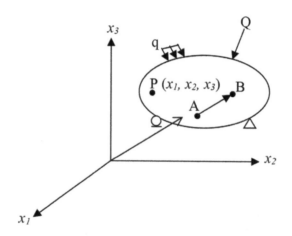

Figure 2.2-1. Reference coordinate system

The location of a point P is uniquely defined by a set of coordinates x_1, x_2, x_3 with reference to the chosen axes, which we can write as:

$$\{x\} = [x_i] = \begin{bmatrix} x_1 \\ x_2 \\ x_3 \end{bmatrix} \qquad (2.2.1)$$

Similarly, we can define other points $A(a_i)$ and $B(b_i)$ as:

$$\{a\} = [a_i] = \begin{bmatrix} a_1 \\ a_2 \\ a_3 \end{bmatrix} \qquad (2.2.2a)$$

$$\{b\} = [b_i] = \begin{bmatrix} b_1 \\ b_2 \\ b_3 \end{bmatrix} \qquad (2.2.2b)$$

Note that in the above, we are using both the indicial and matrix notations where the index i takes the values of $i = 1, 2, 3$. A usual convention is that if an index is repeated twice, a summation is implied. For example:

$$a_i b_i = a_1 b_1 + a_2 b_2 + a_3 b_3 \tag{2.2.3}$$

which, in matrix notation, equals:

$$a_i b_i = \{a\}^T \{b\} \tag{2.2.4}$$

For the matrix notation, recall that 'T' stands for transpose and note that:

$$\{a\}^T \{b\} \neq \{b\}\{a\}^T \tag{2.2.5}$$

However,

$$a_i b_i = b_i a_i \tag{2.2.6a}$$

$$\{a\}^T \{b\} = \{b\}^T \{a\} \tag{2.2.6b}$$

Next, we look at the coordinate changes (transformations) with the change of reference coordinate system. First let us consider just a translation of the coordinate system (Figure 2.2-2a). Let x_i represent the coordinates of any point in the old coordinate system, and x_i' represent the coordinates of the same point in the new system. We can write,

$$x_i = x_i' + c_i \tag{2.2.7a}$$

$$x_i' = x_i - c_i \tag{2.2.7b}$$

where c_i represents the translation of the old origin O to new origin O'. Next, let us consider the rotation of the coordinate system (Figure 2.2-2b) where we can write x_i' as a linear function of x_i as:

$$\begin{aligned} x_1' &= L_{11}x_1 + L_{12}x_2 + L_{13}x_3 \\ x_2' &= L_{21}x_1 + L_{22}x_2 + L_{23}x_3 \\ x_3' &= L_{31}x_1 + L_{32}x_2 + L_{33}x_3 \end{aligned} \tag{2.2.8}$$

Using the index notation or the matrix notation we can rewrite the above equations as:

$$x_i' = L_{ij}x_j \tag{2.2.9}$$

or

$$\{x'\} = [L]\{x\}$$

In the above, $[L]$ is a transformation matrix which depends only on the magnitude of rotation. From the geometry alone the transformation matrix can be represented as:

$$[L] = \begin{bmatrix} e_{11} & e_{21} & e_{31} \\ e_{12} & e_{22} & e_{32} \\ e_{13} & e_{23} & e_{33} \end{bmatrix} = \begin{bmatrix} \cos(\theta_{11}) & \cos(\theta_{21}) & \cos(\theta_{31}) \\ \cos(\theta_{12}) & \cos(\theta_{22}) & \cos(\theta_{32}) \\ \cos(\theta_{13}) & \cos(\theta_{23}) & \cos(\theta_{33}) \end{bmatrix} \tag{2.2.10}$$

where θ_{ij} is the angle between x_i' axis and x_j axis (Figure 2.2-2b). For the general case, where both translation and rotation are involved (Figure 2.2-2c), the coordinate transformation can be written as:

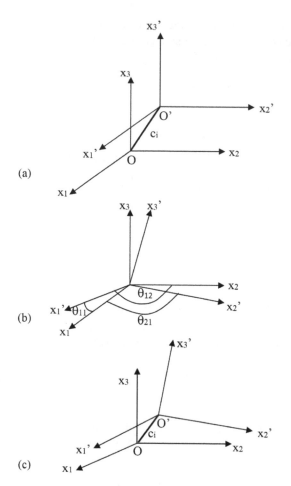

Figure 2.2-2. Coordinate transformation: (a) Translation,
(b) Rotation, (c) Combined translation and rotation

$$x_i' = L_{ij}(x_i - c_i)$$ (2.2.11a)

$$\{x'\} = [L]\big[\{x\} - \{c\}\big]$$ (2.2.11b)

2.3 Strains

The response of a material body subjected to forces is manifested in its deformation. Thus, a measure of the deformation which is independent of any rigid body motion is defined. Such a measure is provided by a *strain tensor* which defines the state of *deformation* of the body completely. Figure 2.3-1a illustrates a typical deformed body, which experiences strain under several forces.

2.3.1 Strain Tensor: Mathematical Definition of Strains

Let us consider a material body in both undeformed (reference configuration) and deformed state in the fixed coordinate system as shown in Figure 2.3-1b. The line segment PQ of length

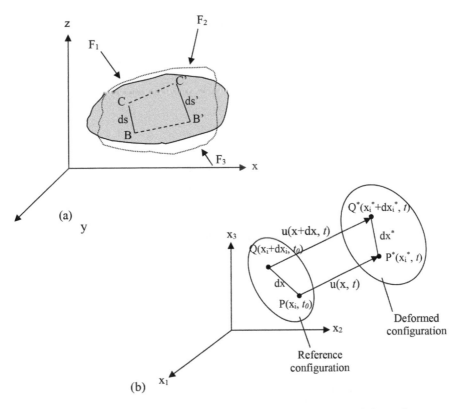

Figure 2.3-1. (a) Solid body under forces, (b) Deformed and non-deformed states

ds is in a non-deformed configuration which moves to P^*Q^* with length ds^* in the deformed configuration. We identify the points P, Q, P^* and Q^* as:

$$P: \; x_i = \{x\} = \{x_1 \quad x_2 \quad x_3\}^T \tag{2.3.1a}$$

$$Q: \; x_i + dx_i = \{x + dx\} = \{x_1 + dx_1 \quad x_2 + dx_2 \quad x_3 + dx_3\}^T \tag{2.3.1b}$$

$$P^*: \; x_i^* = x_i + u_i = \{x + u\} = \{x_1 + u_1 \quad x_2 + u_2 \quad x_3 + u_3\}^T \tag{2.3.1c}$$

In Eq. 2.3.1 the displacements $u_i = u_i(x_i, t)$ are referred to the undeformed reference (Lagrangian) configuration, x_i and time t. We have:

$$Q^*: \; x_i^* + dx_i^* = x_i + dx_i + u_i \{x_i + dx_i, t\} \tag{2.3.1d}$$

$$dx_i^* = dx_i + du_i = dx_i + \frac{\partial u_i}{\partial x_j} dx_j \tag{2.3.1e}$$

$$dx_i^* = \left(\delta_{ij} + \frac{\partial u_i}{\partial x_j} \right) dx_j \tag{2.3.1f}$$

where

$$\delta_{ij} = \begin{bmatrix} 1, \, i = j \\ 0, \, i \neq j \end{bmatrix} \tag{2.3.1g}$$

is the Kronecker delta. The above relation can be equivalently written in matrix notation as:

$$\{dx^*\} = \{[I] + [J]\}\{dx\} \tag{2.3.2}$$

with,

$$\delta_{ij} = [I] = \begin{bmatrix} 1 & 0 & 0 \\ 0 & 1 & 0 \\ 0 & 0 & 1 \end{bmatrix} \tag{2.3.3}$$

$$\frac{\partial u_i}{\partial x_j} = \frac{\partial u_j}{\partial x_i} = [J] = \begin{bmatrix} \dfrac{\partial u_1}{\partial x_1} & \dfrac{\partial u_1}{\partial x_2} & \dfrac{\partial u_1}{\partial x_3} \\[2mm] \dfrac{\partial u_2}{\partial x_1} & \dfrac{\partial u_2}{\partial x_2} & \dfrac{\partial u_2}{\partial x_3} \\[2mm] \dfrac{\partial u_3}{\partial x_1} & \dfrac{\partial u_3}{\partial x_2} & \dfrac{\partial u_3}{\partial x_3} \end{bmatrix} \tag{2.3.4}$$

In the above the vector dx_i changes to dx_i^* due to deformation and the length of these vectors are ds $(\overline{PQ})$ and ds^* $(\overline{P^*Q^*})$ We can then write:

$$ds^2 = dx_j dx_j = \delta_{ij} dx_i dx_j = \{dx\}^T \{dx\} \tag{2.3.5}$$

$$ds^{*2} = dx_k^* dx_k^* = \left(\delta_{kj} + \frac{\partial u_k}{\partial x_j}\right)\left(\delta_{ki} + \frac{\partial u_k}{\partial x_i}\right) dx_j dx_i \tag{2.3.6}$$

$$ds^{*2} - ds^2 = \left(\frac{\partial u_i}{\partial x_j} + \frac{\partial u_j}{\partial x_i} + \frac{\partial u_k}{\partial x_i}\frac{\partial u_k}{\partial x_j}\right) dx_i dx_j \tag{2.3.7a}$$

or

$$ds^{*2} - ds^2 = 2 dx_i E_{ij} dx_j \tag{2.3.7b}$$

In (2.3.7a), for the term in the parenthesis, E_{ij} is introduced which is a set of six linearly independent quantities collectively representing the measure of strains in a general sense. This 'strain tensor' (and the 'stress tensor' to be discussed in the next section) undergoes a transformation in a special manner due to the changes in reference axes. E_{ij} is called the strain tensor here and written as:

$$E_{ij} = \frac{1}{2}\left(\frac{\partial u_i}{\partial x_j} + \frac{\partial u_j}{\partial x_i} + \frac{\partial u_k}{\partial x_i}\frac{\partial u_k}{\partial x_j}\right) \tag{2.3.8}$$

E_{ij} is symmetric, i.e. $E_{ij} = E_{ji}$. This strain tensor was introduced by Green and St. Venant and is often called the *Green strain tensor*. The above can be written in matrix form as:

$$ds^{*2} - ds^2 = \{dx_1 \ dx_2 \ dx_3\}\left[[J] + [J]^T + [J]^T[J]\right]\begin{Bmatrix} dx_1 \\ dx_2 \\ dx_3 \end{Bmatrix} \tag{2.3.9a}$$

$$E_{ij} = \frac{1}{2}\left[[J] + [J]^T + \left[J^T\right][J]\right] \tag{2.3.9b}$$

2.3.2 Small Strain Tensor

We will consider only those situations where the displacement gradients u_{ij} are small:

$$|u_{i,j}| \ll 1 \qquad (2.3.10)$$

In this case the quadratic terms in equations (2.3.7) and (2.3.8) can be ignored and the *Lagrange Green Strain* tensor E_{ij} defined above can be approximated by small strain tensor ε_{ij} as:

$$\varepsilon_{ij} = \frac{1}{2}(u_{i,j} + u_{j,i}) \qquad (2.3.11)$$

or in matrix notation,

$$\left[\varepsilon_{ij}\right] = \frac{1}{2}\left[[J] + [J]^T\right] \qquad (2.3.12)$$

$$\left[\varepsilon_{ij}\right] = \begin{bmatrix} \varepsilon_{11} & \varepsilon_{12} & \varepsilon_{13} \\ \varepsilon_{21} & \varepsilon_{22} & \varepsilon_{23} \\ \varepsilon_{31} & \varepsilon_{32} & \varepsilon_{33} \end{bmatrix} = \begin{bmatrix} \varepsilon_{xx} & 0.5\gamma_{xy} & 0.5\gamma_{xz} \\ 0.5\gamma_{yx} & \varepsilon_{yy} & 0.5\gamma_{yz} \\ 0.5\gamma_{zx} & 0.5\gamma_{zy} & \varepsilon_{zz} \end{bmatrix}$$

$$\begin{bmatrix} \dfrac{\partial u_1}{\partial x_1} & \dfrac{1}{2}\left(\dfrac{\partial u_1}{\partial x_2} + \dfrac{\partial u_2}{\partial x_1}\right) & \dfrac{1}{2}\left(\dfrac{\partial u_1}{\partial x_3} + \dfrac{\partial u_3}{\partial x_1}\right) \\[3ex] \dfrac{1}{2}\left(\dfrac{\partial u_1}{\partial x_2} + \dfrac{\partial u_2}{\partial x_1}\right) & \dfrac{\partial u_2}{\partial x_2} & \dfrac{1}{2}\left(\dfrac{\partial u_2}{\partial x_3} + \dfrac{\partial u_3}{\partial x_2}\right) \\[3ex] \dfrac{1}{2}\left(\dfrac{\partial u_1}{\partial x_3} + \dfrac{\partial u_3}{\partial x_1}\right) & \dfrac{1}{2}\left(\dfrac{\partial u_3}{\partial x_2} + \dfrac{\partial u_2}{\partial x_3}\right) & \dfrac{\partial u_3}{\partial x_3} \end{bmatrix} \qquad (2.3.13)$$

It is observed that the small strain tensor is symmetric (i.e. $\varepsilon_{ij} = \varepsilon_{ji}$). It can be easily demonstrated that in the case of only a rigid body motion (i.e. $ds^{*2} - ds^2 = 0$)

$$E_{ij} = \varepsilon_{ij} = 0 \qquad (2.3.14)$$

This implies that a rigid body motion does not affect Green-Lagrange strain tensor. In (2.3.13), ε_{ii} are the normal strains with ε_{11} being the unit elongation for an element originally in the x_1 direction, ε_{22} being the unit elongation for an element originally in the x_2 direction and ε_{33} being the unit elongation for an element originally in the x_3 direction. Similarly, ε_{ij} represent the shear strains with $2\,\varepsilon_{12}$ being the decrease in angle between two elements initially in the x_1 and x_2 directions, $2\,\varepsilon_{13}$ being the decrease in angle between two elements initially in the x_1 and x_3 directions and finally $2\,\varepsilon_{23}$ being the decrease in angle between two elements initially in the x_2 and x_3 directions.

2.3.3 Physical Meaning of the Strain Tensor

For small displacement gradients (i.e. $E_{ij} = \varepsilon_{ji}$), we can write equation (2.3.7) as:

$$\frac{ds^{*2} - ds^2}{2ds^2} = \frac{dx_i}{ds}\,\varepsilon_{ij}\,\frac{dx_j}{ds} \qquad (2.3.15)$$

$$\frac{(ds^* + ds)(ds^* - ds)}{2ds^2} = n_i\varepsilon_{ij}n_j \qquad (2.3.16)$$

where n_i are the components of the unit vector along $\overline{PQ}$. Note the following:

$$\overline{PQ} = dx_1\hat{i}_1 + dx_2\hat{i}_2 + dx_3\hat{i}_3 \tag{2.3.17a}$$

$$\frac{\overline{PQ}}{\left|\overline{PQ}\right|} = \frac{dx_1}{dj}\hat{i}_1 + \frac{dx_2}{dj}\hat{i}_2 + \frac{dx_3}{dj}\hat{i}_3 \tag{2.3.17b}$$

$$\hat{i}_s = n_1\hat{i}_1 + n_2\hat{i}_2 + n_3\hat{i}_3 \tag{2.3.17c}$$

For small strains, $ds^* = ds$, therefore we can write,

$$\frac{ds^* - ds}{ds} = n_i \varepsilon_{ij} n_j \tag{2.3.18}$$

We identify the left hand side term the "relative elongation", ε, of vector $\overline{PQ}$ which can be seen as the change in length (of the line element PQ) per unit length. Therefore we can now write:

$$\varepsilon = n_i \varepsilon_{ij} n_j \tag{2.3.19}$$

or in matrix form,

$$\varepsilon = \{n\}^T [\varepsilon]\{n\} \tag{2.3.20}$$

To interpret the components of the strain tensor, ε_{ij}, let us choose the line element PQ (of unit length $\overline{PQ} = ds = 1$) aligned with x_2-axis (Figure 2.3-2).
For this case note that,

$$\overline{PQ} = 0\hat{i}_1 + 1\hat{i}_2 + 0\hat{i}_3 \tag{2.3.21}$$

and

$$n_i = \{n\}^T = [0 \quad 1 \quad 0] \tag{2.3.22}$$

Therefore using 2.3.20,

$$\varepsilon = [0 \quad 1 \quad 0] \begin{bmatrix} \varepsilon_{11} & \varepsilon_{12} & \varepsilon_{13} \\ \varepsilon_{21} & \varepsilon_{22} & \varepsilon_{23} \\ \varepsilon_{31} & \varepsilon_{32} & \varepsilon_{33} \end{bmatrix} \begin{Bmatrix} 0 \\ 1 \\ 0 \end{Bmatrix} \tag{2.3.23}$$

and we have:

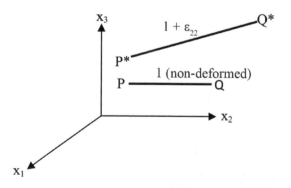

Figure 2.3-2. Definition of strain component ε_{22}

$$\frac{ds^* - ds}{ds} = \varepsilon_{22} \tag{2.3.24a}$$

$$ds^* = 1 + \varepsilon_{22} \tag{2.3.24b}$$

This means that ε_{22} represents the strain component along x_2. Similarly choosing the line segments along x_1 and x_2, we see that the diagonal terms of strain tensor represent the normal strains along the reference directions. Now to interpret the diagonal terms of the strain tensor let us consider two line segments $d_s^{(2)}$ and $d_s^{(3)}$ to evaluate the strain below,

$$ds^{(2)} = 0\hat{i}_1 + 1\hat{i}_2 + 0\hat{i}_3 \tag{2.3.25a}$$

$$ds^{(3)} = 0\hat{i}_1 + 0\hat{i}_2 + 1\hat{i}_3 \tag{2.3.25b}$$

$$n_i^{(2)} = \frac{\partial x_i^2}{\partial s^{(2)}} = \{n_i^{(2)}\}^T = [0 \quad 1 \quad 0] \tag{2.3.26a}$$

$$n_i^{(3)} = \frac{\partial x_i^3}{\partial s^{(3)}} = \{n_i^{(3)}\}^T = [0 \quad 0 \quad 1] \tag{2.3.26b}$$

In the reference configuration, the two line segments are orthogonal, therefore we have,

$$n_i^{(2)} . n_i^{(3)} = 0 \tag{2.3.27}$$

In the deformed configuration (Figure 2.3-3), it is,

$$\cos\left(\frac{\pi}{2} - \gamma_{23}\right) = \frac{dx_i^{*(2)}}{ds^{*(2)}} \frac{dx_i^{*(3)}}{ds^{*(3)}} \tag{2.3.28}$$

From equation (2.3.1f),

$$dx_i^{*(2)} = \left(\delta_{ij} + \frac{\partial u_i}{\partial x_j}\right) dx_j^{(2)} \tag{2.3.29a}$$

$$dx_i^{*(3)} = \left(\delta_{ik} + \frac{\partial u_i}{\partial x_k}\right) dx_k^{(3)} \tag{2.3.29b}$$

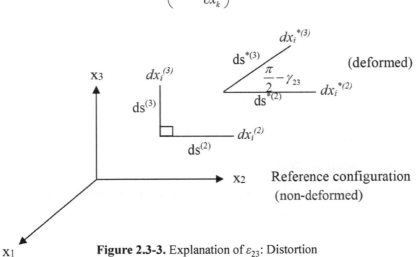

Figure 2.3-3. Explanation of ε_{23}: Distortion

Combining (2.3.28) and (2.3.29), we can write:

$$\sin(\gamma_{23}) = \left(\delta_{jk} + \frac{\partial u_k}{\partial x_j} + \frac{\partial u_j}{\partial x_k} + \frac{\partial u_i}{\partial x_j}\frac{\partial u_i}{\partial x_k} \right) \frac{dx_j^{(2)}}{ds^{*(2)}} \frac{dx_k^{(3)}}{ds^{*(3)}}$$

(2.3.30)

For small strains and assuming $ds^{*(2)} \approx ds^{(2)}$; $ds^{*(3)} \approx ds^{(3)}$ we get,

$$\sin(\gamma_{23}) = \frac{dx_k^{(2)}}{ds^{(2)}}\frac{dx_k^{(3)}}{ds^{(3)}} + 2\varepsilon_{jk}\frac{dx_j^{(2)}}{ds^{(2)}}\frac{dx_k^{(3)}}{ds^{(3)}}$$

(2.3.31a)

$$\gamma_{23} = 2\varepsilon_{jk}n_j^{(2)}n_j^{(3)}$$

(2.3.31b)

$$\gamma_{23} = 2\begin{bmatrix}0 & 1 & 0\end{bmatrix}\{\varepsilon_{jk}\}\begin{Bmatrix}0\\0\\1\end{Bmatrix} = 2\varepsilon_{23}$$

(2.3.31c)

The above relation implies that ε_{23} is half of the angular change between line elements along x_2 and x_3 caused by distortion. Similarly other off-diagonal terms can be interpreted accordingly. In general, we can write,

$$\gamma_{nm} = 2\varepsilon_{nm} = 2n_i\varepsilon_{ij}m_j$$

(2.3.32)

or in matrix form:

$$\varepsilon_{nm} = \{n\}^T[\varepsilon]\{m\}$$

(2.3.33)

We should note that ε_{nm} are the components of strain tensor as defined earlier while γ_{nm} are engineering strain components representing the decrease of angle between n_i and m_i caused by deformation. Finally, all of the above can be interpolated with reference to a material element as seen in Figure 2.3-4. Strains are written as,

$$\varepsilon_{xx} = \frac{dx^* - dx}{dx}$$

(2.3.34a)

$$\varepsilon_{yy} = \frac{dy^* - dy}{dy}$$

(2.3.34b)

$$\varepsilon_{zz} = \frac{dz^* - dz}{dz}$$

(2.3.34c)

$$\gamma_{xy} = \tan^{-1}\frac{\delta(dx)}{dx}$$

(2.3.34d)

Other shear strains can similarly be interpreted.

2.3.4 Strain Transformation

Now we change the reference axis, x_i to a new x_i' with the transformation below,

$$x_i' = L_{ij}(x_j - c_i)$$

(2.3.35)

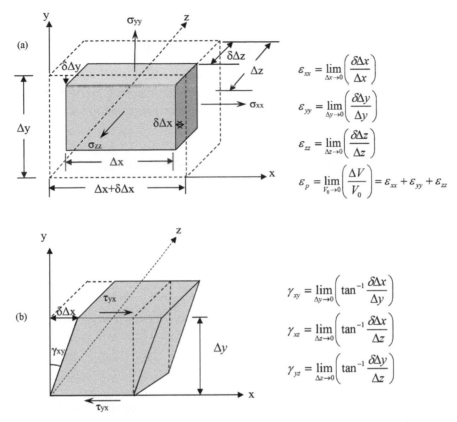

$$\varepsilon_{xx} = \lim_{\Delta x \to 0}\left(\frac{\delta \Delta x}{\Delta x}\right)$$

$$\varepsilon_{yy} = \lim_{\Delta y \to 0}\left(\frac{\delta \Delta y}{\Delta y}\right)$$

$$\varepsilon_{zz} = \lim_{\Delta z \to 0}\left(\frac{\delta \Delta z}{\Delta z}\right)$$

$$\varepsilon_{p} = \lim_{V_0 \to 0}\left(\frac{\Delta V}{V_0}\right) = \varepsilon_{xx} + \varepsilon_{yy} + \varepsilon_{zz}$$

$$\gamma_{xy} = \lim_{\Delta y \to 0}\left(\tan^{-1}\frac{\delta \Delta x}{\Delta y}\right)$$

$$\gamma_{xz} = \lim_{\Delta z \to 0}\left(\tan^{-1}\frac{\delta \Delta x}{\Delta z}\right)$$

$$\gamma_{yz} = \lim_{\Delta z \to 0}\left(\tan^{-1}\frac{\delta \Delta y}{\Delta z}\right)$$

Figure 2.3-4. Stress element under (a) Volumetric strain, (b) Shear strain

or

$$\{x'\} = [L]\Big[\{x\} - \{c\}\Big] \tag{2.3.36}$$

The transformed strain tensor then becomes,

$$\varepsilon'_{ij} = L_{ij}\varepsilon_{kl}L_{jl} \tag{2.3.37}$$

or

$$[\varepsilon'] = [L]^{T}[\varepsilon][L] \tag{2.3.38}$$

Details of the above transformation will be discussed in the next section on the 'stress tensor' as such transformation is similar to this one.

2.3.5 Principal Strains and Directions

As implied in the previous section, the values of the strain components are dependent on the choice of the reference axes. For a special choice of axes the strain tensor takes a very simple form in which only normal strains (diagonal term of $[\varepsilon]$) are non-zero and the shear strains (off diagonal terms of $[\varepsilon]$) become zeros. So,

$$[\varepsilon'] = [L][\varepsilon][L]^T = \begin{bmatrix} \varepsilon_1 & 0 & 0 \\ 0 & \varepsilon_2 & 0 \\ 0 & 0 & \varepsilon_3 \end{bmatrix} \qquad (2.3.39)$$

for $[L]^T = [n_1 \ n_2 \ n_3]$. The details of this special transformation and the process of evaluating ε_1, ε_2, ε_3 is identical to those for principal strains which will be presented in section 2.4.

2.3.6 Deviatoric Strains

In order to characterize material behavior, it is often desirable to work with 'deviatoric strain tensor', e_{ij} defined as,

$$e_{ij} = \varepsilon_{ij} - \frac{1}{3}\varepsilon_{kk}\delta_{ij} \qquad (2.3.40)$$

or

$$[e] = [\varepsilon] - \frac{1}{3}\varepsilon_v[I] \qquad (2.3.41)$$

where $\varepsilon_v = \varepsilon_{kk} = \varepsilon_{11} + \varepsilon_{22} + \varepsilon_{33}$ is the volumetric strain. The relative volume change can be readily shown as:

$$\varepsilon_p = \lim_{V_0 \to 0}\left(\frac{\Delta V}{V_0}\right) = \varepsilon_{xx} + \varepsilon_{yy} + \varepsilon_{zz} \qquad (2.3.42)$$

The deviatoric strain defined above is a measure of only the shearing (distortion) of the material. The principal directions of e_{ij} are the same as ε_{ij}.

2.3.7 Strain Invariants

The following strain quantities are independent of the choice of reference axes and thus, are called "strain invariants":

$$\theta_1 = \varepsilon_{11} + \varepsilon_{22} + \varepsilon_{33} = \varepsilon_1 + \varepsilon_2 + \varepsilon_3 = \varepsilon_{ii} \qquad (2.3.43a)$$

$$\theta_2 = \varepsilon_{11}\varepsilon_{22} + \varepsilon_{22}\varepsilon_{33} + \varepsilon_{11}\varepsilon_{33} - \varepsilon_{23}^2 - \varepsilon_{12}^2 - \varepsilon_{13}^2 = \frac{1}{2}\theta_1^2 - \frac{1}{2}\varepsilon_{ij}\varepsilon_{ji} \qquad (2.3.43b)$$

$$\theta_3 = \varepsilon_{11}\varepsilon_{22}\varepsilon_{33} - \varepsilon_{11}\varepsilon_{23}^2 - \varepsilon_{22}\varepsilon_{13}^2 - \varepsilon_{33}\varepsilon_{12}^2 + 2\varepsilon_{12}\varepsilon_{13}\varepsilon_{23} = \det(\varepsilon_{ij}) = \varepsilon_1\varepsilon_2\varepsilon_3 \qquad (2.3.43c)$$

In terms of principal strains the general strain invariants can also be written as,

$$I_1 = \varepsilon_{ii} = \varepsilon_1 + \varepsilon_2 + \varepsilon_3 \qquad (2.3.44a)$$

$$I_2 = \frac{1}{2}\varepsilon_{ij}\varepsilon_{ji} = \frac{1}{2}\left(\varepsilon_1^2 + \varepsilon_2^2 + \varepsilon_3^2\right) \qquad (2.3.44b)$$

$$I_3 = \frac{1}{2}\varepsilon_{ij}\varepsilon_{jk}\varepsilon_{kl} = \frac{1}{3}\left(\varepsilon_1^3 + \varepsilon_2^3 + \varepsilon_3^3\right) \qquad (2.3.44c)$$

Another set of strain invariants are defined in terms of deviatoric strains,

$$J_1 = e_{ii} = e_1 + e_2 + e_3 \qquad (2.3.45a)$$

$$J_2 = \frac{1}{2} e_{ij} e_{ji} = \frac{1}{2} (e_1^2 + e_2^2 + e_3^2) \tag{2.3.45b}$$

$$J_3 = \frac{1}{2} e_{ij} e_{jk} e_{kl} = \frac{1}{3} (e_1^3 + e_2^3 + e_3^3) = e_1 e_2 e_3 \tag{2.3.45c}$$

2.3.8 Octahedral Normal and Shear Strains

An octahedral plane is defined as a plane where its normal makes equal angles with the three principal directions (Figure 2.3-5). For the normal to the octahedral plane, we have:

$$n = \frac{1}{\sqrt{3}} \begin{bmatrix} 1 \\ 1 \\ 1 \end{bmatrix} \tag{2.3.46}$$

In the reference system collinear with the principal strain directions (see above), the strain tensor takes the form below,

$$[\varepsilon] = \begin{bmatrix} \varepsilon_1 & 0 & 0 \\ 0 & \varepsilon_2 & 0 \\ 0 & 0 & \varepsilon_3 \end{bmatrix} \tag{2.3.47}$$

The normal and shear strains on the octahedral plane are written as:

$$\varepsilon_0 = [n]^T \{q\} = [n]^T [n][n] \tag{2.3.48a}$$

$$\frac{\gamma_0}{2} = \sqrt{q^T q - \varepsilon_0^2} \tag{2.3.48b}$$

where $q = [\varepsilon][n]$. It can further be shown that:

$$\varepsilon_0 = \frac{\tilde{I}_1}{3} \tag{2.3.49}$$

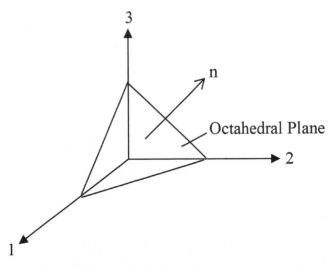

Figure 2.3-5. Octahedral plane

$$\gamma_0 = 2\sqrt{\frac{2}{3}\tilde{J}_2}$$

(2.3.50)

2.3.9 Strain Compatibility

For a strain field to yield a unique displacement field, it must obey the following conditions below. Figure 2.3-6 shows the cases where a strain field is not acceptable.

$$\frac{\partial^2 \varepsilon_{xx}}{\partial y^2} + \frac{\partial^2 \varepsilon_{xy}}{\partial x^2} = 2\frac{\partial^2 \varepsilon_{xy}}{\partial x \partial y}$$

(2.3.51a)

$$\frac{\partial^2 \varepsilon_{y}}{\partial z^2} + \frac{\partial^2 \varepsilon_{z}}{\partial y^2} = 2\frac{\partial^2 \varepsilon_{yz}}{\partial x \partial z}$$

(2.3.51b)

$$\frac{\partial^2 \varepsilon_{z}}{\partial x^2} + \frac{\partial^2 \varepsilon_{x}}{\partial z^2} = 2\frac{\partial^2 \varepsilon_{zx}}{\partial x \partial z}$$

(2.3.51c)

$$\frac{\partial}{\partial x}\left(-\frac{\partial \varepsilon_{yz}}{\partial x} + \frac{\partial \varepsilon_{zx}}{\partial y} + \frac{\partial \varepsilon_{xy}}{\partial z}\right) = \frac{\partial^2 \varepsilon_{x}}{\partial z \partial y}$$

(2.3.51d)

$$\frac{\partial}{\partial y}\left(-\frac{\partial \varepsilon_{zx}}{\partial y} + \frac{\partial \varepsilon_{xy}}{\partial z} + \frac{\partial \varepsilon_{yz}}{\partial x}\right) = \frac{\partial^2 \varepsilon_{y}}{\partial z \partial x}$$

(2.3.51e)

$$\frac{\partial}{\partial z}\left(-\frac{\partial \varepsilon_{xy}}{\partial z} + \frac{\partial \varepsilon_{yz}}{\partial x} + \frac{\partial \varepsilon_{zx}}{\partial y}\right) = \frac{\partial^2 \varepsilon_{z}}{\partial x \partial y}$$

(2.3.51f)

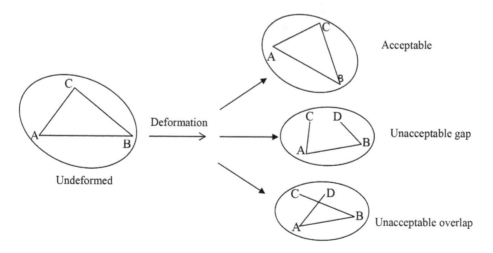

Figure 2.3-6. Various deformation states

2.4 Stresses

After obtaining a description of deformation of a material body, we now describe the loading within the body at an arbitrary point. The body is treated as a continuum (Figure 2.4-1 and as described in the beginning of this chapter) and typically two kinds of forces are recognized: *body forces, b_i,* and *surface forces, f_i.*

2.4.1 Stresses on a Plane

Taking *n* as the outward unit normal at point *P* of a small element of surface ΔA of *S*, let ΔF be the resultant force exerted across ΔA upon the material (Figure 2.4-2). The force element ΔF will depend upon the choice of ΔA and on *n*. The force distribution is, in general, equipollent to a force and a moment at *P*. The average force per unit area on ΔA is given by $\Delta F/\Delta A$. The *Cauchy stress principle* asserts that ratio $\Delta F/\Delta A$ tends to a definite limit dF/dA as ΔA approaches zero at point *P*, while at the same time the moment of ΔF about the point *P* vanishes

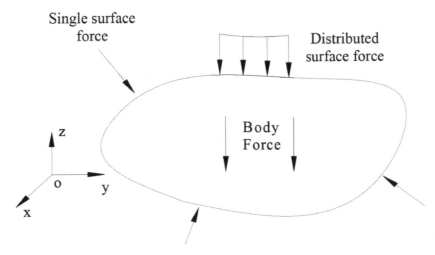

Figure 2.4-1 Body and surface forces

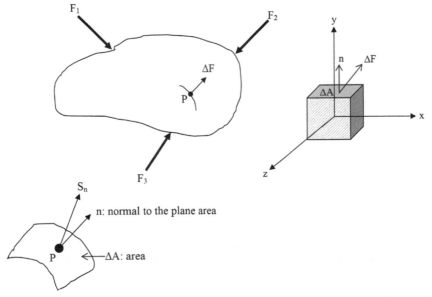

Figure 2.4-2. Definition of stress for a point on a solid body and on the surface of an element of material

in the limiting process. The resulting vector dF/dA is called the stress vector. Mathematically the stress vector is defined by:

$$S_n = \lim_{\Delta A \to 0} \left(\frac{\Delta F}{\Delta A} \right) = \frac{dF}{dA}$$

(2.4.1)

or more definitively for 2-D,

$$\sigma_{xx} = \lim_{\delta A \to 0} \left(-\frac{\delta F_x}{\delta A} \right)$$

(2.4.2a)

$$\sigma_{xy} = \lim_{\delta A \to 0} \left(-\frac{\delta F_y}{\delta A} \right)$$

(2.4.2b)

$$\sigma_{xz} = \lim_{\delta A \to 0} \left(-\frac{\delta F_z}{\delta A} \right)$$

(2.4.2c)

In index notation the stress vector is defined in the x-direction as,

$$S_x = \sigma_{xx} i + \sigma_{xy} j + \sigma_{xz} k$$

(2.4.3)

$$S_x = \left\{ \begin{array}{c} \sigma_{xx} \\ \sigma_{xy} \\ \sigma_{xz} \end{array} \right\} = \left\{ \begin{array}{c} \sigma_{11} \\ \sigma_{12} \\ \sigma_{13} \end{array} \right\}$$

(2.4.4)

and in other directions,

$$S_y = \sigma_{yx} i + \sigma_{yy} j + \sigma_{yz} k$$

(2.4.5)

$$S_y = \left\{ \begin{array}{c} \sigma_{yx} \\ \sigma_{yy} \\ \sigma_{yz} \end{array} \right\} = \left\{ \begin{array}{c} \sigma_{21} \\ \sigma_{22} \\ \sigma_{23} \end{array} \right\}$$

(2.4.6)

$$S_z = \sigma_{zx} i + \sigma_{zy} j + \sigma_{zz} k$$

(2.4.7)

$$S_z = \left\{ \begin{array}{c} \sigma_{zx} \\ \sigma_{zy} \\ \sigma_{zz} \end{array} \right\} = \left\{ \begin{array}{c} \sigma_{31} \\ \sigma_{32} \\ \sigma_{33} \end{array} \right\}$$

(2.4.8)

These stresses are shown on respective planes in Figure 2.4-3.

2.4.2 Stresses at a Point

At an arbitrary point in a continuum, Cauchy's stress principle associates a stress vector with the normal vector in each unit. The totality of all possible pairs of such stress vectors and unit normal vector defines the state of stress of that point. We can describe the state of stress at a given point by giving the stress vector on each of the three mutually perpendicular planes at the point (Figure 2.4-4). Coordinate transformation equations then serve to relate the stress vector on any other plane at the point to the given three planes. Stress vector is,

$$\sigma = [\sigma_{xx} \ \sigma_{xy} \ \sigma_{zz} \ \sigma_{xy} \ \ \sigma_{yz} \ \sigma_{zx}]^T = [\sigma_{xx} \ \sigma_{xy} \ \sigma_{zz} \ \tau_{xy} \ \tau_{yz} \ \tau_{zx}]^T$$

(2.4.9)

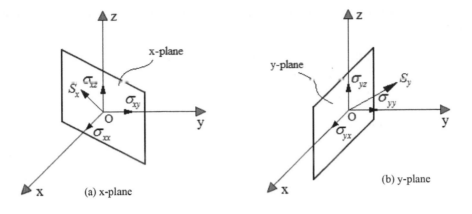

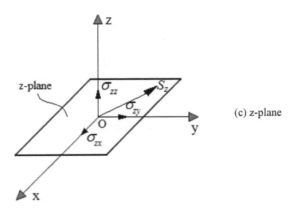

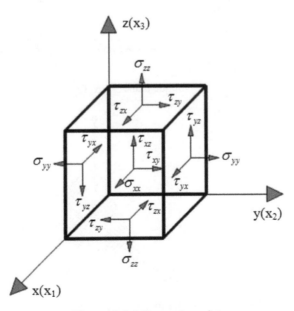

Figure 2.4-3. Stresses on various planes

Figure 2.4-4 Stress at a point

The nine stress vector components are the components of a second-order Cartesian tensor known as the *stress tensor*.

$$\sigma = \begin{bmatrix} \sigma_{xx} & \tau_{xy} & \tau_{xz} \\ \tau_{yx} & \sigma_{yy} & \tau_{yz} \\ \tau_{zx} & \tau_{zy} & \sigma_{zz} \end{bmatrix} = \begin{bmatrix} \sigma_{11} & \sigma_{12} & \sigma_{13} \\ \sigma_{21} & \sigma_{22} & \sigma_{23} \\ \sigma_{31} & \sigma_{32} & \sigma_{33} \end{bmatrix} \qquad (2.4.10)$$

It can be shown from the force equilibrium that:

$$\tau_{xy} = \tau_{yx} \qquad (2.4.11a)$$

$$\tau_{yz} = \tau_{zy} \qquad (2.4.11b)$$

$$\tau_{zx} = \tau_{xz} \qquad (2.4.11c)$$

The components perpendicular to the planes (σ_{11}, σ_{22}, σ_{33}) are called *normal stresses*. Those acting tangents to the planes (σ_{12}, σ_{13}, σ_{21}, σ_{23}, σ_{31}, σ_{32}) are called *shear stresses*. A stress component is positive when it acts in the positive direction of the coordinate axes, and on the plane whose outer normal points in one of the positive coordinate directions. The component σ_{ij} acts in the direction of the jth coordinate axis and on the plane whose outward normal is parallel to the ith coordinate axis.

2.4.3 Stresses on an Oblique Plane

If in case planes rotate, the normal and shear stresses acting on that plane are calculated considering the rotation angles, α, β and γ measured from the coordinate axes. Figure 2.4-5

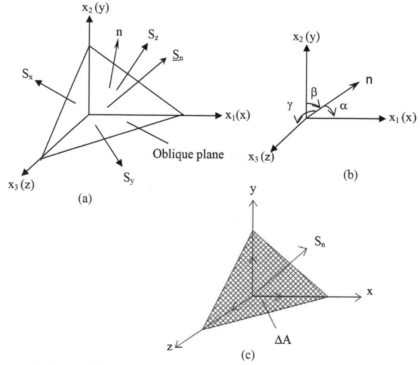

Figure 2.4-5 (a) Oblique plane and its stresses, (b) Direction cosines, (c) Resultant normal stress acting on the oblique plane

shows an oblique plane which has a normal, n, and direction cosines $l_n = \cos \alpha$, $m_n = \cos \beta$, $n_n = \cos \gamma$. If we write the force equilibrium:

$$\sum \vec{F} = 0 = S_n A + S_{-x} l_n A + S_{-y} m_n A + S_{-z} n_n A \tag{2.4.12}$$

we get the normal stress in terms of components along the three directions using the direction cosines:

$$S_n = S_x l_n + S_y m_n + S_z n_n \tag{2.4.13}$$

which is actually,

$$S_n = \begin{bmatrix} S_{nx} & S_{ny} & S_{nz} \end{bmatrix} \tag{2.4.14}$$

with the stress vector written as:

$$\sigma_n = \begin{Bmatrix} \sigma_{nx} \\ \sigma_{ny} \\ \sigma_{nz} \end{Bmatrix} \tag{2.4.15}$$

and using the direction cosines we get:

$$\begin{Bmatrix} \sigma_{nx} \\ \sigma_{ny} \\ \sigma_{nz} \end{Bmatrix} = \begin{Bmatrix} \sigma_{xx} \\ \sigma_{xy} \\ \sigma_{xz} \end{Bmatrix} \cdot l_n + \begin{Bmatrix} \sigma_{yx} \\ \sigma_{yy} \\ \sigma_{yz} \end{Bmatrix} \cdot m_n + \begin{Bmatrix} \sigma_{zx} \\ \sigma_{zy} \\ \sigma_{zz} \end{Bmatrix} \cdot n_n \tag{2.4.16}$$

or

$$\begin{Bmatrix} \sigma_{nx} \\ \sigma_{ny} \\ \sigma_{nz} \end{Bmatrix} = \begin{bmatrix} \sigma_x & \tau_{xy} & \tau_{xy} \\ \tau_{yx} & \sigma_y & \tau_{yz} \\ \tau_{zx} & \tau_{zy} & \sigma_z \end{bmatrix} \cdot \begin{Bmatrix} l_n \\ m_n \\ n_n \end{Bmatrix} \tag{2.4.17}$$

where $\{l_n \, m_n \, n_n\} = \{\cos \alpha \, \cos \beta \, \cos \gamma\}$. The above equation represents the stresses acting on an oblique plane for which we know the direction cosines (Figure 2.4-6).

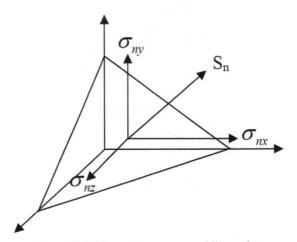

Figure 2.4-6 Stress vector on an oblique plane

2.4.4 Stress Transformation

Let the rectangular Cartesian coordinate systems xyz and $x'y'z'$ be related to another by the direction cosines,

$$\sigma_{x'x'} = \sigma_{xx} \cdot l_{x'} + \sigma_{xy} \cdot m_{x'} + \sigma_{xz} \cdot n_{x'} \tag{2.4.18}$$

$$\sigma_{x'x'} = [l_{x'} \ \ m_{x'} \ \ n_{x'}] \begin{Bmatrix} \sigma_{xx} \\ \sigma_{xy} \\ \sigma_{xz} \end{Bmatrix} \tag{2.4.19}$$

Similarly we can write for $\sigma_{x'y'}$ and $\sigma_{x'z'}$:

$$\begin{Bmatrix} \sigma_{x'x'} \\ \sigma_{y'y'} \\ \sigma_{z'z'} \end{Bmatrix} = \begin{bmatrix} l_{x'} & m_{x'} & n_{x'} \\ l_{y'} & m_{y'} & n_{y'} \\ l_{z'} & m_{z'} & n_{z'} \end{bmatrix} \cdot \begin{Bmatrix} \sigma_{nx} \\ \sigma_{ny} \\ \sigma_{nz} \end{Bmatrix} = [L] \cdot \begin{Bmatrix} \sigma_{nx} \\ \sigma_{ny} \\ \sigma_{nz} \end{Bmatrix} \tag{2.4.20}$$

Transformation of stresses on an oblique plane can be seen in Figure 2.4-7. Table 2.4-1 shows all the direction cosines for the transformed coordinates. Equation 2.4.20 can be rewritten as,

$$\begin{Bmatrix} \sigma_{x'x'} \\ \sigma_{x'y'} \\ \sigma_{x'z'} \end{Bmatrix} = [L] \cdot [\sigma] \cdot \begin{Bmatrix} l_{x'} \\ m_{x'} \\ n_{x'} \end{Bmatrix} \tag{2.4.21}$$

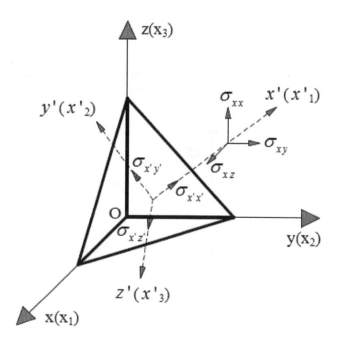

Figure 2.4-7 Transformation of stresses on an oblique plane

Table 2.4-1 Direction cosines of rectangular Cartesian coordinate systems *xyz* and *x'y'z'*

Coordinates	*x*	*y*	*z*
x'	$l_{x'} = \cos(x', x)$	$m_{x'} = \cos(x', y)$	$n_{x'} = \cos(x', z)$
y'	$l_{y'} = \cos(y', x)$	$m_{y'} = \cos(y', y)$	$n_{y'} = \cos(y', z)$
z'	$l_{z'} = \cos(z', x)$	$m_{z'} = \cos(z', y)$	$n_{z'} = \cos(z', z)$

Repeating the above steps with $y' = n$ and $z' = n$ we get,

$$\begin{bmatrix} \sigma_{x'x'} & \sigma_{x'y'} & \sigma_{x'z'} \\ \sigma_{y'x'} & \sigma_{y'y'} & \sigma_{y'z'} \\ \sigma_{z'x'} & \sigma_{z'y'} & \sigma_{z'z'} \end{bmatrix} = [L] \cdot [\sigma] \cdot \begin{bmatrix} l_{x'} & l_{y'} & l_{z'} \\ m_{x'} & m_{y'} & m_{z'} \\ n_{x'} & n_{y'} & n_{z'} \end{bmatrix} = [L] \cdot [\sigma] \cdot [L]^T \quad (2.4.22)$$

So finally the "transformed" stress vector becomes:

$$[\sigma'] = [L] \cdot [\sigma] \cdot [L]^T \quad (2.4.23)$$

2.4.5 Principal Stresses and Invariants

Since the stress components of a point change with the coordinates, we can find a coordinate system in which the point has only normal stresses but no shear stress. At that point, we call these normal stresses *principal stresses* (Figure 2.4-8). From the previous section, we know that:

$$\begin{Bmatrix} \sigma_{nx} \\ \sigma_{ny} \\ \sigma_{nz} \end{Bmatrix} = \sigma \cdot \begin{Bmatrix} l_n \\ m_n \\ n_n \end{Bmatrix} \quad (2.4.24)$$

$$\begin{bmatrix} \sigma_{xx} - \sigma & \sigma_{xy} & \sigma_{xz} \\ \sigma_{yx} & \sigma_{yy} - \sigma & \sigma_{yz} \\ \sigma_{zx} & \sigma_{zy} & \sigma_{zz} - \sigma \end{bmatrix} \begin{Bmatrix} l_n \\ m_n \\ n_n \end{Bmatrix} = 0 \quad (2.4.25)$$

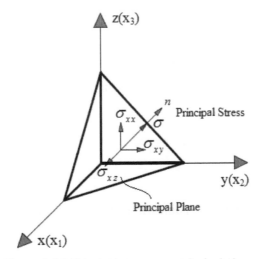

Figure 2.4-8 Principal stresses on principal planes

Let

$$
\begin{bmatrix}
\sigma_{xx} - \sigma & \sigma_{xy} & \sigma_{xz} \\
\sigma_{yx} & \sigma_{yy} - \sigma & \sigma_{yz} \\
\sigma_{zx} & \sigma_{zy} & \sigma_{zz} - \sigma
\end{bmatrix} = A
\tag{2.4.26}
$$

and let $det(A) = 0$ yielding:

$$
\sigma^3 - I_1\sigma^2 + I_2\sigma + I_3 = 0
\tag{2.4.27}
$$

in which the stress invariants are,

$$
I_1 = \sigma_{xx} + \sigma_{yy} + \sigma_{zz}
\tag{2.4.28a}
$$

$$
I_2 = \sigma_{xx}\sigma_{yy} + \sigma_{yy}\sigma_{zz} + \sigma_{zz}\sigma_{xx} - \sigma_{xy}^2 - \sigma_{yz}^2 - \sigma_{zx}^2
\tag{2.4.28b}
$$

$$
I_3 = \sigma_{xx}\sigma_{yy}\sigma_{zz} - \sigma_{xx}\sigma_{yz}^2 - \sigma_{yy}\sigma_{zx}^2 - \sigma_{zz}\sigma_{xy}^2 - 2\sigma_{xy}\sigma_{yz}\sigma_{zx}
\tag{2.4.28c}
$$

These invariants can also be written as:

$$
I_1 = \sigma_{xx} + \sigma_{yy} + \sigma_{zz} = \sigma_{x'x'} + \sigma_{y'y'} + \sigma_{z'z'}
\tag{2.4.29a}
$$

$$
I_2 = \begin{vmatrix} \sigma_{xx} & \sigma_{xy} \\ \sigma_{yx} & \sigma_{yy} \end{vmatrix} + \begin{vmatrix} \sigma_{yy} & \sigma_{yz} \\ \sigma_{zy} & \sigma_{zz} \end{vmatrix} + \begin{vmatrix} \sigma_{zz} & \sigma_{zx} \\ \sigma_{xz} & \sigma_{xx} \end{vmatrix}
\tag{2.4.29b}
$$

$$
I_3 = \begin{vmatrix} \sigma_{xx} & \sigma_{xy} & \sigma_{xz} \\ \sigma_{yx} & \sigma_{yy} & \sigma_{yz} \\ \sigma_{zx} & \sigma_{zy} & \sigma_{zz} \end{vmatrix}
\tag{2.4.29c}
$$

2.4.6 Evaluation of Principal Stresses

Solution of equation 2.4-26 yields three values of σ satisfying $\sigma_1 > \sigma_2 > \sigma_3$. Now to get the direction cosines of the normal to the principal plane, $\sigma = \sigma_1$ is substituted in the relation below:

$$
\begin{bmatrix}
\sigma_{xx} - \sigma & \sigma_{xy} & \sigma_{xz} \\
\sigma_{yx} & \sigma_{yy} - \sigma & \sigma_{yz} \\
\sigma_{zx} & \sigma_{zy} & \sigma_{zz} - \sigma
\end{bmatrix}
\begin{Bmatrix} l_1 \\ m_1 \\ n_1 \end{Bmatrix} = 0
\tag{2.4.30}
$$

and gives out the solution for l_1, m_1 and n_1 satisfying,

$$
l_1^2 + m_1^2 + n_1^2 = 1
\tag{2.4.31}
$$

Then, we repeat the above with $\sigma = \sigma_2$ and $\sigma = \sigma_3$ and evaluate l_2, m_2, n_2 and l_3, m_3, n_3 in a similar way.

Example 2.4-1: State of stress at a point within a body is given below as, $\sigma = \begin{bmatrix} 3 & 1 & 1 \\ 1 & 0 & 2 \\ 1 & 2 & 0 \end{bmatrix}$,

find the principal planes and the corresponding principal stresses.

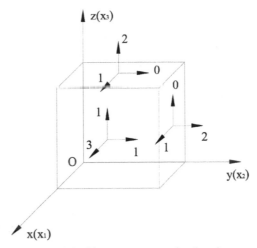

Figure 2.4-9. Given stress state for the element

Solution:

We can write $A = \begin{bmatrix} 3-\sigma & 1 & 1 \\ 1 & 0-\sigma & 2 \\ 1 & 2 & 0-\sigma \end{bmatrix}$ following 2.4.29 leading to:

$$\begin{bmatrix} 3-\sigma & 1 & 1 \\ 1 & 0-\sigma & 2 \\ 1 & 2 & 0-\sigma \end{bmatrix} \begin{Bmatrix} l \\ m \\ n \end{Bmatrix} = 0$$

whose

$$det(A) = -(\sigma+2)(\sigma-4)(\sigma-1) = 0 \rightarrow \sigma_{1,2,3} = 4, 1, -2, \; \sigma_1 = 4, \quad \sigma_2 = 1, \quad \sigma_3 = -2$$

Let $\sigma = \sigma_1 = 4$ and substitute it into above and get (Figure 2.4-10):

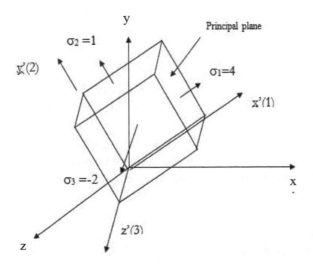

Figure 2.4-10. Normal stresses on rotated plane

$$\begin{bmatrix} 3-\sigma_1 & 1 & 1 \\ 1 & 0-\sigma_1 & 2 \\ 1 & 2 & 0-\sigma_1 \end{bmatrix} \begin{Bmatrix} l_1 \\ m_1 \\ n_1 \end{Bmatrix} = \begin{bmatrix} -1 & 1 & 1 \\ 1 & -4 & 2 \\ 1 & 2 & -4 \end{bmatrix} \begin{Bmatrix} l_1 \\ m_1 \\ n_1 \end{Bmatrix} = 0$$

Then we find the direction cosines using,

$$l_1^2 + m_1^2 + n_1^2 = 1, \ -l_1 + m_1 + n_1 = 0, \ l_1 - 4m_1 + 2n_1 = 0, \quad l_1 + 2m_1 - 4n_1 = 0 \ \text{and} \ l_1 = \pm\frac{2}{\sqrt{6}},$$

$$m_1 = \pm\frac{1}{\sqrt{6}}, \ n_1 = \pm\frac{1}{\sqrt{6}}.$$

If we repeat the above with $\sigma = \sigma_2$ and $\sigma = \sigma_3$, we get $l_2 = \pm\frac{1}{\sqrt{3}}$, $l_2 = \pm 0$; $m_2 = \pm\frac{1}{\sqrt{3}}$, $m_2 = \pm\frac{1}{\sqrt{2}}$; $n_2 = \pm\frac{1}{\sqrt{3}}$, $n_2 = \pm\frac{1}{\sqrt{2}}$. The stress tensor values at a point P are given by the matrix:

$$[\sigma_{ij}] = \begin{bmatrix} 7 & 0 & -2 \\ 0 & 5 & 0 \\ -2 & 0 & 4 \end{bmatrix}$$

Example 2.4-2: Determine the stress vector on the plane at P whose unit normal is $\{2/3, -2/3, 1/3\}$.

Solution:

Using the equation:

$$\begin{Bmatrix} \sigma_{nx} \\ \sigma_{ny} \\ \sigma_{nz} \end{Bmatrix} = \begin{bmatrix} \sigma_{xx} & \tau_{xy} & \tau_{xz} \\ \tau_{yx} & \sigma_{yy} & \tau_{yz} \\ \tau_{zx} & \tau_{zy} & \sigma_{zz} \end{bmatrix} \begin{Bmatrix} l_n \\ m_n \\ n_n \end{Bmatrix}$$

we have:

$$\begin{Bmatrix} \sigma_{nx} \\ \sigma_{ny} \\ \sigma_{nz} \end{Bmatrix} = \begin{bmatrix} 7 & 0 & -2 \\ 0 & 5 & 0 \\ -2 & 0 & 4 \end{bmatrix} \begin{Bmatrix} 2/3 \\ 2/3 \\ 1/3 \end{Bmatrix} = \begin{Bmatrix} 4 \\ -3/10 \\ 0 \end{Bmatrix}$$

Example 2.4-3: From the stress vector of problem 2.4.2, determine (a) the component perpendicular to the plane, (b) the magnitude of stress vector, and (c) the angle between the stress vector and the unit normal of the plane.

Solution:

(a) $[4 \ -3/10 \ 0] \begin{bmatrix} 2/3 \\ -2/3 \\ 1/3 \end{bmatrix} = 44/9$

(b) $\sqrt{4^2 + (-3/10)^2} = 5.2$

(c) Since $\cos\theta = (44/9)/5.2 = 0.94$, the angle between the stress vector and the unit normal can be evaluated as $\theta = 20°$.

Example 2.4-4: The state of stress at a point is given with respect to the Cartesian axes x-y-z by the matrix:

$$[\sigma_{ij}] = \begin{bmatrix} 2 & -2 & 0 \\ -2 & \sqrt{2} & 0 \\ 0 & 0 & -\sqrt{2} \end{bmatrix}$$

Determine the stress tensor for the rotated axes x'-y'-z' related to the axes xyz by the transformation tensor given as:

$$[L] = \begin{bmatrix} 0 & 1/\sqrt{2} & 1/\sqrt{2} \\ 1/\sqrt{2} & 1/2 & -1/2 \\ -1/\sqrt{2} & 1/2 & -1/2 \end{bmatrix}$$

Solution:
From the equation:

$$\begin{bmatrix} \sigma'_{11} & \sigma'_{12} & \sigma'_{13} \\ \sigma'_{21} & \sigma'_{22} & \sigma'_{23} \\ \sigma'_{31} & \sigma'_{32} & \sigma'_{33} \end{bmatrix} = \begin{bmatrix} l_{x'} & m_{x'} & n_{x'} \\ l_{y'} & m_{y'} & n_{y'} \\ l_{z'} & m_{z'} & n_{z'} \end{bmatrix} \begin{bmatrix} \sigma_{11} & \sigma_{12} & \sigma_{13} \\ \sigma_{21} & \sigma_{22} & \sigma_{23} \\ \sigma_{31} & \sigma_{32} & \sigma_{33} \end{bmatrix} \begin{bmatrix} l_{x'} & l_{y'} & l_{z'} \\ m_{x'} & m_{y'} & m_{z'} \\ n_{x'} & n_{y'} & n_{z'} \end{bmatrix}$$

we have:

$$[\sigma'_{ij}] = \begin{bmatrix} 0 & 1/\sqrt{2} & 1/\sqrt{2} \\ 1/\sqrt{2} & 1/2 & -1/2 \\ -1/\sqrt{2} & 1/2 & -1/2 \end{bmatrix} \begin{bmatrix} 2 & -2 & 0 \\ -2 & \sqrt{2} & 0 \\ 0 & 0 & -\sqrt{2} \end{bmatrix} \begin{bmatrix} 0 & 1/\sqrt{2} & -1/\sqrt{2} \\ 1/\sqrt{2} & 1/2 & 1/2 \\ 1/\sqrt{2} & -1/2 & -1/2 \end{bmatrix}$$

$$= \begin{bmatrix} 0 & 0 & 2 \\ 0 & 1-\sqrt{2} & -1 \\ 2 & -1 & 1+\sqrt{2} \end{bmatrix}$$

Example 2.4-5: The state of stress at a point is given as follows: $\sigma_x = -800$ kPa, $\sigma_y = -400$ kPa, $\tau_{xy} = 400$ kPa, $\tau_{yz} = -600$ kPa, and $\tau_{zx} = 500$ kPa. Determine: (a) the stresses on a plane whose normal has direction cosines $l_1 = 1/4$ and $l_2 = 1/2$, and (b) the normal and shearing stresses on that plane.

Solution:

From the relation, $l_1^2 + l_2^2 + l_3^2 = 1$, we can get $(1/4)^2 + (1/2)^2 + l_3^2 = 1$, $l_3 = \sqrt{11}/4$. Then,

(a) $\begin{bmatrix} \sigma_{nx} \\ \sigma_{ny} \\ \sigma_{nz} \end{bmatrix} = \begin{bmatrix} -800 & 400 & 500 \\ 400 & 1200 & -600 \\ 500 & -600 & -400 \end{bmatrix} \begin{bmatrix} 1/4 \\ 1/2 \\ \sqrt{11}/4 \end{bmatrix} = \begin{bmatrix} 414.6 \\ 202.5 \\ -506.7 \end{bmatrix}$

(b) $\sigma_n = \sigma_{n1} \cdot l_1 + \sigma_{n2} \cdot l_2 + \sigma_{n3} \cdot l_3$

$\qquad = (414.6)(1/4) + (202.5)(1/2) - (506.7)(\sqrt{11}/4) = -214.73\,\text{kPa}$

$\sigma_R^2 = \sigma_{n1}^2 + \sigma_{n2}^2 + \sigma_{n3}^2 = 471306.7$, $\sigma_R^2 = \sigma_n^2 + \tau^2, \tau^2 = 425197.73 \rightarrow \tau = 652.1$ kPa

2.4.7 Principal Normal Stresses

Principal normal stresses can be written as (see Figure 2.4-11),

$$\sigma_{n1} = \sigma_1 \cdot l_1 \qquad\qquad (2.4.32a)$$

$$\sigma_{n2} = \sigma_2 \cdot l_2 \qquad\qquad (2.4.32b)$$

$$\sigma_{n3} = \sigma_3 \cdot l_3 \qquad\qquad (2.4.32c)$$

From equilibrium and further manipulation,

$$\sigma_n = \sigma_{n1} \cdot l + \sigma_{n2} \cdot m + \sigma_{n3} \cdot n \qquad\qquad (2.4.33)$$

which can be written as:

$$\sigma_n = \sigma_1 \cdot l^2 + \sigma_2 \cdot m^2 + \sigma_3 \cdot n^2 \qquad\qquad (2.4.34)$$

If we also write,

$$\sigma_R^2 = \sigma_{n1}^2 + \sigma_{n2}^2 + \sigma_{n3}^2 \qquad\qquad (2.4.35)$$

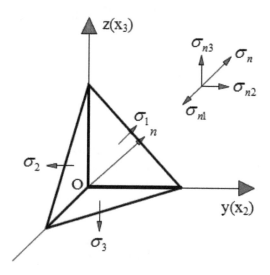

Figure 2.4-11. Principal normal stresses

or

$$\sigma_R^2 = \sigma_1^2 l^2 + \sigma_2^2 m^2 + \sigma_3^2 n^2 \tag{2.4.36}$$

we can get,

$$\tau_n^2 = \sigma_R^2 - \sigma_n^2 \tag{2.4.37}$$

leading to,

$$\tau_n^2 = (\sigma_1^2 l^2 + \sigma_2^2 m^2 + \sigma_3^2 n^2) - (\sigma_1 \cdot l^2 + \sigma_2 \cdot m^2 + \sigma_3 \cdot n^2)^2 \tag{2.4.38}$$

We now want to find the extreme values of τ_n^2 subjected to the constant relation that satisfies (2.4.31). This is obtained by solving the following,

$$F = \tau_n^2 - \lambda(l_1^2 + m_1^2 + n_1^2 - 1) \tag{2.4.39}$$

and requiring $\dfrac{\partial F}{\partial l} = 0$, $\dfrac{\partial F}{\partial m} = 0$ and $\dfrac{\partial F}{\partial n} = 0$ which gives,

$$2l[m^2(\sigma_1 - \sigma_2)^2 + n^2(\sigma_3 - \sigma_1)^2 - \lambda] = 0 \tag{2.4.40a}$$

$$2m[n^2(\sigma_2 - \sigma_3)^2 + l^2(\sigma_1 - \sigma_2)^2 - \lambda] = 0 \tag{2.4.40b}$$

$$2n[l^2(\sigma_3 - \sigma_1)^2 + m^2(\sigma_2 - \sigma_3)^2 - \lambda] = 0 \tag{2.4.40c}$$

The solution of the above set of equations is listed in Table 2.4-2. Figure 2.4-12 shows the normal and tangential directions along with corresponding stresses.

Table 2.4-2 Principal stresses with direction cosines

l	m	n	$\tau_{n\,max}^2$	λ	σ_n
± 1	0	0	0	0	σ_1
0	± 1	0	0	0	σ_2
0	0	± 1	0	0	σ_3
0	$\pm\dfrac{1}{\sqrt{2}}$	$\pm\dfrac{1}{\sqrt{2}}$	$\dfrac{(\sigma_2 - \sigma_3)^2}{2}$	$2\tau_n^2$	$\dfrac{\sigma_2 + \sigma_3}{2}$
$\pm\dfrac{1}{\sqrt{2}}$	0	$\pm\dfrac{1}{\sqrt{2}}$	$\dfrac{(\sigma_3 - \sigma_1)^2}{2}$	$2\tau_n^2$	$\dfrac{\sigma_3 + \sigma_1}{2}$
$\pm\dfrac{1}{\sqrt{2}}$	$\pm\dfrac{1}{\sqrt{2}}$	0	$\dfrac{(\sigma_1 - \sigma_2)^2}{2}$	$2\tau_n^2$	$\dfrac{\sigma_1 + \sigma_2}{2}$

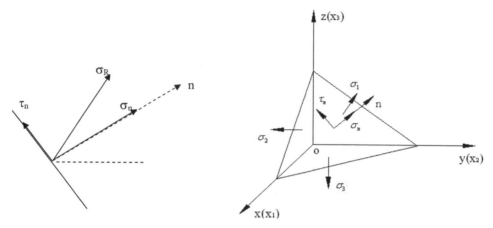

Figure 2.4-12. Normal and tangential directions with related stresses

2.4.8 Octahedral Stresses

The stresses acting on an octahedral plane is represented by the face *ABC* in Figure 2.4-13 with $OA = OB = OC$. The normal to this oblique face has equal direction cosines relative to the principal axes. Since (2.4.31) is satisfied, we have,

$$l = m = n = \frac{\sqrt{3}}{3} \tag{2.4.41}$$

In Figure 2.4-13, plane *ABC* is one of the eight faces of a regular octahedron. Referring to Figure 2.4-12 where the normal and shear stresses acting on an oblique element are seen, we can write,

$$\sigma_n = \sigma_1 \cdot l^2 + \sigma_2 \cdot m^2 + \sigma_3 \cdot n^2 \tag{2.4.42}$$

$$\sigma_n^2 + \tau_n^2 = \sigma_R^2 = \sigma_1^2 \cdot l^2 + \sigma_2^2 \cdot m^2 + \sigma_3^2 \cdot n^2 \tag{2.4.43}$$

when simultaneously solved gives:

$$\tau^2 = \sigma_1^2 l^2 + \sigma_2^2 m^2 + \sigma_3^2 n^2 - \sigma^2 \tag{2.4.44a}$$

$$\tau^2 = \sigma_1^2 l^2 + \sigma_2^2 m^2 + \sigma_3^2 n^2 - \left(\sigma_1 l^2 + \sigma_2 m^2 + \sigma_3 n^2\right)^2 \tag{2.4.44b}$$

Expanding and using the expressions $1 - l^2 = m^2 + n^2$, $1 - m^2 = l^2 + n^2$ and so on, the following result is obtained for the shear stress on an oblique plane:

$$\tau = \left[(\sigma_1 - \sigma_2)^2 l^2 m^2 + (\sigma_2 - \sigma_3)^2 n^2 m^2 + (\sigma_3 - \sigma_1)^2 l^2 n^2\right]^{1/2} \tag{2.4.45}$$

This clearly indicates that if all the principal stresses are equal, the shear stress vanishes, regardless of the direction cosines. Equations (2.4.46) and (2.4.48) are now applied to provide an expression for the *octahedral shear stress*, which is rearranged to form:

$$\tau_{oct} = \frac{1}{3}\sqrt{(\sigma_1 - \sigma_2)^2 + (\sigma_2 - \sigma_3)^2 + (\sigma_3 - \sigma_1)^2} \tag{2.4.46}$$

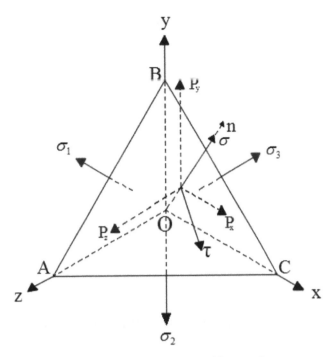

Figure 2.4-13 Octahedral plane with normal stresses

Through the use of Equations (2.4.42) and (2.4.46), we obtain the *octahedral normal stress*:

$$\sigma_{oct} = \frac{\sigma_1 + \sigma_2 + \sigma_3}{3} \tag{2.4.47}$$

The normal stress acting on an octahedral plane is thus the average of the principal stresses, the *mean stress*. Octahedral stresses are shown in another oriented element in Figure 2.4-14.

Example 2.4-6: The stress tensor at a point P is given with respect to axes o-x-y-z by the values,

$$\left[\sigma_{ij}\right] = \begin{bmatrix} 3 & 1 & 1 \\ 1 & 0 & 2 \\ 1 & 2 & 0 \end{bmatrix}$$

Determine the principal stress values and the principal stress directions represented by the axes o-x'-y'-z'.

Solution:

The principal stress values σ_i, $i = 1, 2, 3$ are evaluated using the characteristic equation obtained from:

$$\begin{vmatrix} 3-\sigma & 1 & 1 \\ 1 & -\sigma & 2 \\ 1 & 2 & -\sigma \end{vmatrix} = 0$$

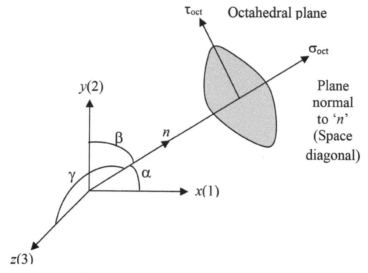

Figure 2.4-14 Octahedral plane and stresses

Upon expansion, we get $(\sigma + 2)(\sigma - 4)(\sigma - 1) = 0$ for which the roots are the principal stress values $\sigma_1 = -2$, $\sigma_2 = 1$, $\sigma_3 = 4$. Let the x' axis be the direction of σ_1, and let l_x, m_x, n_x be the direction cosines of this axis. Then:

$$(3 + 2)\, l_x + m_x + n_x = 0$$

$$l_x + 2\, m_x + 2\, n_x = 0$$

Hence $l_x = 0$; $m_x = n_x$ and since $l_x^2 + m_x^2 + n_x^2 = 1$ therefore, $l_x = 0$, $m_x = 1/\sqrt{2}$, $n_x = -1/\sqrt{2}$. The same way, we can get: $l_y = 1/\sqrt{3}$, $m_y = -1/\sqrt{3}$, $n_y = -1/\sqrt{3}$, $l_z = -2/\sqrt{6}$, $m_z = -1/\sqrt{6}$, $n_z = -1/\sqrt{6}$.

Example 2.4-7: Evaluate the invariants I_1, I_2, I_3 for the stress tensor:

$$\left[\sigma_{ij}\right] = \begin{bmatrix} 6 & -3 & 0 \\ -3 & 6 & 0 \\ 0 & 0 & 8 \end{bmatrix}$$

Determine the principal stress values for this state of stress and show that the diagonal of the stress tensor yields the same values for the stress invariants.

Solution:

$$I_1 = \sigma_{xx} + \sigma_{yy} + \sigma_{zz} = 6 + 6 + 8 = 20.$$

$$I_2 = \sigma_{xx}\sigma_{yy} + \sigma_{yy}\sigma_{zz} + \sigma_{zz}\sigma_{xx} - \sigma_{xy}^2 - \sigma_{yz}^2 - \sigma_{zx}^2 = 36 + 48 + 48 - 9 = 123.$$

$$I_3 = \begin{vmatrix} \sigma_{xx} & \tau_{xy} & \tau_{xz} \\ \tau_{yx} & \sigma_{yy} & \tau_{yz} \\ \tau_{zx} & \tau_{zy} & \sigma_{zz} \end{vmatrix} = 6(48) + 3(-24) = 216$$

The principal stress values of $\left[\sigma_{ij}\right] = \begin{bmatrix} 6 & -3 & 0 \\ -3 & 6 & 0 \\ 0 & 0 & 8 \end{bmatrix}$ are $\sigma_1 = 3$, $\sigma_2 = 8$, $\sigma_3 = 9$. In terms of principal values:

$$I_1 = \sigma_1 + \sigma_2 + \sigma_3 = 3 + 8 + 9 = 20$$

$$I_2 = \sigma_1\sigma_2 + \sigma_2\sigma_3 + \sigma_3\sigma_1 = 24 + 72 + 27 = 123$$

$$I_3 = \sigma_1\sigma_2\sigma_3 = (3)(8)(9) = 216$$

Example 2.4-8: Calculate the principal stresses, principal directions and the principal shear stresses for the given stress tensor σ_{ij} below. Also check if the three principal directions are orthogonal to each other.

$$\left[\sigma_{ij}\right] = \begin{bmatrix} 4 & 1 & 0 \\ 1 & 7 & 3 \\ 0 & 3 & 2 \end{bmatrix}$$

Solution:

The three invariants are calculated as,

$$I_1 = \sigma_{xx} + \sigma_{yy} + \sigma_{zz} = 13$$

$$I_2 = \sigma_{xx}\sigma_{yy} + \sigma_{yy}\sigma_{zz} + \sigma_{zz}\sigma_{xx} - \tau_{xy}^2 - \tau_{yz}^2 - \tau_{zx}^2 = 28 + 14 + 8 - 1 - 0 - 9 = 40$$

$$I_3 = \begin{vmatrix} \sigma_{xx} & \tau_{xy} & \tau_{xz} \\ \tau_{yx} & \sigma_{yy} & \tau_{yz} \\ \tau_{zx} & \tau_{zy} & \sigma_{zz} \end{vmatrix} = 4(5) - (2) = 18$$

The characteristic equation becomes,

$$\begin{vmatrix} 4-\sigma & 1 & 0 \\ 1 & 7-\sigma & 3 \\ 1 & 3 & 2-\sigma \end{vmatrix} \begin{aligned} &= (4-\sigma)[(7-\sigma)(2-\sigma)-9]-(2-\sigma)+3 \\ &= \sigma^3 - 13\sigma^2 + 40\sigma - 21 = 0 \end{aligned}$$

having the three roots as, $\sigma_1 = 0.5412$, $\sigma_2 = 3.8739$, $\sigma_3 = 8.5849$. The principal directions are calculated as:

$$\begin{bmatrix} 4-0.5412 & 1 & 0 \\ 1 & 7-0.5412 & 3 \\ 1 & 3 & 2-0.5412 \end{bmatrix} \begin{Bmatrix} l_1 \\ l_2 \\ l_3 \end{Bmatrix} = \begin{Bmatrix} 0 \\ 0 \\ 0 \end{Bmatrix}$$

leading to,

$$l = \{0.1254, -0.4338, 0.8922\}$$

Similarly,

$$m = \{0.9728, -0.1227, -0.1964\},$$

$$n = \{-0.1947, -0.8926, -0.4067\}$$

As for orthogonality of the directions, it can be seen that, $l \cdot m = 0$, $l \cdot n = 0$, $m \cdot n = 0$ relations will satisfy so the principal directions are orthogonal to each other. The principal shear stresses can be defined as the octahedral stresses, σ_{oct} and τ_{oct} which are determined as,

$$\sigma_{oct} = \frac{I_1}{3} = \frac{13}{3},$$

$$\tau_{oct} = \frac{1}{3}\left[(\sigma_1 - \sigma_2)^2 + (\sigma_2 - \sigma_3)^2 + (\sigma_1 - \sigma_3)^2\right]^{\frac{1}{2}}$$

$$= \frac{1}{3}[11.1 + 22.193 + 64.7]^{\frac{1}{2}} = 3.3$$

2.4.9 Two-Dimensional (2-D) Stress States in Plane

In two dimensions (2-D), there are two basic stress states and one strain state that are used to idealize an engineering problem based on the stresses and strains that are present on adjacent planes. These are called: (i) *Plane stress state*, (ii) *Axi-symmetric stress state* and (iii) *Plane strain* state (Figure 2.4-15).

In the plane stress and plane strain cases where the stresses and strains are respectively in the plane, the independent stresses are,

$$\sigma = \begin{bmatrix} \sigma_{yy} & \tau_{yz} \\ \tau_{zy} & \sigma_{zz} \end{bmatrix} \tag{2.4.48}$$

$$\tau_{yz} = \tau_{zy} \tag{2.4.49}$$

and for the axisymmetric case they are,

$$\sigma = \begin{bmatrix} \sigma_{rr} & \tau_{rz} \\ \tau_{zr} & \sigma_{zz} \end{bmatrix} \tag{2.4.50}$$

$$\tau_{rz} = \tau_{zr} \tag{2.4.51}$$

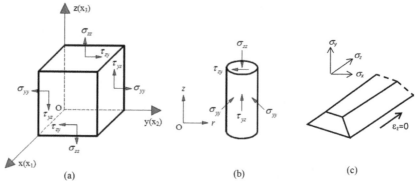

Figure 2.4-15 2-D Stress states of (a) Plane stress, (b) Axisymmetric, (c) Plane strain

2.4.10 Deviatoric Stresses

Mean normal stress is calculated as:

$$\sigma_m - \frac{1}{3}(\sigma_{11} + \sigma_{22} + \sigma_{33}) \tag{2.4.52a}$$

Total stress vector is then written as a summation of mean stress and deviatoric stress:

$$\sigma_{ij} = S_{ij} + \sigma_m \delta_{ij} \tag{2.4.52b}$$

which is written in tensor form as:

$$\begin{bmatrix} \sigma_{11} & \sigma_{12} & \sigma_{13} \\ \sigma_{21} & \sigma_{22} & \sigma_{23} \\ \sigma_{31} & \sigma_{32} & \sigma_{33} \end{bmatrix} = \begin{bmatrix} S_{11} & S_{12} & S_{13} \\ S_{21} & S_{22} & S_{23} \\ S_{31} & S_{32} & S_{33} \end{bmatrix} + \begin{bmatrix} \sigma_m & 0 & 0 \\ 0 & \sigma_m & 0 \\ 0 & 0 & \sigma_m \end{bmatrix} \tag{2.4.52c}$$

In terms of triaxial stress components (p, q), mean and deviatoric stresses become:

$$p = \frac{1}{3}(\sigma_1 + \sigma_2 + \sigma_3) = \frac{1}{3}(\sigma_{xx} + \sigma_{yy} + \sigma_{zz}) \tag{2.4.52d}$$

$$q = \frac{1}{\sqrt{2}}\sqrt{(\sigma_1 - \sigma_2)^2 + (\sigma_2 - \sigma_3)^2 + (\sigma_3 - \sigma_1)^2} \tag{2.4.52e}$$

Corresponding triaxial strain components ε_p and ε_q are written as:

$$\varepsilon_p = \varepsilon_1 + \varepsilon_2 + \varepsilon_3 = \varepsilon_{xx} + \varepsilon_{yy} + \varepsilon_{zz} \tag{2.4.52f}$$

$$\varepsilon_q = \frac{\sqrt{2}}{3}\sqrt{(\varepsilon_1 - \varepsilon_2)^2 + (\varepsilon_2 - \varepsilon_3)^2 + (\varepsilon_3 - \varepsilon_1)^2} \tag{2.4.52g}$$

The invariants of deviatoric stresses are also calculated considering the stress decomposition of (2.4.52b). So we write,

$$[\sigma] = [S] + [p] \tag{2.4.53}$$

from which we get:

$$[S] = [\sigma] - [p] \tag{2.4.54}$$

Using the above, deviatoric stress invariants are calculated as,

$$J_1 = 0 \tag{2.4.55a}$$

$$J_2 = \frac{I_1^2}{3} - I_2 \tag{2.4.55b}$$

$$J_3 = I_3 - I_2\sigma_m + 2\sigma_m^3 \tag{2.4.55c}$$

We can clearly notice that the first deviatoric invariant is zero as it is related only to the mean stress.

2.4.11 Axi-Symmetric Condition

For the axisymmetric stress state as can be seen in Figure 2.4-16, the intermediate principal stress becomes equal to the minor principal stress, hence the stress and strain parameters become:

$$p = \frac{\sigma_1 + 2\sigma_3}{3} \tag{2.4.56a}$$

$$q = \sigma_1 - \sigma_3 \tag{2.4.56b}$$

$$\varepsilon_p = \varepsilon_1 + 2\varepsilon_3 \tag{2.4.56c}$$

$$\varepsilon_q = \frac{2}{3}(\varepsilon_1 - \varepsilon_3) \tag{2.4.56d}$$

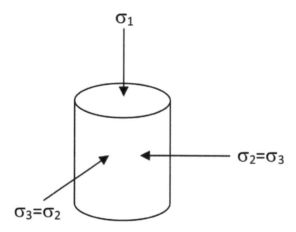

Figure 2.4-16 Axi-symmetric stresses, $\sigma_2 = \sigma_3$

2.4.12 Plane Strain Condition

For the plane strain condition,

$$p = \frac{\sigma_1 + \sigma_2 + \sigma_3}{3} \tag{2.4.57}$$

as the third stress is non-zero, $\sigma_3 \neq 0$. However, for most geotechnical engineering analyses of plain strain, σ_1 and σ_2 are considered. The deviatoric stress is,

$$q = \frac{1}{\sqrt{2}}\sqrt{(\sigma_1 - \sigma_2)^2 + (\sigma_2 - \sigma_3)^2 + (\sigma_3 - \sigma_1)^2} \tag{2.4.58}$$

Similarly we can define the volumetric and shear stresses,

$$\varepsilon_p = \varepsilon_1 + \varepsilon_2 \tag{2.4.59}$$

$$\varepsilon_q = \frac{2}{3}\sqrt{\varepsilon_1^2 + \varepsilon_2^2 + \varepsilon_1\varepsilon_2} \tag{2.4.60}$$

In conclusion, it should be mentioned here that most of the problems encountered in geotechnical engineering that can be simplified in 2-D, can be idealized as either a plane strain or an axisymmetric problem. For some conditions, it is convenient to define mean and deviatoric stresses in terms of variables (p, q).

Computer Implementation

Various calculations for the stress state of an element for which a number of examples are given above are implemented in the MATLAB® program called "**STRESS.m**". Find it at the end of the chapter.

2.5 Mohr's Circle

As briefly presented previously, using the Mohr's Circle method is quite an effective and visually useful way to work with stress states (and similarly strain states). This is particularly true for soils and related problems in soil mechanics. Thus, Mohr's Circle finds a special use in the mechanics of soils not just because it is an effective tool to work with stresses, but it is also necessary to find reliable engineering solutions to problems using the Mohr's Circle. For a given element of soil under 2-D stress state (Figure 2.5-1), one will be interested in the new stress state of a transformed element through rotating at an angle, θ. Although the stress transformation equations are always at our disposal to evaluate the stresses acting on rotated elements, the Mohr Circle as sketched from those equations help return the new stresses on the rotated element in a faster and more effective manner utilizing geometrical relations.

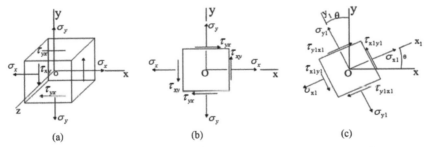

Figure 2.5-1 (a) 3-D Stress element, (b) 2-D representation and (c) Rotated 2-D element

2.5.1 Stress Transformation and Mohr's Circle

In 2-D using Mohr's circle we get two real roots, σ_1 and σ_3 of the equation,

$$\begin{bmatrix} \sigma_{xx} - \sigma & \sigma_{yx} \\ \sigma_{xy} & \sigma_{yy} - \sigma \end{bmatrix} \begin{Bmatrix} l \\ m \end{Bmatrix} = 0 \tag{2.5.1}$$

where $A = \begin{bmatrix} \sigma_{xx} - \sigma & \sigma_{yx} \\ \sigma_{xy} & \sigma_{yy} - \sigma \end{bmatrix}$ and we write $det(A) = 0$. Corresponding direction cosines

of the principal planes are $\begin{bmatrix} l_1 & l_3 \\ m_1 & m_3 \end{bmatrix}$. Figure 2.5-2 shows original and rotated stress elements along with the principal stresses acting on principal planes. We can write for the shear stresses:

$$\tau_{xy} = \tau_{yz} \tag{2.5.2a}$$

$$\tau_{x'y'} = \tau_{y'z'} \tag{2.5.2b}$$

Now we can define the stress transformation for plane stress using the following equations derived from the equilibrium of forces in the rotated directions x'-y' as can be seen in Figure 2.5-3.

$$\sigma_{x'} = \frac{\sigma_x + \sigma_y}{2} + \frac{\sigma_x - \sigma_y}{2}\cos 2\theta + \tau_{xy}\sin 2\theta \tag{2.5.3a}$$

$$\tau_{x'y'} = -\frac{\sigma_x - \sigma_y}{2}\sin 2\theta + \tau_{xy}\cos 2\theta \tag{2.5.3b}$$

$$\sigma_{y'} = \frac{\sigma_x + \sigma_y}{2} - \frac{\sigma_x - \sigma_y}{2}\cos 2\theta - \tau_{xy}\sin 2\theta \tag{2.5.3c}$$

$$\sigma_{x'} + \sigma_{y'} = \sigma_x + \sigma_y \tag{2.5.3d}$$

The last relation can also be used instead of (2.5.3c) to find σ_y. If we sketch the Mohr circle, the center of the circle, $C(\sigma_{ave}, 0)$ will have the coordinate,

$$\sigma_{ave} = \frac{\sigma_x + \sigma_y}{2} \tag{2.5.4}$$

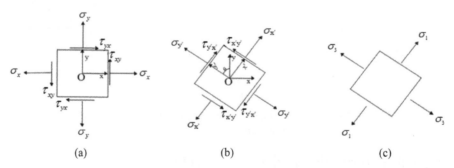

(a)   (b)   (c)

Figure 2.5-2 (a) Original stress element, (b) Rotated element, (c) Principal stresses

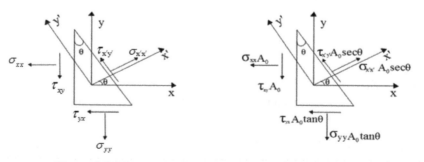

Figure 2.5-3 Stresses and corresponding forces on inclined planes

and radius,

$$R = \sqrt{\left(\frac{\sigma_x - \sigma_y}{2}\right)^2 + \tau_{xy}^2} \qquad (2.5.5)$$

Therefore, the equation of the circle is,

$$(\sigma - \sigma_{ave})^2 + \tau_{xy}^2 = R^2 \qquad (2.5.6)$$

Here, the stresses acting on a plane oriented with an angle θ from the vertical are calculated as:

$$\sigma_{x'} = \frac{\sigma_x + \sigma_y}{2} + R\cos\beta \qquad (2.5.7a)$$

$$\tau_{x'y'} = R\sin\beta \qquad (2.5.7b)$$

where

$$\cos\beta = \frac{1}{R}\left(\frac{\sigma_x - \sigma_y}{2}\cos 2\theta + \tau_{xy}\sin 2\theta\right) \qquad (2.5.8a)$$

$$\cos\beta = \frac{1}{R}\left(\frac{\sigma_x - \sigma_y}{2}\cos 2\theta + \tau_{xy}\sin 2\theta\right) \qquad (2.5.8b)$$

Principal stresses are then calculated:

$$\sigma_1 = \frac{\sigma_x + \sigma_y}{2} + R \qquad (2.5.9a)$$

$$\sigma_3 = \frac{\sigma_x + \sigma_y}{2} - R \qquad (2.5.9b)$$

with the angles,

$$\cos 2\theta = \frac{\sigma_x - \sigma_y}{2R} \qquad (2.5.10a)$$

$$\sin 2\theta = \frac{\tau_{xy}}{R} \qquad (2.5.10b)$$

Deviatoric stresses can also be calculated:

$$\sigma = s + \sigma_m I \qquad (2.5.11a)$$

$$\sigma_{ij} = s_{ij} + \sigma_m \delta_{ij} \qquad (2.5.11b)$$

Invariants of deviatoric stresses were given in (2.4.55). Figure 2.5-4 shows the sign convention for shear stresses. A clear illustration of the Mohr's circle of stresses acting on adjacent planes of the physical element which correspond simply to stress points on the circle is shown in Figure 2.5-5.

Example 2.5-1: For the given element in plane stress, (a) construct the Mohr's circle, and (b) find the stresses acting on a rotated element oriented at an angle of 30° from the horizontal.

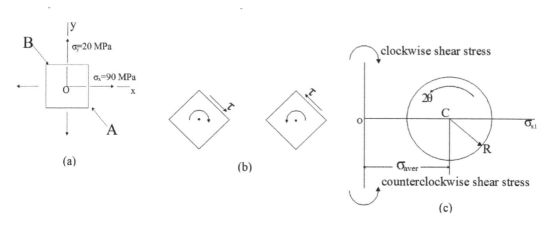

Figure 2.5-4 Given problem and alternative sign convention for shear stresses: (a) Given problem, (b) Clockwise and counterclockwise shear stresses, and (c) Axes for Mohr's circle. (Note that clockwise shear stresses are plotted upward and counterclockwise shear stresses are plotted downward.)

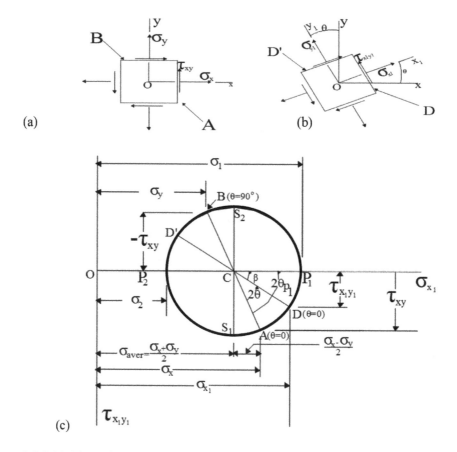

Figure 2.5-5 (a) Given element of stress, (b) Inclined planes and their stresses, (c) Mohr's circle representation of inclined planes and their stresses

Solution:

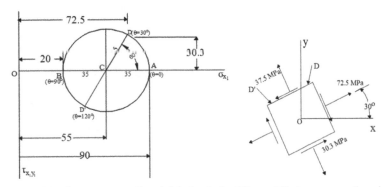

Figure 2.5-6 (a) The corresponding Mohr's circle (*Note:* All stresses on the circle have units of MPa), (b) Rotated element and its stresses

Example 2.5-2: For the given element in plane stress, (a) construct the Mohr's circle, (b) find the stresses acting on a rotated element oriented at an angle of $\theta = 40°$ from the horizontal, (c) show the principal stresses and (d) find the maximum shear stresses.

Solution:

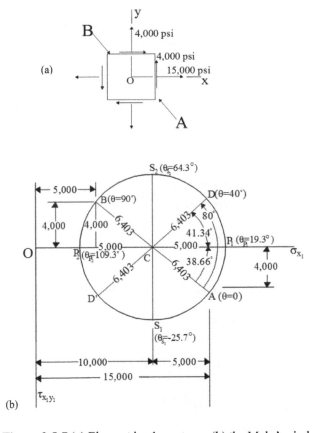

Figure 2.5-7 (a) Element in plane stress, (b) the Mohr's circle

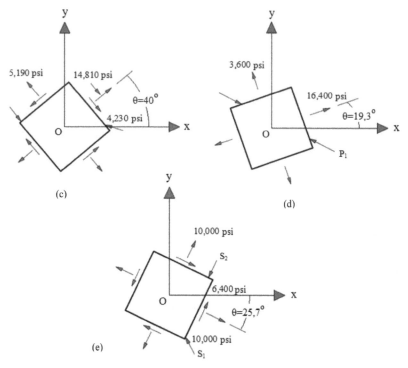

Figure 2.5-7 (c) Stresses acting on an element oriented at 40°, (d) Principal stresses and (e) Maximum shear stresses. (Note: All stresses have units of psi.)

2.5.2 3-D Stresses

If we assume $\sigma_1 > \sigma_2 > \sigma_3$, this system given in (2.4.42) and (2.4.43) is solved for:

$$l^2 = \frac{\tau_n^2 + (\sigma_n - \sigma_2)(\sigma_n - \sigma_3)}{(\sigma_1 - \sigma_2)(\sigma_1 - \sigma_3)} \geq 0 \tag{2.6.1a}$$

$$m^2 = \frac{\tau_n^2 + (\sigma_n - \sigma_3)(\sigma_n - \sigma_1)}{(\sigma_2 - \sigma_3)(\sigma_2 - \sigma_1)} \geq 0 \tag{2.6.1b}$$

$$n^2 = \frac{\tau_n^2 + (\sigma_n - \sigma_1)(\sigma_n - \sigma_2)}{(\sigma_3 - \sigma_1)(\sigma_3 - \sigma_2)} \geq 0 \tag{2.6.1c}$$

The above set can then be manipulated into the following form:

$$\tau_n^2 + (\sigma_n - \frac{\sigma_2 - \sigma_3}{2})^2 \geq (\frac{\sigma_2 - \sigma_3}{2})^2 = \tau_1^2 \tag{2.6.2a}$$

$$\tau_n^2 + (\sigma_n - \frac{\sigma_3 - \sigma_1}{2})^2 \geq (\frac{\sigma_3 - \sigma_1}{2})^2 = \tau_2^2 \tag{2.6.2b}$$

$$\tau_n^2 + (\sigma_n - \frac{\sigma_1 - \sigma_2}{2})^2 \geq (\frac{\sigma_1 - \sigma_2}{2})^2 = \tau_3^2 \tag{2.6.2c}$$

which lead to the Mohr's circles in the stress space seen in Figure 2.6-1. Octahedral stresses can also be evaluated (see Figure 2.4-14),

$$l = m = n = \frac{\sqrt{3}}{3} \tag{2.6.3}$$

$$\sigma_{oct} = \frac{\sigma_1 + \sigma_2 + \sigma_3}{3} \tag{2.6.4}$$

$$\tau_{oct} = \frac{1}{3}\sqrt{(\sigma_1 - \sigma_2)^2 + (\sigma_2 - \sigma_3)^2 + (\sigma_3 - \sigma_1)^2} \tag{2.6.5}$$

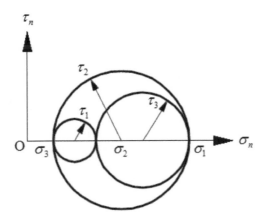

Figure 2.6-1 Mohr's circle

Computer Implementation

In this section, a number of MATLAB scripts along with related functions are developed to calculate the stress transformation as well as sketch Mohr's circle for a given stress state. Below is the given initial plane stress states and related rotation angles measured from the horizontal axis. Mohr circles have been developed using the program "**Mohr_Circle.m**".

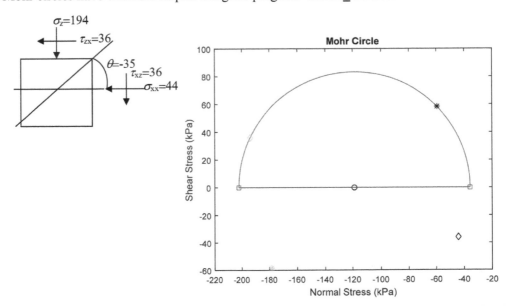

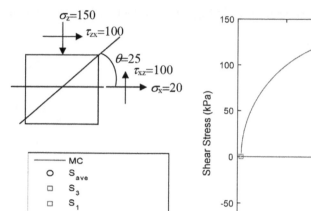

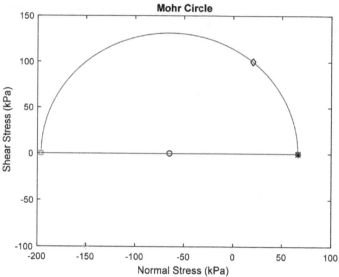

MATLAB® Scripts

Main program "STRESS.m"

```
% "STRESS.m"
% Given the stress tensor at a point, this program performs stress
%transformation
% Stress Matrix at a Point
S11=200;
S12=100;
S13=300;
S21=100;
S22=0;
S23=0;
S31=300;
S32=0;
S33=0;
S = [S11,S12,S13;S21,S22,S23;S31,S32,S33];
%Plane through the point(direction cosines of its normal)
l1 = 2/3;
l2 = -2/3;
l3 = 1/3;
lv=l1^2+l2^2+l3^2;
l = [l1,l2,l3];
%Evaluate the stress vector (Snx, Sny, Snz) on the plane
Sn = S*(l');
Snn = Sn'*l';
%Evaluate the resultant stress on the plane
Sr2 = Sn(1)^2 + Sn(2)^2 + Sn(3)^2;
%Evaluate the normal and shear stresses on the plane
Snn2 = Snn^2;
Sss = sqrt(Sr2-Snn2);
%Evaluate the stress tensor with reference to new set of axes
%(normals to another set of oriented planes - defined by direction cosines)
l11=0;
l12=1/sqrt(2);
l13=1/sqrt(2);
l21=1/sqrt(2);
l22=1/2;
l23=-1/2;
l31=1/sqrt(2);
l32=1/2;
l33=-1/2;
l  = [ l11,l12,l13;l21,l22,l23;l31,l32,l33 ];
%Evaluate the stress vector on second coordinate system at givethe point
ST= l*S*(l');
%Evaluate the first, second and third stress invariants (defined in terms of
%original stress components)
I1=S11+S22+S33;
I2=S11*S22+S22*S33+S33*S11-S12^2-S23^2-S31^2;
I3=S11*S22*S33-S11*S23^2-S22*S31^2-S33*S12^2-2*S12*S23*S31;
%Evaluate the Direction Cosines of Principal Axes and Principal Stresses
[Lp,Sp] = eig(S);
%Another way to evaluate the invariants (in terms of principal stresses)
I1=Sp(1,1)+Sp(2,2)+Sp(3,3);
I2=Sp(1,1)*Sp(2,2)+Sp(2,2)*Sp(3,3)+Sp(3,3)*Sp(1,1);
I3=Sp(1,1)*Sp(2,2)*Sp(3,3);
```

```
%Evaluate the Normal and Shear Stresses on Octahedral Plane
Snoct = (Sp(1,1) + Sp(2,2) + Sp(3,3))/3;
Ssoct = sqrt((Sp(1,1)-Sp(2,2))^2 + (Sp(2,2) - Sp(3,3))^2 + (Sp(3,3)-
Sp(1,1))^2)/3;
```

Main program "Mohr_Circle.m"

```
%Chapter 2: Basic Mechanics
%"Mohr_Circle.m"
%Mohr's Circle of Plane Stresses
%Stress Transformation
%Sign Convention: Tension is + for Normal Stress, CCW is + for Shear
% Rotation Angle
theta=25; % in degrees from the horizontal axis
Sx=20;
Sy=-150;
Sxy=100;
Stress=[Sx Sy Sxy];
n=1000; %Number of stresses to show for the Mohr circle
Save=0.5*(Sx+Sy); %Sigma Average
Rad=sqrt((0.5*(Sx-Sy))^2+Sxy^2); %Radius

%Calculate principal stresses
P1=Save+Rad;
P3=Save-Rad;
S=P3:(P1-P3)/n:P1; %Normal stresses
T=sqrt(Rad.^2-(S-Save).^2); %Shear stresses

%Rotated Stresses
%Call the stress transformation function
S_r=Plane_Stress_Transformation(Stress,theta);
% Sxx=Save+Rad*cos(beta);
S_r_x=S_r(1);
S_r_y=S_r(2);
S_r_xy=S_r(3);

% Plot the Mohr-Circle
plot(S,T)
title('Mohr Circle')
hold on
plot(Save,0,'ko',P3,0,'rs',P1,0,'s',...
    Sx,Sxy,'kd',Sy,-Sxy,'gd',S_r_x,S_r_xy,'k*',S_r_y,-S_r_
xy,'g*',S,zeros(length(S)),'-')
xlabel('Normal Stress (kPa)')
ylabel('Shear Stress (kPa)')
legend('MC','S_a_v_e','S_3','S_1','Stress Point 1','Stress Point 2',...
    'Rotated Stress Point 1','Rotated Stress Point 2','Location',...
    'SouthEast')
```

Sub program "Plane_Stress_Transformation.m"

```
function [Sigma_rotated]=Plane_Stress_Transformation(stress,theta)
%Stress Transformation Equations
%Plane Stress
%Rotation Angle
%theta in degrees
Sx=stress(1);
```

```
Sy=stress(2);
Sxy=stress(3);
Sxx_r=0.5*(Sx+Sy)+0.5*(Sx-Sy)*cos(2*theta*pi/180)+Sxy*sin(2*theta*pi/180);
Syy_r=0.5*(Sx+Sy)-0.5*(Sx-Sy)*cos(2*theta*pi/180)-Sxy*sin(2*theta*pi/180);
Sxy_r=-0.5*(Sx-Sy)*sin(2*theta*pi/180)+Sxy*cos(2*theta*pi/180);
% Rotated Stresses
Sigma_rotated=[Sxx_r Syy_r Sxy_r];
%Verify
if Sx+Sy==Sxx_r+Syy_r
    display('correct transformation')
else
    error('incorrect transformation, check the calculation')
end

end
```

Physical Laws and Governing Equations

3.1 Introduction

In Chapter 2, we considered only a single-phase continuum and presented the basic concepts in relation to strains and stresses. However, we clearly recognize that in general soil is a multi-phase medium with three separate phases of solid (soil skeleton), liquid (pore water), and gas (pore air). The behavior of such a system should then be modeled by considering the interaction of all the phases. This requires an analysis of the coupled flow (of the pore fluid) and deformation (of the solid skeleton).

It has been common to treat the saturated porous medium (i.e. soil) as a combination of separate phases of fluid and solid resulting in a simpler decoupled analysis. The mathematical theory of "Consolidation" describing the time dependent dissipation of pore pressures through the flow of pore water and the resulting compression of the soil skeleton is such an example. Karl Terzaghi in 1923 developed this theory in one-dimensional (1-D) form by adopting '*the principle of effective stress*' (Terzaghi, 1923, 1925). Later, the equations governing the response of saturated porous media incorporating the fluid-solid skeleton interaction was established for the quasi-static (QS) case by Biot (1941), who then presented a general extended set of equations governing the response of a saturated linear elastic porous solid under dynamic conditions (Biot 1955, 1962). Later, Truesdell (1957, 1962) introduced the "mixture theory" to formulate such coupled equations. This formulation has been subsequently extended to consider the nonlinearity of deformation (Prevost 1980, 1982; Zienkiewicz et al. 1999). Depending on the motion of the pore fluid and the solid skeleton as well as the permeability of the porous medium, it is possible to obtain different formulations in this coupled problem. Most of these formulations are obtained by neglecting some or all of the inertial terms while others refer to drained and undrained extremes in the porous medium.

In this chapter, the basic principles and derivations of various general governing equations (with varying degrees of idealizations) are presented using the basic laws of *conservation of momentum* and *conservation of mass*. These, in turn, yield equilibrium and mass balance equations, respectively. In the next few chapters we will present the constitutive laws, which yield the equations governing the stress-strain relations for soils.

The equilibrium relations are written in terms of total stresses including both the fluid and the solid; however, the deformation of the solid phase is described in terms of the "effective stress". Stress-strain relationship is also written in terms of effective stresses and the flow of pore fluid is governed by the mass balance equation. Finally, these governing equations yield a set of coupled algebraic equations that need to be solved simultaneously for the field variables: u, the displacement of the solid skeleton; w, the displacement of the pore fluid relative to that of skeleton and p, pore fluid pressure.

3.2 Idealizations

The following key idealizations are made for the formulations of the governing equations to be used throughout this book:

- Soils are of multi-phase materials with particulate (discrete) nature. However, we still treat such materials as a continuum with each material point as overlapping system of different phases. Figure 3.2-1 depicts such a system.
- For this three-phase continuum, water and the gaseous (air) phases are considered to constitute a single compressible liquid. The effects of gas diffusing through water and movement of water vapor are ignored. This assumption is justifiable only when the degree of saturation is very high (95-100%) which is the case only for the soils with a rather small amount of gas present in their voids which are termed as *fully saturated soils*.
- Consistent with the scope of this book, we will restrict ourselves to static conditions ignoring the inertial effects associated with both the motion of the fluid and that of the solid skeleton.

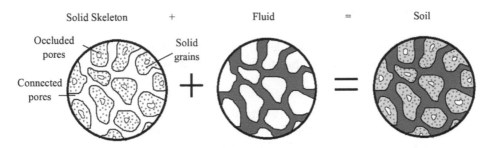

Figure 3.2-1 Soil considered as a multi-phase continuum

3.3 Total and Effective Stresses in Soils

As noted earlier, soil is a multi-phase material and any stress acting on it is supposed to be carried by all the phases, in particular the solid phase that is comprised of the soil grains and the fluid phase. Figure 3.3-1 shows a close-up view of an inter-granular contact area in a soil element. If we write the components of total contact force, F, carried by the phases we get,

$$F = \sigma_s A_s + \sigma_w A_w + \sigma_g A_g \tag{3.3.1}$$

Dividing both sides by the total area gives,

$$\sigma = \frac{F}{A} = \frac{A_s}{A}\sigma_s + \frac{A_w}{A}\sigma_w + \frac{A_g}{A}\sigma_g \tag{3.3.2}$$

which can also be written as,

$$\sigma = a\sigma_s + (1-a)\sigma_w + (1-a-x)(\sigma_g - \sigma_w) \tag{3.3.3}$$

where $a = \dfrac{A_s}{A}$ is the solid area ratio and $\chi = \dfrac{A_w}{A}$ is the water area ratio. If the soil is fully saturated, we have $1 - a - \chi = 0$ leading to,

$$\sigma = a\sigma_s + (1-a)\sigma_w \tag{3.3.4}$$

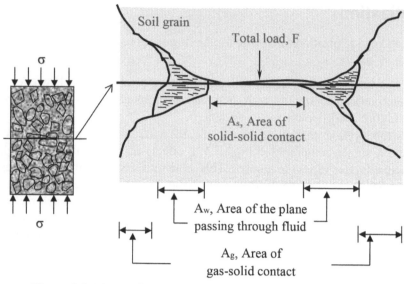

Figure 3.3-1 Transmission of the load through multiphase medium

where the first term is the stress carried solely by *solid skeleton* and the second term is the stress carried by pore water. Therefore, the final form of the equation becomes,

$$\sigma = \sigma' + p_w, \tag{3.3.5}$$

where in soil mechanics terminology, σ' is termed as the "effective stress" and p_w is the "pore water pressure". Here we must note the distinction between the inter-granular contact stress and the effective stress. See the Figure 3.3-2 below. The contact stress (p_s) is a measure of the load (per unit contact area) which develops at grain-to-grain contacts and will usually be very high. The effective stress is a measure of the load (per unit total area) carried by the solid skeleton that is calculated considering a particular cross section of the soil.

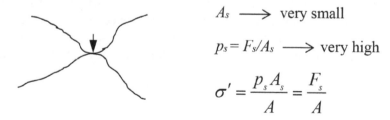

Figure 3.3-2 Contact stress and effective stress

3.3.1 Principle of Effective Stress

We use the *principle of effective stress* for understanding the response of saturated soils everywhere, which is stated as:

- Effective stress is the difference between total stress and the pore water pressure:

$$\sigma' = \sigma - p_w, \tag{3.3.6}$$

- Effective stress controls the deformation of soils.

3.3.2 Total and Effective Stress Invariants

The invariants of total and effective stresses for various stress states are given here. Recalling (2.4.56a) for the triaxial stress state, the mean effective stresses can now be defined as:

$$p' = \frac{\sigma'_1 + 2\sigma'_3}{3} = \frac{\sigma_1 - u + 2(\sigma_3 - u)}{3} = \frac{\sigma_1 + 2\sigma_3 - 3u}{3} = \frac{\sigma_1 + 2\sigma_3}{3} - u \tag{3.3.7}$$

which can be written as:

$$p' = p - u \tag{3.3.8}$$

Now using (2.4.56b), the total and effective deviatoric stresses become:

$$q' = \sigma'_1 - \sigma'_3 = (\sigma_1 - u) - (\sigma_3 - u) = \sigma_1 - \sigma_3 = q \tag{3.3.9}$$

which means,

$$q' = q \tag{3.3.10}$$

3.4 Law of Conservation of Momentum: Equilibrium Equations

Any load applied on a saturated soil mass will cause a stress field and an associated pore water pressure. Under the developed stress field, the total soil mass has to satisfy the law of conservation of momentum. This is alternatively expressed in the form of equilibrium equations as presented below. According to Newton's second law of motion:

$$\sum F = \frac{\partial(mV)}{\partial t} = m\frac{\partial V}{\partial t} + V\frac{\partial m}{\partial t} \tag{3.4.1}$$

$$\sum F = ma \tag{3.4.2}$$

where ΣF represents all the surface tractions and body forces acting on the element of soil with 'a' being the absolute acceleration with respect to the inertial frame of reference. For static problems (3.4.2) reduces to:

$$\sum F = 0 \tag{3.4.3}$$

In this section, the equilibrium equations are developed for the total soil (solid skeleton and the pore fluid) in terms of the effective stresses and the pore pressure.

3.4.1 Static Equilibrium of a Soil Element

Figure 3.4-1 shows a soil element in equilibrium under a stress state.
For the equilibrium in x-direction,

$$\left(\sigma_{xx} + \frac{\partial\sigma_{xx}}{\partial x} \cdot dx\right) \cdot dy \cdot dz - \sigma_{xx} \cdot dy \cdot dz + \left(\tau_{yx} + \frac{\partial\tau_{yx}}{\partial y} \cdot dy\right) \cdot dx \cdot dz - \tau_{yx} \cdot dx \cdot dz$$

$$+ \left(\tau_{zx} + \frac{\partial\tau_{zx}}{\partial z} \cdot dz\right) \cdot dx \cdot dy - \tau_{zx} \cdot dx \cdot dy + f_x \cdot dx \cdot dy \cdot dz = 0 \tag{3.4.4}$$

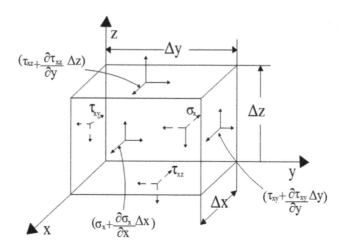

Figure 3.4-1 Equilibrium of a soil element

The above equation can be written as:

$$\frac{\partial \sigma_{xx}}{\partial x} + \frac{\partial \tau_{yx}}{\partial y} + \frac{\partial \tau_{zx}}{\partial z} + f_x = 0 \tag{3.4.5a}$$

Similarly considering the equilibrium of the soil element in y- and z-directions following equations of equilibrium can be readily derived:

$$\frac{\partial \tau_{xy}}{\partial x} + \frac{\partial \sigma_{yy}}{\partial y} + \frac{\partial \tau_{zy}}{\partial z} + f_y = 0 \tag{3.4.5b}$$

$$\frac{\partial \tau_{xz}}{\partial x} + \frac{\partial \tau_{yz}}{\partial y} + \frac{\partial \sigma_{zz}}{\partial z} + f_z = 0 \tag{3.4.5c}$$

Using the effective stress relation in (3.3.5), equations (3.4.5) can also be written in terms of effective normal stresses as:

$$\frac{\partial \sigma'_{xx}}{\partial x} + \frac{\partial \tau_{yx}}{\partial y} + \frac{\partial \tau_{zx}}{\partial z} + f_x + \frac{\partial p}{\partial x} = 0 \tag{3.4.6a}$$

$$\frac{\partial \tau_{xy}}{\partial x} + \frac{\partial \sigma'_{yy}}{\partial y} + \frac{\partial \tau_{zy}}{\partial z} + f_y + \frac{\partial p}{\partial y} = 0 \tag{3.4.6b}$$

$$\frac{\partial \tau_{xz}}{\partial x} + \frac{\partial \tau_{yz}}{\partial y} + \frac{\partial \sigma'_{zz}}{\partial z} + f_z + \frac{\partial p}{\partial z} = 0 \tag{3.4.6c}$$

The above equations may also be written in the following tensorial form:

$$\sigma'_{ij,j} + f_j + \delta_{ij} p_w = 0 \tag{3.4.7}$$

3.4.2 Fluid Equilibrium and D'Arcy's Law

We now write the equilibrium equation for an element of pore fluid as can be seen in Figure 3.4-2. In the figure, the equilibrium under the following forces is being considered;

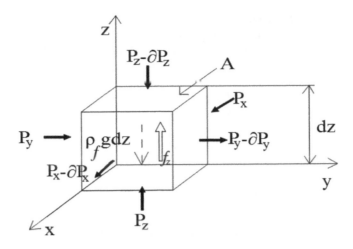

Figure 3.4-2 Representative element for fluid equilibrium

$P(P_x, P_y, P_z)$ are the forces due to the fluid pressure on the boundary surfaces in the x, y, and z directions, $\rho_f g dz$ is the gravitational force in the z-direction, $f(f_x, f_y, f_z)$ are the seepage forces in respective directions. Considering the equilibrium in the z-direction, we have,

$$P_z - (P_z - dP_z) - \rho_f gAdz + f_z = 0 \tag{3.4.8a}$$

$$\frac{\partial p}{\partial z} Adz - \rho_f gAdz + i_z \rho_f gAdz = 0 \tag{3.4.8b}$$

$$-\frac{\partial p}{\partial z} + \rho_f g - \frac{\partial h}{\partial z}\rho_f g = 0 \tag{3.4.9a}$$

where h is the total head which is the sum of the pressure head, velocity head and the elevation head as well as $i_z = dh/dz$ being the gradient of the total head in the z-direction. Similarly, considering the equilibrium of forces in the x- and y-directions the following can be derived:

$$-\frac{\partial p}{\partial x} - \frac{\partial h}{\partial x}\rho_f g = 0 \tag{3.4.9b}$$

$$-\frac{\partial p}{\partial y} - \frac{\partial h}{\partial y}\rho_f g = 0 \tag{3.4.9c}$$

D'Arcy's law represents a constitutive law for the flow according to which the fluxes (i.e. volume of fluid flowing per unit time and per unit total area of the medium) in respective directions are given by,

$$q_x = -k_x \frac{\partial h}{\partial x} \tag{3.4.10a}$$

$$q_y = -k_y \frac{\partial h}{\partial y} \tag{3.4.10b}$$

$$q_z = -k_z \frac{\partial h}{\partial z} \tag{3.4.10c}$$

In the above q_x, q_y and q_z are identified as the superficial velocities of pore fluid (relative to soil skeleton), which can be alternatively represented as time derivatives of displacements u, v and w as $\dot{u}$, $\dot{v}$, and $\dot{w}$. Now combining the above D'Arcy's equations with the fluid equilibrium equations in (3.4.9), we can rewrite them as,

$$-\frac{\partial p}{\partial x} - \frac{u}{k_x}\rho_f g = 0 \qquad (3.4.11a)$$

$$-\frac{\partial p}{\partial y} - \frac{v}{k_y}\rho_f g = 0 \qquad (3.4.11b)$$

$$-\frac{\partial p}{\partial z} + \rho_f g - \frac{\dot{w}}{k_z}\rho_f g = 0 \qquad (3.4.11c)$$

In tensorial form, the above becomes the final equation of the fluid equilibrium neglecting the associated inertial terms,

$$-p_i + \rho_f g_i - \frac{u_i}{k_i}\rho_f g_i = 0 \qquad (3.4.12)$$

3.5 Law of Conservation of Mass

The equation set is completed when the continuity of flow condition is written. We can discuss the continuity of flow by showing its vertical component in an element as in Figure 3.5-1. Here, $\psi_{z-dz/2}$ represents the mass flux or the amount of fluid flowing through the bottom of the element, $\psi_{z+dz/2}$ is the mass flux out of the top of the element and q_z is the rate of flow in the vertical direction. If we write the net mass flux in the z-direction as the difference between the flow into the bottom and the flow out of the top, we find,

$$\psi_{znet} = \psi_{z-dz/2} - \psi_{z+dz/2} \qquad (3.5.1)$$

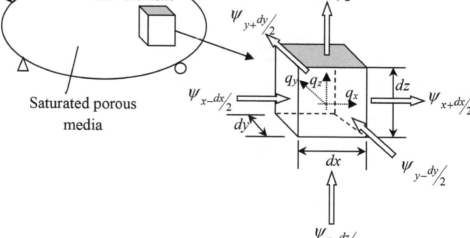

Figure 3.5-1 Element for flow continuity

which is simply,

$$\psi_{znet} = \rho_f \left(q_z - \frac{\partial q_z}{\partial z}\frac{dz}{2} \right) - \rho_f \left(q_z + \frac{\partial q_z}{\partial z}\frac{dz}{2} \right) = -\rho_f \frac{\partial q_z}{\partial z} dz \qquad (3.5.2)$$

We can similarly write the net mass fluxes for other directions using the $\psi_{x-dx/2}$ and $\psi_{x+dx/2}$ for the x-direction and also $\psi_{y-dy/2}$ and $\psi_{y+dy/2}$ for the y-direction. Adding them to find the total mass flux by assuming the unit weight constant and making use of (3.4.10) gives the total mass flux,

$$\psi_T = -\left\{ \frac{\partial}{\partial z}\left[\rho_f \left(-k_z \frac{\partial h}{\partial z} \right) \right] + \frac{\partial}{\partial y}\left[\rho_f \left(-k_y \frac{\partial h}{\partial y} \right) \right] \right.$$
$$\left. + \frac{\partial}{\partial x}\left[\rho_f \left(-k_x \frac{\partial h}{\partial x} \right) \right] \right\} = -\rho_f q_{i,i} V \qquad (3.5.3)$$

Earlier we have considered the porous medium to be rigid (i.e. non-deformable) and fluid to be incompressible. This yields $\psi_T = 0$. Recall also that for a homogeneous and isotropic medium this equation is reduced to the well-known Laplace equation:

$$\nabla^2 h = 0 \qquad (3.5.4)$$

Now, by considering the soil medium to be deformable, we have,

$$\psi_T = \frac{\partial m_f}{\partial t} = \frac{\partial}{\partial t}\rho_f V_f = V_f \frac{\partial \rho_f}{\partial t} + \rho_f \frac{\partial V_f}{\partial t} \qquad (3.5.5)$$

where the first term on the right hand side of the equation represents the compressibility of the fluid leading to density changes and the second, compressibility of the solid skeleton leading to a change in the volume of pores. This equation essentially gives the total change in mass flux of water due to flow. Taking the partial derivative of the first term, it can be written as,

$$V_f \frac{\partial \rho_f}{\partial p}\frac{\partial p}{\partial t} = V_f \frac{\partial}{\partial p}\left(\frac{m_f}{V_f} \right)\frac{\partial p}{\partial t} = -V_f \frac{\partial V_f}{\partial p}\left(\frac{m_f}{V_f^2} \right)\frac{\partial p}{\partial t}$$
$$= -V_f \left(\frac{1}{V_f}\frac{\partial V_f}{\partial p}(\rho_f) \right)\frac{\partial p}{\partial t} = \frac{nV}{K_f}\rho_f \frac{\partial p}{\partial t} \qquad (3.5.6)$$

making use of $V_f = nV$ and of $\frac{1}{V_f}\frac{\partial V_f}{\partial p} = \frac{\partial \varepsilon_v}{\partial p} = \frac{1}{K_f}$. Now if we define 'compressibility' as β, the first term becomes,

$$V_f \frac{\partial \rho_f}{\partial t} = nV\rho_f \beta \frac{\partial p}{\partial t} \qquad (3.5.7)$$

and similarly the second term is,

$$\rho_f \frac{\partial V_f}{\partial t} = \rho_f V \frac{\partial \varepsilon_v}{\partial t} \qquad (3.5.8)$$

Then (3.5.5) simply becomes,

$$\psi_T = \rho_f V \left(\frac{\partial \varepsilon_v}{\partial t} + n\beta \frac{\partial p}{\partial t} \right) \qquad (3.5.9)$$

Now combining the two definitions of total fluxes, (3.5.3) and (3.5.9), we get,

$$-\rho_f q_{i,i} V = \rho_f V \left(\frac{\partial \varepsilon_v}{\partial t} + n\beta \frac{\partial p}{\partial t} \right) \tag{3.5.10}$$

In Cartesian coordinates, (3.5.10) is also written as,

$$k \frac{\rho_f}{\rho_f g} \left(\frac{\partial^2 p}{\partial x^2} + \frac{\partial^2 p}{\partial y^2} + \frac{\partial^2 p}{\partial z^2} \right) V = \rho_f V \left(\frac{\partial \varepsilon_v}{\partial t} + n\beta \frac{\partial p}{\partial t} \right) \tag{3.5.11}$$

or

$$\frac{k}{\rho_f g} \nabla^2 p = \left(\frac{\partial \varepsilon_v}{\partial t} + n\beta \frac{\partial p}{\partial t} \right) \tag{3.5.12}$$

Note that $q_i = -k \left(\frac{\partial h}{\partial x} + \frac{\partial h}{\partial y} + \frac{\partial h}{\partial z} \right)$ and $h = \frac{p}{\rho_w g}$, hence giving $q_{i,i} = \frac{-k}{\rho_f g} \nabla^2 p$. If we use the effective stress principle, we get,

$$\frac{k}{\rho_f g} \nabla^2 p = \frac{1}{D} \frac{\partial}{\partial t} (\sigma_{oct} + p) + n\beta \frac{\partial p}{\partial t} \tag{3.5.13}$$

where D is *the compressibility modulus* of the soil skeleton and σ_{oct} is the total mean stress (octahedral normal stress). This equation can also be written in 3-D indicial notation as,

$$\dot{\varepsilon}_{ii} + \dot{\bar{w}}_{i,i} = -\frac{n}{K_f} \dot{p} \tag{3.5.14}$$

and is called the "Mass Balance Equation". Its complete derivation is omitted here. For only slight unsaturation (i.e. S close to 1), the pore fluid can still be treated as of single phase and the effect can be captured by simply modifying the fluid compressibility (Okusa 1985, Rahman et al. 1994). The compressibility of the pore fluid, β, is modified as,

$$\beta = \frac{1}{K_f} = \frac{1}{K_p} + \frac{1-S}{p_0} \tag{3.5.15}$$

where p_0 represents absolute pore pressure (or the reference pressure) and K_p is the bulk modulus of the fluid itself (i.e. in case of water, $K_p = K_w$). In the 2-D plane strain form, equation (3.5.14) becomes:

$$(\dot{\varepsilon}_x + \dot{\varepsilon}_z) + \left(\frac{\partial \dot{\bar{w}}_x}{\partial x} + \frac{\partial \dot{\bar{w}}_z}{\partial z} \right) = -\frac{n}{K_f} \dot{p} \tag{3.5.16}$$

or written in terms of pore pressure,

$$\frac{-k}{\rho_f g} \left(\frac{\partial^2 p}{\partial x^2} + \frac{\partial^2 p}{\partial z^2} \right) = \frac{1}{D} \frac{\partial}{\partial t} (\sigma_{oct} + p) + n\beta \frac{\partial p}{\partial t} \tag{3.5.17}$$

3.5.1 1-D Consolidation Equation

If we neglect the compressibility of pore fluid (which is mostly accepted in soil mechanics, $\beta = 0$), the last term in equation (3.5.13) drops and the equation becomes,

$$\frac{k}{\rho_f g}\nabla^2 p = \frac{1}{D}\frac{\partial}{\partial t}(\sigma_{oct} + p)$$ (3.5.18)

which can also be written as,

$$\frac{kD}{\rho_f g}\nabla^2 p = \frac{\partial \sigma_{oct}}{\partial t} + \frac{\partial p}{\partial t}$$ (3.5.19)

which clearly shows that the flow of pore fluid and the deformation of the solid skeleton are coupled. If we neglect the first term on the right hand side of (3.5.19) which states that the total mean stress in the soil stays constant (which is the case if load relatively remains constant in time), the equation becomes:

$$\frac{kD}{\rho_f g}\nabla^2 p = \frac{\partial p}{\partial t}$$ (3.5.20)

which is often called the *Rendulic's Theory*. If the change in octahedral stress is taken as the change due to applied outside surface load ($\sigma_{oct} = f(\sigma_z)$), we obtain the Terzaghi's 1-D consolidation theory as,

$$C_v \frac{\partial^2 p}{\partial z^2} = \frac{\partial p}{\partial t}$$ (3.5.21)

where C_v is the coefficient of consolidation written as $C_v = \dfrac{kD}{\gamma_w} = \dfrac{k}{m_v \gamma_w}$. Here m_v is called the

volumetric compressibility coefficient whose inverse is the *elastic constrained modulus*, D (or the compressibility modulus as mentioned before).

PART II

Elemental Response: Constitutive Models

I.   Introduction

In Chapter 2, the effect of loading on a material body was defined in terms of the stress state represented by its components, followed by the kinematics of deformation in terms of strain components. For the general case of 3-D loading and deformation, there are nine components of both stress and strain. However, because of the symmetry there are only six independent components. Additionally in Chapter 2, the strains were defined in terms of three displacement components followed by six strain-displacement compatibility relationships. In Chapter 3, the basic laws governing load-deformation behavior of a material body was presented including three equilibrium equations.

Our ultimate objective is to predict the deformation response of a geotechnical system subjected to loading. For example, consider a common problem of a footing resting on a foundation soil shown below (Figure II.1).

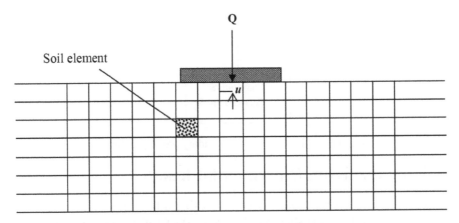

Figure II.1 Load-deformation problem

It is evident that the global behavior of a system (as in the above problem) is intimately related to the behavior of all the small elements of the soil. Thus, in order to evaluate the global deformation response, we must first formulate a constitutive model (stress-strain relationship) for an element of soil (highlighted in Figure II.1). We seek this relationship in the following form:

$$\{\sigma'\} = [D]\{\varepsilon\} \tag{II.1}$$

These constitutive relations together with the compatibility and equilibrium equations provide us with the necessary number of equations to solve a boundary value problem (Figure II.2).

It should be noted here that for soils the stress-strain laws are typically written in terms of 'effective stresses', which we denote as σ'. In Chapters 4, 5 and 6, for the sake of convenience, we will suppress the prime superscript.

Soil as a material is: (i) a particulate system consisting of an assemblage of discrete particles with void spaces between the grains, (ii) a multiphase system with solid grains, and the void spaces filled with air and/or water, (iii) non-homogeneous porous medium with its properties varying spatially, and (iv) anisotropic material with its properties varying with the direction of loading. The actual behavior of soils tends to be quite complex and depends on many factors. Therefore, it becomes necessary to develop idealized constitutive models for explaining various soil behaviors under various conditions. In this part of the book, two broad classes of constitutive models are presented. These are (1) time-independent behavior and (2) time dependent behavior. In Chapter 4, we will present the elastic constitutive laws followed by *the theory of plasticity* based constitutive laws presented in Chapter 5. Finally, the viscoelastic and viscoplastic constitutive models for simulating time-dependent behavior are presented in Chapter 6.

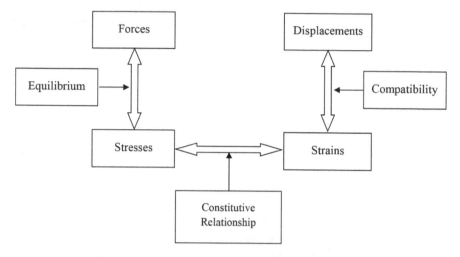

Figure II.2 Equations of a boundary value problem

II.   Soil Behavior: From Experimental Results

We usually study the stress-strain behavior of soils using a triaxial shearing test in which an element (a cylindrical soil sample shown in Figure II.3) of soil is in its first stage of loading subjected to an isotropic stress state (with $\sigma_c = \sigma_1 = \sigma_2 = \sigma_3$) during which soil is allowed to consolidate. This is often followed by an isotropic unloading to σ_0, if desired to cause an "overconsolidation". In the second stage of loading, soil element is subjected to an increasing axial loading (with $\sigma_2 = \sigma_3$), which can be either stress (σ_1) or strain (ε_1) controlled, with the measurement of axial strain (ε_1) or axial stress (σ_1) accordingly. In a drained test, volumetric strain of the soil is monitored, while in an undrained test, with volume kept constant, pore pressure generation is measured. This test allows to control two principal stresses independently. As was mentioned in the previous chapter in terms of octahedral and mean stresses, the stress state in a triaxial system can be expressed as:

$$p = I1/3 = \frac{\sigma_1 + 2\sigma_3}{3}, \quad q = \sqrt{3J_2} = \sigma_1 - \sigma_3 \qquad (II.2)$$

$$\sigma_1; \varepsilon_1$$

$$\sigma_2 = \sigma_3$$
$$\varepsilon_2 = \varepsilon_3$$

$$\sigma_3; \varepsilon_3 \qquad\qquad \sigma_2; \varepsilon_2$$

Figure II.3 Stress and strain states in the triaxial test

where p and q are the mean and deviatoric stresses, which are related to the first invariant of stress tensor, I_1 and the second invariant of deviatoric stress tensor, J_2 (see Chapter 2). Similarly, the corresponding strain state is represented as:

$$\varepsilon_p = \varepsilon_1 + 2\varepsilon_3, \ \varepsilon_q = \frac{2}{3}(\varepsilon_1 - \varepsilon_3) \tag{II.3}$$

where ε_p and ε_q are the volumetric and shear strains that are related to the first invariant of strain tensor and the second invariant of deviatoric strain tensor.

Typical results from a drained test on a slightly overconsolidated clay is shown in Figure II.4. The dashed lines represent the response during the first stage of loading while the solid lines represent the response during the second stage of shearing. The soil element is first loaded to a pre-consolidation stress p'_c followed by an unloading to p'_o. Thereafter the sample is subjected to increasing axial stress (with $\Delta\sigma_1$) while keeping the radial stress constant ($\Delta\sigma_2 = \Delta\sigma_3 = 0$).

Studying the experimental results of Figure II.4, we can readily observe the following:

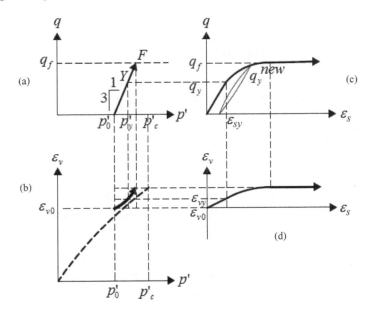

Figure II.4 Typical behavior: (a) Stress path, (b) Volumetric response, (c) Shear response, (d) Strain path (reproduced after Puzrin, 2012)

- The volumetric behavior in general is non-linear and irreversible as seen in (b).
- In the initial part of the second stage of loading, both the volumetric (b) and shear (c) are reversible and linear.
- The above behavior continues until the stress path reaches the yield stress state Y (p'_y, q_y), which depends on pre-consolidation pressure ratio, p_c/p_g'. For normally consolidated soils ($p_c'/p_0' = 1$), yielding will begin right in the beginning of shearing.
- Beyond this yield stress, soil begins to deform plastically and its behavior becomes irreversible. In Figure (II.4c), the descending dotted line representing unloading does not follow the stress-strain curve for initial loading.
- When the soil is reloaded again (see the ascending dotted line in (c)), the new yield stress q_y^{new} is higher than the one reached in initial loading as the soil undergoes strain-hardening. The new yield stress is equal to the maximum stress reached prior to unloading. This

implies that soil retains its "memory", the maximum stress it has experienced in the past.

- Beyond the yield stress, during plastic deformation shearing is accompanied by increasing volumetric strains (see (d)) implying soil contraction. However, with large over-consolidation, the soils exhibit dilation during shearing.
- Plastic deformation continues until the soil element reaches the failure stress state $F(p'_f, q'_f)$. Beyond this point, plastic flow continues at constant stress which we refer to as *the critical state*.
- It is also observed that if the tests are performed at a higher stress or strain rate, both the volumetric and deviatoric behaviors become different demonstrating the rate dependency of the deformation behavior of soils.

The above observations for the soil behavior are made with reference to test results on clays. From similar tests on sands, one may observe that deformation of loose sands is similar to normally and lightly overconsolidated clays while those of dense sands are similar to highly overconsolidated clays.

III. Modeling of Soil Behavior

A mathematical description of soil behavior tends to become complex due to the inherent nature of soils. The following general features of the observed deformation behavior of soils must be kept in mind. The stress-strain relationship in general are:

- Non-linear during loading, unloading and reloading.
- Stress path dependent (with different strains developing at the same stress).
- Stress level dependent (changing with confinement stress).
- Irreversible (with the development of residual strains in a closed stress cycle).
- Rate dependent (different stress-strain curves at different strain rates)

Additionally the deformation in soils exhibit the following:
- Memory (remembers the previous highest stress before unloading)
- Dilatancy (experiences volumetric change during shearing)
- Hardening (yield stress changes with plastic straining)

4

Elasticity

4.1 Elastic Constitutive Law

As observed in Section II above, at the initial stage of shearing, before the soil begins to yield, the stress-strain relationship is reversible which can be modeled by an 'elastic constitutive law'. Furthermore, if the stress-strain relationship is linear then *linear elasticity* may be used as a model. In Figure 4.1-1 a typical stress-strain relationship for soils subjected to a monotonic loading is shown along with an approximating linear elastic stress-strain relationship. As we can see, the linear elastic model may simply be used to approximate the deformation behavior of soils only within a rather small range of strains during the initial phase of the loading.

Fundamentally speaking, an elastic material is defined as one for which the stress depends only on the strain without being influenced by past thermodynamic history. Furthermore, if the stresses are considered as the derivatives of the 'strain energy density function' with respect to strains, the resulting models are called the Green-elastic or the hyperelastic models. We will consider only this type of elastic model in this chapter.

4.1.1 Fundamentals

For an isotropic linear elastic material we have:

$$\varepsilon_{xx} = \frac{\sigma_{xx}}{E}, \ \varepsilon_{yy} = -\upsilon \frac{\sigma_{xx}}{E}, \ \varepsilon_{zz} = -\upsilon \frac{\sigma_{xx}}{E} \tag{4.1.1a}$$

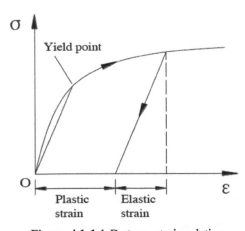

Figure 4.1-1 1-D stress-strain relation

$$\varepsilon_{yy} = \frac{\sigma_{yy}}{E}, \ \varepsilon_{xx} = -\upsilon\frac{\sigma_{yy}}{E}, \ \varepsilon_{zz} = -\upsilon\frac{\sigma_{yy}}{E} \qquad (4.1.1b)$$

$$\varepsilon_{zz} = \frac{\sigma_{zz}}{E}, \ \varepsilon_{xx} = -\upsilon\frac{\sigma_{zz}}{E}, \ \varepsilon_{yy} = -\upsilon\frac{\sigma_{zz}}{E} \qquad (4.1.1c)$$

written in terms of all three stresses in 3-D shown in Figure 4.1-2.

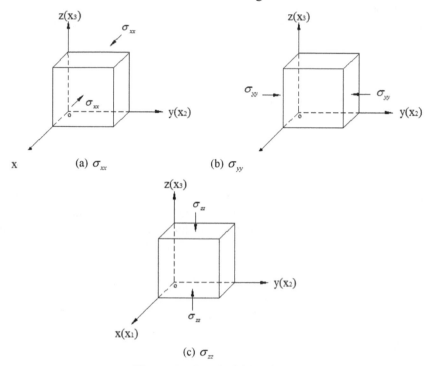

(a) σ_{xx} (b) σ_{yy}

(c) σ_{zz}

Figure 4.1-2 Uniaxial loading

In 3-D, stress-strain relations become:

$$\varepsilon_{xx} = \frac{1}{E}\left[\sigma_{xx} - \upsilon\left(\sigma_{yy} + \sigma_{zz}\right)\right] \qquad (4.1.2a)$$

$$\varepsilon_{yy} = \frac{1}{E}\left[\sigma_{yy} - \upsilon\left(\sigma_{xx} + \sigma_{zz}\right)\right] \qquad (4.1.2b)$$

$$\varepsilon_{zz} = \frac{1}{E}\left[\sigma_{zz} - \upsilon\left(\sigma_{yy} + \sigma_{xx}\right)\right] \qquad (4.1.2c)$$

$$\gamma_{xy} = \frac{\tau_{xy}}{G}, \ \gamma_{yz} = \frac{\tau_{yz}}{G}, \ \gamma_{xz} = \frac{\tau_{xz}}{G} \qquad (4.1.2d)$$

for which the components are shown in Figure 4.1-3.

In (4.1.2), E is the elasticity modulus and υ is Poisson's ratio. $G = \dfrac{E}{2(1+\upsilon)}$ is the shear modulus. The above relationship in terms of strains in matrix form becomes:

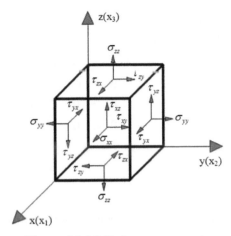

Figure 4.1-3 3-D stress components

$$\begin{Bmatrix} \varepsilon_x \\ \varepsilon_y \\ \varepsilon_z \\ \tau_{xy} \\ \tau_{yz} \\ \tau_{zx} \end{Bmatrix} = \begin{bmatrix} 1/E & -\upsilon/E & -\upsilon/E & 0 & 0 & 0 \\ -\upsilon/E & 1/E & -\upsilon/E & 0 & 0 & 0 \\ -\upsilon/E & -\upsilon/E & 1/E & 0 & 0 & 0 \\ 0 & 0 & 0 & 1/G & 0 & 0 \\ 0 & 0 & 0 & 0 & 1/G & 0 \\ 0 & 0 & 0 & 0 & 0 & 1/G \end{bmatrix} \begin{Bmatrix} \sigma_x \\ \sigma_y \\ \sigma_z \\ \tau_{xy} \\ \tau_{yz} \\ \tau_{zx} \end{Bmatrix} \qquad (4.1.3)$$

Alternatively it can be written in terms of stresses as:

$$\begin{Bmatrix} \sigma_x \\ \sigma_y \\ \sigma_z \\ \tau_{xy} \\ \tau_{yz} \\ \tau_{zx} \end{Bmatrix} = \begin{bmatrix} D_1 & D_2 & D_2 & 0 & 0 & 0 \\ D_2 & D_1 & D_2 & 0 & 0 & 0 \\ D_2 & D_2 & D_1 & 0 & 0 & 0 \\ 0 & 0 & 0 & D_3 & 0 & 0 \\ 0 & 0 & 0 & 0 & D_3 & 0 \\ 0 & 0 & 0 & 0 & 0 & D_3 \end{bmatrix} \begin{Bmatrix} \varepsilon_x \\ \varepsilon_y \\ \varepsilon_z \\ \gamma_{xy} \\ \gamma_{yz} \\ \gamma_{zx} \end{Bmatrix} \qquad (4.1.4)$$

where $D_1 = K + (4/3)G$, $D_2 = K - (2/3)G$ and $D_3 = G$ with $K = \dfrac{E}{3(1-2\upsilon)}$ being the bulk modulus.

Thus, equation (4.1.4) can also be written as:

or

$$\begin{Bmatrix} \sigma_x \\ \sigma_y \\ \sigma_z \\ \tau_{xy} \\ \tau_{yz} \\ \tau_{zx} \end{Bmatrix} = \begin{bmatrix} K+\dfrac{4}{3}G & K-\dfrac{2}{3}G & K-\dfrac{2}{3}G & 0 & 0 & 0 \\[2mm] K-\dfrac{2}{3}G & K+\dfrac{4}{3}G & K-\dfrac{2}{3}G & 0 & 0 & 0 \\[2mm] K-\dfrac{2}{3}G & K-\dfrac{2}{3}G & K+\dfrac{4}{3}G & 0 & 0 & 0 \\[2mm] 0 & 0 & 0 & G & 0 & 0 \\ 0 & 0 & 0 & 0 & G & 0 \\ 0 & 0 & 0 & 0 & 0 & G \end{bmatrix} \begin{Bmatrix} \varepsilon_x \\ \varepsilon_y \\ \varepsilon_z \\ \gamma_{xy} \\ \gamma_{yz} \\ \gamma_{zx} \end{Bmatrix} \qquad (4.1.5)$$

$$\begin{Bmatrix} \sigma_x \\ \sigma_y \\ \sigma_z \\ \tau_{xy} \\ \tau_{yz} \\ \tau_{zx} \end{Bmatrix} = \begin{bmatrix} \lambda+2G & \lambda & \lambda & 0 & 0 & 0 \\ \lambda & \lambda+2G & D_2 & 0 & 0 & 0 \\ \lambda & \lambda & \lambda+2G & 0 & 0 & 0 \\ 0 & 0 & 0 & G & 0 & 0 \\ 0 & 0 & 0 & 0 & G & 0 \\ 0 & 0 & 0 & 0 & 0 & G \end{bmatrix} \begin{Bmatrix} \varepsilon_x \\ \varepsilon_y \\ \varepsilon_z \\ \gamma_{xy} \\ \gamma_{yz} \\ \gamma_{zx} \end{Bmatrix} \tag{4.1.6}$$

or

$$\sigma_{ij} = \lambda \varepsilon_v \delta_{ij} + 2G\varepsilon_{ij} \tag{4.1.7}$$

where $\lambda = \dfrac{\upsilon E}{(1+\upsilon)(1-2\upsilon)}$ and $G = \dfrac{E}{2(1+\upsilon)}$ are Lame's constants and ε_v is the volumetric strain written as $\varepsilon_v = \varepsilon_{xx} + \varepsilon_{yy} + \varepsilon_{zz}$.

4.1.2 Principal Stresses and Strains

The elastic stress-strain relations can also be written in terms of principal stresses given that they are readily available. Therefore:

$$\varepsilon_1 = \frac{1}{E}\left[\sigma_1 - \upsilon\left(\sigma_2 + \sigma_3\right)\right] \tag{4.1.8a}$$

$$\varepsilon_2 = \frac{1}{E}\left[\sigma_2 - \upsilon\left(\sigma_1 + \sigma_3\right)\right] \tag{4.1.8b}$$

$$\varepsilon_3 = \frac{1}{E}\left[\sigma_3 - \upsilon\left(\sigma_2 + \sigma_1\right)\right] \tag{4.1.8c}$$

or in matrix form the above equations will be:

$$\begin{Bmatrix} \varepsilon_1 \\ \varepsilon_2 \\ \varepsilon_3 \end{Bmatrix} = \frac{1}{E} \begin{bmatrix} 1 & -\upsilon & -\upsilon \\ -\upsilon & 1 & -\upsilon \\ -\upsilon & -\upsilon & 1 \end{bmatrix} \begin{Bmatrix} \sigma_1 \\ \sigma_2 \\ \sigma_3 \end{Bmatrix} \tag{4.1.9}$$

We can easily derive the principal stresses from here as:

$$\sigma_1 = \frac{E}{(1+\upsilon)(1-2\upsilon)}\left[(1-\upsilon)\varepsilon_1 + \upsilon\left(\varepsilon_2 + \varepsilon_3\right)\right] \tag{4.1.10a}$$

$$\sigma_2 = \frac{E}{(1+\upsilon)(1-2\upsilon)}\left[(1-\upsilon)\varepsilon_2 + \upsilon\left(\varepsilon_1 + \varepsilon_3\right)\right] \tag{4.1.10b}$$

$$\sigma_3 = \frac{E}{(1+\upsilon)(1-2\upsilon)}\left[(1-\upsilon)\varepsilon_3 + \upsilon\left(\varepsilon_1 + \varepsilon_2\right)\right] \tag{4.1.10c}$$

or in matrix form:

$$\begin{Bmatrix} \sigma_1 \\ \sigma_2 \\ \sigma_3 \end{Bmatrix} = \frac{E}{(1+\upsilon)(1-2\upsilon)} \begin{bmatrix} 1-\upsilon & \upsilon & \upsilon \\ \upsilon & 1-\upsilon & \upsilon \\ \upsilon & \upsilon & 1-\upsilon \end{bmatrix} \begin{Bmatrix} \varepsilon_1 \\ \varepsilon_2 \\ \varepsilon_3 \end{Bmatrix} \tag{4.1.11}$$

4.1.3 Plain Strain Condition

For many problems of geotechnical engineering, it is mostly plausible to simplify the domain in 2-D, such as the case for plane strain stress state where $\varepsilon_{xx} = 0$, $\varepsilon_{xy} = 0$ and $\varepsilon_{xz} = 0$ as can be seen for a retaining wall structure in Figure 4.1-4. Then the stress-strain relations are:

$$\begin{Bmatrix} \sigma_{yy} \\ \sigma_{zz} \\ \tau_{yz} \end{Bmatrix} = \frac{E(1-\upsilon)}{(1+\upsilon)(1-2\upsilon)} \begin{bmatrix} 1 & \dfrac{\upsilon}{1-\upsilon} & 0 \\ \dfrac{\upsilon}{1-\upsilon} & 1 & 0 \\ 0 & 0 & \dfrac{1-2\upsilon}{2(1-\upsilon)} \end{bmatrix} \begin{Bmatrix} \varepsilon_{yy} \\ \varepsilon_{zz} \\ \varepsilon_{yz} \end{Bmatrix} \qquad (4.1.12)$$

We need to keep in mind that the third stress (the one perpendicular to the plane of interest) is also non-zero and calculated as:

$$\sigma_{xx} = \upsilon\left(\sigma_{zz} + \sigma_{yy}\right) \qquad (4.1.13)$$

But the other shear stresses vanish:

$$\tau_{xz} = \tau_{xy} = 0 \qquad (4.1.14)$$

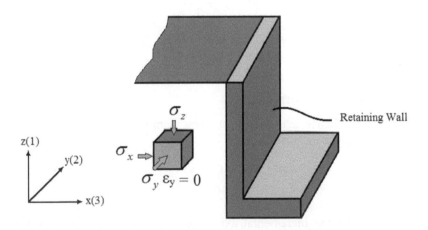

Figure 4.1-4 Plane strain problem of retaining wall

4.1.4 Axi-Symmetric Condition

Another 2-D idealization for several geotechnical problems is the axisymmetric condition. For example, a circular tank in Figure 4.1-5 is in such a stress state for which if we use the principal values of stress and strain we get:

$$\begin{Bmatrix} \varepsilon_1 \\ \varepsilon_3 \end{Bmatrix} = \frac{1}{E} \begin{bmatrix} 1 & -2\upsilon \\ -\upsilon & 1-\upsilon \end{bmatrix} \begin{Bmatrix} \sigma_1 \\ \sigma_3 \end{Bmatrix} \qquad (4.1.15)$$

or in terms of stresses,

$$\begin{Bmatrix} \sigma_1 \\ \sigma_3 \end{Bmatrix} = \frac{E}{(1+\upsilon)(1-2\upsilon)} \begin{bmatrix} 1-\upsilon & 2\upsilon \\ 1-\upsilon & 1 \end{bmatrix} \begin{Bmatrix} \varepsilon_1 \\ \varepsilon_3 \end{Bmatrix} \qquad (4.1.16)$$

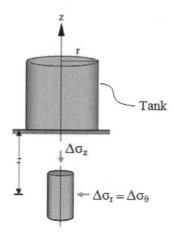

Figure 4.1-5 Axi-symmetric stress state under a circular tank

We can rewrite the same state using the individual stress components with:

$$\begin{Bmatrix} \sigma_{11} \\ \sigma_{22} \\ \sigma_{33} \\ \sigma_{12} \end{Bmatrix} = \begin{bmatrix} K+\frac{4}{3}G & K-\frac{2}{3}G & K-\frac{2}{3}G & 0 \\ K-\frac{2}{3}G & K+\frac{4}{3}G & K-\frac{2}{3}G & 0 \\ K-\frac{2}{3}G & K-\frac{2}{3}G & K+\frac{4}{3}G & 0 \\ 0 & 0 & 0 & G \end{bmatrix} \begin{Bmatrix} \varepsilon_{11} \\ \varepsilon_{22} \\ \varepsilon_{33} \\ \varepsilon_{12} \end{Bmatrix} \tag{4.1.17}$$

4.1.5 Other Considerations

If we use the triaxial (hence axisymmetric) stress parameters given in (II.2) to define the necessary stress-strain relationship, we get:

$$\begin{Bmatrix} p \\ q \end{Bmatrix} = \begin{bmatrix} K & 0 \\ 0 & 3G \end{bmatrix} \begin{Bmatrix} \varepsilon_p \\ \varepsilon_q \end{Bmatrix} \tag{4.1.18}$$

Let $\delta\sigma_a$ and $\delta\sigma_r$ be the incremental axial and radial stress components for the triaxial stress state (Figure 4.1-6). Then the mean and deviatoric stress increments and corresponding volumetric and deviatoric strain increments become:

$$\delta p = \frac{\delta\sigma_a + 2\delta\sigma_r}{3} \tag{4.1.19}$$

$$\delta q = \delta\sigma_a - \delta\sigma_r \tag{4.1.20}$$

$$\delta\varepsilon_p = \delta\varepsilon_a + 2\delta\varepsilon_r \tag{4.1.21}$$

$$\delta\varepsilon_q = \frac{2}{3}(\delta\varepsilon_a - \delta\varepsilon_r) \tag{4.1.22}$$

Incremental work done by the stresses and strains is then calculated as:

$$\delta W = p'\delta\varepsilon_p + q\delta\varepsilon_q = \delta\sigma_a'\delta\varepsilon_a + 2\delta\sigma_r'\delta\varepsilon_r \tag{4.1.23}$$

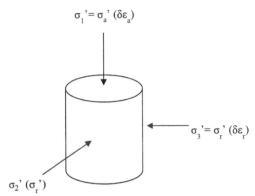

$\sigma_1' = \sigma_a' \, (\delta\varepsilon_a)$

$\sigma_3' = \sigma_r' \, (\delta\varepsilon_r)$

$\sigma_2' \, (\sigma_r')$

Figure 4.1-6 Triaxial stress state

The above equations can be written in the following matrix forms:

$$\begin{Bmatrix} \delta p' \\ \delta q' \end{Bmatrix} = \begin{bmatrix} 1/3 & 2/3 \\ 1 & -1 \end{bmatrix} \begin{Bmatrix} \delta\sigma_a' \\ \delta\sigma_r' \end{Bmatrix} \tag{4.1.24}$$

$$\begin{Bmatrix} \delta\sigma_a' \\ \delta\sigma_r' \end{Bmatrix} = \begin{bmatrix} 1 & 2/3 \\ 1 & -1/3 \end{bmatrix} \begin{Bmatrix} \delta p' \\ \delta q' \end{Bmatrix} \tag{4.1.25}$$

$$\begin{Bmatrix} \delta\varepsilon_p \\ \delta\varepsilon_q \end{Bmatrix} = \begin{bmatrix} 1 & 2 \\ 2/3 & -2/3 \end{bmatrix} \begin{Bmatrix} \delta\varepsilon_a \\ \delta\varepsilon_r \end{Bmatrix} \tag{4.1.26}$$

$$\begin{Bmatrix} \delta\varepsilon_a \\ \delta\varepsilon_r \end{Bmatrix} = \begin{bmatrix} 1/3 & 1 \\ 1/3 & -1/2 \end{bmatrix} \begin{Bmatrix} \delta\varepsilon_p \\ \delta\varepsilon_q \end{Bmatrix} \tag{4.1.27}$$

$$\begin{Bmatrix} \delta\varepsilon_a \\ \delta\varepsilon_r \end{Bmatrix} = \frac{1}{E} \begin{bmatrix} 1 & -2\upsilon \\ -\upsilon & 1-\upsilon \end{bmatrix} \begin{Bmatrix} \delta\sigma_a \\ \delta\sigma_r \end{Bmatrix} \tag{4.1.28}$$

$$\begin{Bmatrix} \delta\varepsilon_p \\ \delta\varepsilon_q \end{Bmatrix} = \frac{1}{E} \begin{bmatrix} 1 & 2/3 \\ 1 & -1/3 \end{bmatrix} \begin{bmatrix} 1 & -2\upsilon \\ -\upsilon & 1-\upsilon \end{bmatrix} \begin{bmatrix} 1 & 2/3 \\ 1 & -1/3 \end{bmatrix} \begin{Bmatrix} \delta p' \\ \delta q' \end{Bmatrix}$$

$$= \frac{1}{E} \begin{bmatrix} 3(1-2\upsilon) & 0 \\ 0 & 2(1+\upsilon)/3 \end{bmatrix} \begin{Bmatrix} \delta p' \\ \delta q' \end{Bmatrix} \tag{4.1.29}$$

$$\begin{Bmatrix} \delta\varepsilon_p \\ \delta\varepsilon_q \end{Bmatrix} = \begin{bmatrix} \dfrac{1}{K} & 0 \\ 0 & \dfrac{1}{3G} \end{bmatrix} \begin{Bmatrix} \delta p' \\ \delta q' \end{Bmatrix} \tag{4.1.30}$$

Figure 4.1-7 shows the relationships between the above stress and strain parameters. It is nice to observe that the E and ν can be obtained by the K and G moduli also as: $E = \dfrac{9KG}{G+3K}$, $G/K = \dfrac{3(1-2\upsilon)}{2(1+\upsilon)}$, $\upsilon = \dfrac{3K-2G}{2(G+3K)}$. In a conventional drained triaxial test, we have the following

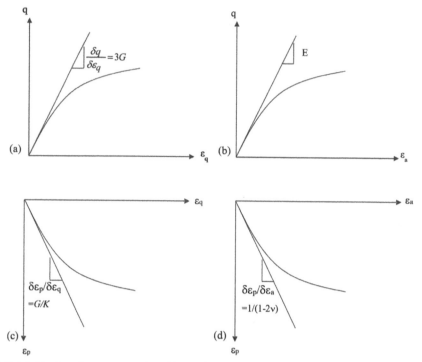

Figure 4.1-7 (a) Deviatoric stress-strain, (b) Deviatoric stress-axial strain, (c) Mean strain-deviatoric strain, (d) Mean strain-axial strain relationships and related moduli

conditions hold: $\delta\sigma_r = 0$, $\delta q = \delta\sigma_a$ $\delta p' = \dfrac{1}{3}\delta\sigma_a$. Thus we get: $\dfrac{\delta q}{\delta p'} = 3 \Rightarrow \dfrac{\delta\varepsilon_p}{\delta\varepsilon_q} = \dfrac{G}{K}$. If $\dfrac{\delta q}{\delta p'} = \lambda$, then $\dfrac{\delta\varepsilon_p}{\delta\varepsilon_q} = \dfrac{3G}{\lambda K}$. In a conventional undrained triaxial test, these conditions hold: $\delta\varepsilon_p \to 0$, $\dfrac{\delta p}{K_u} \to 0$, $K_u = \dfrac{E_u}{2(1-\upsilon_u)} \to \infty$, $\upsilon_u \to 0.5$ where K_u is the undrained bulk modulus.

Constrained Loading Condition

In the case of an oedometer (consolidation) test (Figure 4.1-8) where the soil element is constrained in lateral direction with $\delta\varepsilon_r = 0$ yields,

$$\frac{\delta\sigma'_r}{\delta\sigma'_a} = \frac{\upsilon}{1-\upsilon} = \frac{3K-2G}{3K+4G} \qquad (4.1.31)$$

Since:

$$\delta\varepsilon_a = \frac{1}{E}\delta\sigma_a - \frac{2\upsilon}{E}\frac{\upsilon}{1-\upsilon}\delta\sigma_a \qquad (4.1.32)$$

$$\frac{\delta\sigma_a}{\delta\varepsilon_a} = \frac{E(1-\upsilon)}{(1+\upsilon)(1-2\upsilon)} \qquad (4.1.33)$$

is also obtained. The oedometer modulus, also called *the constrained modulus* is then calculated as:

$$E_{oed} = M = \frac{\delta\sigma'_a}{\delta\varepsilon_a} = \frac{E(1-\upsilon)}{(1+\upsilon)(1-2\upsilon)} = K + \frac{4}{3}G \qquad (4.1.34)$$

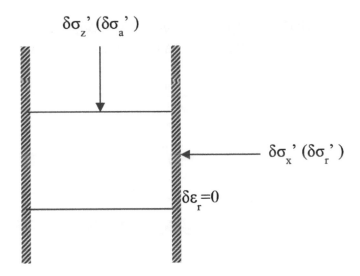

$\delta\sigma_z{}'\,(\delta\sigma_a{}')$

$\delta\sigma_x{}'\,(\delta\sigma_r{}')$

$\delta\varepsilon_r = 0$

Figure 4.1-8 Oedometer test stress state

Pore Pressure Changes: Undrained Loading

It is important that the pore pressure changes (Δp_w) in geotechnical loading conditions (i.e. stress states) are also calculated. In the isotropic loading case:

$$\Delta V_w = nV_0 C_w \Delta p_w \tag{4.1.35}$$

$$\Delta V_s = V_0 C_c \Delta\sigma' \tag{4.1.36}$$

are the volume change in water part of the soil and the overall soil structure, respectively. Here, C_w and C_c are the water and soil compressibilities respectively and V_0 is the original volume of the soil and n is the porosity. If the compressibility in isotropic state will be due to the compressibility of pore water (neglecting the solid grain compressibility), we have:

$$\Delta V_{sk} = \Delta V_w \tag{4.1.37}$$

$$V_0 C_c \Delta\sigma' = nV_0 C_w \Delta p_w \tag{4.1.38}$$

giving:

$$B = \frac{\Delta p_w}{\Delta\sigma} = \frac{1}{(1 + n\dfrac{C_w}{C_c})} \tag{4.1.39}$$

or

$$\Delta p_w = B\Delta\sigma \tag{4.1.40}$$

In the above, B is the "first Skempton pore pressure parameter". Assuming full saturation, the compressibility of pore water will be much less than the soil structure, hence $C_w \ll C_c \Rightarrow B \approx 1$. In the case of axial loading, we have:

$$\Delta\sigma_2' = \Delta\sigma_3' = 0 - \Delta p_w = -\Delta p_w \tag{4.1.41}$$

and with (4.1.37), we can write:

$$nV_0C_w\Delta p_w = V_0C_c(\Delta\sigma_1 - \Delta p_w) + V_0C_c(-\Delta p_w) + V_0C_c(-\Delta p_w)$$
$$= V_0C_c(\Delta\sigma_1 - 3\Delta p_w) = V_0C_c\Delta\sigma_1 - 3V_0C_c\Delta p_w \quad (4.1.42)$$

giving:

$$\frac{\Delta p_w}{\Delta\sigma} = \frac{V_0C_c}{V_0(nC_w + 3C_c)} \quad (4.1.43)$$

$$\tilde{B} = \frac{\Delta p_w}{\Delta\sigma} = \frac{1}{(n\dfrac{C_w}{C_c} + 3)} \quad (4.1.44)$$

In the case of triaxial loading:

$$\Delta p_w = \frac{\Delta\sigma_3}{1 + n\dfrac{C_w}{C_c}} + \frac{\Delta\sigma_1 - \Delta\sigma_3}{n\dfrac{C_w}{C_c} + 3} \quad (4.1.45)$$

yielding:

$$\Delta p_w = B\Delta\sigma_3 + \tilde{B}(\Delta\sigma_1 - \Delta\sigma_3) \quad (4.1.46)$$

or

$$\Delta p_w = B[\Delta\sigma_3 + A(\Delta\sigma_1 - \Delta\sigma_3)] \quad (4.1.47)$$

where the A is the "second Skempton pore pressure parameter". The parameter A should be determined in tests as the soil is mostly a nonlinear material. Again if the soil is saturated,

$$\Delta p_w = \Delta\sigma_3 + A(\Delta\sigma_1 - \Delta\sigma_3) \quad (4.1.48)$$

For the undrained condition, $\Delta\varepsilon_v = 0$, $\Delta p' = 0$ and we can write:

$$\frac{(\Delta\sigma_1 - \Delta p_w) + 2(\Delta\sigma_3 - \Delta p_w)}{3} = 0 \quad (4.1.49)$$

Hence,

$$\Delta p_w = \Delta\sigma_3 + \frac{1}{3}(\Delta\sigma_1 - \Delta\sigma_3) \quad (4.1.50)$$

Again for undrained loading, but in terms of effective stress, the elastic volumetric strain is:

$$\Delta\varepsilon_p^e = \frac{\Delta p'}{K'} = 0 \quad (4.1.51)$$

yielding again $\Delta p' = 0$ and $K' = \dfrac{E'}{3(1 - 2\upsilon')}$. In terms of total stress, it is:

$$\Delta\varepsilon_p^e = \frac{\Delta p}{K} = 0 \quad (4.1.52)$$

but this time we should observe that $\Delta p \neq 0$ and therefore $K_u = \dfrac{E_u}{3(1 - 2\upsilon_u)}$ with E_u and υ_u being

the elastic undrained parameters. In the case of incompressibility limit, we write: $K \to K_u = \infty$ $\Rightarrow 3(1 - 2\upsilon_u) = 0$ resulting in $\upsilon_u = 1/2$. The same can be written for the shear with $G = G_u = G'$ giving:

$$E_u = \frac{1.5E'}{(1+\upsilon')} \qquad (4.1.53)$$

If for example, we use $\upsilon = 0.33$, then $E_u = 1.1\ E'$, which means that undrained modulus is about 10% larger than the drained modulus.

In the case of anisotropy, we speak of two cases for soils: (i) Structural anisotropy, and (ii) Stress-induced anisotropy. As for elastic parameters, they vary in the three spatial directions (i.e. E_z, E_x, υ_{xx}, υ_{zx}, υ_{zz}). In radial and axial directions the following relationship holds:

$$\begin{Bmatrix} \delta\varepsilon_z \\ \delta\varepsilon_r \end{Bmatrix} = \begin{bmatrix} \dfrac{1}{E_z} & -\dfrac{2\upsilon_{rz}}{E_r} \\ -\dfrac{\upsilon_{zr}}{E_z} & \dfrac{1-\upsilon_{rr}}{E_r} \end{bmatrix} \begin{Bmatrix} \delta\sigma_z \\ \delta\sigma_r \end{Bmatrix} \qquad (4.1.54)$$

$$\frac{\upsilon_{rz}}{\upsilon_{zr}} = \frac{E_r}{E_z} \qquad (4.1.55)$$

5

Plasticity Theory: Nonlinear Deformation of Soils

5.1 Introduction

As mentioned earlier, soil is a nonlinear material experiencing a significant amount of inelastic deformation under applied loads. In this chapter, we will summarize the fundamentals of the *Theory of Plasticity*, which provides a framework for the development of a constitutive model to understand the stress-strain relationship of soils. For many geotechnical problems, evaluation of response of soils requires development of such theoretical models.

The classical plasticity is based on defining the stress-strain relationships of materials for a given state of stress. As for soils, the idea is to be able to capture the mechanical behavior under given drainage and loading conditions. An ideal situation is to simulate the actual stress-strain relationships as observed in laboratory tests following various realistic stress paths. The essential features of soil behavior relevant to the problems to be studied in geotechnical engineering are briefly summarized in the following points:

- In general, strain increment caused by a stress increment consists of both a recoverable (elastic) and an irrecoverable (inelastic/plastic) part.
- Slope of the stress-strain curve depends upon the stress level, direction of stress increment (i.e. loading, reloading and unloading), and the past state history (stress, strain, modification of material micro-structure etc.) which changes as the plastic (or elastic-plastic) loading continues.
- As observed in soil tests, behavior of soils depends more on the level of stress, pore pressure, past history, direction of stress increment and material state than on time. The observed time dependence is associated with the pore fluid flow.
- Specifically for soils, there exists a correspondence between undrained and fully drained behavior. With respect to volumetric behavior, if the soil is fully drained then the loading will result in change in volume. In the case of undrained condition, (the tendency of) this volume change which is not permitted, results in a corresponding (tendency of) change in pore fluid pressure. In the case of partial drainage, during a load increment both will take place, i.e. volume change as well as the change in pore pressure. Shearing behavior is not affected by the drainage conditions.

For these, the classical plasticity theory provides a suitable framework for the development of a constitutive model, which reproduces the soil behavior under monotonic loading. These models employ a plastic potential and a yield surface, the latter being allowed to expand or contract depending on whether the material is hardening or softening. However in these models, soil remains elastic within the yield surface where no plastic deformation is allowed to develop. As a consequence, such models fail to reproduce the phenomena occurring during cyclic loading, such as the progressive generation of pore fluid pressure or densification. In order to overcome this limitation, several different approaches have been developed. These are briefly mentioned in the following categories: (i) Classical isotropic and kinematic hardening plastic models, (ii) Bounding surface models, (iii) Generalized plasticity, (iv) Hypoplasticity and incrementally nonlinear models, (iv) Densification models, and (v) Endochronic theory. In this chapter, we focus our attention on soils and soil behavior although the first few sections present the essential elements of the plasticity theory applicable to many other materials as well.

5.2 Nonlinear Deformation of Soils

Soils are naturally nonlinear materials. Figure 5.2-1 shows a typical one-way cyclic stress-strain relationship. Material is loaded until point A which continues in a linear fashion as long as it behaves linear-elastically. Strains exhibited by the soil at this stage are entirely *recoverable*. However, since soils are not actually linear elastic materials, particularly above a certain small stress level, the behavior is such that the stress will decrease until the material is loaded until point B. At that point, if we unload the soil it follows the initial straight line suggesting a linear unloading behavior which results in *irrecoverable* or plastic strains. Upon reloading however, we see that there is some amount of permanent strain occurring in the soil which eventually, beyond point B, returns to the actual virgin loading curve.

Another way of seeing this behavior is Figure 5.2-2 where a clay soil is sheared in a drained triaxial test. In the initial part of the test, soil shearing behavior is *reversible* until the stress path reaches its yield stress at point $Y(p_y, q_y)$ (Figure 5.2-2a). From this point onwards, the material yields plastically and its behavior becomes *irreversible*. Plastic yielding continues until soil reaches the *failure* stress $F(p_f, q_f)$. At failure, plastic flow continues at constant stress. M is the slope of the failure line in the stress path (Figure 5.2-2c). Plastic flow occurs after the material has yielded essentially producing corresponding shear strains (Figure 5.2.2-b).

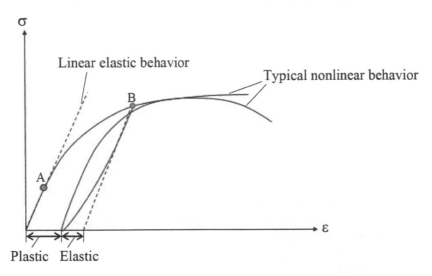

Figure 5.2-1 Typical soil behaviors in terms of stress-strain relationships

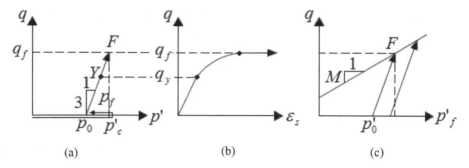

Figure 5.2-2 Drained behavior of clay soil, (a) Stress path, (b) Deviatoric behavior, (c) Failure envelope

5.3 Elements of Plasticity

The theory of plasticity has three essential elements:

- Yield Criterion
- Flow Rule
- Post-Yield Deformation

The "yield criterion" defines the stress state, which will cause yielding or the onset of plastic deformation. "Flow rule" defines the development of plastic strains during yielding. It may or may not be associated with the way material yields. "Post-yield deformation" deals with how the material is deformed following the first yield. Following yielding, an *incremental stress-strain relationship* in conjunction with a *hardening law* define the nature of straining as well as the change in the location and/or size of the *yield surface*. Within this described general framework, several specific constitutive models have been and could still be developed for soils. These models vary with respect to how these main elements are taken into consideration. In the following sections, only the essential aspects of these components are presented. The focus is to introduce the formulation of the 'incremental stress-strain relationship' in general, and then to illustrate its use in the context of few commonly used soil models.

5.4 Yielding Criteria

5.4.1 General Definition

Using the soil test results (as mentioned above), we need to be able to model the failure state of soils regardless of their initial effective mean stress, p_0'. As mentioned before, soils and many natural materials have a so called *yielding stress level* or *yield stress* that distinguishes elastic and elastic-plastic behavior, the latter of which contains both elastic and plastic strains. Such a stress value in a 1-D loading becomes a line in 2-D and a surface in 3-D which is then also called a *yield surface*. Throughout the course of loading, it is necessary to define the onset of plasticity. The soil material initially deforms elastically under tension, compression or shear stresses. There, however, exists a surface in multi-dimensions (line or a curve in 1-D) beyond which permanent deformations start to exhibit. These include both recoverable and irrecoverable deformations as stated. Irrecoverable deformations are defined as plastic strains whereas stress paths inside the yield surface result in solely recoverable deformations also called *elastic strains* as also discussed previously. This 'yield surface' is represented by a mathematical function called the *yield function* which represents the *yield criterion*. Figure 5.4-1(a) shows a 1-D elastic-perfectly plastic behavior. While the actual realistic behavior has a curvilinear transition from the elastic regime to the perfectly plastic regime, it is possible and perfectly reasonable to idealize it as a combination of two linear segments for all practical purposes. In general, the initial yield function, f is described as,

$$f(\sigma) = f_c \tag{5.4.1}$$

with

$$\sigma = \sigma(\sigma_{xx}, \sigma_{yy}, \sigma_{zz}, \sigma_{xy}, \sigma_{yz}, \sigma_{xz}) \tag{5.4.2}$$

where f_c is a constant for perfectly plastic material but a variable for strain or work hardening materials. In a uniaxial situation it becomes,

$$|\sigma| - \sigma_y = 0 \tag{5.4.3}$$

where σ_y is the yield normal stress. In the case of the stress vector (or point in 1-D) being within the yield surface which is also the elastic domain, we write,

$$f < f_c \tag{5.4.4}$$

A 2-D and a 3-D surface of a representative yield function in the stress-space can be seen in Figure 5.4-1(b) and 5.4-1(c), respectively. In 3-D, the material encapsulates the possible existing stress states in a cone-shaped volume where the size of it decreases towards extension and increases towards compression. This behavior accurately models the actual observed soil response since soils' tensile strengths are much lower than their compressive strengths.

There are many different yield functions available in the related literature such as the Tresca criterion, von Mises criterion, Mohr-Coulomb function etc. In this section, we limit ourselves to the Mohr-Coulomb and the Drucker-Prager yield criteria which are both commonly used for modeling soil static behavior.

5.4.2 Mohr-Coulomb Criterion

For several elastic-perfectly plastic models, Mohr-Coulomb criterion is used to define the yield function as,

$$\tau = c + \sigma_n \tan\phi \tag{5.4.5}$$

Here, from the geometry of the Mohr circle and the failure envelope (Figure 5.4-2), we get,

$$\tau = \frac{\sigma_1 - \sigma_3}{2}\cos\phi \tag{5.4.6}$$

$$\sigma_n = \frac{\sigma_1 + \sigma_3}{2} - \frac{\sigma_1 - \sigma_3}{2}\cos\phi \tag{5.4.7}$$

From (5.4.6) and (5.4.7) we obtain,

$$\frac{\sigma_1 - \sigma_3}{2}\cos\phi = -2c\cos\phi + (\sigma_1 + \sigma_3)\sin\phi \tag{5.4.8}$$

which is written in terms of the invariants as,

$$f(\sigma) = \sigma_m \sin\phi + \bar{\sigma}F(\theta) - c\cos\phi = 0 \tag{5.4.9}$$

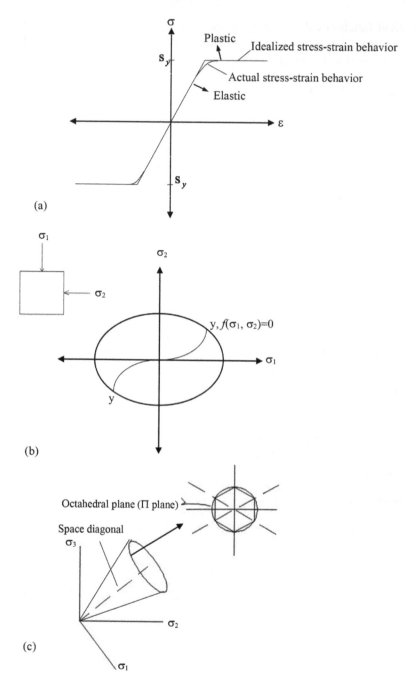

Figure 5.4-1 (a) 1-D perfectly plastic stress-strain behavior, (b) 2-D yield surface in stress-space, (c) 3-D yield cone in stress-space and its cross section, the Π-plane

where *c* is the *cohesion*, ϕ is the *internal friction angle* and we have,

$$\sigma_m = \frac{1}{3}(\sigma_x + \sigma_y + \sigma_z) \tag{5.4.10}$$

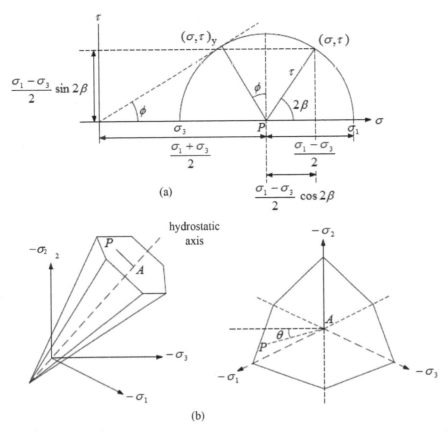

Figure 5.4-2 (a) Mohr-Coulomb yield function, (b) Mohr-Coulomb prismatic cone and Π-plane

$$\bar{\sigma} = \sqrt{\frac{1}{2}(S_x^2 + S_y^2 + S_z^2) + \tau_{xy}^2 + \tau_{yz}^2 + \tau_{zx}^2} = \sqrt{J_2} \tag{5.4.11}$$

$$S_x = \sigma_x - \sigma_m, \quad S_y = \sigma_y - \sigma_m, \quad S_z = \sigma_z - \sigma_m \tag{5.4.12}$$

and also,

$$F(\theta) = \cos\theta - \frac{1}{\sqrt{3}}\sin\phi\sin\theta \tag{5.4.13}$$

with the Lode angle,

$$\theta = \frac{1}{3}\sin^{-1}\left(-\frac{3\sqrt{3}}{2}\frac{J_3}{\bar{\sigma}^3}\right), \quad -\frac{\pi}{6} \le \theta \le \frac{\pi}{6} \tag{5.4.14}$$

where the third deviatoric invariant is calculated as,

$$J_3 = S_x S_y S_z + 2\tau_{xy}\tau_{yz}\tau_{zx} - S_x\tau_{yz}^2 - S_y\tau_{zx}^2 - S_z\tau_{xy}^2 \tag{5.4.15}$$

It should be mentioned here that when viewed in the octahedral plane where the mean stress, σ_m is constant, such a yield surface has sharp vertices leading to gradient discontinuities. When implemented in a numerical code, these discontinuities may (and often times) lead to singularities because the derivatives of the yield function with respect to the stress vector are undefined at those points. Unfortunately, the stress states at or near those vertices are communicable in a boundary value problem in geotechnical engineering and must be carefully addressed. Therefore, it is necessary to "smoothen" the yield function around the corners and such a treatment is taken care of with a new hyperbolic yield function proposed by Abbo and Sloan (1995). Here the new yield function is written as,

$$f(\sigma) = \sigma_m \sin\phi + \sqrt{\overline{\sigma}^2 [F(\theta)]^2 + a^2 \sin^2\phi} - c\cos\phi = 0 \tag{5.4.16}$$

where if $a \leq 0.25c\cot\phi$ is satisfied then the new surface closely represents the actual Mohr-Coulomb surface. In the simulations presented in this section, this yield function is used. In order for the octahedral cross-section to be similar to the Mohr-Coulomb cross-section as stated by Abbo and Sloan (1995), $F(\theta)$ is defined as,

$$F(\theta) = \begin{cases} A(\theta) - B(\theta)\sin 3\theta, & \text{if } |\theta| > \theta_T \\ \cos\theta - \dfrac{1}{\sqrt{3}}\sin\phi\sin\theta, & \text{if } |\theta| \leq \theta_T \end{cases} \tag{5.4.17}$$

where

$$A(\theta) = \frac{1}{3}\cos\theta_T \left(3 + \tan\theta_T \tan 3\theta_T + \frac{1}{\sqrt{3}} sign(\theta)(\tan 3\theta_T - 3\tan\theta_T)\sin\phi \right) \tag{5.4.18a}$$

$$B(\theta) = \frac{1}{3\cos 3\theta_T}\left(sign(\theta)\sin\theta_T + \frac{1}{\sqrt{3}}\sin\phi\cos\theta_T \right) \tag{5.4.18b}$$

$$sign(\theta) = \begin{cases} +1 & \text{for } \theta \geq 0 \\ -1 & \text{for } \theta < 0 \end{cases} \tag{5.4.19}$$

Thus, such an option for the Lode angle function is also made available in the related code written to demonstrate the yield function of the "modified Mohr-Coulomb model".

5.4.3 Drucker-Prager Criterion

Drucker-Prager model defines the yield surface as a cone with circular cross section on the Π-plane (Figure 5.4-3). In the classical elastic-perfectly plastic form of Drucker-Prager model, the yield surface is defined as,

$$f(\sigma) = \sqrt{J_2} - \alpha I_1 - k = 0 \tag{5.4.20}$$

where I_1 is the first invariant of the Cauchy stress and J_2 is the second invariant of the deviatoric part of the Cauchy stress with,

$$J_2 = \frac{1}{6}\left[(\sigma_{11} - \sigma_{22})^2 + (\sigma_{22} - \sigma_{33})^2 + (\sigma_{33} - \sigma_{11})^2 \right] + \sigma_{12}^2 + \sigma_{23}^2 + \sigma_{13}^2 \tag{5.4.21a}$$

$$I_1 = \sigma_{11} + \sigma_{22} + \sigma_{33} \tag{5.4.21b}$$

In (5.4.20), α and k are the model parameters that are generally represented in terms of cohesion and friction angle as,

$$\alpha = \frac{2\sin\phi}{\sqrt{3}(3+\sin\phi)} \quad, \quad k = \frac{6c\cos\phi}{\sqrt{3}(3+\sin\phi)} \tag{5.4.22a}$$

$$\alpha = \frac{2\sin\phi}{\sqrt{3}(3-\sin\phi)} \quad, \quad k = \frac{6c\cos\phi}{\sqrt{3}(3-\sin\phi)} \tag{5.4.22b}$$

In the above, the first couple is used when the Drucker-Prager yield surface is assumed to circumscribe the Mohr-Coulomb yield surface and the second couple is used when the Drucker-Prager yield surface inscribes the Mohr-Coulomb yield surface as can be seen in Figure 5.4-4 showing the cross-sections of the Mohr-Coulomb and Drucker-Prager cones on the Π-plane.

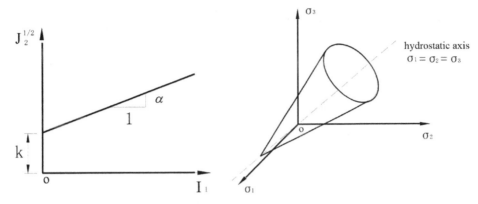

Figure 5.4-3 Drucker-Prager yield criterion

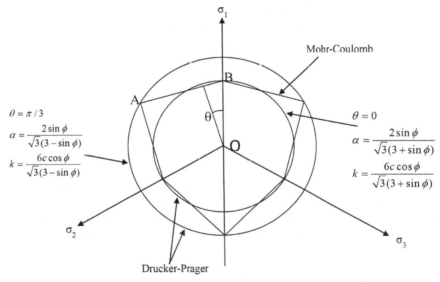

Figure 5.4-4 Drucker-Prager and Mohr-Coulomb yield surfaces

5.5 Post-Yield Behavior

Post-yield behavior, as self-explanatory, defines the soil elasto-plastic behavior under the loads applied right when the material experiences yielding. Typically, it is governed by the fact whether or not the material hardens (or softens) under ongoing total stresses (or total strains). That being said, we will begin our discussion with the perfectly plastic case with no hardening.

5.5.1 Plastic Flow

Plastic flow occurs when the stress vector reaches the yield surface. Material starts to yield and plastic strains start to develop in the direction normal to the tangent line drawn to the yield curve at the stress point which is also on the curve. Figure 5.5-1 shows this in the stress

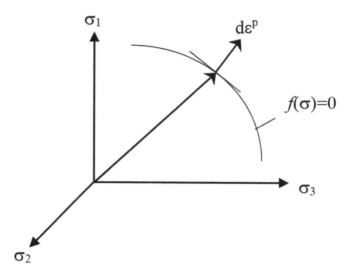

Figure 5.5-1 Yield surface and plastic flow

space. Such a state where the plastic flow takes place along the normal direction to the yield surface is called an *associated plastic behavior*. Here by *associated*, we mean the post-yield behavior to be associated or *consistent* with the pre-yield or the elastic behavior. As for soils, such a behavior is not typically observed (except in some cases for some normally consolidated clays) and the monotonic stress-strain behavior of soils can be well simulated by considering *non-associated plasticity*. This distinction will be made clear in the definition of the flow rule in the next section.

5.5.2 Flow Rule

The flow rule is necessary to define the relationship between subsequent increments of the plastic strain $d\varepsilon^p$ and the present state of stress, σ_{ij}, for the material under yielding. It simply describes how the plastic strain increments should be calculated during the course of loading. We should note here that only the normal component of the stress increment to the yield surface, $d\sigma_{ij}$, produces plastic strain increments which are also proportional to this stress increment. So we write,

$$d\varepsilon_{ij}^p = d\lambda \frac{\partial f}{\partial \sigma_{ij}} \qquad (5.5.1)$$

where $d\lambda$ is the "slip rate" or the "proportionality factor" representing the magnitude of plastic strain increment and $\partial f/\partial \sigma_{ij}$ represents the direction. The magnitude of the plastic strain increment can also be defined such that,

$$d\lambda = \frac{1}{H} \frac{\partial f}{\partial \sigma_{kl}} d\sigma_{kl} \qquad (5.5.2)$$

where H is called the "hardening modulus" which could be a function of the stress, strain or the loading history. If we substitute (5.5.2) into (5.5.1) we get,

$$d\varepsilon_{ij}^p = \frac{1}{H} \frac{\partial f}{\partial \sigma_{kl}} d\sigma_{kl} \frac{\partial f}{\partial \sigma_{ij}} \qquad (5.5.3)$$

This form of the flow rule is called "the associated flow rule" and in some sources presented as the "normality condition". Although it is simpler to define the tangent of the stress vector to the yield surface at the current stress point, this method has been proven not to work properly for porous materials such as soils, rock or concrete. Therefore the rule (5.5.3) is redefined as,

$$d\varepsilon_{ij}^p = d\lambda \frac{\partial g}{\partial \sigma_{ij}}$$ (5.5.4)

where g is called the *plastic potential function* (Figure 5.5-2). For some materials, yield function and plastic potential function can be taken as equal, however for most soils, generally a potential function that is different than the yield function is used to calculate the plastic strains more accurately. Equation (5.5.4) is called the "non-associated flow rule". A generalized approach for determining plastic stress-strain relations for any yield criterion is suggested by Drucker (1949) and the details are provided subsequently.

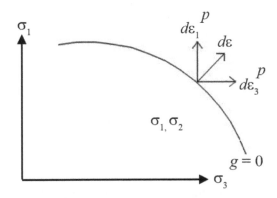

Figure 5.5-2 Potential function and the normality rule of plastic strain increment

5.5.3 Plastic Loading and Consistency Condition

For the perfectly plastic case, if the stress vector lies on the yield surface and we have plastic deformation, we need to define a state called "consistency condition" to represent the fact that the current stress vector should stay on the same yield surface during the course of loading. That is:

$$df = \frac{\partial f}{\partial \sigma_{ij}} d\sigma_{ij} = 0 , \ f = f_c$$ (5.5.5)

For the special case of stress path originating on the same surface, the loading condition for the elastic behavior is defined as:

$$df = \frac{\partial f}{\partial \sigma_{ij}} d\sigma_{ij} < 0 , \ f = f_c$$ (5.5.6)

which is sufficient to satisfy that the stress vector moves within the yield surface.

5.5.4 Incremental Stress-Strain Relation

The stress-strain relationship is written in its incremental form to be able to predict the actual soil behavior accurately by integrating the rate equations. That is to say that the actual stress-strain relationship is integrated within the soil domain. We write the strain decomposition in an incremental form as,

$$d\varepsilon = d\varepsilon^e + d\varepsilon^p \tag{5.5.7}$$

where $d\varepsilon^e$ is the elastic strain increment and $d\varepsilon^p$ is the plastic strain increment. Taking the elastic strains from this relation and using in the stress-strain relationship, we get:

$$d\sigma = D^e(d\varepsilon - d\varepsilon^p) \tag{5.5.8}$$

where D^e is the elastic constitutive matrix. It is always possible to define this stress-strain relationship in terms of elastic strains. Let us substitute the non-associated flow rule of (5.5.4) into (5.5.8) giving:

$$d\sigma = D^e\left(d\varepsilon - d\lambda\frac{\partial g}{\partial \sigma}\right) \tag{5.5.9}$$

At this point we need to talk more about the *consistency condition*. With the help of the consistency condition which states basically that stresses must remain on the yield surface at all times during the course of loading, we write,

$$df = \left(\frac{\partial f}{\partial \sigma}\right)^T d\sigma = 0 \tag{5.5.10}$$

assuming no hardening behavior. Using (5.5.3) in this equation gives,

$$\left(\frac{\partial f}{\partial \sigma}\right)^T D^e(d\varepsilon - d\lambda\frac{\partial g}{\partial \sigma}) = 0 \tag{5.5.11}$$

with leaving the only unknown as the plastic strain magnitude (or also called *slip rate*), $d\lambda$, provided that $d\varepsilon$ is given. A bit of vector algebra yields the final form of $d\lambda$:

$$d\lambda = \frac{\left(\dfrac{\partial f}{\partial \sigma}\right)^T D^e}{\left(\dfrac{\partial f}{\partial \sigma}\right)^T D^e \dfrac{\partial g}{\partial \sigma}} d\varepsilon \tag{5.5.12}$$

Next, (5.5.12) is substituted in (5.5.9) giving the final incremental stress-strain relationship as,

$$d\sigma = D^e\left[I - \frac{\dfrac{\partial g}{\partial \sigma}\left(\dfrac{\partial f}{\partial \sigma}\right)^T D^e}{\left(\dfrac{\partial f}{\partial \sigma}\right)^T D^e \dfrac{\partial g}{\partial \sigma}}\right] d\varepsilon \tag{5.5.13}$$

where I is the identity matrix and the entire term in the brackets along with the elastic matrix is called the "elasto-plastic matrix", D^{ep}.

5.6 Perfect Plasticity

Perfect plasticity is a property of materials that undergo irreversible deformations without any increase in stresses or loads. As self-implied, it describes a "perfect" condition that refers to an idealized stress-strain relationship, that is, when the stress level exceeds the yield stress of the material further deformations do not result in further increase in stress (see Figure 5.6-1). For soils, perfect plasticity can be used as a preliminary estimate of more complex stress-strain behavior that can be captured better with the inclusion of some type of hardening rules. Therefore, for practical purposes, we can say that perfect plasticity is a good enough model to utilize in static conditions. In a perfectly plastic material where the size of the yield surface does not change we can write,

$$\dot{\sigma}\dot{\varepsilon}^p = 0 \tag{5.6.1}$$

where $\dot{\sigma}$ is the rate of stress and $\dot{\varepsilon}^p$ is the rate of plastic strain. Although the theory of plasticity dates back to 1864 when Tresca proposed the *maximum shear stress criterion* (Tresca criterion) for metals, the concept of perfect plasticity was first used within the context of geotechnical engineering with applications to earth pressures and retaining walls by Coulomb and Rankine. In early analyses, soil was considered to be a rigid-perfectly plastic material and calculations led to an estimate of the maximum load that the soil-structure system could sustain before it would collapse. Interested readers can refer to Drucker (1966) for some of the problems that arise in using perfect plasticity for soils and the justifications made for its practical use. In this chapter, we discuss the most commonly used perfectly plastic soil models, Mohr-Coulomb and Drucker-Prager.

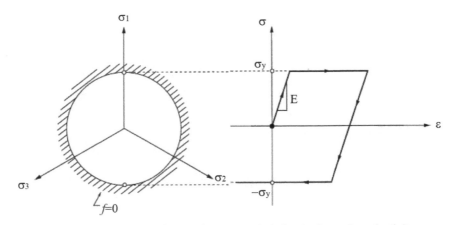

Figure 5.6-1 Yield surface and stress-strain behavior in perfect plasticity

5.6.1 Mohr-Coulomb Model with Triaxial Formulation

Within the context of triaxial loading conditions (i.e. $\sigma_1 > \sigma_2 = \sigma_3$), it is common to present the stresses in the p', q stress space during triaxial testing. Here $p' = \dfrac{\sigma_1' + 2\sigma_3'}{3}$ is the effective mean stress and $q = q' = \sigma_1' - \sigma_3'$, the deviatoric stress. The strain counterparts are ε_p and ε_q, respectively. As far as the Mohr-Coulomb yield criterion is concerned, associated plasticity with $f(\sigma) = g(\sigma)$ gives:

$$f(p',q) = q - Mp' - d = 0 \tag{5.6.2}$$

where M and d are the inclination and the intercept of the failure surface, respectively. Figure 5.6-2a shows the equation of this line as well as the stress-strain relationship as idealized as elastic-perfectly plastic. As the stress point reaches the failure surface at (p_f, q_f), failure takes place and continuous increase in magnitude occurs for strains at constant stresses (Figure 5.6-2b). This type of deformation is called the *"plastic flow"* which is idealized as the elastic-perfectly plastic model (Figure 5.6-2c).

Plastic Loading and Consistency Condition

In associated plasticity, the stress increment vector and the plastic strain increment vector both reside on the yield surface (or a similar failure surface). The unit normal is drawn from the yield curve and the direction of loading or unloading is determined based on the direction of the unit vector on the yield function (Figure 5.6-3). An immediate condition occurs here that the stress vector must stay on the yield curve upon ongoing loading which is written as,

$$f(p', q) = f(p' + dp', q + dq) = 0 \tag{5.6.3}$$

$$df = f(p' + dp', q + dq) - f(p', q) = 0 \tag{5.6.4}$$

The above becomes:

$$df = \frac{\partial f}{\partial p'} dp' + \frac{\partial f}{\partial q} dq = 0 \tag{5.6.5}$$

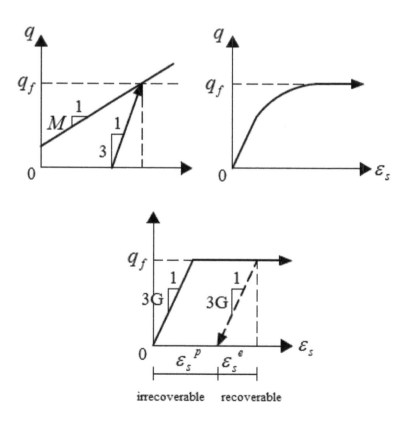

Figure 5.6-2 (a) Mohr-Coulomb yield function, (b) Actual stress strain response, (c) Idealized response (reproduced after Puzrin, 2012)

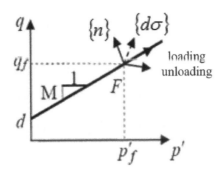

Figure 5.6-3 Loading/unloading directions and unit normal

when plastic loading occurs. Here, $df < 0$ is the condition for the material to behave elastically during unloading. Therefore, we can write: $n^T d\sigma < 0$. The normal vector to the failure surface at the current failure stress state shown as in Figure 5.6-4 is,

$$n = \left\{ \begin{array}{c} \dfrac{\partial f}{\partial p'} \\[2mm] \dfrac{\partial f}{\partial q} \end{array} \right\} = \left\{ \begin{array}{c} -M \\ 1 \end{array} \right\} \tag{5.6.6}$$

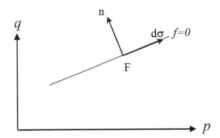

Figure 5.6-4 Stress increment and the normal on the yield surface

Drucker's Postulate for Associated Plasticity

Drucker (1949) states that in order for a work-hardening material to be stable at failure, the work done by the plastic strains, the plastic work, must be non-negative. That is:

$$d\sigma^T d\varepsilon^P = dp'd\varepsilon_p^p + dq d\varepsilon_q^p \geq 0 \tag{5.6.7}$$

The *second order plastic work* (dW^p) describes the stability postulate more accurately for the post-failure behavior. If $dW^p > 0$, material absorbs energy and hence the plastic behavior is considered stable. That is, further deformation requires additional work to be done on the material plastic behavior. If, however $dW^p < 0$, material releases energy and the behavior is unstable which means further deformation does not require additional external work. These states are analogous to the position of a ball placed at the bottom and at the top of a bowl (Figure 5.6-5).

While the Drucker's stability postulate explains why a material could be stable at failure, there is by no means a universal rule to dictate that all materials must be stable at failure. Thus,

Stable behavior   Unstable behavior

Figure 5.6-5 Ball position describing the stability state

it is more of a reasonable explanation to a certain physical phenomenon as opposed to a general law.

Convexity and Normality

If $df = 0$, we get,

$$n^T d\sigma = -Mdp' + dq = 0 \rightarrow dq/dp' = M \qquad (5.6.8)$$

which implies that n is normal to the $d\sigma$ at the failure surface which is the $f(\sigma)$. This is called the "normality condition". For a convex yield surface, f, to satisfy both the Drucker's Postulate and the consistency condition, the incremental plastic strain vector must be parallel to the normal, n, of that $f(s)$ surface. This is called the "convexity rule" (Figure 5.6-6). Therefore assuming the associated flow rule we write:

$$d\varepsilon^P = d\lambda.n = d\lambda \left\{ \frac{\partial f}{\partial p'}, \frac{\partial f}{\partial q} \right\}^T \qquad (5.6.9)$$

where $d\lambda$ is the plastic multiplier which must always satisfy $d\lambda > 0$. At failure, this becomes:

$$\begin{Bmatrix} d\varepsilon_p^P \\ d\varepsilon_q^P \end{Bmatrix} = d\lambda \begin{Bmatrix} -M \\ 1 \end{Bmatrix} \qquad (5.6.10)$$

Equation (5.6.9) implies that the direction of incremental plastic strain coincide with that of the principal direction of the stress increment. That said, associated flow rule describes only the direction of the incremental plastic strain vector but not the magnitude. For that we need to use the incremental stress-strain relationship stated earlier for the elastic case. Thus we have,

$$\begin{Bmatrix} dp' \\ dq \end{Bmatrix} = \begin{bmatrix} K & 0 \\ 0 & 3G \end{bmatrix} \begin{Bmatrix} d\varepsilon_p^e \\ d\varepsilon_q^e \end{Bmatrix} \qquad (5.6.11)$$

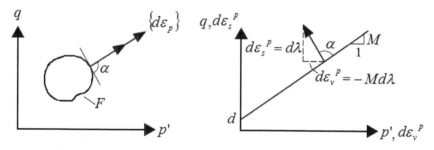

Figure 5.6-6 Convexity, normality and plastic flow (after Puzrin, 2012)

where $\begin{Bmatrix} d\varepsilon_p^e \\ d\varepsilon_q^e \end{Bmatrix}$ is the elastic strain vector. Using now the consistency condition (5.6.5):

$$df = Kd\varepsilon_p \frac{\partial f}{\partial p'} + 3Gd\varepsilon_q \frac{\partial f}{\partial q} - d\lambda \left[K\left(\frac{\partial f}{\partial p'} \right)^2 + 3G\left(\frac{\partial f}{\partial q} \right)^2 \right] = 0 \qquad (5.6.12)$$

which is now solved for $d\lambda$ as,

$$d\lambda = \frac{Kd\varepsilon_p \dfrac{\partial f}{\partial p'} + 3Gd\varepsilon_q \dfrac{\partial f}{\partial q}}{K\left(\dfrac{\partial f}{\partial p'} \right)^2 + 3G\left(\dfrac{\partial f}{\partial q} \right)^2} \qquad (5.6.13)$$

resulting in the volumetric and deviatoric plastic strains respectively as,

$$d\varepsilon_p^p = \frac{Kd\varepsilon_p \left(\dfrac{\partial f}{\partial p'} \right)^2 + 3Gd\varepsilon_q \dfrac{\partial f}{\partial q} \dfrac{\partial f}{\partial p'}}{K\left(\dfrac{\partial f}{\partial p'} \right)^2 + 3G\left(\dfrac{\partial f}{\partial q} \right)^2} \qquad (5.6.14a)$$

$$d\varepsilon_q^p = \frac{Kd\varepsilon_p \dfrac{\partial f}{\partial q} \dfrac{\partial f}{\partial p'} + 3Gd\varepsilon_q \left(\dfrac{\partial f}{\partial q} \right)^2}{K\left(\dfrac{\partial f}{\partial p'} \right)^2 + 3G\left(\dfrac{\partial f}{\partial q} \right)^2} \qquad (5.6.14b)$$

At Mohr-Coulomb failure, (5.6.14) relations become,

$$d\varepsilon_p^p = \frac{KM^2 d\varepsilon_p - 3GMd\varepsilon_q}{KM^2 + 3G} \qquad (5.6.15a)$$

$$d\varepsilon_q^p = \frac{-MKd\varepsilon_p + 3Gd\varepsilon_q}{KM^2 + 3G} \qquad (5.6.15b)$$

presenting essentially that the plastic strain increments do not depend on the stress increments but the total strains only. Substituting (5.6.14) into the triaxial form of (5.5.8) gives:

$$\begin{Bmatrix} dp' \\ dq \end{Bmatrix} = \frac{3KG}{K\left(\dfrac{\partial f}{\partial p'} \right)^2 + 3G\left(\dfrac{\partial f}{\partial q} \right)^2} \begin{bmatrix} \left(\dfrac{\partial f}{\partial q} \right)^2 & -\dfrac{\partial f}{\partial q} \dfrac{\partial f}{\partial p'} \\ -\dfrac{\partial f}{\partial q} \dfrac{\partial f}{\partial p'} & \left(\dfrac{\partial f}{\partial p'} \right)^2 \end{bmatrix} \begin{Bmatrix} d\varepsilon_p \\ d\varepsilon_q \end{Bmatrix} \qquad (5.6.16)$$

We should note here that zero stress increments do not lead to zero strain increments as long as the following holds:

$$\frac{\partial f}{\partial p'} d\varepsilon_q - \frac{\partial f}{\partial q} d\varepsilon_p = 0 \qquad (5.6.17)$$

The matrix in (5.6.16) is termed as the "elasto-plastic constitutive matrix" or sometimes called the "elasto-plastic material stiffness matrix" as,

$$D^{ep} = \frac{3KG}{K\left(\frac{\partial f}{\partial p'}\right)^2 + 3G\left(\frac{\partial f}{\partial q}\right)^2} \begin{bmatrix} \left(\frac{\partial f}{\partial q}\right)^2 & -\frac{\partial f}{\partial q}\frac{\partial f}{\partial p'} \\ -\frac{\partial f}{\partial q}\frac{\partial f}{\partial p'} & \left(\frac{\partial f}{\partial p'}\right)^2 \end{bmatrix} \tag{5.6.18}$$

This is a singular matrix as it is not invertible with the natural restriction on the strains essentially satisfying:

$$3Gd\varepsilon_q \frac{\partial f}{\partial q} + Kd\varepsilon_p \frac{\partial f}{\partial p'} \geq 0 \tag{5.6.19}$$

We also need to know the stresses in the case of failure, so (5.6.16) becomes:

$$\begin{Bmatrix} dp' \\ dq \end{Bmatrix} = \frac{3KG}{KM^2 + 3G} \begin{bmatrix} 1 & M \\ M & M^2 \end{bmatrix} \begin{Bmatrix} d\varepsilon_p \\ d\varepsilon_q \end{Bmatrix} \tag{5.6.20}$$

Having derived the entire plasticity formulation of a soil element defined by the Mohr-Coulomb yield criterion in triaxial stress state assuming an associated flow rule, it is important that we present how the formulation is modified in the case of *non-associated flow rule*. The latter is both more general and more applicable to a wider range of soils the reason of which is basically the need to be able to model *dilatancy*. That is, accurate modeling of dilatancy requires that the incremental plastic strain vector is not normal to the failure (i.e. yield) surface. Otherwise, such a normality condition overestimates dilatancy. Thus, in the case of non-associativity, we have to have a "plastic potential function" defined in the way stated in (5.5.4). In this case, incremental plastic strain vector will be normal to the plastic potential surface written now in triaxial stress-state as:

$$d\varepsilon^P = d\lambda n_p = d\lambda \left\{ \frac{\partial g}{\partial p'}, \frac{\partial g}{\partial q} \right\}^T \tag{5.6.21}$$

where n_p is the normal vector to the plastic potential surface. This *non-associated flow rule* again does not define the magnitude of the plastic strains for which $g(p', q)$ needs to be defined first. In order to be similar to the relation of (5.6.2), we can use the following form for the g function (Figure 5.6.7):

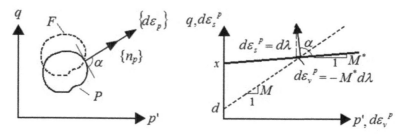

Figure 5.6-7 Non-associated flow rule and plastic potential function (reproduced after Puzrin, 2012)

$$g(p', q) = q - M^* p' - d^* = 0 \tag{5.6.22}$$

where d^* is derived considering that for each failure point (p_f, q_f) on the f surface, d^* should be such that plastic potential passes through this point as:

$$d^*(M - M^*)p' + d \tag{5.6.23}$$

Equation (5.6.10) now becomes:

$$\begin{Bmatrix} d\varepsilon_p^p \\ d\varepsilon_q^p \end{Bmatrix} = d\lambda \begin{Bmatrix} -M^* \\ 1 \end{Bmatrix} \tag{5.6.24}$$

Drucker's Postulate for Non-Associated Plasticity

In non-associated plasticity, Drucker's stability postulate changes in that, the second order plastic work can now be negative. For the incremental stress vector directed downwards along the failure surface, we have,

$$\begin{Bmatrix} dp' \\ dq \end{Bmatrix} = d\lambda \begin{Bmatrix} -1 \\ -M \end{Bmatrix} \tag{5.6.25}$$

satisfying the plastic loading condition:

$$df = -Mdp' + dq = 0 \tag{5.6.26}$$

Once this is substituted into the plastic work of (5.6.7), we get,

$$d\sigma^T d\varepsilon^p = dp' d\varepsilon_p^p + dq d\varepsilon_q^p = d\lambda^2(M^* - M) < 0 \tag{5.6.27}$$

since $M^* < M$. The take on from this is that the material is now unstable and the stability postulate is violated. However, this is not an undesirable fact for all materials as mentioned before and in order to be able to identify the behavior of some soil-like pressure-dependent frictional materials in the most accurate way, this is indeed desired.

Stress-Strain Formulation

Turning our attention back to the stress-strain formulation with non-associated flow rule, we repeat the above process by first taking the decomposition of strains for both shear and volumetric components as:

$$\varepsilon_p = \varepsilon_p^e + \varepsilon_p^p \tag{5.6.28a}$$

$$\varepsilon_s = \varepsilon_s^e + \varepsilon_s^p \tag{5.6.28b}$$

and then rewriting (5.6.12) such that we get,

$$d\lambda = \frac{Kd\varepsilon_p \dfrac{\partial f}{\partial p'} + 3Gd\varepsilon_q \dfrac{\partial f}{\partial q}}{K \dfrac{\partial f}{\partial p'} \dfrac{\partial g}{\partial p'} + 3G \dfrac{\partial f}{\partial q} \dfrac{\partial g}{\partial p'}} \tag{5.6.29}$$

This results in the plastic strain increments respectively as:

$$d\varepsilon_p^p = \frac{Kd\varepsilon_p \dfrac{\partial f}{\partial p'}\dfrac{\partial g}{\partial p'} + 3Gd\varepsilon_q \dfrac{\partial f}{\partial q}\dfrac{\partial g}{\partial p'}}{K\dfrac{\partial f}{\partial p'}\dfrac{\partial g}{\partial p'} + 3G\dfrac{\partial f}{\partial q}\dfrac{\partial g}{\partial q}} \tag{5.6.30a}$$

$$d\varepsilon_q^p = \frac{Kd\varepsilon_p \dfrac{\partial f}{\partial p'}\dfrac{\partial g}{\partial q} + 3Gd\varepsilon_q \dfrac{\partial f}{\partial q}\dfrac{\partial g}{\partial q}}{K\dfrac{\partial f}{\partial p'}\dfrac{\partial g}{\partial p'} + 3G\dfrac{\partial f}{\partial q}\dfrac{\partial g}{\partial q}} \tag{5.6.30b}$$

Again for the Mohr-Coulomb model, (5.6.30) relations become:

$$d\varepsilon_p^p = \frac{KMM^* d\varepsilon_p - 3GM^* d\varepsilon_q}{KMM^* + 3G} \tag{5.6.31a}$$

$$d\varepsilon_q^p = \frac{-MKd\varepsilon_p + 3Gd\varepsilon_q}{KMM^* + 3G} \tag{5.6.31b}$$

Equation (5.6.16) now becomes:

$$\begin{Bmatrix} dp' \\ dq \end{Bmatrix} = \frac{3KG}{K\dfrac{\partial f}{\partial p'}\dfrac{\partial g}{\partial p'} + 3G\dfrac{\partial f}{\partial q}\dfrac{\partial g}{\partial q}} \begin{bmatrix} \dfrac{\partial f}{\partial q}\dfrac{\partial g}{\partial q} & -\dfrac{\partial f}{\partial q}\dfrac{\partial g}{\partial p'} \\ -\dfrac{\partial g}{\partial q}\dfrac{\partial f}{\partial p'} & \dfrac{\partial f}{\partial p'}\dfrac{\partial g}{\partial p'} \end{bmatrix} \begin{Bmatrix} d\varepsilon_p \\ d\varepsilon_q \end{Bmatrix} \tag{5.6.32}$$

And finally at failure (5.6.32) becomes:

$$\begin{Bmatrix} dp' \\ dq \end{Bmatrix} = \frac{3KG}{KMM^* + 3G} \begin{bmatrix} 1 & M^* \\ M & MM^* \end{bmatrix} \begin{Bmatrix} d\varepsilon_p \\ d\varepsilon_q \end{Bmatrix} \tag{5.6.33}$$

5.6.2 Drucker-Prager Model with Triaxial Formulation

For the triaxial stress state, Drucker-Prager yield surface (5.4.20) takes the form,

$$f(\sigma) = \frac{q}{\sqrt{3}} - 3\alpha p - k = 0 \tag{5.6.34}$$

where

$$p = \frac{\sigma_{11} + 2\sigma_{33}}{3}, \quad q = \sigma_{11} - \sigma_{33}, \quad \sigma_{22} = \sigma_{33} \tag{5.6.35}$$

Based on section 5.4.3, if we follow the same steps described above to obtain the elasto-plastic formulation and use,

$$\frac{\partial f}{\partial \sigma} = \begin{pmatrix} \partial f / \partial p \\ \partial f / \partial q \end{pmatrix} = \begin{pmatrix} -3\alpha \\ 1/\sqrt{3} \end{pmatrix} \tag{5.6.36}$$

then for associated flow in p'-q space we get,

$$D^{en} = \begin{bmatrix} K & 0 \\ 0 & 3G \end{bmatrix} \left(\begin{pmatrix} 1 & 0 \\ 0 & 1 \end{pmatrix} - \frac{\begin{bmatrix} 9\alpha^2 K & -3\sqrt{3}\alpha G \\ -\sqrt{3}\alpha K & G \end{bmatrix}}{(9\alpha^2 K + G)} \right) \tag{5.6.37}$$

which, with manipulation leads to the following matrix which can also be derived in general stress state,

$$D^{ep} = \begin{bmatrix} \dfrac{GK}{(9\alpha^2 K + G)} & \dfrac{3\sqrt{3}\alpha GK}{(9\alpha^2 K + G)} \\ \dfrac{3\sqrt{3}\alpha GK}{(9\alpha^2 K + G)} & \dfrac{27\alpha^2 GK}{(9\alpha^2 K + G)} \end{bmatrix} \tag{5.6.38}$$

5.7 Hardening Plasticity

During a uniaxial compression (or tension) test, stress-strain relationship is observed to reach certain stress levels and corresponding strain responses. The elasto-plastic behavior of engineering materials can be drastically different from one another under applied loads. This is also the case for soils as we know different types of soils may behave completely differently under similar loading or drainage conditions. A general stress-strain behavior is shown in Figure 5.7-1. We can see that at the end of purely elastic behavior, the material yields at point A and the behavior changes from elastic to elastic-plastic (or elasto-plastic). From that point permanent deformations or irrecoverable strains are exhibited on the material. The material is then unloaded at point B where the response is purely elastic and the relationship is linear. It is, then, reloaded through lines CB, ED and FE where the behavior is fully elastic. Then the material keeps on yielding through points D and E and finally fails at point G with a failure stress, σ_f. If we continue the loading-unloading cycles, we observe that the point of yielding at every cycle in tensile and compression regions shifts (i.e. increases in absolute value) suggesting a separate law to describe this behavior. In classical plasticity, this behavior is called "*hardening*". Due to plastic flow, work or strain hardening can occur in many materials including soils. One theory proposed by Hill (1950) assumes that hardening depends only on the plastic work and is independent of the strain path. This implies that the resistance to further plastic deformations depends only on the total plastic work that has been done on the material. So the yield criterion is then written as,

$$f = f(\sigma, W^p) \tag{5.7.1}$$

where W^p is the plastic work. Another theory is based on the plastic strain as a measure of hardening. Then the yield function is rewritten as,

$$f = f(\sigma, \varepsilon^p) \tag{5.7.2}$$

In classical plasticity, the former definition is termed as "work hardening" and the latter as "strain hardening". In strain hardening, there is usually a relation to define the hardening as a function of either the plastic strain increment or some form of plastic distortional strain. There are several hardening rules that have been proposed to describe the growth of yield surface in the case of strain-hardening. The choice of the hardening rule depends on the actual material behavior observed in laboratory tests.

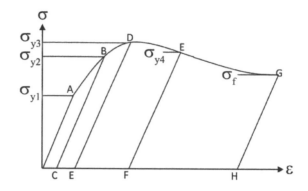

Figure 5.7-1 Fundamental behavior of an elasto-plastic material

In general, there are three types of hardening rules that are commonly used. These are:

(i) Isotropic hardening,
(ii) Kinematic hardening,
(iii) Combined isotropic and kinematic hardening or also called "mixed hardening".

As shown in Figure 5.7-2, the isotropic hardening rule presents only an expansion in the size of the yield surface whereas a kinematic hardening rule implies a kinematic movement of the yield surface as a translation or a rotation (the latter sometimes being called *rotational hardening*) in the stress space without a change in the size of the yield surface (Figure 5.7-2). A mixed hardening rule, however, includes both isotropic and kinematic hardening and is more flexible than the two being used independently in describing the post-yield behavior of the material. While yield surface is the main surface used to calculate plastic strains for associated plasticity, plastic potential surface is necessary to modify the way plastic strains are calculated as figured in Figure 5.7-3 (see the flow rule above). In the last decade, many models with quite complex hardening laws have been developed, implemented and applied to solve complex geotechnical engineering problems. These models tend to capture actual soil behavior and require modeling of elemental stress-strain relationship. We will discuss mainly the Mohr-Coulomb and Drucker-Prager models in this chapter. However, our discussion begins with introducing the three main hardening types mentioned above.

5.7.1 Isotropic Hardening

Constitutive material models that involve isotropic hardening have yield surfaces that expand through (mostly) the direction of the plastic strain increment. A hardening type is isotropic if the evolution of the yield surface is such that, at any state of hardening, it corresponds to a uniform (isotropic) expansion of the initial yield surface without any translation or rotation. Mathematically, this is described in the yield function by using a state variable, κ, that could either be a function (linear or nonlinear) of plastic strain variables (i.e. deviatoric plastic strain increment) or simply a scalar, that is:

$$f = f(\sigma, \kappa(\varepsilon^P)) \tag{5.7.3}$$

or in the case of work hardening:

$$f = f(\sigma, \kappa(W^P)) \tag{5.7.4}$$

Isotropic hardening could be treated within the concept of both strain hardening and work hardening which will be elaborated more subsequently. Isotropic hardening within the scope of this book is classified into *linear isotropic hardening* and *nonlinear isotropic hardening*. Here the yield function is altered such that,

$$f(\sigma, \kappa) = f(\sigma) - \kappa = 0 \tag{5.7.5}$$

and κ is taken either a constant (as in linear isotropic) or some function of the plastic strain history or stress (as in nonlinear isotropic). In this section, κ is defined as:

$$\kappa = \sigma_0 + h\eta_y \tag{5.7.6}$$

where σ_0 is the initial yield stress, h is the hardening parameter determined from laboratory tests and η_y is the strain history parameter keeping track of the plastic strain history. The nonlinear stress dependent variation of κ can be written as:

$$\kappa = \eta_y \sigma_m \tag{5.7.7}$$

which depends on the mean stress. The change in η_y as a function of plastic strain increment defines the isotropic hardening rule. Although, a simple linear dependency of η_y to the plastic shear strain increment can be used, a common approach is to define a parameter called "incremental plastic distortion" that is used to store the history of plastic deformation as:

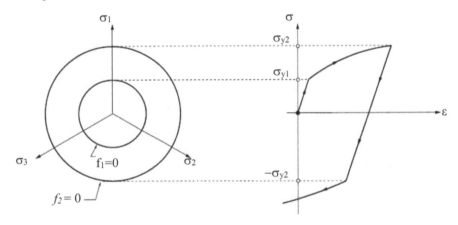

Figure 5.7-2 Evolution of yield surface and related stress-strain behavior in isotropic hardening

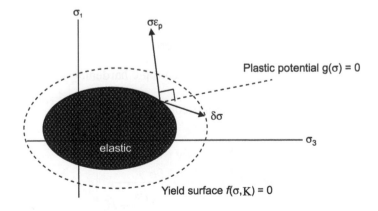

Figure 5.7-3 Yield and plastic potential surfaces in 2-D stress space

$$\eta_y = \sqrt{de^p \, de^p} \tag{5.7.8}$$

where de^p is the plastic shear strain increment. We then rewrite the consistency condition including the hardening term as:

$$\left(\frac{\partial f}{\partial \sigma}\right)^T d\sigma + \left(\frac{\partial f}{\partial \kappa}\right)^T d\kappa = 0 \tag{5.7.9}$$

If we use the chain rule, we obtain the consistency condition as,

$$\left(\frac{\partial f}{\partial \sigma}\right)^T d\sigma + \left(\frac{\partial f}{\partial \kappa}\right)^T \frac{\partial \kappa}{\partial \varepsilon^p} d\varepsilon^p = 0 \tag{5.7.10}$$

If the flow rule and the stress-strain relationship of (5.5.3) is substituted in the above relation, we get,

$$\left(\frac{\partial f}{\partial \sigma}\right)^T D\left(d\varepsilon - d\lambda \frac{\partial g}{\partial \sigma}\right) + \left(\frac{\partial f}{\partial \kappa}\right)^T \frac{\partial \kappa}{\partial \varepsilon^p} d\lambda \frac{\partial g}{\partial \sigma} = 0 \tag{5.7.11}$$

where the unknown will be the scalar parameter, the slip rate, $d\lambda$ which is obtained as:

$$d\lambda = \frac{\left(\dfrac{\partial f}{\partial \sigma}\right)^T D d\varepsilon}{\left[\left(\dfrac{\partial f}{\partial \sigma}\right)^T D \dfrac{\partial g}{\partial \sigma} - \left(\dfrac{\partial f}{\partial \kappa}\right)^T \dfrac{\partial \kappa}{\partial \varepsilon^p} \dfrac{\partial g}{\partial \sigma}\right]} \tag{5.7.12}$$

Here, the second term in the denominator is typically referred to as the *hardening modulus*, H and so the slip rate takes the final form,

$$d\lambda = \frac{\left(\dfrac{\partial f}{\partial \sigma}\right)^T D d\varepsilon}{\left[\left(\dfrac{\partial f}{\partial \sigma}\right)^T D \dfrac{\partial g}{\partial \sigma} + H\right]} \tag{5.7.13}$$

with

$$H = -\left(\frac{\partial f}{\partial \kappa}\right)^T \frac{\partial \kappa}{\partial \varepsilon^p} \frac{\partial g}{\partial \sigma} \tag{5.7.14}$$

where $\dfrac{\partial \kappa}{\partial \varepsilon^p}$ differential is determined from the isotropic hardening rule. Therefore, the final stress-strain relationship in incremental form becomes,

$$d\sigma = D\left(I - \frac{\dfrac{\partial g}{\partial \sigma}\left(\dfrac{\partial f}{\partial \sigma}\right)^T D}{\left(\dfrac{\partial f}{\partial \sigma}\right)^T D \dfrac{\partial g}{\partial \sigma} + H}\right) d\varepsilon \tag{5.7.15}$$

More on isotropic hardening can be found in the subsequent sections within the context of Mohr-Coulomb model.

5.7.2 Kinematic Hardening

If the hardening type is kinematic, typically the yield surface moves in the stress-space with no change in its size or shape, as shown in Figure 5.7-4. In this case, the center of the yield surface which is called the "back stress", changes its location by either translating or by translating and rotating together. This ensures another hardening variable to be placed in the yield function that is mostly a function of stress tensor used to model the variation of the yield surface center during loading. Now the yield function takes the form,

$$f(\sigma, \bar{\alpha}) = f(\sigma - \bar{\alpha}) = 0 \tag{5.7.16}$$

where $\bar{\alpha}$ denotes the translation of the center of yield surface in stress space which is a function of the plastic strain tensor, ε_{ij}^p. Consistency condition then takes the form,

$$\left(\frac{\partial f}{\partial \sigma}\right)^T d\sigma - \left(\frac{\partial f}{\partial \alpha}\right)^T \frac{\partial \bar{\alpha}}{\partial \varepsilon^p} d\varepsilon^p = 0 \tag{5.7.17}$$

which makes the hardening modulus H as,

$$H = \left(\frac{\partial f}{\partial \alpha}\right)^T \frac{\partial \bar{\alpha}}{\partial \varepsilon^p} \frac{\partial g}{\partial \sigma} \tag{5.7.18}$$

As a special case, Ziegler's kinematic hardening rule (Ziegler, 1959) is introduced here and implemented for the elemental level modeling of a soil material. Ziegler's rule assumes that the rate of translation takes place in the direction of the reduced stress or the back stress, $(\sigma - \bar{\alpha})$, in the form,

$$d\bar{\alpha} = d\mu(\sigma - \bar{\alpha}) \tag{5.7.19}$$

where $d\mu$ is a multiplier that depends on the history of the deformation which can be assumed as,

$$d\mu = md\varepsilon^p \tag{5.7.20}$$

where m is a positive material constant. The plastic hardening modulus now becomes,

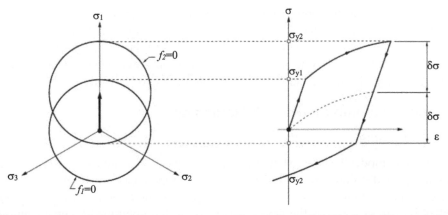

Figure 5.7-4 Evolution of yield surface and corresponding stress-strain behavior in kinematic hardening

$$H = mC\left(\frac{\partial f}{\partial \sigma}\right)^T \sqrt{\frac{\partial g}{\partial \sigma}\frac{\partial g}{\partial \sigma}}(\sigma - \bar{\alpha}) \tag{5.7.21}$$

where parameter C is depending on the yield function and defined within the context of "effective plastic strain" expressed as:

$$d\varepsilon_e^p = \frac{dW_p}{\sigma_e} = \frac{\sigma d\varepsilon^p}{\sigma_e} = \frac{\sigma d\lambda \frac{dg}{d\sigma}}{\sigma_e} = \frac{\sigma \frac{dg}{d\sigma} \frac{\sqrt{d\varepsilon^p d\varepsilon^p}}{\sqrt{\frac{\partial g}{\partial \sigma}\frac{\partial g}{\partial \sigma}}}}{\sigma_e} = C\sqrt{d\varepsilon^p d\varepsilon^p} \tag{5.7.22}$$

Here, we adopt the work hardening and write it in incremental form. So, C needs to be derived for a specific model (i.e. Mohr-Coulomb). Therefore, we should define the concept of effective stress (σ_e) and effective strain (ε_e) which are found to be quite useful in formulating the hardening behavior of soils. We should note immediately that the "effective" here does not mean the Terzaghi effective stress of soil solids but is related to the plasticity theory. Effective stress and effective strain are used to correlate the test results obtained by different loading programs. The relation between the two is easily defined for the uniaxial state. If we rewrite the yield function in its general form as,

$$f(\sigma,\kappa) = F(\sigma) - \kappa(\varepsilon^p) = 0 \tag{5.7.23}$$

where $F(\sigma)$ is used to define the effective stress such that,

$$F(\sigma) = C\sigma_e^n \tag{5.7.24}$$

We limit our discussion on kinematic hardening here since for more, it requires a special model to be utilized to finalize the formulation and adapt it to the solution of a boundary value problem in geotechnical engineering.

5.7.3 Mixed Hardening

Mixed hardening is generally a combination of some form of isotropic hardening and kinematic hardening. The yield function includes the state variables that are utilized to define both types of hardening. That is,

$$f = f(\sigma - \bar{\alpha},\kappa) \tag{5.7.25}$$

where the yield surface is allowed both to expand (and/or change in shape) and translate simultaneously in the stress space. This way, more realistic and sophisticated constitutive models can be developed, especially for complex materials like soil. Here, the hardening modulus, H becomes a combination of the one obtained for both isotropic and kinematic hardening models. Again more detailed discussion can be found in coming sections.

5.7.4 Mohr-Coulomb Model with Hardening

Isotropic Hardening

When we want to model the stress-strain behavior of soils as measured through classical triaxial compression (or extension) tests, it is desirable to write the constitutive relations in p'-q form (as before we prefer to use the prime (') here to represent the effective stress) where p' is the mean effective stress and q is the deviatoric stress, which combined together, represent the stress state of the soil. Similarly, the corresponding components of deformation are ε_p and ε_q as per volumetric and deviatoric strains, respectively. If we rewrite the yield function of the Mohr-Coulomb model,

$$f(p',q) = q - \eta_y p' = 0 \tag{5.7.26}$$

which, in the case of perfectly plastic model, becomes,

$$f(p',q) = q - Mp' = 0 \tag{5.7.27}$$

with M being the slope of the line in the p'-q space separating the inaccessible and elastic regions (Figure 5.7-5a) and is a function of the angle of internal friction, ϕ, in triaxial compression,

$$M = \frac{6\sin\phi}{3 - \sin\phi} \tag{5.7.28}$$

In (5.7.26), η_y is the hardening parameter indicating the current size of the yield surface and is proposed to take a hyperbolic relationship (Wood, 2004) with the distortional strain (deviatoric plastic strain) as:

$$\frac{\eta_y}{\eta_p} = \frac{\varepsilon_q^p}{A + \varepsilon_q^p} \tag{5.7.29}$$

where A is a scaling soil constant and η_p is the limiting value of the stress ratio (see Figure 5.7-5b). The incremental stress-strain relationship in p'-q can be written as,

$$\begin{Bmatrix} dp' \\ dq \end{Bmatrix} = \begin{bmatrix} K & 0 \\ 0 & 3G \end{bmatrix} \begin{Bmatrix} d\varepsilon_p^e \\ d\varepsilon_q^e \end{Bmatrix} \tag{5.7.30}$$

with $d\varepsilon_p^e$ and $d\varepsilon_q^e$ being the elastic strain components. The non-associated flow rule which works well for soils or soil-like materials is,

$$\begin{Bmatrix} d\varepsilon_p^p \\ d\varepsilon_q^p \end{Bmatrix} = \mu \begin{Bmatrix} \partial g/\partial p' \\ \partial g/\partial q \end{Bmatrix} = \mu \begin{Bmatrix} M - \eta \\ 1 \end{Bmatrix} \tag{5.7.31}$$

Naturally, plastic potential function is written accordingly as,

$$g(\sigma) = q - Mp' \ln\frac{p_r'}{p'} = 0 \tag{5.7.32}$$

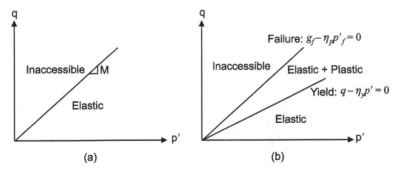

Figure 5.7-5 (a) Yield function locus for elastic-perfectly plastic Mohr-Coulomb model, (b) Yield function and failure loci for elastic-isotropic hardening plastic Mohr-Coulomb model (reproduced after Wood, 2004)

where p_r' is an arbitrary value introduced to be sure that the potential curve passes through the current stress state. The nonlinear isotropic hardening rule based on incremental distortional plastic strain $d\varepsilon_q^p$ is written according to (5.7.29) as:

$$d\eta_p = \frac{(\eta_p - \eta_y)^2}{A\eta_p} d\varepsilon_q^p$$

(5.7.33)

leading to the final form of the stress-strain relationship given below,

$$\begin{Bmatrix} dp' \\ dq \end{Bmatrix} = \left(\begin{bmatrix} K & 0 \\ 0 & 3G \end{bmatrix} - \frac{\begin{bmatrix} -K^2 \eta_y (M - \eta_y) & 3GK(M - \eta_y) \\ -3GK\eta_y & 9G^2 \end{bmatrix}}{3G - K\eta_y (M - \eta_y) + \dfrac{p'(\eta_p - \eta_y)^2}{A\eta_p}} \right) \begin{Bmatrix} d\varepsilon_p \\ d\varepsilon_q \end{Bmatrix}$$

(5.7.34)

Following the classical plasticity notion, we can also define a plastic potential function in order to make use of a non-associated flow rule and carry out an isotropic hardening model with the Mohr-Coulomb yield criterion. Such a potential function satisfying the non-associated flow rule can be written as:

$$g(\sigma) = \sigma_m \sin \phi_d + \sqrt{\bar{\sigma}^2 [F(\theta)]^2 + a^2 \sin^2 \phi_d} - c \cos \phi_d = 0$$

(5.7.35)

where σ_m is the mean stress, $\bar{\sigma} = \sqrt{J_2}$ and instead of the friction angle, ϕ, we now use a so called "dilation angle", ϕ_d as measured in direct shear tests for granular materials that exhibit the tendency to increase in volume under shearing, also called dilatancy. The dilation angle is, in general, smaller than the friction angle for soils. As we can see, isotropic hardening directly affects the stress-strain relationship through the elasto-plastic matrix.

Drained Triaxial Test

In drained triaxial test, it requires that the change in all-around pressure (or cell or radial pressure) is zero as a result of no change in pore pressure. That is,

$$\delta \sigma_r = \delta \sigma_3 = 0$$

(5.7.36a)

giving

$$\delta p' = \delta q / 3$$

(5.7.36b)

which, for the linear elastic case, leads to,

$$\frac{\delta q}{\delta p'} = 3 \Rightarrow \frac{\delta \varepsilon_p}{\delta \varepsilon_q} = \frac{\delta p / K}{\delta q / 3G} = \frac{G}{K}$$

(5.7.36c)

If we write,

$$\frac{\delta q}{\delta p'} = M$$

(5.7.36d)

then

$$\frac{\delta \varepsilon_p}{\delta \varepsilon_q} = \frac{3G}{MK}$$

(5.7.36e)

where $\delta\varepsilon_p$ and $\delta\varepsilon_q$ are the incremental volumetric and deviatoric strains, respectively. In 3-D stress state, the drained loading constrained under strain controlled loading results in the following equation,

$$\begin{pmatrix} \delta\sigma_1 \\ 0 \end{pmatrix} = \begin{bmatrix} D_{11}^{ep} & D_{12}^{ep} + D_{13}^{ep} \\ D_{21}^{ep} & D_{22}^{ep} + D_{23}^{ep} \end{bmatrix} \begin{pmatrix} \delta\varepsilon_1 \\ \delta\varepsilon_3 \end{pmatrix} \tag{5.7.36f}$$

where the terms in the constitutive matrix are the respective entries of the elasto-plastic rigidity matrix.

Undrained Triaxial Test

In an undrained test, there is no volume change in the material during shearing, so we write,

$$\delta\varepsilon_p = \frac{\delta p}{K_u} = 0 \tag{5.7.36g}$$

with the undrained bulk modulus being,

$$K_u = \frac{E_u}{2(1 - 2v_u)} \to \infty \tag{5.7.36h}$$

along with undrained Poisson's ratio $v_u \to 0.5$. For undrained loading, in terms of effective stress,

$$\delta\varepsilon_p^e = \frac{\Delta p'}{K'} = 0, \; \delta p' = 0 \tag{5.7.36i}$$

and $K' = \dfrac{E'}{3(1 - 2v')}$. In terms of total stress we write,

$$\delta\varepsilon_p^e = \frac{\delta p}{K} = 0, \; \delta p \neq 0 \tag{5.7.36j}$$

and $K = \dfrac{E_u}{3(1 - 2v_u)}$ since $K \to K_u = \infty \Rightarrow 3(1 - 2v_u) = 0$. We then have, $v_u = 0.5$ and $G = G_u$ $= G'$; so,

$$E_u = \frac{1.5E'}{(1 + v')} \tag{5.7.36k}$$

Computer Implementation

The triaxial formulation of the Mohr-Coulomb model presented above is implemented into a MATLAB® program, "**Extended_MC_Plasticity.m**". It is then used to simulate drained and undrained triaxial test results for the elastic-isotropic hardening plastic Mohr-Coulomb model previously studied also by Wood (2004). Undrained results are presented in Figure 5.7-6 and the drained results are in Figure 5.7-7.

Combined Isotropic-Kinematic Hardening

In combined isotropic-kinematic hardening, the yield surface (or potential surface in the case of non-associated flow rule) translates along the direction of the plastic strain and expands in size simultaneously during monotonic loading. Here, the hardening modulus, H becomes a

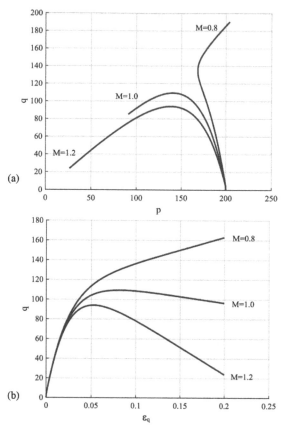

Figure 5.7-6 Elastic-nonlinear isotropic hardening plastic Mohr-Coulomb model undrained triaxial behavior, (a) *p-q* relationship for three M values, (b) Shear stress-shear strain relationship for each *M*, *G* = 3 MPa, *K* = 2 MPa, *A* = 0.02, p_0 = 200 kPa, η_y = 0.02, η_p = 1 (reproduced after Wood, 2004)

combination of the one obtained for both isotropic and kinematic hardening models. For Mohr-Coulomb model with Ziegler's kinematic hardening rule, it takes the form,

$$H = mC(1-M_w)\left(\frac{\partial f}{\partial \sigma}\right)^T \sqrt{\frac{\partial g}{\partial \sigma}\frac{\partial g}{\partial \sigma}}(\sigma-\bar{\alpha})-CM_w\left(\frac{\partial f}{\partial \kappa}\right)^T\frac{\partial \kappa}{\partial \varepsilon^p}\sqrt{\frac{\partial g}{\partial \sigma}\frac{\partial g}{\partial \sigma}} \tag{5.7.37}$$

where M_w is a weighting parameter (also called mixed hardening parameter) that decides how much isotropic hardening there will be in proportion to kinematic hardening. *C* is again the material parameter defined using the yield function. If we write such a yield criterion stated earlier in (5.7.24) using the Mohr-Coulomb model, we get,

$$\sigma_m \sin\phi + \sqrt{\bar{\sigma}^2\left[F(\theta)\right]^2+a^2\sin^2\phi} = C\sigma_e^n \tag{5.7.38}$$

For the simplest case of a uniaxial test, we have $\sigma_e = \sigma_1$, $\sigma_2 = \sigma_3 = 0$. Taking *n* = 1 we have,

$$\sigma_e = \frac{\frac{I_1}{3}\sin\phi + \sqrt{J_2}\cos\theta - \sqrt{\frac{J_2}{3}}\sin\phi\sin\theta}{\frac{\sin\phi}{3}+\frac{\cos\theta}{\sqrt{3}}-\frac{\sin\phi\sin\theta}{3}} \tag{5.7.39}$$

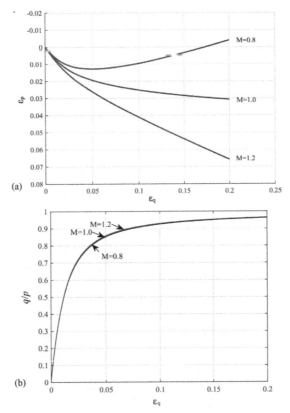

Figure 5.7-7 Elastic-nonlinear isotropic hardening plastic Mohr-Coulomb model drained triaxial behavior, (a) Volumetric strain behavior, (b) Shear stress-shear strain relationship, G = 3 MPa, K = 5 MPa, A = 0.005, p_0 = 100 kPa, η_p = 1 (reproduced after Wood, 2004)

Here we obtain C for the uniaxial stress state as:

$$C = \frac{\sin\phi}{3} + \frac{\cos\theta}{\sqrt{3}} - \frac{\sin\phi\sin\theta}{3} \tag{5.7.40}$$

which can also be derived for multi-dimensions (Chen and Mizuno, 1990):

$$C = \frac{\sigma \dfrac{dg}{d\sigma}}{\sigma_e \sqrt{\dfrac{\partial g}{\partial\sigma}\dfrac{\partial g}{\partial\sigma}}} \tag{5.7.41}$$

As an example we derive the form below for the Mohr-Coulomb model with non-associated flow rule,

$$C = \frac{\dfrac{I_1}{3}\sin\phi_d + \sqrt{J_2}\cos\theta - \sqrt{\dfrac{J_2}{3}}\sin\phi_d\sin\theta}{\left\{ \dfrac{\dfrac{I_1}{3}\sin\phi + \sqrt{J_2}\cos\theta - \sqrt{\dfrac{J_2}{3}}\sin\phi\sin\theta}{\dfrac{\sin\phi_d}{3} + \dfrac{\cos\theta}{\sqrt{3}} - \dfrac{\sin\phi_d\sin\theta}{3}} \right\} \sqrt{\dfrac{\sin^2\phi_d}{3} + \dfrac{1}{2}\left(\cos\theta - \dfrac{1}{\sqrt{3}}\sin\phi_d\sin\theta\right)^2}} \tag{5.7.42}$$

Computer Implementation

Combined isotropic-kinematic hardening formulation of the Mohr-Coulomb model is implemented into the MATLAB program called "**MC_DP_Hardening_Plasticity.m**" which is a quite versatile code allowing the user to control the following simultaneously:
 (i) Type of plasticity model (Mohr-Coulomb (MC) or Drucker-Prager (DP)),
(ii) Stress state (3-D state with a possibility to simulate any stress or strain state, i.e. triaxial),
(iii) Drainage conditions (drained or undrained test simulation),
(iv) Selection of the type of hardening (in terms of linear or nonlinear hardening in isotropic, kinematic or in combined fashion),
 (v) Selection of the type of flow rule (associated or non-associated),
(vi) Selection of Mohr-Coulomb yield surface (smoothed corners with a hyperbolic function or the regular Mohr-Coulomb prismatic cone).

It should be noted here, however, that the user must be very cautious in entering the input material parameters especially in the case of hardening as the results are rather sensitive to the physically meaningful ranges of those parameters. That is, they may end up being 'no result' to very high magnitudes of stresses and plastic strains. Thus, in order to demonstrate the flexibility of the code, some of the results presented herein may refer more to a general material behavior than a regular soil one.

Figures 5.7-8, 5.7-9 and 5.7-10 present nonlinear isotropic and combined isotropic-kinematic hardening plastic response of a soil element under undrained triaxial stress conditions using the Mohr-Coulomb yield function. The effect of hardening parameter A can be seen in all the response plots for both kinds of hardening along with the material parameter m. The drained behavior is plotted in Figure 5.7-11 and Figure 5.7-12 for varying η_y and M values.

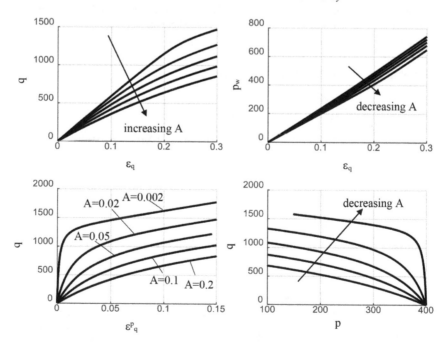

Figure 5.7-8 Elastic-non-linear isotropic hardening Mohr-Coulomb undrained behavior ($G = 2$ MPa, $K = 3.8$ Mpa, $c = 5$ kPa, $\phi = 30°$, $\phi_d = 22°$, $\theta_i = 15$, $h = 100$, $p_0 = 400$, $\eta_y = 0.02$, $H_p = 100$, $m = 10$)

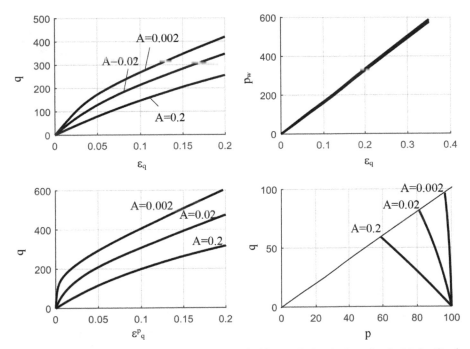

Figure 5.7-9 Elastic-combined nonlinear isotropic-kinematic hardening plastic Mohr-Coulomb model, undrained behavior, effect of Λ ($G = 1.5$ MPa, $K = 2.8$ Mpa, $c = 0$, $\phi = 22°$, $\phi_d = 18°$, $\theta_t = 15$, $h = 100$, $\sigma_0 = 70$ kPa, $p_0 = 100$ kPa, $H_p = 100$, $m = 10$, $M_w = 0.5$, values kPa)

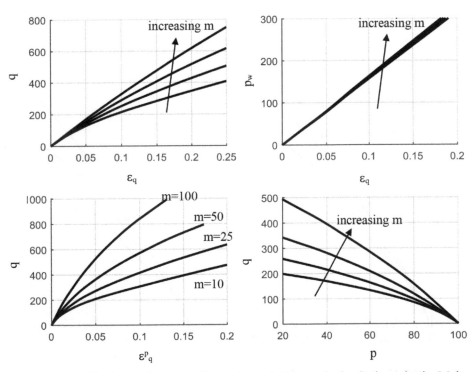

Figure 5.7-10 Elastic-combined nonlinear isotropic-kinematic hardening plastic Mohr-Coulomb model, undrained behavior, effect of m ($G = 1.5$ MPa, $K = 2.8$ MPa, $c = 0$, $\phi = 22°$)

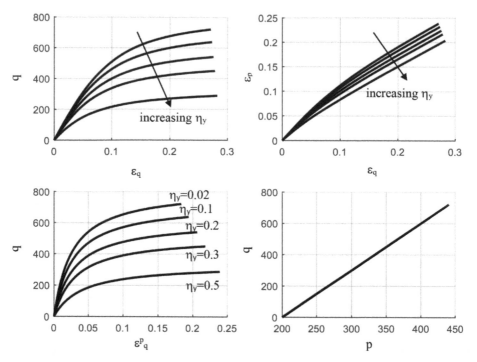

Figure 5.7-11 Elastic-nonlinear isotropic hardening plastic Mohr-Coulomb model, drained behavior ($G = 3$ MPa, $K = 2$ MPa, $c = 0$, $\phi = 30°$, $\phi_d = 18°$, $A = 0.02$, $p_0 = 200$ kPa, η_y = varying, $\eta_p = 1$, $H_p = 100$, $m = 10$, in kPa)

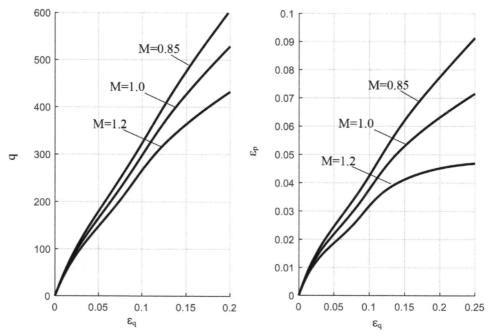

Figure 5.7-12 Elastic-nonlinear isotropic hardening plastic Mohr-Coulomb model, drained behavior, effect of J_3 (same parameters of previous figure, $\eta_y = 0.02$, values kPa)

5.7.5 Drucker-Prager Model with Hardening

Isotropic Hardening

In the case of isotropic hardening for the Drucker-Prager model, the yield function takes the form,

$$f(\sigma) = \sqrt{J_2} - \alpha(\varepsilon_e^p)I_1 - k(\varepsilon_e^p) = 0 \tag{5.7.43}$$

where α and k are functions of the effective plastic strain which can both also be functions of the distortional plastic strain. Once again the isotropic hardening can be assumed linear or nonlinear. We can write the last term of the yield function in (5.7.43) as,

$$k(\varepsilon_e^p) = \frac{1 + \sqrt{3}\alpha_1 H_p \varepsilon_e^p}{\sqrt{3}} \tag{5.7.44}$$

where H_p is the nonlinear plastic hardening modulus associated with the rate of expansion of the yield surface. However, we will use the relation derived by Chen and Mizuno, (1990) and assume that k changes as,

$$k = \frac{1 + \sqrt{3}\alpha_1}{\sqrt{3}}\sigma_e \tag{5.7.45}$$

where $\alpha = \alpha_1$ is taken as constant for the sake of some simplicity (for other forms in terms of effective strain, see Chen and Mizuno, 1990). The effective stress, σ_e is defined as,

$$\sigma_e = \frac{\sqrt{3}(\alpha_1 I_1 + \sqrt{J_2})}{1 + \sqrt{3}\alpha_1} \tag{5.7.46}$$

The plastic potential function is then written as,

$$g(\sigma) = \sqrt{J_2} - \alpha_2 I_1 = 0 \tag{5.7.47}$$

where $0 \le \alpha_2 \le \alpha_1$ relation holds with constant α_1 and now α_2 is a function of the dilation angle. One suggestion is,

$$\alpha_2 = \frac{2\sin\phi_d}{\sqrt{3}(3 - \sin\phi_d)} \tag{5.7.48}$$

Although the linear isotropic hardening can still be used with the Drucker-Prager model, hardening modulus can still be a function of plastic strain in terms of the following form:

$$H = -\left(\frac{\partial f}{\partial k}\right)^T \frac{\partial k}{\partial \varepsilon^p} C \sqrt{\frac{\partial g}{\partial \sigma}\frac{\partial g}{\partial \sigma}} \tag{5.7.49}$$

where C is,

$$C = \frac{(1 + \sqrt{3}\alpha_1)}{\sqrt{\frac{3}{2} + 9\alpha_1^2}} \tag{5.7.50}$$

Using the yield function and k we get,

$$H = \frac{(\alpha_2 I_1 + \sqrt{J_2})(1 + \sqrt{3}\alpha_1)^2 H_p}{3k} \qquad (5.7.51)$$

Kinematic Hardening

In the case of kinematic hardening for the Drucker-Prager model, the yield function is modified such that,

$$f(\sigma - \bar{\alpha}) = \sqrt{\bar{J_2}} - \alpha \bar{I_1} - k = 0 \qquad (5.7.52)$$

where $\bar{J_2}$ and $\bar{I_1}$ are the modified invariants defined using $\bar{\sigma} = \sigma - \bar{\alpha}$ with $\bar{\alpha}$ being the back stress. The readers should be cautious not to confuse it with the model parameter α. The consistency condition and the hardening modulus defined previously for the Mohr-Coulomb model hold also for the Drucker-Prager model. The plastic hardening modulus is still defined as in (5.7.49) with the parameter C being now:

$$C = \frac{(1 + \sqrt{3}\alpha_1)(\sqrt{\bar{J_2}} + \alpha_2 \bar{I_1})}{(\sqrt{\bar{J_2}} + \alpha_1 \bar{I_1})\sqrt{\left(\frac{1}{2} + 3\alpha_2{}^2\right)}} \qquad (5.7.53)$$

The plastic potential function takes the form,

$$\frac{\partial g}{\partial \sigma} = \frac{1}{2} + 3\alpha_2{}^2 \qquad (5.7.54)$$

and the incremental back stress, $d\bar{\alpha}$, is updated using Ziegler's rule as:

$$d\bar{\alpha} = mCd\varepsilon_q^p(\sigma - \bar{\alpha}) \qquad (5.7.55)$$

Combined Isotropic-Kinematic Hardening

As is the case for the Mohr-Coulomb model, the combined hardening rule for the Drucker-Prager model allows the translation (or rotation) and expansion of the yield surface during the course of elasto-plastic loading. The yield function now becomes,

$$f(\bar{\sigma}, \varepsilon_e^p) = \sqrt{\bar{J_2}} - \alpha(\varepsilon_e^p)\bar{I_1} - k(\varepsilon_e^p) = 0 \qquad (5.7.56)$$

which can easily be seen that this form is a combination of (5.7.43) and (5.7.52). If we use Ziegler's kinematic hardening rule and include isotropic hardening, hardening modulus is derived as,

$$H = \underbrace{H_{kin}(1 - M_w)}_{kinematic} + \underbrace{\frac{(\alpha_2 I_1 + \sqrt{J_2})(1 + \sqrt{3}\alpha_1)^2 H_p}{3k}}_{isotropic} M_w \qquad (5.7.57)$$

where H_{kin} is the plastic hardening modulus for the kinematic hardening part and calculated as in (5.7.21):

$$H_{kin} = mC\left(\frac{\partial f}{\partial \sigma}\right)^T \sqrt{\frac{\partial g}{\partial \sigma}\frac{\partial g}{\partial \sigma}}(\sigma - \bar{\alpha}) \qquad (5.7.58)$$

Computer Implementation

Using the same code "**MC_DP_Hardening_Plasticity.m**", Drucker-Prager model is simulated in various stress and drainage states. Figure 5.7-13 presents the undrained behavior with linear isotropic hardening. An example for the behavior of combined hardening soil material under undrained loading is plotted in Figure 5.7-14 and the drained loading is shown in Figure 5.7-15.

Plane Strain Formulation

As in many geotechnical engineering problems, the stress state associated with the problem in 2-D is plane strain. Therefore, the constitutive formulation of plasticity must be modified to suit that stress state at the material level. As for the Drucker-Prager model, we start off by writing the yield function again:

$$f(\sigma,k) = \sqrt{J_2} - \alpha I_1 - k = 0 \tag{5.7.59}$$

where

$$\frac{\partial f}{\partial I_1} = -\alpha; \qquad \frac{\partial f}{\partial J_2} = \frac{1}{2\sqrt{J_2}} \tag{5.7.60}$$

with

$$I_1 = \sigma_{11} + \sigma_{22} + \sigma_{33}$$

and

$$J_2 = \frac{1}{6}[(\sigma_{11}-\sigma_{22})^2 + (\sigma_{22}-\sigma_{33})^2 + (\sigma_{11}-\sigma_{33})^2] + \sigma_{12}^2$$

As usual, if we take the partial derivatives with respect to stress, we get,

$$\frac{\partial f}{\partial \sigma_{11}} = \frac{\partial f}{\partial I_1}\frac{\partial I_1}{\partial \sigma_{11}} + \frac{\partial f}{\partial J_2}\frac{\partial J_2}{\partial \sigma_{11}} = -\alpha + \frac{1}{2\sqrt{J_2}}S_{11} \tag{5.7.61a}$$

$$\frac{\partial f}{\partial \sigma_{22}} = \frac{\partial f}{\partial I_1}\frac{\partial I_1}{\partial \sigma_{22}} + \frac{\partial f}{\partial J_2}\frac{\partial J_2}{\partial \sigma_{22}} = -\alpha + \frac{1}{2\sqrt{J_2}}S_{22} \tag{5.7.61b}$$

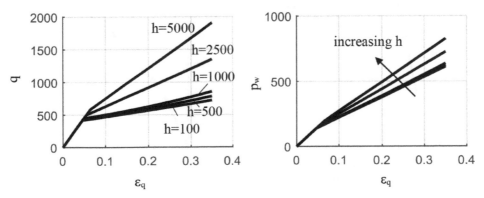

Figure 5.7-13 Elastic-linear isotropic hardening Drucker-Prager undrained response (parameters same as previous analysis, σ_0 = 150 kPa, p_0 = 200 kPa, η_y = 0.02, h = varying, m = 10, values kPa), $H = h\eta_y$

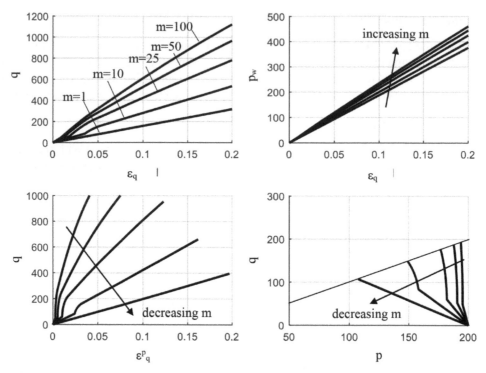

Figure 5.7-14 Elastic-combined nonlinear isotropic-kinematic hardening plastic Drucker-Prager model, undrained behavior, varying m ($G = 2.5$ MPa, $K = 2$ MPa, $c = 50$ kPa, $p_0 = 200$ kPa, $\phi = 25°$, $h = 150$, $\sigma_0 = 100$ kPa, $\eta_y = 0.02$, $\eta_p = 1$, $H_p = 80$, $M_w = 0.5$, values kPa)

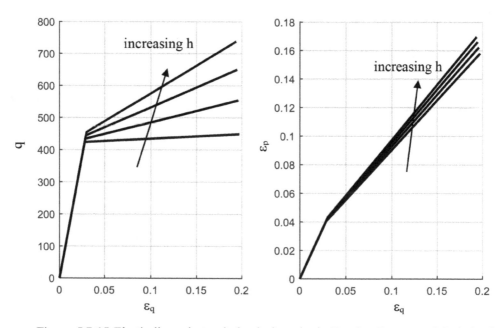

Figure 5.7-15 Elastic-linear isotropic hardening plastic Drucker-Prager model, drained behavior, varying h ($G = 5$ MPa, $K = 3.5$ MPa, $c = 50$ kPa, $p_0 = 200$ kPa, $\phi = 25°$, $\sigma_0 = 50$ kPa, h varying, $\eta_y = 0.02$, $\eta_p = 1$, $H_p = 80$, $m = 1$, values kPa)

$$\frac{\partial f}{\partial \sigma_{33}} = \frac{\partial f}{\partial I_1}\frac{\partial I_1}{\partial \sigma_{33}} + \frac{\partial f}{\partial J_2}\frac{\partial J_2}{\partial \sigma_{33}} = -\alpha + \frac{1}{2\sqrt{J_2}}S_{33} \qquad (5.7.61c)$$

$$\frac{\partial f}{\partial \sigma_{12}} = \frac{\partial f}{\partial I_1}\frac{\partial I_1}{\partial \sigma_{12}} + \frac{\partial f}{\partial J_2}\frac{\partial J_2}{\partial \sigma_{12}} = -\alpha + \frac{1}{2\sqrt{J_2}}S_{12} \qquad (5.7.61d)$$

which is written in the vector form as:

$$\left(\frac{\partial f}{\partial \underline{\sigma}}\right)^T = \left[-\alpha + \frac{1}{2\sqrt{J_2}}S_{11} \quad -\alpha + \frac{1}{2\sqrt{J_2}}S_{22} \quad -\alpha + \frac{1}{2\sqrt{J_2}}S_{33} \quad \frac{1}{\sqrt{J_2}}S_{12}\right] \qquad (5.7.62)$$

The stress-strain relationship then becomes:

$$\begin{Bmatrix} d\sigma_{11} \\ d\sigma_{22} \\ d\sigma_{12} \\ d\sigma_{33} \end{Bmatrix} = \left[D_{ep}\right]\begin{Bmatrix} d\varepsilon_{11} \\ d\varepsilon_{22} \\ d\varepsilon_{12} \end{Bmatrix} \qquad (5.7.63)$$

where the elasto-plastic constitutive matrix is,

$$\left[D_{ep}\right] = \begin{bmatrix} K+\frac{4}{3}G & K-\frac{2}{3}G & 0 \\ K-\frac{2}{3}G & K+\frac{4}{3}G & 0 \\ 0 & 0 & G \\ K-\frac{2}{3}G & K-\frac{2}{3}G & 0 \end{bmatrix} - \frac{1}{H}\begin{bmatrix} H_{11}^2 & H_{11}H_{22} & H_{11}H_{22} \\ H_{22}H_{11} & H_{22}^2 & H_{22}H_{12} \\ H_{12}H_{11} & H_{33}H_{22} & H_{33}H_{12} \end{bmatrix} \qquad (5.7.64)$$

with the terms,

$$H = 9K\alpha^2 + G; \quad H_{11} = 3K\alpha + \frac{G}{\sqrt{J_2}}S_{11}; \quad H_{22} = 3K\alpha + \frac{G}{\sqrt{J_2}}S_{22};$$

$$H_{33} = 3K\alpha + \frac{G}{\sqrt{J_2}}S_{33}; \quad H_{12} = \frac{G}{\sqrt{J_2}}S_{12} = \frac{G}{\sqrt{J_2}}\sigma_{12}.$$

Example 5.7-1: Develop the stress-strain relationship for associated Drucker-Prager model in triaxial stress state and simulate a strain-controlled drained static triaxial shear test.

Solution:
We start with the stress and strain vectors defined as:

$$d\varepsilon = \begin{bmatrix} d\varepsilon_1 & 0 & 0 \end{bmatrix}^T \qquad (5.7.65a)$$

$$d\sigma = \begin{bmatrix} d\sigma_1 & d\sigma_3 & d\sigma_3 \end{bmatrix}^T \qquad (5.7.65b)$$

leading to the stress variables dp and dq and the corresponding strain variables $d\varepsilon_p$ and $d\varepsilon_q$ as $dp' = \dfrac{d\sigma_1' + 2d\sigma_3'}{3}$, $dq = d\sigma_1' - d\sigma_3'$ and $d\varepsilon_q = d\varepsilon_1 + 2d\varepsilon_3$, $d\varepsilon_q = \dfrac{2}{3}(d\varepsilon_1 - d\varepsilon_3)$. Elastic stress-strain relationship is given as usual as, $d\sigma_1' = \left(K + \dfrac{4}{3}G\right)d\varepsilon_1^e$ or in p-q space as in (5.7.30),

$$\begin{pmatrix} dp' \\ dq \end{pmatrix} = \begin{bmatrix} K & 0 \\ 0 & 3G \end{bmatrix} \begin{pmatrix} d\varepsilon_p^e \\ d\varepsilon_q^e \end{pmatrix} \tag{5.7.66}$$

For the Drucker-Prager model we have the following yield function in triaxial state:

$$f(p,q,k) = \frac{q}{\sqrt{3}} + 3\alpha p - k = 0 \tag{5.7.67}$$

If we use associated flow rule and develop an elasto-plastic stress-strain relationship whose constitutive matrix is given in (5.6.38), we have a relationship such as the one below:

$$\begin{pmatrix} dp' \\ dq \end{pmatrix} = \begin{bmatrix} \dfrac{GK}{(9\alpha^2 K + G)} & \dfrac{3\sqrt{3}\alpha GK}{(9\alpha^2 K + G)} \\ \dfrac{3\sqrt{3}\alpha GK}{(9\alpha^2 K + G)} & \dfrac{27\alpha^2 GK}{(9\alpha^2 K + G)} \end{bmatrix} \begin{pmatrix} d\varepsilon_p \\ d\varepsilon_q \end{pmatrix} \tag{5.7.68}$$

Relations in (5.6.35) can now be used to derive the following:

$$\frac{d\sigma_1 + 2d\sigma_3}{3} = -K\left(\frac{9\alpha^2 K}{9\alpha^2 K + G} - 1\right)(d\varepsilon_1) + \left(-\frac{3\sqrt{3}\alpha GK}{9\alpha^2 K + G}\right)\left(\frac{2d\varepsilon_1}{3}\right) \tag{5.7.69a}$$

$$(d\sigma_1 - d\sigma_3) = \left(-\frac{3\sqrt{3}\alpha GK}{9\alpha^2 K + G}\right)(d\varepsilon_1) - 3G\left(\frac{G}{9\alpha^2 K + G} - 1\right)\left(\frac{2d\varepsilon_1}{3}\right) \tag{5.7.69b}$$

Solving the system gives the vertical stress increment $d\sigma_1$:

$$(d\sigma_1) = \left(K + \frac{4}{3}G\right)(d\varepsilon_1) - \left(\frac{2\dfrac{\sqrt{3}}{3}G + 3K\alpha}{9\alpha^2 K + G}\right)(d\varepsilon_1) \tag{5.7.70}$$

where the first term on the right is the elastic part and the second term is the plastic part. A simple separate MATLAB program "**DP_Assoc_Plas.m**" where the above Drucker-Prager model for a 1-D case is implemented and provided at the end of the chapter which gives the result in Figure 5.7-16.

5.8 Loading/Unloading Criterion

In order for the stress path to follow the applied loads in the material, the loading-unloading condition must be well defined. The simplest way to distinguish between a loading and an unloading condition is to define in the stress space a normalized direction, n, for any given state of stress, σ, such that all increments of stress are separated into two cases: loading and unloading, such that,

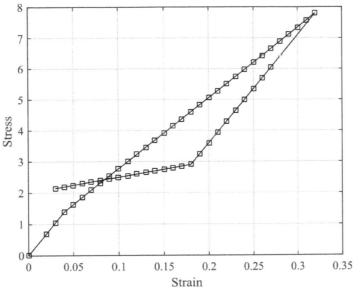

Figure 5.7-16 Axial stress-strain relationship from Drucker-Prager model

$$d\varepsilon_L = C_L d\sigma_L \quad \text{for} \quad n:d\sigma > 0 \quad \rightarrow \quad \text{Loading} \tag{5.8.1a}$$

$$d\varepsilon_U = C_U d\sigma_U \quad \text{for} \quad n:d\sigma < 0 \quad \rightarrow \quad \text{Unloading} \tag{5.8.1b}$$

where C_L and C_U are the constitutive matrices in loading and unloading, respectively. There is also the neutral loading written as,

$$n:d\sigma = 0 \tag{5.8.2}$$

where loading increments take place tangential to the yield surface and do not create plastic deformation (Figure 5.8-1). These are defined within the so called "Generalized Theory of Plasticity" introduced by Zienkiewicz and Mroz (1985) which was later extended by Pastor et al. (1985) and Pastor et al. (1990). Zienkiewicz et al. (1990) suggest that the set of surfaces that are defined as a result of this distinction between loading and unloading are equivalent to those used in classical plasticity and need not be defined explicitly. Chen and Mizuno (1990) express loading-unloading criterion in terms of the complementary energy density function, Ω. Unloading is indicated by the condition $d\Omega < 0$ where $d\Omega = \varepsilon_{ij}d\sigma_{ij}$ is the incremental change in Ω. So we write,

$$\Omega = \Omega_{\max} , \ d\Omega > 0 \rightarrow \text{Loading} \tag{5.8.3a}$$

$$\Omega \leq \Omega_{\max} , \ d\Omega < 0 \rightarrow \text{Unloading} \tag{5.8.3b}$$

$$\Omega < \Omega_{\max} , \ d\Omega > 0 \rightarrow \text{Reloading} \tag{5.8.3c}$$

where $\Omega_{\max}$ is the maximum previous value of Ω at the material point. Anandarajah (2010) defines a scalar quantity, L, called the loading index which is a scalar product between the rate of stress, $\dot{\sigma}$ and the unit normal, n to the yield surface at the current stress point as,

$$L = \dot{\sigma}_{kl} n_{kl} \tag{5.8.4}$$

where

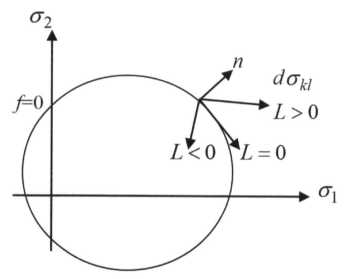

Figure 5.8-1 Loading-unloading conditions represented by the direction of loading vector on the yield surface

$$n_{kl} = \frac{1}{g} \frac{\partial \phi}{\partial \sigma_{kl}}$$
(5.8.5)

and $g = \left| \dfrac{\partial \phi}{\partial \sigma} \right|$. This form is called the "unit normal formulation" but the $\dfrac{\partial \phi}{\partial \sigma_{kl}}$ can also be used instead of the unit normal. So the loading-unloading criterion is defined as (see Figure 5.8-1),

$$L > 0 \rightarrow \quad \text{Loading (elasto-plastic)} \tag{5.8.6a}$$

$$L = 0 \rightarrow \text{Neutral-loading (elastic)} \tag{5.8.6b}$$

$$L < 0 \rightarrow \text{Unloading (elastic)} \tag{5.8.6c}$$

The above relations are useful for the elemental level calculations of stress-strain behavior. However, in finite element applications, the stress increment is calculated for a given strain increment therefore it is necessary to determine whether the strain increment will cause elasto-plastic behavior or elastic behavior. Here the variables in the rate equations must be converted into incremental quantities and as Anandarajah (2010) suggested, (5.8.6) are not useful. Therefore, for multi-dimensional plasticity, the best way to decide whether the loading step is an elasto-plastic or purely an elastic one is to define a trial *elastic stress*,

$$\sigma_{n+1}^{tri} = \sigma_n + C^e \Delta \varepsilon \tag{5.8.7}$$

where $n + 1$ represents the forward integration point and C^e is the elastic constitutive matrix. If σ_{n+1}^{tri} lies outside the yield surface the step is elasto-plastic, if it is still inside the yield surface the step will then be elastic.

5.9 Exercise Problems

5.9.1 For the elastic-perfectly plastic constitutive model, Mohr-Coulomb, obtain the final form of the elasto-plastic matrix, D^{ep} using the yield functions given in the text and assuming non-associated flow rule where yield function (f) is not equal to the plastic potential function (g). (Hint: Recall the flow rule, consistency condition and the stress-strain relationships in the process of developing the formulation).

    a. Obtain the final form of D^{ep} for Mohr-Coulomb for the plane strain case using the following material parameters and stresses:

        $E = 10$ MPa, $v = 0.3$, $c = 100$ kPa, $\phi = 300$, $\sigma_{11} = 100$ kPa, $\sigma_{33} = 50$ kPa, $\sigma_{12} = 75$ kPa

    b. Recalculate the elasto-plastic matrix by considering a triaxial stress state.

5.9.2 Repeat the above problem with the Drucker-Prager model considering the material parameters as well as the stress state below. Then comment on the differences between the results in terms of the use of both models, Mohr-Coulomb and Drucker-Prager. Obtain the final form of D^{ep} in the case of triaxial stress state: a. Plane strain, b. Triaxial stress state

        $E = 10$ MPa, $v = 0.3$, $k = 20$, $\alpha = 0.1$, $\sigma_{11} = 100$ kPa, $\sigma_{33} = 50$ kPa, $\sigma_{12} = 50$ kPa

5.9.3 Formulate the undrained constrain in terms of stress-strain relationship for the triaxial stress state in a stress-controlled test simulation.

5.9.4 Redo the previous problem for a 3-D stress state. Recall the drained constrain presented in the text namely, Eq. 5.7.36f.

5.9.5 Derive the parameter C, for the Mohr-Coulomb model with an associated flow rule using Eq. 5.7.41.

5.9.6 Redo the previous problem for the Drucker-Prager model and consider both non-associated and associated formulations.

5.9.7 Describe how a plasticity formulation developed in 3-D can be adapted to simulate a triaxial stress state (axi-symmetric condition) in an undrained situation under both strain and stress controlled tests. You can keep the yield function in its general definition given in the book and present your steps for the following cases:

    a. Elastic-perfectly plastic material

    b. Elastic isotropic hardening material

    c. Elastic mixed hardening material

MATLAB® Scripts

Main program "Extended_MC_Plasticity.m"

```matlab
%% Extended Mohr-Coulomb Model with Non-linear Hardening
% Strain-controlled triaxial test simulation
clear all
% close all
Drain='D';                          % Drained Test: "D", Undrained Test: "U"
Spec_Strain=0.3;                    % Shear strain loading

L=length(Spec_Strain);
N=1000;                             % # of sub increments
ns=N*ones(1,L);
eps=zeros(1,6);
for j=1:L
    if Drain=='D'
        eps(j,1)=Spec_Strain;
        eps(j,2)=0;
        eps(j,3)=0;
    elseif Drain=='U'
        eps(j,1)=Spec_Strain;
        eps(j,2)=-Spec_Strain/2;
        eps(j,3)=-Spec_Strain/2;
    else
        % do nothing
    end
end
s=size(eps);
for j=1:s(1)                        % Create specified sub-strain matrix
    for k=1:s(2)
        deps(j,k)=eps(j,k)./ns(j);
    end
end
deps_p=deps(1,1)+2*deps(1,3);
deps_q=2/3*(deps(1,1)-deps(1,3));
deps=[deps_p, deps_q];
% MATERIAL PARAMETERS
Mat_Par.G=3000;                     % Shear modulus
Mat_Par.K=5000;                     % Bulk modulus
Mat_Par.phi=30;% 25.378;% 20.67;    % Friction angle
Mat_Par.A=0.005;                    % Isotropic MC Hardening Parameter
eta_y=0.02;
eta_p=1.0;
G=Mat_Par.G;
K=Mat_Par.K;
phi=Mat_Par.phi; % as degrees
A=Mat_Par.A;
M=6*sin(phi*pi/180)/(3-sin(phi*pi/180));    % Critical State Stress Ratio for
% Triaxial Compression
% Elastic_Matrix=[K 0
%                 0 3*G];
Sigma=100;%50;
% INITIALIZATION
Stress=[Sigma 0];                   % Stress State P,Q
Total_Strain=[0 0];                 % Eps-P, Eps-Q
Plas_Strain=[0 0];
count=1;
```

```
Plot_Vol_Stress(count) = Sigma;
for j=1:s(1)                          % Start doing the calculations for each
% entry of max. total strain tensor
    Stress_Inc = [0 0],
    Plas_Strain_Inc = [0 0];          % Initial plastic strain Inc
    Total_Strain_Inc = deps(j,:);
    for i=1:ns(j)                      % For each strain increment
        P=Stress(1);
        Q=Stress(2);
        f1 = Q-eta_y*P;
        % Elastic Predictor
        if Drain=='D'
            Mult=(3*Elastic_Matrix(1,2)-
            Elastic_Matrix(2,2))/(Elastic_Matrix(2,1)-3*Elastic_Matrix(1,1));
            Total_Strain_Inc(1)=Mult*Total_Strain_Inc(2); % Adjusted dEps_p
        elseif Drain=='U'
            % do nothing
        end
        Stress_Inc = Elastic_Matrix*Total_Strain_Inc';
        Stress = Stress + Stress_Inc';
        P=Stress(1);
        Q=Stress(2);
        f2 = Q-eta_y*P;
        if ( f1 < 0 && f2 <= 0 )        % If we are in elastic region
            Plas_Strain_Inc = [0 0];
        elseif ( ( f1<0 || f1==0 || f1>0 ) && f2>0 ) % If we are in plastic
        % region which must be on the yield surf.
            Stress = Stress - Stress_Inc';
            % Gradients
            dF_by_dSig=[-eta_y 1];
            dG_by_dSig=[M-eta_y 1];
            % Plastic Strain
            numer = dF_by_dSig*Elastic_Matrix;
            denom = dF_by_dSig*Elastic_Matrix*dG_by_dSig';
            H = P*(eta_p-eta_y)^2/(A*eta_p);
            del_lambda = numer*Total_Strain_Inc'/(denom+H);
            if del_lambda<0
                display ('del_lambda < 0')  % The scalar must be positive
                del_lambda=0;
            else
                % do nothing
            end
            Plas_Strain_Inc = del_lambda*dG_by_dSig;
            % Stress Increment
            numer = Elastic_Matrix*dG_by_dSig'*dF_by_dSig*Elastic_Matrix;
            Plastic_Matrix = numer/(denom+H);
            Elasto_Plastic_Matrix = Elastic_Matrix - Plastic_Matrix;
            if Drain=='U'
                % do nothing
            elseif Drain=='D'
                Mult=(3*Elasto_Plastic_Matrix(1,2)-
                Elasto_Plastic_Matrix(2,2))/(Elasto_Plastic_Matrix(2,1)-
                3*Elasto_Plastic_Matrix(1,1));
                Total_Strain_Inc(1)=Mult*Total_Strain_Inc(2); % Adjusted
                % dEps_p
            end
            Stress_Inc = Elasto_Plastic_Matrix*Total_Strain_Inc';
            % Update Isotropic Hardening Parameter
```

```
            del_eta_y = (eta_p-eta_y)^2*Plas_Strain_Inc(2)/(A*eta_p);
            eta_y = eta_y + del_eta_y;

        elseif ( ( f1<0 || f1==0 || f1>0 ) && f2<0 ) % Elastic Unloading
            Stress = Stress - Stress_Inc';
            Stress_Inc = Elastic_Matrix*Total_Strain_Inc';
            Plas_Strain_Inc = [0 0];
        end
    Stress = Stress + Stress_Inc';
    Plas_Strain = Plas_Strain + Plas_Strain_Inc;
    Total_Strain = Total_Strain + Total_Strain_Inc;
    count = count + 1;
    Plot_Plastic_Vol_Strain(count) = Plas_Strain(1);
    Plot_Plastic_Shear_Strain(count) = Plas_Strain(2);
    Plot_Vol_Stress(count) = Stress(1);
    Plot_Shear_Stress(count) = Stress(2);
    Plot_Total_Vol_Strain(count) = Total_Strain(1);
    Plot_Total_Shear_Strain(count) = Total_Strain(2);
    Plot_eta(count) = eta_y;

    end
end
%PLOTS
figure
plot(Plot_Total_Shear_Strain,Plot_eta,'k', 'LineWidth',0.5)
xlabel('Eps-q')
ylabel('q/p')
grid on
figure
subplot(2,2,1)
hold on
plot(Plot_Total_Shear_Strain,Plot_Shear_Stress,'b', 'LineWidth',2)
xlabel('Eps-q')
ylabel('q')
grid on
subplot(2,2,2)
plot(Plot_Total_Shear_Strain,Plot_Total_Vol_Strain,'b', 'LineWidth',2)
xlabel('Eps-q')
ylabel('Eps-p')
grid on
subplot(2,2,3)
plot(Plot_Plastic_Shear_Strain,Plot_Shear_Stress,'b', 'LineWidth',2)
xlabel('Eps-plas-q')
ylabel('q')
grid on
subplot(2,2,4)
figure
hold on
plot(Plot_Vol_Stress,Plot_Shear_Stress,'b', 'LineWidth',2)
xlabel('p')
ylabel('q')
grid on
```

Main program "DP_Assoc_Plas.m"

```
% 1-D Associated Plasticity Formulation for Triaxial Loading of
% Drucker-Prager Model
clear all
```

```
close all
% Input Parameters
nu=0.3;                                % Poisson's ratio
G=10;                                  % Shear Modulus
k=3;                                   % DP parameter, y-intercept of sqrt(J2)-I1
                                         plot
alpha=0.8;                             % DP Model parameter, slope of sqrt(J2)-I1
                                         plot

K=2*G*(1+nu)/(3*(1-2*nu));             % Bulk modulus
M=K+4*G/3;                             % Constrained modulus

% Initial values
Stress(1)=0;
Strain(1)=0;

% Strain Controlled Loading
for j=2:32
    Strain(j)=j/100;
end
for j=33:57
    Strain(j)=(60-j)/100;
end

sigma_yield1=sqrt(3)*M*k/(2*G+3*sqrt(3)*K*alpha); % Yield stress in
% compression
sigma_yield2=sqrt(3)*M*k/(-2*G+3*sqrt(3)*K*alpha);% Yield stress in tension
const_1=(2*sqrt(3)/3*G-3*K*alpha)^2/(9*K*alpha^2+G);
const_2=(2*sqrt(3)/3*G+3*K*alpha)^2/(9*K*alpha^2+G);

for i=1:56
    delta_Strain=Strain(i+1)-Strain(i)
    if delta_Strain > 0
        delta_stress=M*delta_Strain
        if (Stress(i)+delta_stress) <= sigma_yield1
            Stress(i+1)=Stress(i)+delta_stress
        else
            Stress(i+1)=Stress(i)+delta_stress-const_1*delta_Strain
        end
    elseif delta_Strain < 0
        delta_stress=M*delta_Strain
        if (Stress(i)+delta_stress) >= sigma_yield2
            Stress(i+1)=Stress(i)+delta_stress
        else
            Stress(i+1)=Stress(i)+delta_stress-const_2*delta_Strain
        end
    end
end
% Plot Stress-Strain Relationship
plot(Strain, Stress, '-ks')
title('Stress-Strain Relationship using the Drucker-Prager Model')
ylabel('Stress')
xlabel('Strain')
```

Main program "MC_DP_Hardening_Plasticity.m"

```
%% 3-D Formulation of Mohr-Coulomb and Drucker-Prager Soil Plasticity Models
% "MC_DP_Hardening_Plasticity.m"
```

```
% Primarily monotonic loading with elastic unloading option
% Drainage control option
% Strain or stress-induced loading control option
% Hardening control option
%%%%%%%%%%%%%%%%%%%%%%%%%%%%%%%%%%%%%%%%%%%%%%%%%%%%%%%%%%%%%%%%%%%%%%%%%%%%%%%
clear all
% close all
global FLAG FLAG_H FLAG_HT FLAG_IHT FLAG_DRN FLAG_YF FLAG_FLOW FLAG_ETA_Y
FLAG_TEST

FLAG_TEST = '3AXI'              % 3AXI: Triaxial Test, 3DIM: 3-D State
FLAG = 'MC';                    % MC: Mohr-Coulomb, DP: Drucker Prager
FLAG_H = 'H';                   % H: Hardening, NH: No-Hardening
FLAG_HT = 'M';                  % I: Isotropic Hardening, K: Kinematic
                                  Hardening, M:Mixed Hardening
FLAG_IHT = 'NL';                % Use with (FLAG_HT=I), L: Linear
                                  Isotropic Hardening, NL: Non-linear
                                  Isotropic Hardening
FLAG_DRN = 'UND';               % DRN: Drained, UND: Undrained
FLAG_YF = 'SH';                 % PR: Regular Prismatic MC yield surface,
                                  SH: Smoothed Hyperbolic MC yield
                                  function as in Abbo and Sloan (1995)
FLAG_FLOW = 'NONAS';            % ASSOC: Associated Flow Rule, NONAS: Non-
                                  associated Flow Rule
% Specified Total Strain INC. Tensors
deps_vert=0.005*ones(1,50);
N=100;                          % # of sub increments
L=length(deps_vert);
ns=N*ones(1,L);
if FLAG_DRN == 'DRN'
    for jj=1:L
        eps(jj,1)=deps_vert(jj);
        eps(jj,2)=0;% 0.5*deps_vert(jj);
        eps(jj,3)=0;% 0.5*deps_vert(jj);
        eps(jj,4)=0;
        eps(jj,5)=0;
        eps(jj,6)=0;
    end
elseif FLAG_DRN == 'UND'
    for jj=1:L
        eps(jj,1)=deps_vert(jj);
        eps(jj,2)=-deps_vert(jj)/2;
        eps(jj,3)=-deps_vert(jj)/2;
        eps(jj,4)=0;
        eps(jj,5)=0;
        eps(jj,6)=0;
    end
end
s=size(eps);
for j=1:s(1)                    % Create specified sub-strain matrix
    for k=1:6
        deps(j,k)=eps(j,k)./ns(j);
    end
end


% MATERIAL PARAMETERS
Mat_Par.G=1500;                 % Shear modulus
Mat_Par.K=2800;                 % Bulk modulus
Mat_Par.c=0;                    % Cohesion
```

```
Mat_Par.phi=22;                    % Friction angle
Mat_Par.phi_dil=18;                % Dilation angle if equal to friction
                                     angle, becomes associative

Mat_Par.Theta_T=15;                % Transition Lode angle
Mat_Par.h=100;                     % Linear Hardening Modulus
Mat_Par.A=0.2;                     % Isotropic Hardening Param.
Mat_Par.S0=70;                     % Constant initial yield stress
Mat_Par.eta_y=0.02;                % Initial hardening parameter for the size
                                     of yield surface

Mat_Par.eta_p=1;                   % Limit value of the stress ratio
Mat_Par.Hp=100;                    % Nonlinear plastic hardening modulus
                                     associated with the rate of expansion of
                                     the potential/yield surface

Mat_Par.const_m=10;                % Positive material parameter determined
                                     from the tests to be used in Ziegler's
                                     Kin. Hard. Rule

Mat_Par.MW=0.5;                    % Weighting parameter -1<MW<1 to determine
                                     what % is Kin. Hard (MW=0) and what % is
                                     Isot. Hard. (MW=1)

Scell=100;                         % Consolidation cell pressure

phi=Mat_Par.phi;
phi_dil=Mat_Par.phi_dil;
A=Mat_Par.A;
eta_p=Mat_Par.eta_p;
const_m=Mat_Par.const_m;
MW=Mat_Par.MW;


M=6*sin(phi*pi/180)/(3-sin(phi*pi/180));      % Critical State Stress Ratio
                                                for Triaxial Compression
Elastic_Matrix = elastic_matrix( Mat_Par );   % Call the elastic matrix

% INITIALIZATION
Stress=[Scell Scell Scell 0 0 0];  % Initial Stress State, Default: Triaxial
                                     Compression
Total_Strain=[0 0 0 0 0 0];
Plas_Strain=[0 0 0 0 0 0];
Back_Alpha=[0 0 0 0 0 0];          % Back stress of Kinematic Hardening
Eff_Plas_Strain=0;
Pore_Pressure=0;
del_Sig_Cons=0;                    % Change in consolidation pressure
count=1;
Plot_Vol_Stress(count) = Scell;
FLAG_ETA_Y=0;                      % Flag for hardening eta_y
for j=1:s(1)                       % Start doing the calculations for each
                                     element of max. total strain tensor
    Stress_Inc = [0 0 0 0 0 0];
    Plas_Strain_Inc = [0 0 0 0 0 0];        % Initial plastic strain inc
    Strain_Inc = deps(j,:);
    for i=1:ns(j)                           % For each strain increment
        [I1, Shear] = components( Stress );
        if FLAG == 'MC'  %Mohr Coulomb
            f1 = yield_func_MC( Stress, Mat_Par, Back_Alpha );
        elseif FLAG == 'DP' %Drucker Prager
            f1 = yield_func_DP( Stress, Mat_Par, Back_Alpha );
        else
            % do nothing
        end
```

```
% Elastic Predictor
% Apply the Drainage Constraint
if FLAG_DRN == 'DRN'
    den_el=Elastic_Matrix(2,2)+Elastic_Matrix(2,3);
    Strain_Inc(3)= -Elastic_Matrix(2,1)*Strain_Inc(1)/den_el;
    Mod_DE =(
    Elastic_Matrix(1,1)*(Elastic_Matrix(2,2)+Elastic_Matrix(2,3))-...

    Elastic_Matrix(2,1)*(Elastic_Matrix(1,2)+Elastic_Matrix(1,3)) )/
    den_el;
    Stress_Inc(1) = Mod_DE*Strain_Inc(1);
    Strain_Inc(2)=Strain_Inc(3);
    Stress = Stress + Stress_Inc;
elseif FLAG_DRN == 'UND'
    Stress_Inc = Elastic_Matrix*Strain_Inc';
    Stress = Stress + Stress_Inc';
end
[I1, Shear] = components( Stress );
[J2, J3, Theta] = invariants( Shear );

if FLAG=='MC'   %Mohr Coulomb
    f2 = yield_func_MC( Stress, Mat_Par, Back_Alpha );
elseif FLAG == 'DP' %Drucker Prager
    f2 = yield_func_DP( Stress, Mat_Par, Back_Alpha );
else
    % do nothing
end

if ( f1 < 0 && f2 <= 0 )              % If we are in elastic region
    if FLAG_DRN == 'UND'
        Stress = Stress - Stress_Inc'; % Back to current stress
    elseif FLAG_DRN == 'DRN'
        Stress = Stress - Stress_Inc;
    end
    Plas_Strain_Inc = [0 0 0 0 0 0];
    Total_Strain_Inc = Strain_Inc;
    [ del_Vol_Stress, del_Shear_Stress ] = triaxial_stress_state(
    Stress_Inc );
    del_Pore_Pressure = del_Shear_Stress/3 - del_Vol_Stress;
elseif ( ( f1<0 || f1==0 || f1>0 ) && f2>0 ) % If we are in plastic
% region
    if FLAG_DRN == 'UND'
        Stress = Stress - Stress_Inc'; % Back to current stress
    elseif FLAG_DRN == 'DRN'
        Stress = Stress - Stress_Inc;
    end
    [I1, Shear] = components( Stress );
    [J2, J3, Theta] = invariants( Shear );
    Total_Strain_Inc = Strain_Inc;
    % Calculate Gradients of F and G
    dF_by_dSig=Gradient_of_Surface(Stress, phi, Mat_Par);
    if FLAG_FLOW == 'ASSOC'
        dG_by_dSig=dF_by_dSig;
    elseif FLAG_FLOW == 'NONAS'
        dG_by_dSig=Gradient_of_Surface(Stress, phi_dil, Mat_Par);
    end
    numer = dF_by_dSig*Elastic_Matrix;
    denom = dF_by_dSig*Elastic_Matrix*dG_by_dSig';
```

```
            if FLAG_H == 'H'
                % Calculate the Hardening Modulus H
                [ H_Mod, param_C ] = get_hardening_modulus( Stress, Mat_Par,
                Back_Alpha, f2 );
            elseif FLAG_H == 'NH'
                H_Mod = 0.0; % No 'h'ardening, No H
            else
                error 'Is Hardening Specified?'
            end
            % Calculate the slip rate
            del_lambda = numer*Total_Strain_Inc'/(denom + H_Mod);
            if del_lambda<0
            display ('del_lambda < 0')
% The scalar has to be greater than zero
                del_lambda=0;
            else
                % do nothing
            end
            % Calculate Plastic Strain
            Plas_Strain_Inc = del_lambda*dG_by_dSig; % Non-associated flow
            rule
            % Calculate Stress Increment
            numer = Elastic_Matrix*dG_by_dSig'*dF_by_dSig*Elastic_Matrix;
            Plastic_Matrix = numer/(denom + H_Mod);
            Elasto_Plastic_Matrix = Elastic_Matrix - Plastic_Matrix;
            % Apply the Drainage Constraint
            if FLAG_DRN == 'DRN'
                den=Elasto_Plastic_Matrix(2,2)+Elasto_Plastic_Matrix(2,3);
                Total_Strain_Inc(3)= -
                Elasto_Plastic_Matrix(2,1)*Total_Strain_Inc(1)/den;
                Mod_DEP = (
                Elasto_Plastic_Matrix(1,1)*(Elasto_Plastic_
                Matrix(2,2)+Elasto_Plastic_Matrix(2,3))-...

                Elasto_Plastic_Matrix(2,1)*(Elasto_Plastic_
                Matrix(1,2)+Elasto_Plastic_Matrix(1,3)) )/den;
                Stress_Inc (1) = Mod_DEP*Total_Strain_Inc(1);
                Total_Strain_Inc(2)=Total_Strain_Inc(3);
            elseif FLAG_DRN == 'UND'
                Stress_Inc = Elasto_Plastic_Matrix*Total_Strain_Inc';
            end
            % Calculate pore pressure inc.
            %%%%%%%%%%%%%%% Triaxial
            if FLAG_TEST == '3AXI'
                [ del_Vol_Stress, del_Shear_Stress ] = triaxial_stress_state(
                Stress_Inc );
                del_Pore_Pressure = del_Shear_Stress/3 + del_Sig_Cons -
                del_Vol_Stress;
            elseif FLAG_TEST == '3DIM'
                [del_Vol_Stress, del_Shear_Stress] = components( Stress_Inc
                );
                del_Shear_Stress = sqrt(
                del_Shear_Stress(1)^2+del_Shear_Stress(2)^2+del_Shear_
                Stress(3)^2+...

                2*del_Shear_Stress(4)^2+2*del_Shear_Stress(5)^2+2*del_Shear_
                Stress(6)^2 );
                del_Pore_Pressure = del_Shear_Stress/3 + del_Sig_Cons -
                del_Vol_Stress;
```

```
            end
            %%%%%%%%%%%%%% Triaxial
            % Update the Hardening Parameters
            if FLAG_H == 'H'
                [ Back_Alpha ] = update_hardening( Plas_Strain_Inc, Stress,
                param_C, Mat_Par, Back_Alpha );
            elseif FLAG_H == 'NH'
                % do nothing
            else
                error 'Is Hardening Specified?'
            end
        elseif ( ( f1<0 || f1==0 || f1>0 ) && f2<0 ) % Elastic Unloading
            if FLAG_DRN == 'UND'
                Stress = Stress - Stress_Inc'; % Back to current stress
            elseif FLAG_DRN == 'DRN'
                Stress = Stress - Stress_Inc;
            end
            Total_Strain_Inc = Strain_Inc;
            % Apply the Drainage Constraint
            if FLAG_DRN == 'DRN'
                den=Elastic_Matrix(2,2)+Elastic_Matrix(2,3);
                Total_Strain_Inc(3)= -
                Elastic_Matrix(2,1)*Total_Strain_Inc(1)/den;
                Mod_DEP = (
                Elastic_Matrix(1,1)*(Elastic_Matrix(2,2)+Elastic_
                Matrix(2,3))-...

                Elastic_Matrix(2,1)*(Elastic_Matrix(1,2)+Elastic_Matrix(1,3))
                )/den;
                Stress_Inc (1) = Mod_DEP*Total_Strain_Inc(1);
                Total_Strain_Inc(2)=Total_Strain_Inc(3);
            elseif FLAG_DRN == 'UND'
                Stress_Inc = Elastic_Matrix*Total_Strain_Inc';
            end
            Plas_Strain_Inc = [0 0 0 0 0 0];
        end
if FLAG_DRN == 'UND'
    Stress = Stress + Stress_Inc';
elseif FLAG_DRN == 'DRN'
    Stress = Stress + Stress_Inc;
end
Pore_Pressure=Pore_Pressure+del_Pore_Pressure;
Plas_Strain = Plas_Strain + Plas_Strain_Inc;
Total_Strain = Total_Strain + Total_Strain_Inc;
count = count + 1;

Plot_Plastic_Vert_Strain(count) = abs(Plas_Strain(1));
[ Plas_Vol_Strain, Plas_Shear_Strain ] = triaxial_strain_state(
Plas_Strain );
Plot_Plastic_Shear_Strain(count) = abs(Plas_Shear_Strain);
Plot_Plastic_Vol_Strain(count) = abs(Plas_Vol_Strain);
Plot_Total_Vert_Strain(count) = abs(Total_Strain(1));
[ Total_Vol_Strain, Total_Shear_Strain ] = triaxial_strain_state(
Total_Strain );
Plot_Total_Shear_Strain(count) = abs(Total_Shear_Strain);
Plot_Total_Vol_Strain(count) = abs(Total_Vol_Strain);

Plot_Vert_Stress(count) = Stress(1);
```

```
    [ Vol_Stress, Shear_Stress ] = triaxial_stress_state( Stress );
    Plot_Shear_Stress(count) = Shear_Stress;
    Plot_Vol_Stress(count) = Vol_Stress;
    Plot_Pore_Pressure(count) = abs(Pore_Pressure);

    end
end
% figure
subplot(2,2,1)
hold on
plot(Plot_Total_Shear_Strain,Plot_Shear_Stress,'k', 'LineWidth',2)
xlabel('Eps-q')
ylabel('q')
grid on
subplot(2,2,2)
hold on
if FLAG_DRN == 'DRN'
    plot(Plot_Total_Shear_Strain,Plot_Total_Vol_Strain,'k', 'LineWidth',2)
    xlabel('Eps-q')
    ylabel('Eps-p')
elseif FLAG_DRN == 'UND'
    plot(Plot_Total_Shear_Strain,Plot_Pore_Pressure,'k', 'LineWidth',2)
    xlabel('Eps-q')
    ylabel('pw')
end
grid on
subplot(2,2,3)
hold on
plot(Plot_Plastic_Shear_Strain,Plot_Shear_Stress,'k', 'LineWidth',2)
xlabel('Eps-plas-q')
ylabel('q')
grid on
subplot(2,2,4)
hold on
plot(Plot_Vol_Stress,Plot_Shear_Stress,'k', 'LineWidth',2)
xlabel('p')
ylabel('q')
grid on

%-------------------------------------------------------------
```

Subprogram "components.m"

```
function [ mean, devia ] = components( stress )
% Calculates the mean (volumetric) and deviatoric (shear)
% components of stress
I1=stress(1)+stress(2)+stress(3);
mean=I1/3;
devia=[stress(1)-mean stress(2)-mean stress(3)-mean stress(4) stress(5)
stress(6)];
end
%-------------------------------------------------------------
```

Subprogram "yield_func_MC.m"

```
function [ yield_f ] = yield_func_MC( stress, mat_par, back_stress )
% Mohr-Coulomb model yield function
```

```
global FLAG_H FLAG_HT FLAG_IHT FLAG_YF FLAG_ETA_Y ETA_Y FLAG_TEST
cohesion=mat_par.c;
phi=mat_par.phi; % in degrees
h=mat_par.h;
S0=mat_par.S0;
if FLAG_ETA_Y == 0
    ETA_Y=mat_par.eta_y;
elseif FLAG_ETA_Y == 1
    % do nothing
end

if FLAG_H == 'H'
    if FLAG_HT == 'I'
        [sig_m, Sij] = components( stress );
        [J2,~,Theta] = invariants(Sij);
        if FLAG_TEST == '3AXI'
            Theta = pi/6; % Triaxial compression
        elseif FLAG_TEST == '3DIM'
            % do nothing
        end
        if FLAG_YF == 'SH' % As in Abbo and Sloan (1995)
            [K_theta,~]=LodeAngle(Theta, mat_par, phi);
            if phi == 0
                param_a=0.05*cohesion;
            else
                param_a=0.05*cohesion*cot(phi*pi/180);
            end
            yield_f =
            sig_m*sin(phi*pi/180)+sqrt(J2*K_theta^2+param_
            a^2*sin(phi*pi/180)^2)-cohesion*cos(phi*pi/180);
        elseif FLAG_YF == 'PR' % Regular MC cone
            yield_f = sig_m*sin(phi*pi/180)+sqrt(J2)*(cos(Theta)-
            sin(phi*pi/180)*sin(Theta)/sqrt(3))-cohesion*cos(phi*pi/180);
        end
        if FLAG_IHT == 'L'
            yield_f = yield_f - (S0+h*ETA_Y);
        elseif FLAG_IHT == 'NL'
            yield_f = yield_f - ETA_Y*sig_m;
        else
            error 'What is the Isotropic Hardening Type?'
        end
    elseif FLAG_HT == 'K' || FLAG_HT == 'M'
    [sig_m_bar,J2_bar,Theta_bar]=invariants_translated(stress,back_stress);

        if FLAG_YF == 'SH' % As in Abbo and Sloan (1995)
            [K_theta, ~]=LodeAngle(Theta_bar, mat_par, phi);
            param_a=0.05*cohesion*cot(phi*pi/180);
            yield_f =
            sig_m_bar*sin(phi*pi/180)+sqrt(J2_bar*K_theta^2+param_
            a^2*sin(phi*pi/180)^2)-cohesion*cos(phi*pi/180);
        elseif FLAG_YF == 'PR' % Regular MC
            yield_f = sig_m_bar*sin(phi*pi/180)+sqrt(J2_bar)*(cos(Theta_bar)-
            sin(phi*pi/180)*sin(Theta_bar)/sqrt(3))-cohesion*cos(phi*pi/180);
        end
        if FLAG_HT == 'M'
            if FLAG_IHT == 'L'
                yield_f = yield_f - (S0+h*ETA_Y);
```

```
            elseif FLAG_IHT == 'NL'
                yield_f = yield_f - ETA_Y*sig_m_bar;
            else
                error 'What is the Isotropic Hardening Type?'
            end
        elseif FLAG_HT == 'K'
            % do nothing
        end
    else
        error 'What is the Hardening Type?'
    end

elseif FLAG_H == 'NH'
    [sig_m, Sij] = components( stress );
    [J2,~,Theta] = invariants(Sij);
    %%%%%%%%%%%%%%%% Triaxial
    if FLAG_TEST == '3AXI'
        Theta = pi/6; % Triaxial compression
    elseif FLAG_TEST == '3DIM'
        % do nothing
    end
    %%%%%%%%%%%%%%%% Triaxial
    if FLAG_YF == 'SH' % Abbo and Sloan (1995)
        [K_theta,~]=LodeAngle(Theta, mat_par, phi);
        param_a=0.05*cohesion*cot(phi*pi/180);
        yield_f =
        sig_m*sin(phi*pi/180)+sqrt(J2*K_theta^2+param_a^2*sin(phi*pi/180)^2)-
        cohesion*cos(phi*pi/180);
    elseif FLAG_YF == 'PR' % Regular MC cone
        yield_f = sig_m*sin(phi*pi/180)+sqrt(J2)*(cos(Theta)-
        sin(phi*pi/180)*sin(Theta)/sqrt(3))-cohesion*cos(phi*pi/180);
    end
else
    error 'Is Hardening Specified?'
end

end


%-----------------------------------------------------------------
```

Subprogram "yield_func_DP.m"

```
function [ yield_f ] = yield_func_DP( stress, mat_par, back_alpha )
%Drucker-Prager model yield function
global FLAG_H FLAG_HT FLAG_IHT FLAG_ETA_Y ETA_Y

cohesion=mat_par.c;
phi=mat_par.phi; % in degrees
h=mat_par.h;
S0=mat_par.S0;
Hp=mat_par.Hp;

if FLAG_ETA_Y == 0
    ETA_Y=mat_par.eta_y;
elseif FLAG_ETA_Y == 1
    % do nothing
end
```

```
[alpha1, ~] = Param_DP( mat_par ); % change this for other stress states and/
% or for nonlinear forms
[sig_m, Shear] = components( stress );
I1=3*sig_m;
[J2, ~, ~] = invariants( Shear );

if FLAG_H == 'H'
    if FLAG_HT == 'I'
        if FLAG_IHT == 'L'
            param_k_eff = S0+h*ETA_Y;
        elseif FLAG_IHT == 'NL'
            Sigma_eff = sqrt(3)*(alpha1*I1+sqrt(J2))/(1+sqrt(3)*alpha1); %
            effective stress
            param_k_eff = (1+sqrt(3)*alpha1*Sigma_eff)/sqrt(3);
        else
            error 'What is the Isotropic Hardening Type for DP Model?'
        end
        yield_f = sqrt(J2) - alpha1*I1 - param_k_eff;

    elseif FLAG_HT == 'K' % For other forms see the separate code
        param_k = 6*cohesion*cos(phi*pi/180)/(sqrt(3)*(3-sin(phi*pi/180)));
        [sig_m_bar,J2_bar,~]=invariants_translated(stress,back_alpha);
        I1_bar=3*sig_m_bar;
        yield_f = sqrt(J2_bar) - alpha1*I1_bar - param_k;
    elseif FLAG_HT == 'M'
        [sig_m_bar,J2_bar,~]=invariants_translated(stress,back_alpha);
        I1_bar=3*sig_m_bar;
        if FLAG_IHT == 'L'
            param_k_eff = S0+h*ETA_Y;
        elseif FLAG_IHT == 'NL'
            Sigma_eff = sqrt(3)*(alpha1*I1+sqrt(J2))/(1+sqrt(3)*alpha1);
            param_k_eff = (1+sqrt(3)*alpha1*Sigma_eff)/sqrt(3);
        else
            error 'What is the Isotropic Hardening Type for DP Model?'
        end
        yield_f = sqrt(J2_bar) - alpha1*I1_bar - param_k_eff;
    else
        error 'What is the Hardening Type?'
    end
elseif FLAG_H == 'NH'
    param_k = 6*cohesion*cos(phi*pi/180)/(sqrt(3)*(3-sin(phi*pi/180)));
    yield_f = sqrt(J2) - param_k - alpha1*I1;
else
    error 'Is Hardening Specified?'
end

end

%------------------------------------------------------------------
```

Subprogram "invariants.m"

```
function [J2,J3,Theta]=invariants(Shear)
% Shear Stress Invariants
J2=0.5*(Shear(1)^2+Shear(2)^2+Shear(3)^2+2*Shear(4)^2+2*Shear(5)^2+2*Shear
(6)^2);
```

```
J3=Shear(1)*Shear(2)*Shear(3)+2*Shear(4)*Shear(5)*Shear(6)-
Shear(1)*Shear(5)^2-Shear(2)*Shear(6)^2-Shear(3)*Shear(4)^2;
if J2 == 0
    Theta = 0;
else
    Theta=asin(-1.5*sqrt(3)*J3/(J2^1.5))/3;
end

end
```

%---

Subprogram "LodeAngle.m"

```
function [func_theta, dK_by_dT]=LodeAngle(Angle, mat_par, fric_angle)
% Calculates the Lode angle as in % Abbo & Sloan 1995
phi=fric_angle; % Again for the right selection of phi/phi_dil
Theta_T=mat_par.Theta_T;
if Angle<0.0
    Sgn=-1;
else
    Sgn=1;
end
A=cos(Theta_T*pi/180)*(3+tan(Theta_T*pi/180)*tan(3*Theta_T*pi/180)+...
    Sgn*(tan(3*Theta_T*pi/180)-
3*tan(Theta_T*pi/180))*sin(phi*pi/180)/sqrt(3))/3;
B=(Sgn*sin(Theta_T*pi/180)+sin(phi*pi/180)*cos(Theta_T*pi/180)/sqrt(3))/
3*cos(3*Theta_T*pi/180));
if abs(Angle) > (Theta_T*pi/180)
    func_theta=A-B*sin(3*Angle);
    dK_by_dT=-3*B*cos(3*Angle);
else
    func_theta=cos(Angle)-sin(phi*pi/180)*sin(Angle)/sqrt(3);
    dK_by_dT=-sin(Angle)-sin(phi*pi/180)*cos(Angle)/sqrt(3);
end
end
```
%---

Subprogram "invariants_translated.m"

```
function [I1_bar,J2_bar,Theta_bar]=invariants_translated(stress, back_stress)
% Translates the yield surface based upon the center location in PI-plane
% Used only in Kinematic Hardening

Stress_bar=stress-back_stress;
[ sig_m_bar, ~ ] = components( Stress_bar );
Alpha=[back_stress(1)-sig_m_bar back_stress(2)-sig_m_bar back_stress(3)-
sig_m_bar back_stress(4) back_stress(5) back_stress(6)];
[~, Sij] = components( stress );
Shear_bar=Sij-Alpha;
[J2_bar,~,Theta_bar]=invariants(Shear_bar);

end
```

%---

Subprogram "Param_DP.m"

```
function [ alpha1, alpha2 ] = Param_DP( mat_par )
% Calculates the hardening versions of two alpha parameters of DP Model
phi=mat_par.phi; % in degrees
phi_dil=mat_par.phi_dil;
alpha1 = 2*sin(phi*pi/180)/(sqrt(3)*(3-sin(phi*pi/180)));
alpha2 = 2*sin(phi_dil*pi/180)/(sqrt(3)*(3-sin(phi_dil*pi/180)));

end

%----------------------------------------------------------------
```

Subprogram "elastic_matrix.m"

```
function [ elastic_const_matrx ] = elastic_matrix( mat_par )
% Gives the 3D elastic constitutive matrix
G=mat_par.G;
K=mat_par.K;

e1=K+4*G/3;
e2=K-2*G/3;
elastic_const_matrx=[e1 e2 e2 0 0 0;
                     e2 e1 e2 0 0 0;
                     e2 e2 e1 0 0 0;
                      0  0  0 G 0 0;
                      0  0  0 0 G 0;
                      0  0  0 0 0 G ];

end

%----------------------------------------------------------------
```

Subprogram "Gradient_of_Surface.m"

```
function [delSurface_by_delStress]=Gradient_of_Surface(stress, fric_angle,
mat_par)
global FLAG FLAG_YF FLAG_TEST

c=mat_par.c;
Theta_T=mat_par.Theta_T;
phi=fric_angle; % This is here to be able to use the same function for both f
% and g

% Get the Invariants
[ ~, Sij ] = components( stress );
[J2,J3,Angle]=invariants(Sij);
% Calculate the derivatives of invariants wrt stress
[ dI1_dS, dJ2_dS, dJ3_dS ] = invariant_derivative( Sij, J2 );
%%%%%%%%%%%%%%% Triaxial
if FLAG_TEST == '3AXI'
    Angle = pi/6; % Triaxial compression
elseif FLAG_TEST == '3DIM'
    % do nothing
end
%%%%%%%%%%%%%%% Triaxial
if FLAG=='MC'   %Mohr Coulomb
    Term1=sin(phi*pi/180)*dI1_dS;
```

```
    if FLAG_YF == 'SH' % As in Abbo and Sloan (1995)
        param_a=0.05*c*cot(phi*pi/180);
        % Assuming Hyperbolic Yield Surface
        if J2 == 0
            delSurface_by_delStress=Term1;
        else
            [K, dK_by_dT]=LodeAngle(Angle, mat_par, fric_angle);
            %%%%%%%%%%%%%%% Triaxial
            if FLAG_TEST == '3AXI'
                dK_by_dT = 0; % Triaxial compression
            elseif FLAG_TEST == '3DIM'
                % do nothing
            end
            %%%%%%%%%%%%%%% Triaxial
            Alpha=sqrt(J2)*K/sqrt(J2*K^2+param_a^2*sin(phi*pi/180)^2);
            Term21=Alpha*(K-tan(3*Angle)*dK_by_dT);
            Term2=Term21*dJ2_dS;

            Term31=-Alpha*sqrt(3)*dK_by_dT/(2*cos(3*Angle)*J2);
            Term3=Term31*dJ3_dS;

            delSurface_by_delStress = Term1 + Term2 + Term3;
        end
    elseif FLAG_YF == 'PR'
        Term211 = (cos(Angle)-
        sin(Angle)*sin(phi*pi/180)/sqrt(3))/(2*sqrt(J2));
        %%%%%%%%%%%%%%% Triaxial
        if FLAG_TEST == '3AXI'
            Term2 = Term211*dJ2_dS;
            Term3 = 0.0;
        elseif FLAG_TEST == '3DIM'
            der_angle_der_J2 = (3*sqrt(3)*J3/2)/((J2^2.5)*sqrt(4-
            27*J3^2/(J2^3)));
            Term212 = sqrt(J2)*(-sin(Angle)-'
            cos(Angle)*sin(phi*pi/180)/sqrt(3));
            Term213 = Term212*der_angle_der_J2;
            Term21 = Term211 + Term213;
            Term2 = Term21*dJ2_dS;

            der_angle_der_J3 = -sqrt(3) / ((J2^1.5)*sqrt(4-27*J3^2/(J2^3)));
            Term31 = Term212*der_angle_der_J3;
            Term3 = Term31*dJ3_dS;
        end
        %%%%%%%%%%%%%%% Triaxial
        delSurface_by_delStress = Term1 + Term2 + Term3;
    end

elseif FLAG=='DP' %Drucker Prager
    Alpha = 2*sin(phi*pi/180)/(sqrt(3)*(3-sin(phi*pi/180)));
    if J2 == 0
        delSurface_by_delStress=[Alpha Alpha Alpha 0 0 0];
    else
        delSurface_by_delStress= [ Alpha + ( 2*stress(1)-stress(2)-stress(3)
                                    )/(6*sqrt(J2))
                                    Alpha + ( -stress(1)+2*stress(2)-stress(3)
                                    )/(6*sqrt(J2))
                                    Alpha + ( -stress(1)-stress(2)+2*stress(3)
                                    )/(6*sqrt(J2))
```

```
                              stress(4)/sqrt(J2)
                              stress(5)/sqrt(J2)
                              stress(6)/sqrt(J2) ];
        delSurface_by_delStress=delSurface_by_delStress';
    end
else
    % do nothing
end
end


%--------------------------------------------------------------
```

Subprogram "get_hardening_modulus.m"

```
function [ Hard_Mod, param_C ] = get_hardening_modulus( stress, mat_par,
back_stress, f2 )
% Calculates the hardening modulus (H) for any hardening type
% Also calculates the C parameter
global FLAG FLAG_H FLAG_HT FLAG_IHT FLAG_FLOW ETA_Y FLAG_TEST
cohesion=mat_par.c;
phi=mat_par.phi;
phi_dil=mat_par.phi_dil;
Theta_T=mat_par.Theta_T;
h=mat_par.h;
A=mat_par.A;
eta_p=mat_par.eta_p;
Hp=mat_par.Hp;
const_m=mat_par.const_m;
MW=mat_par.MW;

[sig_m, shear] = components( stress );
I1=3*sig_m;
[J2, ~, Theta] = invariants( shear );
%%%%%%%%%%%%%%%% Triaxial
if FLAG_TEST == '3AXI'
    Theta = pi/6; % Triaxial compression
elseif FLAG_TEST == '3DIM'
    % do nothing
end
%%%%%%%%%%%%%%%% Triaxial
if FLAG_HT == 'I'
    if FLAG_IHT == 'L'
        Hard_Mod = h*ETA_Y;
    elseif FLAG_IHT == 'NL'
        % For Nonlinear Isotropic Hardening H is different for MC and DP
        if FLAG == 'MC'
            Hard_Mod = sig_m*(eta_p-ETA_Y)^2/(A*eta_p); % MC Isotropic
            % Hardening
        elseif FLAG == 'DP'
            % Only for triaxial compression, change it for plane strain or
            % triaxial extension
            [ alpha1, alpha2 ] = Param_DP( mat_par ); % Get the two alpha
            % parameters of DP Model
            param_k = 6*cohesion*cos(phi*pi/180)/(sqrt(3)*(3-
            sin(phi*pi/180)));
            Hard_Mod =
            (alpha2*I1+sqrt(J2))*(1+sqrt(3)*alpha1)^2*Hp/(3*param_k);
        else
```

```
                % do nothing
            end
        else
            error 'What is the Isotropic Hardening Type?'
        end
        param_C = 0; % Irrelevant for isotropic hardening
    elseif FLAG_HT == 'K' || FLAG_HT == 'M'
        % Handle Kinematic Hardening first
        [sig_m_bar,J2_bar,Theta_bar]=invariants_translated(stress, back_stress);
        I1_bar=3*sig_m_bar;
        if FLAG == 'MC'
            if FLAG_FLOW == 'ASSOC'
                df_dS = Gradient_of_Surface(stress, phi, mat_par);
                dG_dS = df_dS;
                dG_dS_Amp =
                sqrt(dG_dS(1)^2+dG_dS(2)^2+dG_dS(3)^2+2*dG_dS(4)^2+2*dG_
                dS(5)^2+2*dG_dS(6)^2);
                Eff_Stress = (f2 +
                cohesion*cos(phi*pi/180))/(sin(phi*pi/180)/3+cos(Theta)/sqrt(3)-
                sin(phi*pi/180)*sin(Theta)/3);
                param_C = (sig_m*sin(phi*pi/180) + sqrt(J2)*(cos(Theta)-
                sin(phi*pi/180)*sin(Theta)/sqrt(3)))/(Eff_Stress*dG_dS_Amp);
            elseif FLAG_FLOW == 'NONAS'
                dG_dS_Amp = sqrt(sin(phi_dil*pi/180)^2/3+(cos(Theta)-
                sin(phi_dil*pi/180)*sin(Theta)/sqrt(3))^2/2); % sqrt(dG/dS.dG/dS)
                Eff_Stress = (f2 +
                cohesion*cos(phi*pi/180))/(sin(phi_dil*pi/180)/3+cos(Theta)/
                sqrt(3)-
                sin(phi_dil*pi/180)*sin(Theta)/3);
                param_C = (sig_m_bar*sin(phi_dil*pi/180) +
                sqrt(J2_bar)*(cos(Theta_bar)-
                sin(phi_dil*pi/180)*sin(Theta_bar)/sqrt(3)))/(Eff_Stress*dG_dS_
                Amp);
                df_dS = Gradient_of_Surface(stress, phi, mat_par);
            end
        elseif FLAG == 'DP'
            [ alpha1, alpha2 ] = Param_DP( mat_par ); % Get the two alpha
            % parameters of DP Model
            param_C = (sqrt(3)*alpha1+1)*(sqrt(J2_bar) +
            alpha2*I1_bar)/((sqrt(J2_bar) + alpha1*I1_bar)*sqrt(3*alpha2^2+0.5));
            df_dS = Gradient_of_Surface(stress, phi, mat_par);
            dG_dS_Amp = 3*alpha2^2+0.5;
        else
            error 'What model is this?'
        end
        H_kin = const_m*param_C*dG_dS_Amp*df_dS*(stress-back_stress)';
        if FLAG_HT == 'K'
            Hard_Mod = H_kin;
        elseif FLAG_HT == 'M'
            if FLAG_IHT == 'L'
                Hard_Mod = H_kin + h*ETA_Y;
            elseif FLAG_IHT == 'NL'
                if FLAG == 'MC'
                    Hard_Mod = H_kin*(1-MW) + (sig_m*(eta_p-
                    ETA_Y)^2/(A*eta_p))*MW; % MC Mixed Hardening
                elseif FLAG == 'DP'
                    % Only for triaxial compression, change it for plane strain
                    % or triaxial extension
```

```
            [ alpha1, alpha2 ] = Param_DP( mat_par );
            param_k = 6*cohesion*cos(phi*pi/180)/(sqrt(3)*(3-
            sin(phi*pi/180)));
            Hard_Mod = H_kin*(1-MW) + ((alpha2*I1+sqrt(J2))*(1+sqrt(3)*al
            pha1)^2*Hp/(3*param_k))*MW;
        else
            error 'What model is this?'
        end
    else
        error 'What is the Isotropic Hardening Type?'
    end
    else
        % do nothing
    end
else
    error 'What is the hardening type?'
end

end
```

%---

Subprogram "triaxial_strain_state.m"

```
function [ vol, shear ] = triaxial_strain_state( strain )
% Give the triaxial stress components
shear = ( strain(1) - strain(3) )*2/3;
vol = strain(1)+strain(2)+strain(3);
end
```

%---

Subprogram "triaxial_stress_state.m"

```
function [ p_stress, q_stress ] = triaxial_stress_state( stress )
% Give the triaxial stress components
q_stress = stress(1) - stress(3);
p_stress = ( stress(1)+stress(2)+stress(3) )/3;
end
```

%---

Subprogram "invariant_derivative.m"

```
function [ der_sigm_der_sig, der_sigbar_der_sig, der_J3_der_sig ] =
invariant_derivative( Shear, J2 )
% Calculates the derivatives of stress invariants wrt stress
sig_bar=sqrt(J2);
% Derivative of I1
der_sigm_der_sig=(1/3)*[1 1 1 0 0 0];
% Derivative of J2
if sig_bar==0
    der_sigbar_der_sig=[0 0 0 0 0];
else
    der_sigbar_der_sig(1)=Shear(1)/(2*sig_bar);
    der_sigbar_der_sig(2)=Shear(2)/(2*sig_bar);
    der_sigbar_der_sig(3)=Shear(3)/(2*sig_bar);
    der_sigbar_der_sig(4)=Shear(4)/sig_bar;
```

```
        der_sigbar_der_sig(5)=Shear(5)/sig_bar;
        der_sigbar_der_sig(6)=Shear(6)/sig_bar;
end
% Derivative of J3
Sik_sk=j=Shear(1)^2+Shear(2)^2+Shear(3)^2+2* Shear(4)^2+2*Shear(5)^2+2*
Shear(6)^2+...
(2*Shear (1)+Shear(5))*(Shear(4)+Shear(6))+(2*Shear(2)+Shear(4))*(Shear(5)
+Shear(6))+...
      (2*Shear(3)+Shear(6))*(Shear(4)+Shear(5));
term=Sik_Skj - 2*J2/3;
der_J3_der_sig(1) = term;
der_J3_der_sig(2) = term;
der_J3_der_sig(3) = term;
der_J3_der_sig(4) = Sik_Skj;
der_J3_der_sig(5) = Sik_Skj;
der_J3_der_sig(6) = Sik_Skj;

end


%-----------------------------------------------------------------
```

Subprogram "update_hardening.m"

```
function [ back_alpha ] = update_hardening( plas_strain_inc, stress, param_C,
mat_par, back_alpha )
% "update_hardening.m" updates the parameters del_eta_y and del_back_alpha
global FLAG_HT FLAG_IHT ETA_Y FLAG_ETA_Y
A=mat_par.A;
eta_p=mat_par.eta_p;
const_m=mat_par.const_m;
MW=mat_par.MW;
[plas_vol_str_inc, ~] = triaxial_strain_state( plas_strain_inc );
plas_shear_str_inc= [plas_strain_inc(1)-plas_vol_str_inc/3
                     plas_strain_inc(2)-plas_vol_str_inc/3
                     plas_strain_inc(3)-plas_vol_str_inc/3
                     plas_strain_inc(4)
                     plas_strain_inc(5)
                     plas_strain_inc(6)]';
plas_dist_inc=sqrt(
plas_shear_str_inc(1)^2+plas_shear_str_inc(2)^2+plas_shear_str_inc(3)^2+...
2*plas_shear_str_inc(4)^2+2*plas_shear_str_inc(5)^2+2*plas_shear_str_inc(6)^2
) ;
if FLAG_HT == 'I'
    if FLAG_IHT == 'L'
        del_eta_y = plas_dist_inc;  % could also be plastic shear strain
    elseif FLAG_IHT == 'NL'
        del_eta_y = (eta_p-ETA_Y)^2*plas_dist_inc/(A*eta_p);
    else
        error 'What is the Isotropic Hardening Type?'
    end
    del_back_alpha = 0.0;            % No Kin. Hard.
    FLAG_ETA_Y=1;
elseif FLAG_HT == 'K'
    PlasStrainAmp = sqrt(
    plas_strain_inc(1)^2+plas_strain_inc(2)^2+plas_strain_inc(3)^2+...
    2*plas_strain_inc(4)^2+2*plas_strain_inc(5)^2+2*plas_strain_inc(6)^2 );
    del_back_alpha = const_m*param_C*PlasStrainAmp*(stress-back_alpha);
    % Ziegler's Rule
```

```
        del_eta_y = 0.0;       % No iso. hard.
elseif FLAG_HT == 'M'
    if FLAG_IHT == 'L'
        del_eta_y = plas_dist_inc;
    elseif FLAG_IHT == 'NL'
        del_eta_y = (eta_p-ETA_Y)^2*plas_dist_inc/(A*eta_p);
    end
    PlasStrainAmp = sqrt(
    plas_strain_inc(1)^2+plas_strain_inc(2)^2+plas_strain_inc(3)^2+...
    2*plas_strain_inc(4)^2+2*plas_strain_inc(5)^2+2*plas_strain_inc(6)^2 );
    del_back_alpha = const_m*param_C*PlasStrainAmp*(stress-back_alpha)*(1-
    MW); % Ziegler's Rule
    FLAG_ETA_Y=1;
else
    error 'What is the hardening type?'
end
ETA_Y = ETA_Y + del_eta_y;                 % Update Isot. Hard.
back_alpha = back_alpha + del_back_alpha;  % Update Kin. Hard.
end
```

6

Viscoelasticity and Viscoplasticity

6.1 Introduction

Some materials such as polymer plastics, biological materials, some soils and metals at high temperatures etc. exhibit gradual deformation and recovery when they are subjected to loading and unloading cycles. The response of such materials is time-dependent with the deformation being dependent upon the rate at which the loads are applied. Modeling of such time-dependent material is available in the framework of the *theory of viscoelasticity*. Viscoelasticity consists of two components: viscosity and elasticity. Viscosity is a measure of resistance to flow, which is a fluid property while elasticity is a characteristic of solids. In general, those materials whose mechanical response to external loads combines the features of both elastic solids and viscous fluids are called *viscoelastic*.

The models presented thus far are rate-independent. That is, stress-strain relationship as well as material strength are not functions of time or the rate of change of them over time is not included in the formulation. It is indeed possible and sometimes preferable to include rate-dependent behavior of engineering materials. Their time-dependent stress-strain relationships matter in determining the overall response to loads which means in this context that the deformation of the material depends on the rate at which loads are applied.

Viscoplasticity based models can incorporate the inelastic strains. In a sense, viscoplasticity is the rate-dependent version of plasticity where material exhibits permanent deformations but over time. Viscoplastic constitutive equations are used to model the behavior of materials that are subjected to stresses at high temperatures and to model the behavior of materials that are deformed at high strain rates.

In this chapter, the focus is on understanding the basic elements of viscoelasticity. First, a number of rheological models consisting of simple springs and dashpots are presented. Some composite forms of these fundamental models are then presented to formulate stress-strain relationships for a rate-dependent material. This is followed by the integral form of arbitrarily changing load in time, where stress-strain form and relaxation modulus are written in the form of convolution integral. Following this, the differential form of the equations leading to the differential operators used to define the rate-dependent stress-strain relationship is developed. Then, an example problem of the response of layered viscoelastic system to vertical circular loading is presented. Lastly, the basics of viscoplasticity is introduced in a 1-D problem where rutting of asphalt concrete is investigated.

To start off, if we take a look at triaxial compression test results, we see that they are performed at a constant deviatoric strain rate as in Figure 6.1-1a. The faster the rate, the higher the strength and stiffer the stress-strain response which is shown in Figure 6.1-1b. This is called *rate dependency*.

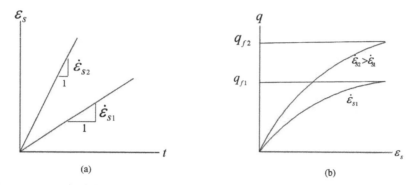

(a)   (b)

Figure 6.1-1 Triaxial compression tests at a constant strain rate: (a) the strain history,
(b) deviatoric stress-strain curves

Another behavior to observe is, when a triaxial test ends at a certain deviatoric stress (q_c < q_f) that is kept constant, corresponding deviatoric strain keeps increasing at three possible rates based upon q_c/q_f ratio (Figure 6.1-2). The phenomenon that the strain is time-dependent is called *creep*. A similar behavior is observed when a triaxial test is terminated at a certain deviatoric strain ε_{sc} which is kept constant afterwards (Figure 6.1-3). In this case, the deviatoric stress decreases partially or fully at a decaying rate. This time-dependency of stresses is called *stress relaxation*.

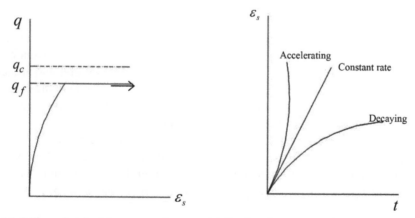

Figure 6.1-2 Creep in triaxial compression test: (a) Deviatoric stress-strain curve, (b) Strain history

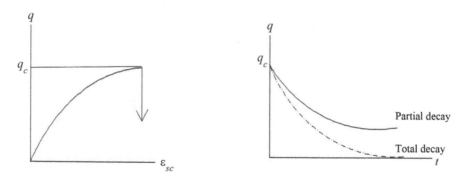

Figure 6.1-3 Stress relaxation in triaxial compression test: (a) Deviatoric stress-strain curve;
(b) Strain history

This section of the chapter explains how such observed time or rate-dependent behaviors of materials can be modeled using various viscous-elastic and viscous-plastic rheological models. Mechanical aspects are investigated and thermomechanical aspects are left for the reader to cover from other related more comprehensive sources.

6.2 Viscoelastic Behavior: Fundamental Rheological Models

6.2.1 Basic Components: Springs and Dashpots

Linear springs and linear viscous dashpots are the basic components for linear viscoelastic models. The linear spring, as indicated in Figure 6.2-1 obeys the Hooke's law when it is subjected to a force:

$$\sigma = E\varepsilon \tag{6.2.1}$$

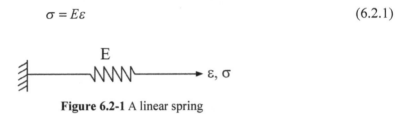

Figure 6.2-1 A linear spring

in which E is the Young's modulus. As indicated in Figure 6.2-2, the dashpot is the ideal viscous component whose stress is proportional to the strain rate. By Newton's law of viscosity, we have

$$\sigma = \eta \frac{d\varepsilon}{dt} = \eta \dot{\varepsilon} \tag{6.2.2}$$

where the constant η is called the coefficient of viscosity which is defined as the ratio of shearing stress to velocity gradient and the dot here and elsewhere denotes the derivative with respect to time. The behavior of viscoelastic material is a kind of combination of the springs and dashpots. Viscoelastic models can be built up by putting the springs and dashpots in series, parallel, or various combinations of these.

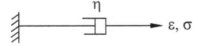

Figure 6.2-2 A linear dashpot

The discussion of viscoelasticity should begin with a number of basic rheological models that are developed to model fundamental material behavior. That is, main rheological models should be capable of explaining the three fundamental aspects of mechanical behavior. These basic features are the *perfectly elastic, perfectly plastic* and *perfectly viscous* behavior that we present in the following. More detailed discussion can be found elsewhere (Flügge, 1975; Puzrin, 2012; Christensen, 2003).

6.2.2 Perfectly Elastic Model (Hookean Element)

The perfectly elastic model can best be represented by a so called *Hookean Element* or *Hookean*

Model where there is an elastic spring (Figure 6.2-1) with a spring constant $k = 3G$ giving elastic deviatoric strain as:

$$\varepsilon_S = \frac{q}{3G} \tag{6.2.3}$$

where q is the deviatoric stress.

6.2.3 Perfectly Viscous Model (Newtonian Element)

In the perfectly viscous model, a viscosity constant η is used to explain the Newtonian model of a viscous fluid or a dashpot (Figure 6.2-2). Here the deviatoric strain is defined as:

$$\dot{\varepsilon}_S = q/\eta \tag{6.2.4}$$

It should be noted here that if $\varepsilon_S = \varepsilon_{SC}$ = constant, the model explains the rate-dependency of strength as:

$$q = \eta \varepsilon_{sc} \tag{6.2.5}$$

or if $\varepsilon_S = 0$ at $t = 0$, $q = q_C$ = constant, then it described the constant rate creep as:

$$\varepsilon_S = \frac{q_C}{\eta} t \tag{6.2.6}$$

6.2.4 Perfectly Plastic Model (Yield Element)

The perfectly plastic model can be described by using the yield stress model where there is a sliding block which starts its movement as soon as a sliding stress, q_y is reached (Figure 6.2-3). The related deviatoric strain is defined as:

$$\varepsilon_S = 0 \quad \text{when} \quad q < q_y$$
$$\dot{\varepsilon}_S > 0 \quad \text{when} \quad q = q_y \tag{6.2.7}$$

where

$$\dot{\varepsilon}_S = d\varepsilon_S/dt \tag{6.2.8}$$

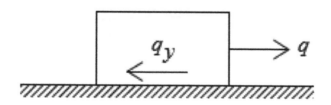

Figure 6.2-3 The yield stress model

6.3 Viscoelastic Behavior: Composite Rheological Models

There are a number of rheological models used to explain the mechanical aspects of viscoelasticity. While individual models are sufficient to explain a material's time-dependent stress-strain relationship, some of the composite models that are made up of multiple individual components are also presented. Given below is a summary of these composite models and more details can be found in related texts like Puzrin (2012).

6.3.1 St. Venant Model

The St. Venant Model is composed of a Hookean Model and a Yield Stress Model which are connected in series as can be seen in Figure 6.3-1a. Deviatoric stress-strain response is linear elastic until the material reaches the yield stress, q_y. Then it becomes perfectly plastic (Figure 6.3-1b).

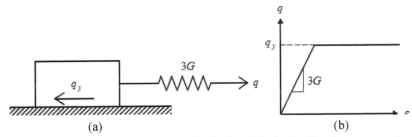

(a)   (b)

Figure 6.3-1 St. Venant model: (a) The mechanical model; (b) Deviatoric stress-strain response

6.3.2 The Linear Hardening Model

The linear hardening model consists of the Hookean and Yield Stress models that are connected in parallel to which another Hookean model is added in series (Figure 6.3-2a). Again the deviatoric stress-strain response is linear elastic until the material reaches the yield stress q_y after which it becomes plastic with linear hardening (Figure 6.3-2b). Following strain-stress relationships hold:

$$\varepsilon_s = \frac{q}{3G} + \frac{q - q_y}{H} \text{ and } \varepsilon_{sy} = \frac{q}{3G}, \frac{q - q_y}{\varepsilon_s - \varepsilon_{sy}} = \frac{3GH}{3G + H} \qquad (6.3.1)$$

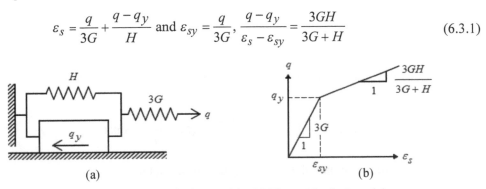

(a)   (b)

Figure 6.3-2 Linear hardening models: (a) The mechanical model,
(b) Deviatoric stress-strain response

6.3.3 The Kelvin (Voigt) Model

The Kelvin model consists of the Hookean and Newtonian models connected in parallel as in Figure 6.3-3a. Its shear strain response to the deviatoric stress step loading shows the decaying creep and the shear strains are entirely recoverable.

The response can be derived by considering the following equation for stresses and strains of the Hookean (H) and Newtonian (N) elements as can be written as:

$$q = q_N + q_H ; \varepsilon_{sN} = \varepsilon_{sH} = \varepsilon_s \qquad (6.3.2)$$

Substitution of the constitutive relations above gives,

$$q = 3G\varepsilon_s + \eta\dot{\varepsilon} \qquad (6.3.3)$$

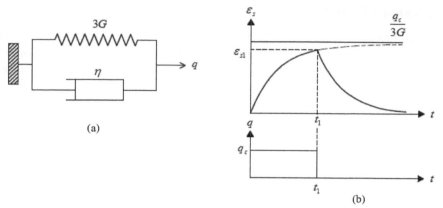

Figure 6.3-3 Kelvin (Voigt) model: (a) Mechanical model, (b) Stress-strain response to step loading

or

$$\frac{q}{\eta} = \frac{3G}{\eta}\varepsilon_s + \dot{\varepsilon}_s$$

(6.3.4)

The latter can be represented in terms of a differential equation as:

$$H(t) = D(t)y(t) + \frac{dy}{dt}$$

(6.3.5)

with the solution,

$$y(t) = e^{-\int D(t)dt}\left[\int_0^t H(\tau)e^{-\int D(t)dt}d\tau + C\right]$$

(6.3.6)

where C is the integration constant yielding $C = 0$ for $y(0) = 0$. For the Kelvin-Voigt model we have,

$$y(t) = \varepsilon_s(t); \ H(t) = \frac{q(t)}{\eta}; \ D(t) = \frac{3G}{\eta} = \text{const}$$

(6.3.7)

and assuming zero immediate shear strain $\varepsilon_s(0) = 0$, we obtain:

$$\varepsilon_s(t) = e^{\frac{3Gt}{\eta}}\int_0^t \frac{q(\tau)}{\eta}e^{\frac{3G\tau}{\eta}}d\tau$$

(6.3.8)

For the constant deviatoric stress $q = q_c$:

$$\varepsilon_s(t) = \frac{q_c}{3G}\left(1 - e^{-\frac{3G}{\eta}t}\right)$$

(6.3.9)

so that $\lim_{t\to\infty}\varepsilon_s = \frac{q_c}{3G}$ i.e. decaying creep. At some $t = t_1$,

$$\varepsilon_{s_1} = e^{\frac{-3Gt_1}{\eta}}\int_0^{t_1}\frac{q_c}{\eta}e^{\frac{3G\tau}{\eta}}d\tau = \frac{q_c}{3G}\left(1 - e^{\frac{-3Gt_1}{\eta}}\right)$$

(6.3.10)

If at this moment $t = t_1$, stress is instantaneously reduced to $q = 0$, for $t > t_1$, we can decompose the integral into two parts such that:

$$\varepsilon_s = e^{\frac{3Gt_1}{\eta}}\left[\int_0^{t_1}\frac{q(t)}{\eta}e^{\frac{3G\tau}{\eta}}d\tau + \int_{t_1}^{t}\frac{q(t)}{\eta}e^{\frac{3G\tau}{\eta}}d\tau\right] \qquad (6.3.11)$$

and taking into account the loading history, we obtain:

$$\varepsilon_s = \varepsilon_{s_1}e^{\frac{3G(t-t_1)}{\eta}} \qquad (6.3.12)$$

So that $\lim_{t\to\infty}\varepsilon_s = 0$ and the strain is fully recoverable.

6.3.4 The Maxwell Model

The Maxwell model consists of the Hookean and Newtonian models connected in series (Figure 6.3-4a). Its shear strain response to the deviatoric stress-step loading exhibits the constant rate creep with shear strain being partly irrecoverable (Figure 6.3-4b). At constant strain, this model results in *total stress relaxation* (Figure 6.3-5a) while at constant shear strain rate, the Maxwell model simulates *the rate dependency of the shear strength* (Figure 6.3-5b).

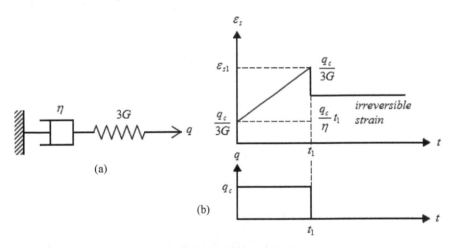

Figure 6.3-4 Maxwell model: (a) Mechanical model, (b) Shear strain response to the deviatoric stress step loading

These responses can be derived by considering the following equations for the stresses and strains resulting from the combination in series of the Hookean (H) and Newtonian (N) elements:

$$q = q_N + q_H \ ; \ \dot{\varepsilon}_{sN} + \dot{\varepsilon}_{sH} = \dot{\varepsilon}_s \qquad (6.3.13)$$

Substitution of the relations (6.2.2)-(6.2.4) into (6.3.13) gives:

$$\dot{\varepsilon}_s = \frac{\dot{q}}{3G} + \frac{q}{\eta} \quad \text{or} \quad 3G\dot{\varepsilon}_s = \frac{3G}{\eta}q + \dot{q} \qquad (6.3.14)$$

From the above equation it follows that for constant stress $q = q_c$, $\dot{q} = 0$ and $\dot{\varepsilon}_s = \frac{q_c}{\eta}$ so

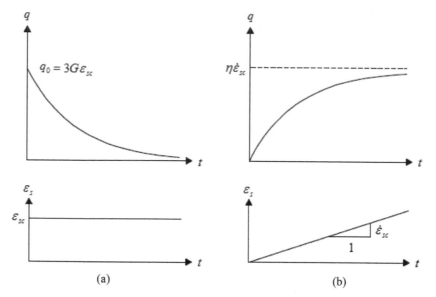

Figure 6.3-5 The deviatoric stress response of the Maxwell model: (a) Constant shear strain loading, (b) Constant shear strain rate loading

that $\varepsilon_s = \dfrac{q_c}{\eta}t + \dfrac{q_c}{3G}$ and for example the creep takes place at a constant rate after the immediate

strain $\dfrac{q_c}{3G}$. At some $t = t_1$:

$$\varepsilon_{s_1} = \frac{q_c}{\eta}t_1 + \frac{q_c}{3G} \tag{6.3.15}$$

If at this moment $t = t_1$, the stress is suddenly reduced to $q = 0$ and there is a release of the immediate strain $\dfrac{q_c}{3G}$. So for $t > t_1$ $\dot\varepsilon_s = 0$ so that:

$$\varepsilon_s = \varepsilon_{s_1} - \frac{q_c}{3G} = \frac{q_c}{\eta}t_1 = \text{constant} \tag{6.3.16}$$

and the strain in Figure 6.3-4b is partially irrecoverable. This equation is also a particular case of Eq. (6.3.4) with the following definitions of the Maxwell model:

$$y(t) = q(t); \; H(t) = 3G\dot\varepsilon_s \; ; \; D(t) = \frac{3G}{\eta} = \text{constant} \tag{6.3.17}$$

Assuming $q(0) = q_0$ yields:

$$q(t) = e^{-\frac{3G}{\eta}t}\left[\int_0^t 3G\dot\varepsilon_s\, e^{\frac{3G}{\eta}\tau}\, d\tau + q_0\right] \tag{6.3.18}$$

For constant shear strain, $\varepsilon_s = \varepsilon_{sc}$, we have $\dot\varepsilon_s = 0$, $q_0 = 3G\varepsilon_{sc}$ and:

$$q(t) = 3G\varepsilon_{sc}e^{-\frac{3G}{\eta}t} \tag{6.3.19}$$

so that $\lim_{t\to\infty} q = 0$ and the stress relaxation is total. For constant shear strain rate $\dot\varepsilon_s = \dot\varepsilon_{sc}$, we have $q_0 = 0$ and,

$$q(t) = e^{-\frac{3G}{\eta}t} \int_0^t 3G\dot{\varepsilon}_{sc} e^{\frac{3G}{\eta}\tau} d\tau = \eta\dot{\varepsilon}_{sc}\left(1 - e^{-\frac{3G}{\eta}t}\right) \tag{6.3.20}$$

so that now $\lim_{t\to\infty} q = \eta\dot{\varepsilon}_{sc}$ and the shear strength is rate independent (see Figure 6.3-5b).

6.3.5 The Bingham Model

The Bingham model has an additional sliding body compared to the Maxwell model. It is located in between the Hookean and Newtonian models which are connected in series (Figure 6.3-6).

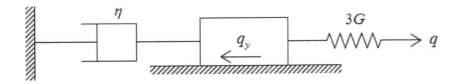

Figure 6.3-6 Bingham model

As in the Maxwell model, the shear stress-strain response exhibits constant rate creep with the strain being partly irrecoverable (Figure 6.3-7a). However, the irrecoverable shear strain can only be controlled by changing the yield stress, q_y. Partial stress relaxation is generated by the Maxwell model at constant shear strain (Figure 6.3-7b).

Considering the following equations for stresses and strains resulting from the combination in series of the Hookean (H) and Newtonian (N) elements (for $q > q_y$):

$$q = q_H = q_N + q_y \; ; \; \dot{\varepsilon}_{sN} + \dot{\varepsilon}_{sH} = \dot{\varepsilon}_s \tag{6.3.21}$$

If we once again substitute (6.2.2) and (6.2.4) into these relations, we get:

$$\dot{\varepsilon}_s = \frac{\dot{q}}{3G} + \frac{q - q_y}{\eta} \tag{6.3.22}$$

or

$$3G\varepsilon_s^* = \frac{3G}{\eta}(q - q_y) + (q - q_y) \tag{6.3.23}$$

As is similar to the previous model, for a constant stress $\dot{q} = 0$, $\dot{\varepsilon}_s = \frac{q_c - q_y}{\eta}$ so that $\varepsilon_s = \frac{q_c - q_y}{\eta}t + \frac{q_c}{3G}$ and the actual creep behavior becomes constant after the immediate strain $\frac{q_c}{3G}$. For $t = t_1$:

$$\varepsilon_{s_1} = \frac{q_c - q_y}{\eta}t_1 + \frac{q_c}{3G} \tag{6.3.24}$$

If at this moment $t = t_1$, the stress is instantaneously reduced to $q = 0$. Following the strain $\frac{q_c}{3G}$, for $t > t_1$ $\dot{\varepsilon}_s = 0$ which gives:

$$\varepsilon_s = \varepsilon_{s_1} - \frac{q_c}{3G} = \frac{q_c - q_y}{\eta} t_1 = \text{constant} \tag{6.3.25}$$

The strain is partially irrecoverable (Figure 6.3-7a). For the Bingham model, we write what is given below into (6.3.4) to obtain the theory,

$$y(t) = q(t) - q_y, \quad H(t) = 3G\dot{\varepsilon}_s; \quad D(t) = \frac{3G}{\eta} = \text{constant} \tag{6.3.26}$$

Taking now $q(0) = q_0 > q_y$ gives,

$$q(t) = q_y + e^{-\frac{3G}{\eta}t}\left[\int_0^t 3G\varepsilon_s \cdot e^{\frac{3G}{\eta}\tau} d\tau + q_0\right] \tag{6.3.27}$$

For $\varepsilon_s = \varepsilon_{sc}$,

$$\dot{\varepsilon}_s = 0, \quad q_0 = 3G\varepsilon_{sc} \tag{6.3.28}$$

and

$$q(t) = q_y + 3G\varepsilon_{sc}e^{-\frac{3G}{\eta}t} \tag{6.3.29}$$

so that $\lim_{t\to\infty} q = q_y$ and the stress relaxation is partial (see Figure 6.3-7b).

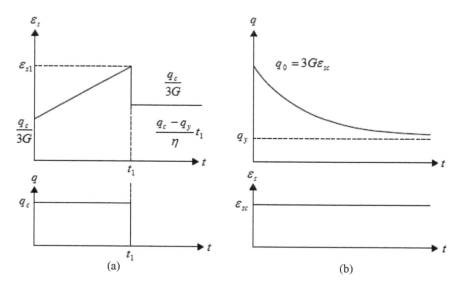

Figure 6.3-7 Bingham model: (a) Shear strain response to the deviatoric stress step loading, (b) Deviatoric stress response to the constant shear strain loading

6.4 Formulation Methods in Viscoelasticity

In understanding viscoelasticity, three major approaches are used to derive stress-strain relationship:

- Integral form

- Derivative form
- Molecular form

The first two forms are equivalent in that, one form uses integrals to define the constitutive relations, while the other relates stresses and strains by means of differentials. The molecular form works well with soft biological tissues. In the molecular form, it is more convenient not to consult with the stress-relaxation or creep experiments but, rather, to use dynamic tests in oscillatory forms. The molecular form of viscoelasticity is out of the scope of this book and is not discussed in this section.

6.4.1 Integral Form of Constitutive Relations

If there is an initial stress σ_0 applied suddenly at $t = 0$, followed by $\sigma(t)$ varying arbitrarily as time elapses, as shown in Figure 6.4-1, stress is thought of being composed of a sequence of infinitesimal step functions with each subsequent step adding an incremental amount to the previous one as,

$$\sigma(t) \cong \sigma_0 + \sum_{i=1}^{N} \Delta\sigma_i'(t) = D(t)\Delta\varepsilon_0 + \sum_{i=1}^{N} D(t-t')\Delta\varepsilon_i, \ t > t' \tag{6.4.1}$$

where Δ is the unit step function and $\Delta\sigma' = \dfrac{d\sigma}{dt}\bigg|_{t=t'} dt'$. This is called the "Boltzmann superposition principle". It should be noted here that (') is used to distinguish the viscoelastic stress associated with time t' and should not be confused with the effective stress. It is now possible to evaluate the system's viscoelastic response. The total strain at time t is then the sum of the strain caused by the entire step loading for the part $t' < t$, which can be written as,

$$\varepsilon(t) = \frac{\sigma_0}{D(t)} + \int_{-\infty}^{t} \frac{1}{D(t-t')} \frac{d\sigma'}{dt'} dt' \tag{6.4.2}$$

or also written more conveniently as,

$$\varepsilon(t) = \sigma_0 J(t) + \int_{-\infty}^{t} J(t-t') \frac{d\sigma'}{dt'} dt' \tag{6.4.3}$$

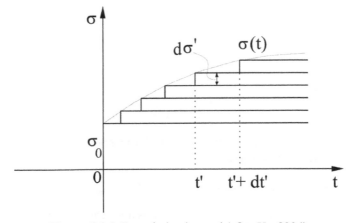

Figure 6.4-1 Convolution integral (after Xu, 2004)

where

$$J(t) = \frac{1}{D(t)} \tag{6.4.4}$$

is called the "compliance" of the material and $J(t - t')$ is the "creep compliance function" which introduces time into these equations as an extra variable.

We can see that the strain response of the material (i.e. soil) depends entirely on the past stress history. Doing integration by parts, (6.4.3) becomes,

$$\varepsilon(t) = \sigma(t)J(0) + \int_{-\infty}^{t} \sigma(t') \frac{dJ(t - t')}{d(t - t')} dt' \tag{6.4.5}$$

These integral forms in (6.4.3) and (6.4.5) are called *the hereditary integrals*. We can also write the stress in terms of the strain and the stress relaxation modulus in the convolution integral form as,

$$\sigma(t) = \varepsilon_0 D(t) + \int_{-\infty}^{t} D(t - t') \frac{d\varepsilon'}{dt'} dt' \tag{6.4.6}$$

or

$$\sigma(t) = \varepsilon(t)D(0) + \int_{-\infty}^{t} \varepsilon(t') \frac{dD(t - t')}{d(t - t')} dt' \tag{6.4.7}$$

Example 6.4-1: The 1-D viscoelastic response of a soil column to a constant strain-rate loading, $\varepsilon(t) = Rt$ can be expressed as: $\sigma(t) = D_{eff}(t) \varepsilon(t)$. Derive an expression for $D_{eff}(t)$, the constant-rate effective modulus, for a viscoelastic soil.

Solution:

If we take the relaxation modulus as $D(t)$ and evaluate its stress response with (6.4.6) using that $d\varepsilon(s)/ds = d(Rs)/ds = R$ and introduce a variable change as $(t - t') = v$, we get,

$$\sigma(t) = D(t)\varepsilon_0 + \int_{-\infty}^{t} D(t - t')dt' = R \int_{-\infty}^{t} D(v)dv$$

We should recall that $\varepsilon(t) = Rt$ and further manipulation gives,

$$\sigma(t) = t\left(\frac{1}{t} \int_{-\infty}^{t} D(v)dv \right) \varepsilon(t) \equiv D_{eff}(t)\varepsilon(t)$$

We can see that here,

$$D_{eff}(t) \equiv \frac{1}{t} \int_{-\infty}^{t} D(v)dv \tag{6.4.8}$$

which can be used to compute the stress response of a viscoelastic soil to constant strain-rate loading by means of $\sigma(t) = D_{eff}(t) \varepsilon(t)$.

6.4.2 Relationship between Relaxation Modulus and Creep Compliance

Expressions (6.4.5) and (6.4.6) relate stresses to strains and strains to stresses through the creep compliance and the relaxation modulus, respectively. It is plausible to expect that these are the reciprocals of each other as $J(t).D(t) \sim 1$. However, when we look at Figure (6.4-2), we see that in general that is not the case and they are not the reciprocals of each other unlike elastic solids. Thus, such a relationship between the two should be derived from the equation (6.4.6). If we make use of step-strain history as $\varepsilon(t) = \varepsilon_0 H(t)$ along with $dH(t)/dt = \delta(t)$, we get.

$$\sigma(t) = \int_{-\infty}^{t} D(t-t')\varepsilon_0 \delta(t) dt' \equiv M(t)\varepsilon_0 \tag{6.4.9}$$

Substituting the above into (6.4.3) with the step-strain function and neglecting the initial stress, we get,

$$\varepsilon(t) = \left(\int_{-\infty}^{t} J(t-t')\frac{dD'}{dt'}dt' \right)\varepsilon_0 = \varepsilon_0 H(t) \tag{6.4.10}$$

This yields,

$$\int_{-\infty}^{t} J(t-t')\frac{dD(t')}{dt'}dt' = H(t) \tag{6.4.11}$$

Following the same steps, we can also evaluate a reverse relation in the form below,

$$\int_{-\infty}^{t} D(t-t')\frac{dJ(t')}{dt'}dt' = H(t) \tag{6.4.12}$$

The last two relations show that the creep compliance and the relaxation modulus are not the reciprocals of one another. Additional insight can be gained by calculating these relations for $t = 0$ and $t = \infty$ as well as invoking the "fading memory principle" that can be found in Gutierrez-Lemini (2014).

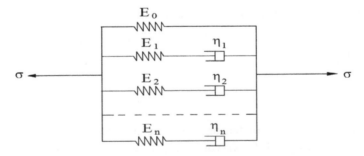

Figure 6.4-2 Generalized spring-dashpot (Maxwell) model

6.4.3 Differential Operator Form of Constitutive Relations

The creep and relaxation integral equations of the stress-strain relationships are only one type of formulation to define the constitutive equations in viscoelasticity. The governing equation of any model of the Kelvin or Maxwell type described previously can also be expressed in terms of differential with respect to time. Such a relationship can be written as,

$$\sigma + p_1 \frac{d\sigma}{dt} + p_2 \frac{d^2\sigma}{dt^2} + \dots + p_n \frac{d^n\sigma}{dt^n} = q_0\varepsilon + q_1 \frac{d\varepsilon}{dt} + q_2 \frac{d^2\varepsilon}{dt^2} + \dots q_n \frac{d^n\varepsilon}{dt^n} \tag{6.4.13}$$

When we look closely at (6.4.13) and equations (6.3.3) & (6.3.12) which can be generalized as,

$$\frac{\dot{\sigma}}{E}+\frac{\sigma}{\eta}=\dot{\varepsilon}_1+\dot{\varepsilon}_2=\dot{\varepsilon} \tag{6.4.14a}$$

$$\sigma = E\varepsilon + \eta\dot{\varepsilon} \tag{6.4.14b}$$

respectively, we see that by choosing $p_0 =1$, $p_1 =\dfrac{\eta}{E}$, $q_1 =\eta$ and the other coefficients being zero, (6.4.13) corresponds to the Maxwell model. By choosing $p_0 = 1$, $q_0 = E$, $q_1 = \eta$ and the other coefficients to be zero, (6.4.13) represents a constitutive relation described by the Kelvin (Voigt) model. This can also be written as,

$$\sum_{k=0}^{n}p_k\frac{d^k\sigma}{dt^k}=\sum_{k=0}^{n}q_k\frac{d^k\varepsilon}{dt^k} \tag{6.4.15}$$

which may further be written as,

$$\partial_p\sigma=\partial_q\varepsilon \tag{6.4.16}$$

where ∂_p and ∂_q are the differential operators defined as,

$$\partial_p=\sum_{k=0}^{n}p_k\frac{d^k}{dt^k} \tag{6.4.17a}$$

$$\partial_q=\sum_{k=0}^{n}q_k\frac{d^k}{dt^k} \tag{6.4.17b}$$

More detailed discussion is in Gutierrez-Lemini (2014).

6.4.4 Constitutive Relations in Time Domain

According to *Boltzman's superposition principle* as in (6.4.1), the stress-strain relationship can be expressed as follows:

$$\sigma_{ij}(t)=\int_{-\infty}^{t}\varepsilon_{kl}(t-\tau)\frac{dD_{ijkl}(\tau)}{d\tau}d\tau \tag{6.4.18}$$

where σ_{ij} = stress tensor; ε_{kl} = strain tensor, t and τ are the actual and pseudo time variables. It is assumed that $\varepsilon_{kl}(t) = 0$ for $t < 0$. $D_{ijkl}(t)$ can be expressed in the following form:

$$D_{ijkl}(t)=\frac{1}{3}[3K(t)-2G(t)]\delta_{ij}\delta_{kl}+G(t)(\delta_{ik}\delta_{jl}+\delta_{il}\delta_{jk}) \tag{6.4.19}$$

where $K(t)$ and $G(t)$ are independent relaxation bulk and shear moduli which are associated with dilation and shear, respectively. δ_{ij} is the Kronecker delta. Introducing the deviatoric components of stress s_{ij} and strain e_{ij} leads to,

$$s_{ij}=2\int_{-\infty}^{t}e_{ij}(t-\tau)\frac{dG(\tau)}{d\tau}d\tau \tag{6.4.20}$$

whereas the volumetric stress σ_{kk} and strain ε_{kk} are related as,

$$\sigma_{kk}=3\int_{-\infty}^{t}\varepsilon_{kk}(t-\tau)\frac{dK(\tau)}{d\tau}d\tau \tag{6.4.21}$$

where

$$s_{ij} = \sigma_{ij} - \frac{1}{3}\delta_{ij}\sigma_{kk} \qquad (6.4.22)$$

$$e_{ij} = \varepsilon_{ij} - \frac{1}{3}\delta_{ij}\varepsilon_{kk} \qquad (6.4.23)$$

and the repeated indices imply summation. From (6.4.20)-(6.4.23), the following equation is derived:

$$\sigma_{ij} = 2\int_{-\infty}^{t} e_{ij}(t-\tau)\frac{dG(\tau)}{d\tau}d\tau + \int_{-\infty}^{t}\delta_{ij}\varepsilon_{kk}(t-\tau)\frac{dK(\tau)}{d\tau}d\tau \qquad (6.4.24)$$

$K(t)$ and $G(t)$ can be obtained from the visco-elastic model composed of springs and dashpots as in Figure 6.4-2.

6.4.5 Constitutive Relations in Fourier Transformed Domain

The Fourier transform is defined as,

$$\bar{f}(\omega) = \int_{-\infty}^{+\infty} f(t)e^{-i\omega t}dt \qquad (6.4.25)$$

and the inverse transform is,

$$f(t) = \frac{1}{2\pi}\int_{-\infty}^{+\infty} \bar{f}(\omega)e^{i\omega t}d\omega \qquad (6.4.26)$$

By using (6.4.25), the Fourier transform of the deviatoric stress-strain relation expressed in (6.4.20) can be written as,

$$\bar{S}_{ij}(\omega) = 2\int_{-\infty}^{t}\left[\int_{-\infty}^{t} e_{ij}(t-\tau)\frac{dG(\tau)}{d\tau}d\tau\right]e^{-i\omega t}dt \qquad (6.4.27)$$

which can be further reduced to the simple form as,

$$\bar{S}_{ij}(\omega) = 2\bar{G}(\omega)\bar{e}_{ij}(\omega) \qquad (6.4.28)$$

Similarly, applying the Fourier transform to the volumetric part of the stress strain relation given in (6.4.21) gives,

$$\bar{\sigma}_{kk}(\omega) = 3\bar{K}(\omega)\bar{\varepsilon}_{kk}(\omega) \qquad (6.4.29)$$

Here, $\bar{G}(\omega)$ and $\bar{K}(\omega)$ are often referred to as *the complex shear modulus* and *complex bulk modulus*, respectively. We see here that in a more general sense, given the Fourier transform of the relaxation modulus $D(t)$, we can obtain the complex modulus,

$$\bar{D}(\omega) = i\omega\int_{0}^{+\infty} D(t)e^{-i\omega t}dt \qquad (6.4.30)$$

6.4.6 Constitutive Relations in Laplace Transformed Domain

The Laplace transform of a function $f(t)$, which is zero for $t < 0$, is given as,

$$\tilde{f}(s) = \int_{0}^{+\infty} f(t)e^{-st}dt \qquad (6.4.31)$$

The inverse Laplace transform is,

$$f(t) = \frac{1}{2\pi i} \int_{\gamma-i\infty}^{\gamma+i\infty} \tilde{f}(s)e^{st}\, ds \tag{6.4.32}$$

where $i = \sqrt{-1}$, s is a complex number and γ is an arbitrary real number but greater than the real parts of all the singularities of $\tilde{f}(s)$. The Laplace transforms of equations (6.4.20) and (6.4.21) give, for $\varepsilon_{ij} = 0$ and $t < 0$,

$$\tilde{S}_{ij} = 2s\tilde{G}\tilde{e}_{ij} \tag{6.4.33}$$

$$\tilde{\sigma}_{kk} = 3s\tilde{K}\tilde{\varepsilon}_{kk} \tag{6.4.34}$$

6.5 1-D Viscoelastic Analysis of Soil Layers under Vertical Circular Loading

Problems of soils and soil-structure systems modeled by using a viscoelastic model are common in geotechnical engineering. A 1-D analysis for the response of viscoelastic soil layers subjected to an axi-symmetric vertical circular loading is presented in this section. As it is presented in the previous section, there are essentially three approaches (direct time integration, Fourier transform, and Laplace transform) to deal with the time factor associated with the numerical methods of analysis of layered viscoelastic systems. In this section, all three approaches are used to evaluate the time-dependent response of the soil layer. A linear visco-elastic layered system subjected to axi-symmetric vertical circular loading on the surface is analyzed for quasi-static case. The method originally developed by Xu and Rahman (2008) is presented here. The method utilizes the generalized Maxwell model or the Kelvin model to describe the viscoelastic soil behavior. A Hankel transform is used to simplify the problem involving a 3-D loading to that involving only a single spatial dimension (1-D). Then the resulting transformed problem is solved by the Finite Element Method (FEM) with three approaches of handling time:

(a) Direct time analysis in which the time interval is discretized by the Finite Difference Method (FDM) the details of which are given in Chapter 9;
(b) Fourier transform applied to the temporal (time based) terms;
(c) Laplace transform applied to the temporal terms.

The solution flow chart is presented in the Appendix. For the latter two approaches, numerical inversion of the Fourier transform or the Laplace transform is required. For that, an inverse FFT (IFFT) and numerical inversion of the Laplace transform which is based on the method developed by Honig and Hirdes (1984) is used. Subsequently, the numerical results for the quasi-static response of all three approaches are compared.

6.5.1 Method 1: Direct Time Integration

The equilibrium equations for a soil layer subjected to axisymmetric loading in cylindrical coordinates are expressed as,

$$\frac{\partial \sigma_r}{\partial r} + \frac{\partial \tau_{rz}}{\partial z} + \frac{\sigma_r - \sigma_\theta}{r} = 0 \tag{6.5.1}$$

$$\frac{\partial \tau_{rz}}{\partial r} + \frac{\partial \sigma_z}{\partial z} + \frac{\tau_{rz}}{r} = 0 \tag{6.5.2}$$

where stress components are time-dependent. Stress-strain relationship for a viscoelastic material in time domain again is,

$$\sigma(t) = D_A(0)\varepsilon(t) - \int_0^t \frac{d D_A(t-\tau)}{d\tau} \varepsilon(\tau) d\tau \tag{6.5.3}$$

where $\sigma(t)$ and $\varepsilon(t)$ are the vectors of time-dependent stress and strain components and $D_A(t)$ is the stress-strain relationship matrix for axial-symmetric problems.

$$D_A(t) = \begin{bmatrix} K(t)+\frac{4}{3}G(t) & K(t)-\frac{2}{3}G(t) & K(t)-\frac{2}{3}G(t) & 0 \\ K(t)-\frac{2}{3}G(t) & K(t)+\frac{4}{3}G(t) & K(t)-\frac{2}{3}G(t) & 0 \\ K(t)-\frac{2}{3}G(t) & K(t)-\frac{2}{3}G(t) & K(t)+\frac{4}{3}G(t) & 0 \\ 0 & 0 & 0 & G(t) \end{bmatrix} \tag{6.5.4}$$

Kinematic equations giving the strain-displacement relationship are as follows:

$$\begin{bmatrix} \varepsilon_r \\ \varepsilon_\theta \\ \varepsilon_z \\ \gamma_{rz} \end{bmatrix} = - \begin{bmatrix} \dfrac{\partial}{\partial r} & 0 & 0 \\ \dfrac{1}{r} & 0 & 0 \\ 0 & 0 & \dfrac{\partial}{\partial z} \\ \dfrac{\partial}{\partial z} & 0 & \dfrac{\partial}{\partial r} \end{bmatrix} \begin{bmatrix} u_r \\ u_\theta \\ u_z \end{bmatrix} \tag{6.5.5}$$

If we perform the Hankel transformation with respect to the spatial coordinate r onto the displacement components, we get,

$$U_z = \int_0^\infty r u_z J_0(\alpha r) dr \tag{6.5.6}$$

$$(U_r, U_\theta) = \int_0^\infty r(u_r, u_\theta) J_1(\alpha r) dr \tag{6.5.7}$$

where J_0 and J_1 are zero order and first order Bessel functions of the first type respectively. The corresponding inverse transforms are,

$$u_z = \int_0^\infty \alpha U_z J_0(\alpha r) d\alpha \tag{6.5.8}$$

$$(u_r, u_\theta) = \int_0^\infty \alpha U_r J_1(\alpha r) d\alpha \tag{6.5.9}$$

Substituting (6.5.8) and (6.5.9) into the kinematics of (6.5.5) gives,

$$\varepsilon_{rr} = \int_0^\infty \alpha (\alpha J_0(\alpha r) - \frac{1}{r} J_1(\alpha r)) U_r d\alpha \tag{6.5.10}$$

$$\varepsilon_{\theta\theta} = \int_0^\infty \alpha \frac{1}{r} J_1(\alpha r) U_r d\alpha \tag{6.5.11}$$

$$\varepsilon_{zz} = \int_0^\infty \alpha J_0(\alpha r) \frac{\partial U_z}{\partial z} d\alpha \tag{6.5.12}$$

$$\gamma_{rz} = \int_0^\infty \alpha \left(\frac{\partial U_r}{\partial z} - \alpha U_z \right) J_1(\alpha r) d\alpha \tag{6.5.13}$$

Now that substituting the above strain components into the stress-strain law of (6.5.3), the stress components can be expressed in terms of transformed displacement components. If the stresses are used in the equilibrium equations, (6.5.1) and (6.5.2), we get,

$$-\alpha M - \frac{\partial}{\partial z} T = 0 \tag{6.5.14}$$

$$-\alpha T + \frac{\partial}{\partial z} N = 0 \tag{6.5.15}$$

where M, T and N are substitutions for the following expressions:

$$M = \alpha \left(a(0) U_r(t) - \int_0^t \frac{da(t-\tau)}{d\tau} U_r(\tau) d\tau \right) + c(0) \frac{\partial U_z}{\partial z}(t) - \int_0^t \frac{dc(t-\tau)}{d\tau} \frac{\partial U_z}{\partial z}(\tau) d\tau \tag{6.5.16}$$

$$T = \alpha \left(f(0) U_z(t) - \int_0^t \frac{df(t-\tau)}{d\tau} U_z(\tau) d\tau \right) - \left(f(0) \frac{\partial U_r}{\partial z}(t) - \int_0^t \frac{df(t-\tau)}{d\tau} \frac{\partial U_r}{\partial z}(\tau) d\tau \right) \tag{6.5.17}$$

$$N = \alpha \left(c(0) U_r(t) - \int_0^t \frac{dc(t-\tau)}{d\tau} U_r(\tau) d\tau \right) + a(0) \frac{\partial U_z}{\partial z}(t) - \int_0^t \frac{da(t-\tau)}{d\tau} \frac{\partial U_z}{\partial z}(\tau) d\tau \tag{6.5.18}$$

where $a(t)$, $c(t)$ and $f(t)$ are substitutions for the following expressions,

$$a(t) = K(t) + \tfrac{4}{3} G(t) \tag{6.5.19}$$

$$c(t) = K(t) - \tfrac{2}{3} G(t) \tag{6.5.20}$$

$$f(t) = G(t) \tag{6.5.21}$$

It is observed that equations (6.5.14)-(6.5.15) can also be written in matrix form,

$$R_{\approx} D(0) V_{\approx} W(t) - R_{\approx} \int_0^t \frac{d D(t-\tau)}{d\tau} V_{\approx} W(\tau) d\tau = 0 \tag{6.5.22}$$

where

$$R_{\approx} = \begin{bmatrix} \alpha & 0 & -\dfrac{\partial}{\partial z} \\ 0 & -\dfrac{\partial}{\partial z} & -\alpha \end{bmatrix} \tag{6.5.23a}$$

$$D(t) = \begin{bmatrix} a(t) & c(t) & 0 \\ c(t) & a(t) & 0 \\ 0 & 0 & f(t) \end{bmatrix} \qquad (6.5.23b)$$

$$V = \begin{bmatrix} \alpha & 0 & \dfrac{\partial}{\partial z} \\[2mm] 0 & \dfrac{\partial}{\partial z} & -\alpha \end{bmatrix}^T \qquad (6.5.23c)$$

$$W = (U_r, U_z)^T \qquad (6.5.23d)$$

In the above, "~" underneath the variable means the variable is a vector and "≈" underneath the variable means the variable is a matrix. Consistent with the associated boundary and continuity equations, the solution of equation (6.5.22) can be obtained by the finite element approximation (see Ch.10 for details). From the *principle of virtual work* used to derive the related equilibrium relations in their "weak form" (see Ch. 10), we can get the following relationship,

$$\int_\Omega \delta W^T(t) V^T D(0) V W(t)dz - \int_0^t \int_\Omega \delta W^T(t) V^T \frac{d D(t-\tau)}{d\tau} V W(\tau)dzd\tau$$

$$- \int_\Omega \delta W^T(t) p \, dz - \int_\Omega \delta W^T(t) q \, dz = 0 \qquad (6.5.24)$$

where Ω is the region along the z-axis because the problem is reduced to that of 1-D after the spatial transform; p is the body force; q is the surface traction. Dividing the layered system along vertical direction into finite elements, displacements within each element can be expressed as

$$U_r = \sum_{i=1}^{n} N_i U_{ri} \qquad (6.5.25a)$$

$$U_z = \sum_{i=1}^{n} N_i U_{zi} \qquad (6.5.25b)$$

where U_{ri} and U_{zi} are the corresponding radial and vertical displacements at nodes i and N_i are the shape functions associated with node i. The displacement vector W^e for each element can then be expressed in matrix form as,

$$W^e = N\{W^e\} \qquad (6.5.26)$$

By substituting equation (6.5.26) into equation (6.5.24), the following expression is arrived,

$$\sum_{e=1}^{N_e} \delta\{W^{e^T}(t)\} \left\{ \int_{\Omega^e} N^T V^T D(0) V N\{W^e(t)\}dz - \int_0^t \int_{\Omega^e} N^T V^T \frac{d D(t-\tau)}{d\tau} \right.$$

$$\left. V N\{W^e(\tau)\}dzd\tau - \int_{\Omega^e} N^T pdz - \int_{\Omega^e} N^T qdz \right\} = 0 \qquad (6.5.27)$$

For arbitrary $\delta(W^{e^T}(t))$ in (6.5.27), the following elemental equation is obtained,

$$K^e(0)\{W^e(t)\} - \int_0^t \frac{d K^e(t-\tau)}{d\tau}\{W^e(\tau)\}d\tau = F^e \qquad (6.5.28)$$

where

$$K^e(t) = \int_{\Omega^e} \underset{\approx}{N}^T \underset{\approx}{V}^T D(t) \underset{\approx}{V} \underset{\approx}{N} dz \tag{6.5.29}$$

$$\underset{\sim}{F}^e = \int_{\Omega^e} \underset{\approx}{N}^T p dz - \int_{\Omega^e} \underset{\approx}{N}^T \underset{\sim}{q} dz \tag{6.5.30}$$

For the assembly process, it is assumed that the stresses and displacements are continuous at layer interfaces, which yields the following set of equations:

$$K(0)\{W(t)\} - \int_0^t \frac{d\,K(t-\tau)}{d\tau}\{W(\tau)\}d\tau = F \tag{6.5.31}$$

For the problem in consideration, the body force is ignored and no surface traction is applied. Therefore, most components of the global force vector F are zero except for the node on the layer surface where there is a nodal vertical force. The global force vector, $\underset{\sim}{F}$, can be written as,

$$\underset{\sim}{F} = (0,\ F_{zz},\ 0,\ 0,\ \dots ,0)^T \tag{6.5.32}$$

If a uniform pressure f_{zz} is applied on the layer surface over a circular region, then

$$F_{zz} = \int_0^\infty rf_{zz}J_0(\alpha r)dr \tag{6.5.33}$$

A full discretization is achieved in equation (6.5.31) by discretizing the convolution integral using the FDM (see Ch. 9 for the detailed discussion) as follows,

$$\int_t^{t+\Delta t} \frac{d\,K(t-\tau)}{d\tau}\{W(\tau)\}d\tau = (1-\theta)\Delta t \left.\frac{d\,K(t-\tau)}{d\tau}\right|_{\tau=t}$$

$$\{W(t)\} + \theta\Delta t \left.\frac{d\,K(t-\tau)}{d\tau}\right|_{\tau=t+\Delta t} \{W(t+\Delta t)\} \tag{6.5.34}$$

The above equation leads to,

$$\left(K(0) - \theta\Delta t \left.\frac{d\,K(t-\tau)}{d\tau}\right|_{\tau=t+\Delta t}\right)\{W(t+\Delta t)\} = F(t+\Delta t) + (1-\theta)\Delta t \left.\frac{d\,K(t-\tau)}{d\tau}\right|_{\tau=t} \{W(t)\} \tag{6.5.35}$$

Therefore, solution at any time can be obtained by a forward marching process. However, it is observed that there is an accumulation of time-history terms on the right-hand-side of equation (6.5.35) as time increases. Thus, the demand on computer storage and computation time will increase as time increases. This problem can be overcome if each component of $D(t)$ is written in terms of a Prony series and therefore bypasses the need to store the entire history of displacements, strains and stresses. In this study, the spatial domain is discretized by FEM using linear shape functions, and the time domain is discretized by the FDM using the *trapezoidal rule*.

For an element lying between coordinates z_i and z_{i+1}, using linear interpolation for displacements, U_r and U_z results in,

$$U_r = U_{r(i)} \frac{z_{i+1} - z}{z_{i+1} - z_i} + U_{r(i+1)} \frac{z - z_i}{z_{i+1} - z_i} \qquad (6.5.36)$$

$$U_z = U_{z(i)} \frac{z_{i+1} - z}{z_{i+1} - z_i} + U_{z(i+1)} \frac{z - z_i}{z_{i+1} - z_i} \qquad (6.5.37)$$

Then, the vector of displacements can be written as,

$$\underset{\sim}{W^e} = \frac{1}{z_{i+1} - z_i} \begin{bmatrix} (z_{i+1} - z) & 0 & (z - z_i) & 0 \\ 0 & (z_{i+1} - z) & 0 & (z - z_i) \end{bmatrix} \{ \underset{\sim}{W^e} \} \qquad (6.5.38)$$

where

$$\{ \underset{\sim}{W^e} \} = \begin{bmatrix} U_{r(i)} & U_{z(i)} & U_{r(i+1)} & U_{z(i+1)} \end{bmatrix}^T \qquad (6.5.39)$$

Equations (6.5.36)-(6.5.39) lead to the following expression:

$$\underset{\approx}{B} = \underset{\approx}{V} \underset{\approx}{N} = \frac{1}{h} \begin{bmatrix} \alpha(z_{i+1} - z) & 0 & \alpha(z - z_i) & 0 \\ 0 & -1 & 0 & 1 \\ -1 & -\alpha(z_{i+1} - z) & 1 & -\alpha(z - z_i) \end{bmatrix} \qquad (6.5.40)$$

which is used to get the stiffness matrix as,

$$\underset{\approx}{K^e}(t) = \int_{z_i}^{z_{i+1}} \underset{\approx}{B}^T \underset{\approx}{D}(t) \underset{\approx}{B} \, dz = \begin{bmatrix} \frac{\alpha^2 a(t)h}{3} + \frac{f(t)}{h} & -\frac{\alpha c(t)}{2} + \frac{f(t)\alpha}{2} & \frac{\alpha^2 a(t)h}{6} - \frac{f(t)}{h} & \frac{\alpha c(t)}{2} + \frac{f(t)}{2}\alpha \\ -\frac{\alpha c(t)}{2} + \frac{f(t)\alpha}{2} & \frac{a(t)}{h} + \frac{f(t)\alpha^2 h}{3} & -\frac{c(t)\alpha}{2} - \frac{f(t)\alpha}{2} & -\frac{a(t)}{h} + \frac{f(t)\alpha^2 h}{6} \\ \frac{\alpha^2 a(t)h}{6} - \frac{f(t)}{h} & -\frac{c(t)\alpha}{2} - \frac{f(t)\alpha}{2} & \frac{\alpha^2 a(t)h}{3} + \frac{f(t)}{h} & \frac{\alpha c(t)}{2} - \frac{\alpha f(t)}{2} \\ \frac{\alpha c(t)}{2} + \frac{f(t)}{2}\alpha & -\frac{a(t)}{h} + \frac{f(t)\alpha^2 h}{6} & \frac{\alpha c(t)}{2} - \frac{\alpha f(t)}{2} & \frac{a(t)}{h} + \frac{\alpha^2 f(t)h}{3} \end{bmatrix} \qquad (6.5.41)$$

where

$$h = z_{i+1} - z_i \qquad (6.5.42)$$

Following the assembly of elemental stiffness matrix and elemental load vectors to get the global matrices and vectors, time domain discretization is performed. Assuming that both the bulk relaxation modulus $K(t)$ and shear relaxation modulus $G(t)$ comply with the law of generalized Maxwell model or generalized Kelvin model are kept constant, each component of constitutive matrix $\underset{\approx}{D}(t)$ can be expressed as a Prony series. For illustration purposes, the detailed time forward marching algorithm is as follows for a material with $K(t)$ and $G(t)$ drawn from the same viscoelastic models but with different parameters. For that case, $\underset{\approx}{D}(t)$ can be written as,

$$\underset{\approx}{D}(t) = \begin{bmatrix} a & c & 0 \\ c & d & 0 \\ 0 & 0 & f \end{bmatrix} \phi(t) = \underset{\approx}{D_0} \, \phi(t) \qquad (6.5.43)$$

where $\phi(t)$ is given as a Prony series of,

$$\phi(t) = \sum_{1}^{n} \phi_i e^{-\alpha_i t} \tag{6.5.44}$$

Plugging (6.5.44) into (6.5.31) results in,

$$\underset{\approx}{K}(0) \cdot \left[\{\underset{\sim}{W}(t)\} - \frac{1}{\displaystyle\sum_{i=1}^{n} \phi_i} \int_0^t \dot\phi(t-\tau)\{\underset{\sim}{W}(\tau)\}d\tau \right] = \underset{\sim}{F}(t) \tag{6.5.45}$$

where

$$\underset{\approx}{K}(0) = \sum_{1}^{n} \phi_i \int \underset{\approx}{B}^T \underset{\approx}{D}_0 \underset{\approx}{B}\, dz \tag{6.5.46}$$

In order to obtain the fully discrete general formulation at each time step $t = t_i$, we begin considering a fully discrete formulation at $t = t_i > 0$ written as,

$$\underset{\approx}{K}(0) \cdot \left[\{\underset{\sim}{W}(t_1)\} - \frac{1}{\displaystyle\sum_{i=1}^{n} \phi_i} \int_0^{t_1} \dot\phi(t_1-\tau)\{\underset{\sim}{W}(\tau)\}d\tau \right] = \underset{\sim}{F}(t_1) \tag{6.5.47}$$

Let

$$\underset{\sim}{r}(t_1) = \int_0^{t_1} \dot\phi(t_1-\tau)\{\underset{\sim}{W}(\tau)\}d\tau \tag{6.5.48}$$

and since

$$\dot\phi(t_1-\tau) = \sum_{i=1}^{n} \alpha_i \phi_i \cdot e^{-\alpha_i(t_1-\tau)} \tag{6.5.49}$$

then

$$\underset{\sim}{r}(t_1) = \int_0^{t_1} \sum_{j=1}^{n} \alpha_j \phi_j \cdot e^{-\alpha_j(t_1-\tau)}\{\underset{\sim}{W}(\tau)\}d\tau = \sum_{j=1}^{n} \underset{\sim j}{r}(t_1) \tag{6.5.50}$$

where

$$\underset{\sim j}{r}(t_1) = \int_0^{t_1} \alpha_j \phi_j \cdot e^{-\alpha_j(t_1-\tau)}\{\underset{\sim}{W}(\tau)\}d\tau, \quad j = 1, 2, \cdots, n \tag{6.5.51}$$

The integral equation (6.5.48) is approximated by the trapezoidal rule of numerical integration. In this method, the given equation or functions' integral within prescribed boundaries is approximated using small trapezoidal areas. More can be found in related calculus literature. Now, setting $\Delta t = t_1$, we obtain,

$$\underset{\sim}{r}(t_1) = \sum_{j=1}^{n} \underset{\sim j}{r}(t_1) = \frac{\Delta t}{2} \left[\sum_{j=1}^{n} \alpha_j \cdot \phi_j \cdot \left[e^{-\alpha_j t_1} \cdot \{\underset{\sim}{W}(0)\} + \{\underset{\sim}{W}(t_1)\} \right] \right] \tag{6.5.52}$$

Substituting equation (6.5.50) into (6.5.45) and rearranging the terms yields,

$$
K(0) \left[1 - \frac{\Delta t}{2 \cdot \left(\sum\limits_{i=1}^{n} \phi_i \right)} \sum\limits_{j=1}^{n} \alpha_j \cdot \phi_j \right] \cdot \{W(t_1)\} = F(t_1) + \frac{\Delta t}{2 \cdot \left(\sum\limits_{i=1}^{n} \phi_i \right)}
$$

$$
\left(\sum\limits_{j=1}^{n} \alpha_j \cdot \phi_j \cdot e^{-\alpha_j t_1} \right) \cdot K(0) \cdot \{W(0)\}
$$

(6.5.53)

Here, $\left\{W(t_1)\right\}$ can be obtained by solving the above set of linear equations, and then $r_j(t_1), j = 1, 2, \cdots, n$ can be calculated. These values are used in the next time step. For $t = t_2$ we have,

$$
K(0) \cdot \left[\{W(t_2)\} - \frac{1}{\sum\limits_{i=1}^{n} \phi_i} \int_0^{t_2} \dot{\phi}(t_2 - \tau) \{W(\tau)\} d\tau \right] = F(t_2)
$$

(6.5.54)

Now we define,

$$
r(t_2) = \int_0^{t_2} \dot{\phi}(t_2 - \tau) \{W(\tau)\} d\tau = \int_0^{t_1} \dot{\phi}(t_2 - \tau) \{W(\tau)\} d\tau + \int_{t_1}^{t_2} \dot{\phi}(t_2 - \tau) \{W(\tau)\} d\tau
$$

(6.5.55)

and by making use of $r_j(t_1)$ and approximating the second integral by the trapezoidal rule, $r(t_2)$ can be written as,

$$
r(t_2) = \int_0^{t_1} \sum\limits_{j=1}^{n} \alpha_j \phi_j \cdot e^{-\alpha_j(t_2 - \tau)} \{W(\tau)\} d\tau + \int_{t_1}^{t_2} \sum\limits_{j=1}^{n} \alpha_j \phi_j \cdot e^{-\alpha_j(t_2 - \tau)} \{W(\tau)\} d\tau
$$

$$
= \int_0^{t_1} \sum\limits_{j=1}^{n} \alpha_j \phi_j \cdot e^{-\alpha_j(t_1 - \tau)} \cdot e^{-\alpha_j(t_2 - t_1)} \{W(\tau)\} d\tau + \frac{t_2 - t_1}{2} \cdot \sum\limits_{j=1}^{n} \alpha_j \phi_j \cdot (e^{-\alpha_j(t_2 - t_1)} \cdot \{W(t_1)\} + \{W(t_2)\})
$$

$$
= \sum\limits_{j=1}^{n} e^{-\alpha_j \Delta t} \cdot r_j(t_1) + \frac{\Delta t}{2} \left[\sum\limits_{j=1}^{n} \alpha_j \cdot \phi_j \left[e^{-\alpha_j(t_2 - t_1)} \cdot \{W(t_1)\} + \{W(t_2)\} \right] \right]
$$

(6.5.56)

Substituting (6.5.56) into (6.5.54) yields,

$$
K(0) \cdot \left[1 - \frac{\Delta t}{2 \cdot \left(\sum\limits_{i=1}^{n} \phi_i \right)} \sum\limits_{j=1}^{n} \alpha_j \cdot \phi_j \right] \cdot \{W(t_2)\} = F(t_2) +
$$

$$
K(0) \cdot \left\{ \frac{\sum\limits_{j=1}^{n} e^{-\alpha_j \Delta t} \cdot r_j(t_1)}{\sum\limits_{i=1}^{n} \phi_i} + \frac{\Delta t}{2 \cdot \left(\sum\limits_{i=1}^{n} \phi_i \right)} \left(\sum\limits_{j=1}^{n} \alpha_j \cdot \phi_j \cdot e^{-\alpha_j \Delta t} \right) \cdot \{W(t_1)\} \right\}
$$

(6.5.57)

where $r_j(t_1), j = 1, 2, \cdots, n$ are determined by equation (6.5.51). Again, $\{W(t_2)\}$ is determined by solving equation (6.5.57), and then $r_j(t_2)$ in equation (6.5.56) is calculated. These values are stored for the next time step. Similarly, at $t = t_i$, we have,

$$
K(0) \cdot \left[1 - \frac{\Delta t}{2 \cdot \left(\sum_{i=1}^{n} \phi_i \right)} \sum_{j=1}^{n_\phi} \alpha_j \cdot \phi_j \right] \cdot \{W(t_i)\} = F(t_i) + K(0) \cdot \sum_{j=1}^{n}
$$

$$
\left(\frac{e^{-\alpha_j \cdot \Delta t}}{\sum_{i=1}^{n} \phi_i} \cdot \left[r_j(t_{i-1}) + \frac{\Delta t}{2} \alpha_j \cdot \phi_j \cdot \{W(t_{i-1})\} \right] \right)
\tag{6.5.58}
$$

After solving (6.5.58) for $\{W(t_i)\}$, $r_j(t_i)$ is obtained as follows:

$$
r_j(t_i) = e^{-\alpha_j \cdot \Delta t} \cdot r_j(t_{i-1}) + \frac{\Delta t}{2} \cdot \alpha_j \cdot \phi_j \cdot \left[e^{-\alpha_j \cdot \Delta t} \cdot \{W(t_{i-1})\} + \{W(t_i)\} \right]
\tag{6.5.59}
$$

Therefore, the fully discrete formulation in the case of a linear viscoelastic material is derived from (6.5.53) for $t = t_1$ and (6.5.58) for $t > t_1$. The advantage of this formulation is that, only the displacement $\{W\}$ and the state variable $r_j, j = 1, 2, \cdots, n$ at the previous time step are involved in the computation of the displacement at the current time step. Once the displacements are calculated, stresses can also be calculated.

6.5.2 Method 2: Fourier Domain Analysis

Fourier transform should be performed to make the governing equations and boundary conditions free from time variable. Applying Fourier transform to all the field variables, the equations of equilibrium become,

$$
\frac{\partial \bar{\sigma}_r}{\partial r} + \frac{\partial \bar{\tau}_{rz}}{\partial z} + \frac{\bar{\sigma}_r - \bar{\sigma}_\theta}{r} = 0
\tag{6.5.60}
$$

$$
\frac{\partial \bar{\tau}_{rz}}{\partial r} + \frac{\partial \bar{\sigma}_z}{\partial z} + \frac{\bar{\tau}_{rz}}{r} = 0
\tag{6.5.61}
$$

The stress-strain relationship is,

$$
\bar{\sigma} = \bar{D} \bar{\varepsilon}
\tag{6.5.62}
$$

where $\bar{\sigma} = (\bar{\sigma}_r, \bar{\sigma}_\theta, \bar{\sigma}_z, \bar{\tau}_{rz})^T$; $\bar{\varepsilon} = (\bar{\varepsilon}_r, \bar{\varepsilon}_\theta, \bar{\varepsilon}_z, \bar{\gamma}_{rz})^T$ and,

$$
\bar{D}(\omega) = \begin{bmatrix}
\bar{\kappa}(\omega) + \frac{4}{3}\bar{G}(\omega) & \bar{\kappa}(\omega) - \frac{2}{3}\bar{G}(\omega) & \bar{\kappa}(\omega) - \frac{2}{3}\bar{G}(\omega) & 0 \\
\bar{\kappa}(\omega) - \frac{2}{3}\bar{G}(\omega) & \bar{\kappa}(\omega) + \frac{4}{3}\bar{G}(\omega) & \bar{\kappa}(\omega) - \frac{2}{3}\bar{G}(\omega) & 0 \\
\bar{\kappa}(\omega) - \frac{2}{3}\bar{G}(\omega) & \bar{\kappa}(\omega) - \frac{2}{3}\bar{G}(\omega) & \bar{\kappa}(\omega) + \frac{4}{3}\bar{G}(\omega) & 0 \\
0 & 0 & 0 & \bar{G}(\omega)
\end{bmatrix}
\tag{6.5.63}
$$

in which $\bar{\kappa}(\omega)$ and $\bar{G}(\omega)$ are the complex bulk modulus and complex shear modulus for the viscoelastic material, respectively. Strain-displacement relationship is then:

$$
\begin{bmatrix} \bar{\varepsilon}_r \\ \bar{\varepsilon}_\theta \\ \bar{\varepsilon}_z \\ \bar{\gamma}_{rz} \end{bmatrix} = -
\begin{bmatrix} \dfrac{\partial}{\partial r} & 0 & 0 \\[6pt] \dfrac{1}{r} & 0 & 0 \\[6pt] 0 & 0 & \dfrac{\partial}{\partial z} \\[6pt] \dfrac{\partial}{\partial z} & 0 & \dfrac{\partial}{\partial r} \end{bmatrix}
\begin{bmatrix} \bar{u}_r \\ \bar{u}_\theta \\ \bar{u}_z \end{bmatrix}
\tag{6.5.64}
$$

where $\bar{u}_r, \bar{u}_\theta, \bar{u}_z$ are Fourier transformed displacements. By applying Hankel transformations to the displacement variables, we have,

$$
\bar{U}_z = \int_0^\infty r \bar{u}_z J_0(\alpha r) dr
\tag{6.5.65}
$$

$$
(\bar{U}_r, \bar{U}_\theta) = \int_0^\infty r(\bar{u}_r, \bar{u}_\theta) J_1(\alpha r) dr
\tag{6.5.66}
$$

and then by following similar procedure as that of the time domain analysis, we can get

$$
-\alpha \left[\alpha \bar{a} \bar{U}_r + \bar{c} \frac{\partial \bar{U}_z}{\partial z} \right] - \frac{\partial}{\partial z} \left[\alpha \bar{U}_z - \frac{\partial \bar{U}_r}{\partial z} \right] \bar{f} = 0
\tag{6.5.67a}
$$

$$
-\alpha \left[\alpha \bar{U}_z - \frac{\partial \bar{U}_r}{\partial z} \right] \bar{f} + \frac{\partial}{\partial z} \left[\alpha \bar{c} \bar{U}_r + \bar{a} \frac{\partial \bar{U}_z}{\partial z} \right] = 0
\tag{6.5.67b}
$$

Equations (6.5.67a) and (6.5.67b) can be written in matrix form as,

$$
R \underset{\approx}{\bar{D}}(\omega) V \underset{\sim}{\bar{W}}(\omega) = 0
\tag{6.5.68}
$$

where

$$
\underset{\approx}{\bar{D}}(\omega) = \begin{bmatrix} \bar{a}(\omega) & \bar{c}(\omega) & 0 \\ \bar{c}(\omega) & \bar{a}(\omega) & 0 \\ 0 & 0 & \bar{f}(\omega) \end{bmatrix}
\tag{6.5.69}
$$

From the *principle of virtual work*, we arrive at,

$$
\int_\Omega \delta \underset{\sim}{\bar{W}}^T(\omega) V^T \underset{\approx}{\bar{D}}(\omega) V \underset{\sim}{\bar{W}}(\omega) dz - \int_\Omega \delta \underset{\sim}{\bar{W}}^T(\omega) p \, dz - \int_\Omega \delta \underset{\sim}{\bar{W}}^T(\omega) q \, dz = 0
\tag{6.5.70}
$$

where the definition of Ω, $\underset{\sim}{p}$ and $\underset{\sim}{q}$ are the same as those defined in the time domain analysis. Adopting the same finite element discretization algorithm as what was done in the time domain analysis, we get the following relationship:

$$
\underset{\approx}{\bar{K}}^e \left\{ \underset{\sim}{\bar{W}}^e(\omega) \right\} = \underset{\sim}{\bar{F}}^e
\tag{6.5.71}
$$

where

$$\underset{\approx}{\bar{K}}^e(\omega) = \int_{\Omega^e} \underset{\approx}{N}^T \underset{\approx}{V}^T \underset{\approx}{\bar{D}}(\omega) \underset{\approx}{V} \underset{\approx}{N} dz \qquad (6.5.72)$$

Assembling the element equations into the global one and applying the boundary conditions leads to,

$$\underset{\approx}{\bar{K}}(\omega)\left\{\underset{\sim}{\bar{W}}(\omega)\right\} = \underset{\sim}{\bar{F}} \qquad (6.5.73)$$

where

$$\underset{\sim}{\bar{F}} = (0, \ \bar{F}_{zz}, \ 0, \ 0, \ \dots \ , 0)^T \qquad (6.5.74)$$

If a time varying uniform pressure $f_{zz}(t)$ is applied on the layer surface over a circular region, then

$$\bar{F}_{zz} = \int_{-\infty}^{\infty} \int_0^{\infty} r f_{zz} J_0(\alpha r) e^{-i\omega t} dr dt \qquad (6.5.75)$$

For the Fourier transform, the algorithm of DFT is used. The boundary load $\bar{F}_{zz}$ is expanded to a number of terms with each term corresponding to a frequency. The set of equations (6.5.73) is solved for each frequency and the inverse DFT is applied to the resulted solution for all the frequencies to obtain the resulting solution in the transformed space-time domain. Finally, the inverse Hankel transform is applied to get the real solution.

6.5.3 Method 3: Laplace Domain Analysis

The main idea for the Laplace domain analysis is similar to that of the transformed space-Fourier domain analysis. That is, by applying Laplace transform with respect to the time variable, the problem is converted into an associated elastic problem; FEM is then used to discretize the space domain and the yielding equations are solved for the Laplace variable "s". Then, inverse Laplace transform is applied to get the time domain solution. In this section, only the applied load history whose Laplace transform is available in an analytical form is considered. The inverse Laplace transform developed by Honig and Hirdes (1984) is utilized.

Computer Implementation

The above 1-D viscoelastic FE formulation is implemented in a number of MATLAB programs "**Viscoelastic_1D_FE_Freq.m**", "**Viscoelastic_1D_Laplace.m**", "**Viscoelastic_1D_FE_Time.m**". Then, to examine the preceding algorithms, two test problems are chosen. These two test problems are also analyzed by Booker and Small (1985), where a hybrid Laplace transform/finite layer method is used.

Example 6.5-1: Single Layer Soil under Constant Load

The problem is about a uniform vertical load q applied over a circular region with radius a on the surface of a single layer of viscoelastic material. The load is applied at time $t = 0$ and then maintained constant. The bottom of the layer is rough and rigid, which means zero displacements there, and the thickness of the layer $h = 2a$, as indicated in Figure 6.5-1. The viscoelastic material is of a constant Poisson's ratio of $v = 0.3$ and the relaxation modulus is derived from the generalized Maxwell model as shown in Figure 6.5-1. This can be expressed as,

$$E(t) = E_0 + E_1 e^{-\frac{E_1 t}{\eta_1}} + E_2 e^{-\frac{E_2 t}{\eta_2}} + E_3 e^{-\frac{E_3 t}{\eta_3}} \qquad (6.5.76)$$

The related relaxation modulus at $t = 0$ is noted as $E_{initial}$. By applying the Fourier and Laplace transform to the constitutive relations of the generalized Maxwell model, the corresponding moduli are obtained,

$$\bar{E}(\omega) = E_0 + \frac{i \cdot \omega}{i\omega/E_1 + 1/\eta_1} + \frac{i \cdot \omega}{i\omega/E_2 + 1/\eta_2} + \frac{i \cdot \omega}{i\omega/E_3 + 1/\eta_3} \qquad (6.5.77)$$

$$\bar{\bar{E}}(s) = E_0 + \frac{s}{s/E_1 + 1/\eta_1} + \frac{s}{s/E_2 + 1/\eta_2} + \frac{s}{s/E_3 + 1/\eta_3} \qquad (6.5.78)$$

With the modulus and Poisson's ratio being available, it is easy to get the constitutive matrix and the analyses can be carried out according to the aforementioned three algorithms. It should be pointed out that the Fourier domain analysis is performed over a time duration, T. The loading history over the duration T is sampled at N equally spaced time instances. As T and N increase, the required computer storage and computational time increase too. Therefore, both T and N must be within a range. For the other two algorithms, this problem does not exist because the time domain analysis is a marching forward process, where solution at the next step can be obtained based on the solution at the current step. The Laplace domain analysis permits one to get the solution at a desired time without prior knowledge of its values at all previous time.

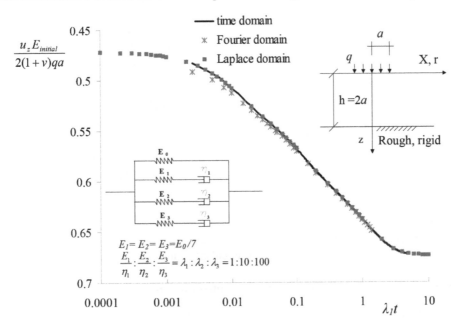

Figure 6.5-1 Time–deflection relationship for constant circular loading on single viscoelastic layer

In Figure 6.5-1, the normalized surface deflection at the loading center u_z vs. time $\lambda_1 t$ is presented based on the results by the three approaches. It is observed that the solutions by the time-domain analysis and the Laplace-domain analysis are in very good agreement. However, there are some minor discrepancies between these two and the Fourier-domain analysis when the time factor is very small. As the time factor increases, results by the three approaches match very well. Stresses and strains beneath the centerline $r = 0$ for $q = 1.0$ and $E_{initial} = 1.0$ are plotted in Figures 6.5-2(a) and 6.5-2(b), respectively. The stress distributions along the vertical depth show no variation with time while the strain distributions vary with the time.

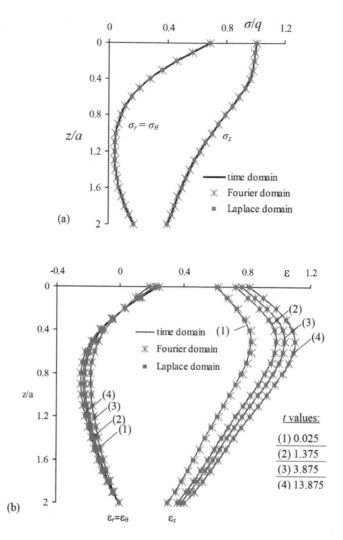

Figure 6.5-2 (a) Stresses at the centerline beneath the circular loading, (b) Strains at the centerline beneath the circular loading

Example 6.5-2: Two-Layered System under Constant Load

In this second problem, a two-layered soil profile with the first layer having viscoelastic properties and the second one with elastic properties is subjected to a vertical circular load. Loading condition and the profile of the layered system are shown in Figure 6.5-3. As far as the viscoelastic properties of the first layer, constant bulk modulus κ is used to describe the volumetric behavior, and the deviatoric behavior is characterized by the viscoelastic model illustrated in Figure 6.5-3. The variable G_0 is defined as the relaxation shear modulus at $t = 0$. The constant bulk modulus K is expressed as a multiple of G_0 as $K = \dfrac{8}{7} G_0$. The material properties for the elastic layer are such that they correspond to the ones of the viscoelastic layer at time $t = 0$, which means the bulk modulus is same as that of the viscoelastic layer and the elastic shear modulus is G_0. In the analysis, the parameter G_0 is chosen to be 2.1, and the load has a unit magnitude. The variations of the results for the surface deflection and stress with the depth as well as the strain-depth distributions are demonstrated in Figures 6.5-3, 6.5-4(a) and 6.5-4(b). It is observed that solutions obtained from the three methods are in a very good match.

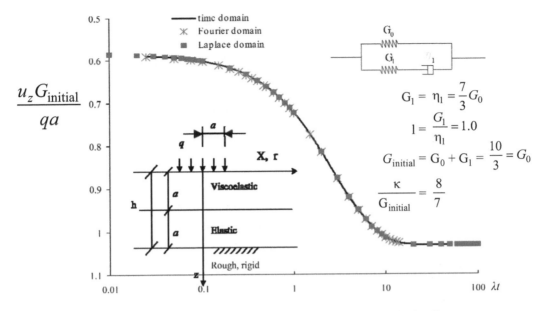

Figure 6.5-3 Time-deflection relationship under constant circular loading on two-layered viscoelastic-elastic system

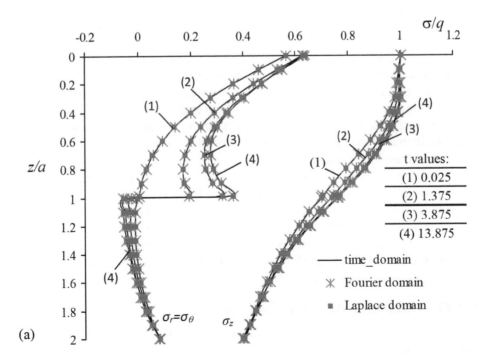

Figure 6.5-4 (a) Stresses on centerline beneath the circular loading for a two-layered system

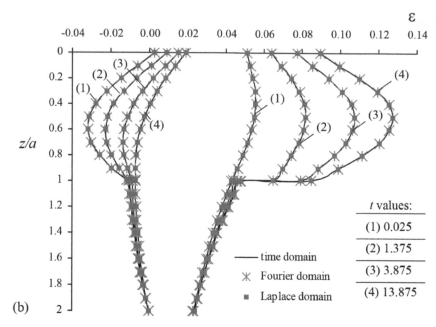

Figure 6.5-4 (b) Strains on centerline beneath the circular loading for a two-layered system

6.6 Viscoplasticity

In simple terms, viscoplasticity pertains to the time-dependent plasticity of material. That is, the material (in this case soil) exhibits permanent deformations while its rigidity moduli and hence, the stress-strain relationship is rate-dependent. One aspect of this time dependency is the phenomenon called *creep* as is discussed previously. Within the context of viscoplasticity, this is shown in Figure 6.6-1. The plots show that the plastic strains evolve over time following the application of loading which is kept constant for a long period of time. The material experiences an ongoing plastic flow which increases at higher stress levels. The high strains corresponding to stresses from high to moderate illustrate a tertiary creep which leads to final rupture of the material. More on this can be found in De Souza Neto et al. (2008).

It is important that one predicts the creep behavior because clays and clay-like soils show such additional time-dependent deformations over time which may induce secondary consolidation and resulting damage to overlying structures. For example, such a need may arise typically in the design and analysis of nuclear reactors and their foundations.

Another aspect of rate dependence as illustrated in Figure 6.6-1, is the phenomenon of stress relaxation which is also discussed in the previous section. The figure shows the evolution of stress in a relaxation test consisting of stretching the specimen (virtually instantaneously) to a prescribed axial strain. Time-dependent response is characterized by the continuous decay of stress in time. Pre-stressed load-carrying components of structures may exhibit stress relaxation which makes it a crucial property to predict.

6.6.1 1-D Formulation of Viscoplasticity

The discussion on viscoplasticity in this section will mainly be about 1-D formulation in terms of modified classical plasticity theory to include time-dependency. Discussion starts off with the presentation of main components of the theory in a hierarchical manner and is valid for

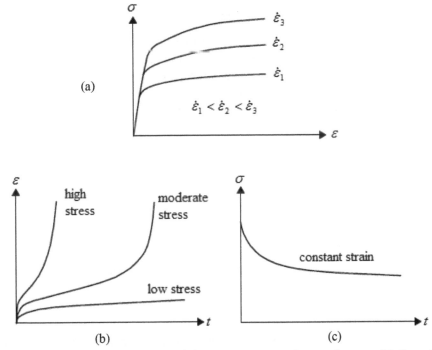

Figure 6.6-1 (a) Uniaxial tensile tests at high temperature at various strain rates, (b) Creep behavior,
(c) Stress relaxation (reproduced after De Souza Neto et al. 2008)

multi-dimensional case also. As is the case for classical rate-independent plasticity, strain decomposition is written as,

$$\varepsilon = \varepsilon^e + \varepsilon^p \qquad (6.6.1)$$

Axial stress is again a linear function of elastic strain through,

$$\sigma = E\varepsilon^e \qquad (6.6.2)$$

As far as the yield function for a 1-D axial loading goes, one can use the conventional form (see 5.4.3) of,

$$f(\sigma) = |\sigma| - \sigma_y \qquad (6.6.3)$$

where σ_y is the yield stress. When the stress is inside the yield surface or the following domain is satisfied, the behavior is purely elastic,

$$\{\sigma | f(\sigma) < 0\} \qquad (6.6.4)$$

It should be mentioned here that some forms of viscoplasticity do not have an elastic domain.

Flow Rule

The main difference in the formulation of the theory of viscoplasticity in comparison to the classical plasticity as described in Ch. 5, is the definition of flow rule. The flow rule is defined solely for evaluation of permanent strains. Going back to section 5.5.3, the definition of (5.5.8) now takes the form,

$$\dot{\varepsilon}^p = \dot{\lambda}\left(\sigma, \sigma_y\right) \text{sign}\left(\sigma\right) \tag{6.6.5}$$

where the plastic strain rate $\dot{\varepsilon}^p$ is now the actual rate of change of plastic strain and which is now affected immensely in time as stress varies. However, that was not the case for rate-independent plasticity where the 'dot' meant a somewhat pseudo time which showed nothing but a change of processes in incremental form and did not have a "real" time meaning. Thus, such a time-dependence was irrelevant. Another main difference lies in the definition of the plastic multiplier $\dot{\lambda}\left(\sigma, \sigma_y\right)$ which, now employs a direct dependence on both stress and its yielding limit. One such definition is given by Peric (1993) as,

$$\dot{\lambda} = \begin{cases} \dfrac{1}{\mu}\left[\left(\dfrac{|\sigma|}{\sigma_y}\right)^{1/\chi} - 1\right] & \xrightarrow{if} f\left(\sigma, \sigma_y\right) \geq 0 \\ \\ 0 & \xrightarrow{if} f\left(\sigma, \sigma_y\right) < 0 \end{cases} \tag{6.6.6}$$

where μ is a 'viscosity-based' time parameter and χ is the non-dimensional 'rate sensitivity' parameter. Both parameters are also temperature dependent.

Hardening Law

Hardening law describes the evolution of yield stress (function or surface in multi-dimensions) as plastic strains develop in rate-independent case. In a viscoplastic model, hardening can be defined in such a manner that the yield stress, σ_y, can be a function of accumulated plastic strain, $\bar{\varepsilon}^p$ as,

$$\sigma_y = \sigma_y\left(\bar{\varepsilon}^p\right) \tag{6.6.7}$$

with

$$\bar{\varepsilon}^p = \int_0^t \left|\dot{\varepsilon}^p\right| dt \tag{6.6.8}$$

Eq. (6.6.6) implies that at a given constant stress σ, an increase in σ_y produces a decrease in the plastic strain rate. As for the classical elastoplastic model, an increase in σ_y is termed as *hardening* whereas its reduction is described as *softening*. If σ_y is constant, the model is referred to as *perfectly viscoplastic*.

Example 6.6-1: What is the total strain of a bar under an axial load that produces a uniform stress $\sigma > \sigma_y$ over the cross-section if the load is applied instantaneously and then stays constant in time? In that case, what is the creep proportional to?

Solution:

Initially ($t = 0$), as the load is applied, the bar deforms purely elastically. Assuming zero initial plastic strain, immediately after the load application, the strain will be,

$$\varepsilon = \varepsilon_0 = \varepsilon_0^e = \frac{\sigma}{E} \tag{6.6.9}$$

Since the stress is kept constant then, elasticity theory dictates that the strain has to be constant as well. Therefore, the following strains will have to be entirely due to viscoplasticity and modeled using the relations above as,

$$\dot{\varepsilon}^p = \frac{1}{\mu}\left[\left(\frac{\sigma}{\sigma_y}\right)^{1/\chi} - 1\right] \tag{6.6.10}$$

Under constant yield stress σ_y, integration of the above relation in time gives the following total strain of the material considering the strain decomposition of (6.6.1) along with (6.6.9) and (6.6.10),

$$\varepsilon(t) = \frac{\sigma}{E} + \frac{1}{\mu}\left[\left(\frac{\sigma}{\sigma_y}\right)^{1/\chi} - 1\right] t \tag{6.6.11}$$

Finally, the creep rate of the material will be constant and proportional to $\left(\dfrac{\sigma}{\sigma_y}\right)^{1/\chi} - 1$.

Stress-Strain Relationship

In viscoplasticity, stress response is clearly dependent on the strain rate, $\dot{\varepsilon}$. For the sake of consistency, if we use the relation (6.6.6) and invert the equation (6.6.11), we get,

$$t(\varepsilon) = \frac{\varepsilon - \dfrac{\sigma}{E}}{\dfrac{1}{\mu}\left[\left(\dfrac{\sigma}{\sigma_y}\right)^{1/\chi} - 1\right]} \tag{6.6.12}$$

which yields the following after some algebraic manipulation,

$$\sigma(t(\varepsilon)) = \sigma_y\left\{1 + \mu\left[\left(\frac{\sigma}{\sigma_y}\right)^{1/\chi} - 1\right]\left[1 - e^{\frac{1}{\mu\alpha}\left(1 - \frac{\varepsilon}{\sigma/E}\right)}\right]\right\} \tag{6.6.13}$$

which can be used to get any stress-strain relationship for an arbitrary strain rate. Figure 6.6-2 shows the response for various β values where

$$\beta = \mu\left[\left(\frac{\sigma}{\sigma_y}\right)^{1/\chi} - 1\right] \tag{6.6.14}$$

In the figure ε^* is the elastic strain and the stresses increase for normalized strain rate β reaching a limit of elastic behavior as is clear. As per limits, $\beta \to 0$ means infinitely slow rates and $\beta \to \infty$ means infinitely fast rates or infinitely viscous material. Such a model seems to capture the experimentally observed behavior illustrated in Figure 6.6-1.

Stress Relaxation at Constant Strain

In order to understand the second phenomenon, stress relaxation under constant strain, we can consider another bar which is under stretch at an infinitely fast strain rate and then kept stretched indefinitely at that constant strain. This way, the material is to experience a stress level above the yield limit. It is expected that the actual stress relaxation behavior of the material as observed in tests (Figure 6.6-1) is captured. Assuming that the plastic strain is zero at $t = 0$ which means during the instantaneous stretching the bar deforms elastically, we write,

$$\sigma_0 = E\varepsilon_0^e = E\varepsilon \tag{6.6.15}$$

Following elastic behavior, stress in the bar is governed by,

$$\sigma = E\left(\varepsilon - \varepsilon^p\right) = \sigma_0 - E\varepsilon^p \tag{6.6.16}$$

where ε^p evolves in time and can be written as,

$$\dot{\varepsilon}^p = \frac{1}{\mu}\left[\left(\frac{\sigma_0 - E\varepsilon^p}{\sigma_y}\right)^{\frac{1}{\chi}} - 1\right] \tag{6.6.17}$$

Example 6.6-2: A bar is under tension and is perfectly viscoplastic with a constant σ_y and $c = 1$. Find a function $\varepsilon^p(t)$ such that $\dot{\varepsilon}^p(t) = c_1 - c_2\varepsilon^p(t)$ with the initial condition $\varepsilon^p(t=0) = 0$.

Solution:

Using (6.6.12), the constants c_1 and c_2 can be written as, $c_1 = \frac{1}{\mu}\left[\frac{\sigma_0}{\sigma_y} - 1\right]$ and $c_2 = \frac{E}{\mu\sigma_y}$. Solution

to the above plastic strain rate function for the given initial condition can be obtained as,

$$\varepsilon^p(t) = \frac{c_1}{c_2}\left(1 - e^{-c_2 t}\right) \tag{6.6.18}$$

Substituting the above relation into (6.6.16) and using the definitions of constants, we get,

$$\sigma(t) = \sigma_0 - \left(\sigma_0 - \sigma_y\right)\left(1 - e^{-\frac{E}{\mu\sigma_y}t}\right) \tag{6.6.19}$$

which describes the stress relaxation process in the bar considering $\sigma = \sigma_0 > \sigma_y$ at $t = 0$.

6.6.2 1-D Rutting Analysis of Concrete Pavement

The above formulation of viscoplasticity can be applied to solve a number of geotechnical engineering problems involving time-dependent behavior, such as the clay soil behavior. It can also be used in the analysis of other engineering materials such as asphalt concrete. For the sake of application of the theoretical arguments, a real-life case study, in this case the rutting phenomenon in asphalt concrete, is investigated. For that, the work of Xu and Rahman (2008) is taken as consideration and the main aspects of the numerical analysis of permanent deformations in the asphalt concrete (AC) are presented.

Rutting is an important factor in flexible pavement design. It has an adverse effect on pavements in terms of influencing the drainage properties leading to reduced frictional strength and thus causing hydroplaning. Rutting stems from the permanent deformation in the pavement layers or the subgrade. Both the top layer of asphalt concrete and the subgrade contribute to rutting in flexible pavements. In the AC layer, the rutting is caused by a combination of densification and shear flow. The initial rutting of the pavement is mainly caused by densification of the granular material in both the AC layer and the subgrade. The subsequent rutting occurs as

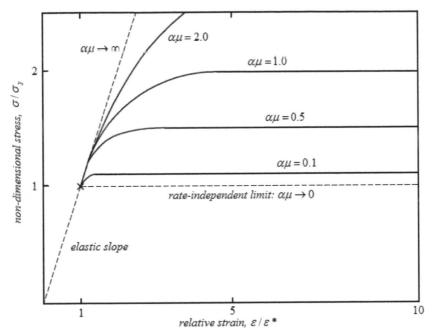

Figure 6.6-2 1-D viscoplastic stress-strain relationships for various β as the strain rate/viscosity parameter (reproduced from De Souza Neto et al. 2008)

a result of shear flow of the mix in the AC layer. In properly compacted pavements, shear flow in the AC layer is the primary cause of rutting, which is analyzed in this section.

Uzan (1996) analyzed the components of the deformation of asphalt concrete based on repeated creep and recovery tests. There, the deformation is divided into four components: plastic and/or viscoplastic components exist under loading only while elastic and viscoelastic components appear under loading/unloading cycles as shown in Figure 6.6-3. The separation of the deformation into components cannot be done directly from the test results. The elastic and plastic components are time independent which develop with the application and removal of the load, without any delay. The elastic one can be obtained directly from the unloading sequence; then the plastic one can be obtained from the loading sequence by subtracting the elastic part from the instantaneous deformation. The viscoelastic and viscoplastic components are time dependent and both develop during the loading; however, only the viscoelastic part recovers during unloading. Uzan (1996) used a viscoelastic model to fit the recovery curve and then subtract the plastic, elastic and viscoelastic component from the loading sequences to get the viscoplastic component. Uzan (1996) proposed a constitutive equation to calculate the deformation under a step loading as,

$$\frac{\varepsilon(t)}{\sigma} = D_e + D_p + D_{ve} \cdot \frac{t^m}{1 + a \cdot t^m} + D_{vp} \cdot t^n \tag{6.6.20}$$

where D_e is the time independent elastic compliance, D_p is the time independent plastic compliance, D_{ve}, a, m are the viscoelastic parameters and D_{vp} and n are the viscoplastic parameters. Here the components of the model can be separated individually. The viscoelastic component is computed by using a power law (Schapery, 1975; Lai, 1976). In the computation of plastic and viscoplastic deformation, it is assumed that both components increase with number of cycles, N, without considering plastic hardening rule or the flow rule.

A 1-D constitutive model is shown in Figure 6.6-3 developed by Xu (2004) for asphalt concrete rutting analysis. In the model, the deformation under a step loading function is expressed as,

$$\frac{\varepsilon(t)}{\sigma} = D_{ve} + D_{ep} + D_{evp} \tag{6.6.21}$$

where D_{ve} is the viscoelastic compliance, D_{ep} is the elastic-plastic compliance and D_{evp} is the elasto-visco-plastic compliance.

As an integrated element, the model in Figure 6.6-3 is highly nonlinear and does not enjoy a numerical solution in its present form. Therefore, a stepwise numerical separation process must be applied in that for the viscoelastic component, and available numerical techniques can be used to solve the recoverable viscoelastic deformation. As for the elastoplastic and elasto-visco-plastic components, iteration algorithms stated in Simo and Hughes (1998) can be utilized to get the deformations provided that the model parameters are determined.

For the model shown in Figure 6.6-4, the total strain is the summation of elastic strain and the visco-plastic strain as,

$$\varepsilon = \varepsilon_e + \varepsilon_{vp} \tag{6.6.22}$$

The stress in the spring can be determined by,

$$\sigma_e = \sigma = E\varepsilon_e \tag{6.6.23}$$

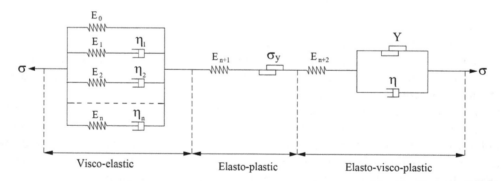

Figure 6.6-3 Schematic model for 1-D pavement analysis

where E is the elasticity modulus. Stress in the slider depends on the value of the yield stress, σ_y. The first time the stress in the slider is greater than the initial yield stress σ_{y0}, viscoplastic deformation starts to develop. Rate of increment depends on the material's strain hardening properties. Here, only the simplest linear isotropic hardening model is considered (Figure 6.6-5). Following yielding, the yield stress is updated as,

$$\sigma_y = \sigma_{y0} + H\varepsilon_{vp} \tag{6.6.24}$$

in which H is the strain hardening parameter and ε_{vp} is the current viscoplastic strain. Therefore, the stress in the slider can be expressed as,

$$\sigma_p = \begin{cases} \sigma \xrightarrow{if} \sigma < \sigma_y \\ \sigma_y \xrightarrow{if} \sigma \geq \sigma_y \end{cases} \tag{6.6.25}$$

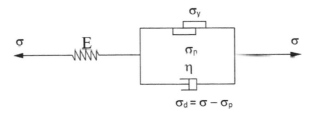

Figure 6.6-4 Elasto-visco-plastic model

In the dashpot, the stress-strain relationship is written as,

$$\sigma_d = \eta \frac{d\varepsilon_{vp}}{dt} \tag{6.6.26}$$

where η is the coefficient of viscosity. Prior to initial yielding, $\varepsilon_{vp} = 0$, therefore the total stress is,

$$\sigma = E\varepsilon \tag{6.6.27}$$

After the initial yielding, the total stress can be written as,

$$\sigma = \sigma_d + \sigma_p \tag{6.6.28}$$

Substituting equations (6.6.25) and (6.6.26) into (6.6.28) yields,

$$\sigma_y + H'\varepsilon_{vp} + \eta \frac{d\varepsilon_{vp}}{dt} = 0 \tag{6.6.29}$$

Substituting equation (6.6.22) into the above equation and using (6.6.23) gives,

$$H'E\varepsilon + \eta E \frac{d\varepsilon}{dt} = H'\sigma + E(\sigma - \sigma_Y) + \eta \frac{d\sigma}{dt} \tag{6.6.30}$$

Introducing the parameter γ as, $\gamma = \dfrac{1}{\eta}$ and plugging it into (6.6.30) results in,

$$\dot{\varepsilon} = \frac{\dot{\sigma}}{E} + \gamma \left[\sigma - \left(\sigma_y + H\varepsilon_{vp} \right) \right] \tag{6.6.31}$$

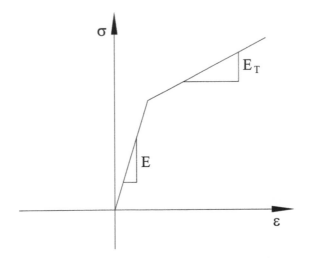

Figure 6.6-5 Elastic linear strain hardening model

which can be rewritten as,

$$\dot{\varepsilon} = \dot{\varepsilon}_e + \dot{\varepsilon}_{vp} \tag{6.6.32}$$

where

$$\dot{\varepsilon}_e = \frac{\dot{\sigma}}{E} \tag{6.6.33}$$

$$\dot{\varepsilon}_{vp} = \gamma \left[\sigma - \left(\sigma_y + H\varepsilon_{vp} \right) \right] \tag{6.6.34}$$

6.7 Exercise Problems

6.7.1 Per Example 6.4-1, consider now a 1-D viscoelastic soil column under a constant stress-rate loading, $\sigma(t) = Rt$ which can be expressed as: $\varepsilon(t) = C_{eff}(t)\,\sigma(t)$. Derive an expression for $C_{eff}(t)$ the constant-rate effective compliance, for a viscoelastic soil.

6.7.2 For a given initial value problem (IVP), obtain the strain and stress relations in time considering a perfectly viscoplastic material (constant σ_y) and $\chi = 1$. Take the total strain as $\varepsilon(t) = \dfrac{\sigma_y}{E} + \alpha t$.

6.7.3 For the example problem of 6.6-2, change the initial plastic strain condition to $\varepsilon^p(t=0) = \varepsilon_0^p$ and redo the problem.

MATLAB® Scripts

Main program ""Viscoelastic_1D_Laplace.m"

```
%-----------------------------------------------------------------------
% Laplace Domain Viscoelastic Analysis of 1-D Soil Layer
% "Viscoelastic_1D_Laplace.m"
%-----------------------------------------------------------------------
clear all
format long;
%-----------------------------------------------------------------------
% Laplace domain time variables
%-----------------------------------------------------------------------
T1= 0.0; % First time value
TN= 2.0; % Last time value
N= 19; % Number of values
IMAN= 0;
ILAPIN= 1;
IKONV= 2;
NS1= 60;
NS2= 30;
ICON= 1;
IKOR= 0;
CON= 1;
IER=0;

for k=1:N
    t(k)=T1+(TN-T1)*k/(N+1);
end

%-----------------------------------------------------------------------
% Laplace Domain Analysis
%-----------------------------------------------------------------------
[H,E]=LAPIN(T1, TN, N, IMAN, ILAPIN, IKONV, NS1, NS2,ICON,IKOR, CON, IER);
y=[t(1:N);H(1,1:N)];

%-----------------------------------------------------------------------
% Print and plot the results
%-----------------------------------------------------------------------
plot (y(2,:), -t(1:N))
title('Laplace transform with respect to t')
xlabel('Strain')
ylabel('z/a')

fid=fopen('out.dat','w+');          % Open a text file in the folder and name
                                    it out.dat
fprintf(fid, '%6.4f   %20.14f\n',y); % Print the outcome variable "y" in that
                                    file
fclose(fid);                        % Close the out.dat file
return
```

Main program "Viscoelastic_1D_FE_Freq.m"

```
%-----------------------------------------------------------------------
% Frequency Domain Viscoelastic Finite Element Analysis of 1-D Soil Layer
% "Viscoelastic_1D_Freq.m"
% Variable descriptions
```

```
% k = element matrix
% f = element vector
% force = node force vector
% kk = system matrix
% ff = system vector
% index = a vector containing system dofs associated with each element
% bcdof = a vector containing dofs associated with boundary conditions
% bcval = a vector containing boundary condition values associated with
%         the dofs in 'bcdof'
%-------------------------------------------------------------------------
clear all
clf
format long
%-----------------------------------
% input data for control parameters
%-----------------------------------
nel=35;                            % number of elements
nnel=2;                            % number of nodes per element
ndof=2;                            % number of dofs per node
nnode=36;                          % total number of nodes in system
sdof=nnode*ndof;                   % total system dofs
%-----------------------------------------
% input data for nodal coordinate values
%-----------------------------------------
gcoord(1)=0;
N=10;
M=15;
P=19;
Q=35;
for i=1:N
    gcoord(i+1)=gcoord(i)+0.01;
end
for i=N+1:M
    gcoord(i+1)=gcoord(i)+0.02;
end
for i=M+1:P
    gcoord(i+1)=gcoord(i)+0.05;
end
for i=P+1:Q
    gcoord(i+1)=gcoord(i)+0.1;
end
%-----------------------------------------------------
% input data for nodal connectivity for each element
%-----------------------------------------------------
for iel=1:nel
    nodes(iel,1)=iel;
    nodes(iel,2)=iel+1;
end
%-----------------------------------------
% input data for material properties
%-----------------------------------------
nmat = 1;                        % total number of materials
mult = 0.1;
E = mult*[1.0*7  1.0   1.0   1.0]; % corresponding to E0, E1, E2, E3 in order
yita = mult*[0    1.0*10 1.0   1.0*0.1];  % first element is 0 for the sake
%of simplicity
v = 0.3;                         % Poisson's ratio
n_component_Maxwell = length(E);
```

```
%-------------------------------------------------
%  assign material property to each element
%-------------------------------------------------
matno=ones(35,1);
%----------------------------------------------------------
%  get Gauss-Quadrature weights and points
%----------------------------------------------------------
[cod,wei]= GAUSS;
%--------------------------------------------------------------------
% set time step parameters
%--------------------------------------------------------------------
dt= 0.025;
loading_t= 13.875;
loading_nt= loading_t/dt+1;
hisload(1:loading_nt)=1.0;

nfft=2048;
mfft= nfft/2;
t=0:dt:nfft*dt;
f=1/dt*(0:mfft)/nfft;              % frequency for Fourier transform
omega= 2*pi*f;
Fload=fft(hisload,nfft);          % Fourier transform of the load
Fiload= ifft(Fload,nfft);
FFload= real(Fiload);
%------------------------------------------------------------------
% input vertical load and output position
%------------------------------------------------------------------
nlds=1;                           % number of loads
vld= [1];                         % vertical load value
xpl= [0.0];                       % x-coord of the load center
ypl= [0.0];                       % y-coord of the load center
radius= [1.0];                    %radius of the load;
nposition=1;                      % number of output positions
xp= [0.0];                        % x-coords of output positions
yp= [0.0];                        % y-coords of output positions
%---------------------------------------------------------------
%  set integration scheme params
%---------------------------------------------------------------
nrho= 10;                         % number of integration sections
drho= 1.495/radius(1);            % length of each section
%-------------------------------------------
% input data for boundary conditions
%-------------------------------------------
ncb=2;                  % number of d.o.f with specified values of 'u'
bcdof= [71 72];         % d.o.f numbers of the d.o.f with specified 'u' values
bcval= [0.0 0.0];       % the corresponding values of 'u'

nfb=1;                  % number of nodes with specified node force
fbdof= [2];             % d.o.f numbers of the d.o.f with specified node force
fbval=[1.0];            % the corresponding values of specified node force
%---------------------------------------------------------------
% initialize stress and displacement
%---------------------------------------------------------------
stressxx= zeros(nnode,nposition,nfft);
stressyy= zeros(nnode,nposition,nfft);
stresszz= zeros(nnode,nposition,nfft);
stressxy= zeros(nnode,nposition,nfft);
stresszy= zeros(nnode,nposition,nfft);
```

```
stressxz= zeros(nnode,nposition,nfft);
dispx= zeros(nnode,nposition,nfft);
dispy= zeros(nnode,nposition,nfft);
dispz= zeros(nnode,nposition,nfft);

stressrr= zeros(nnode,nposition,nfft)
stressoo= zeros(nnode,nposition,nfft);
stresszz= zeros(nnode,nposition,nfft);
strainrr= zeros(nnode,nposition,nfft);
strainoo= zeros(nnode,nposition,nfft);
strainzz= zeros(nnode,nposition,nfft);
stresszz_test= zeros(nnode,nposition,nfft);

%--------------------------------------------------
% march forward in time
%--------------------------------------------------
for iomega=1:mfft+1
    % Associated viscoelastic parameters for certain frequency
    freq = omega(iomega);
    dynamic_modulus=0;
    for jj=1:n_component_Maxwell
        if jj==1
            dynamic_modulus=dynamic_modulus+E(jj);
        else
            dynamic_modulus=dynamic_modulus+complex(E(jj)*(freq*yita(jj))^2,E
            (jj)^2*freq*yita(jj))/...
            (E(jj)^2+(freq*yita(jj))^2);
        end
    end

    a = (1-v)*dynamic_modulus/((1+v)*(1-2*v));
    d = a;
    b = v*dynamic_modulus/((1+v)*(1-2*v));
    c = b;
    f = dynamic_modulus/2/(1+v);
%%%%%%%%%%%%%%%%%%%%%%%%%%%%%%%%%%%%%%%%%%%%%%%%%%%%%%%%%%%%%%%%%%%%%%%%%%
    counter= 0; % the order of the current frequency
    be= 0.0;
    iomega
    for i=1:nrho
        counter=(i-1)*20;
        ay=be;
        be=be+drho;
        for j=1:20
            counter=counter+1;
            rho=0.5*(be-ay)*cod(j)+0.5*(be+ay);
            %----------------------------------------
            % initialization of matrices and vectors
            %----------------------------------------
            force = zeros(sdof,1);  % initialization of node force vector
            ff=zeros(sdof,1);       % initialization of system force vector
            kk=zeros(sdof,sdof);    % initialization of system matrix
            index=zeros(nnel*ndof,1);  % initialization of index vector
            %-----------------------------------------------------------------
            % computation of element matrices and vectors and their assembly
            %-----------------------------------------------------------------
```

```
for iel=1:nel          % loop for the total number of elements
    nprop = matno(iel); % material no. for the element
    nl=nodes(iel,1); nr=nodes(iel,2); % extract nodes for (iel)-
                                      %th element
    xl=gcoord(nl); xr=gcoord(nr);% extract nodal coord values for
                                 %the element
    eleng=xr-xl;           % element length
    index=FEELDOF1(iel,nnel,ndof);% extract system dofs
                                  % associated with element


    [kd]= elematrx_freq(rho,dynamic_modulus,v,eleng)
                    % compute element matrix
    if (nprop==1)
        k=kd;
    end
    f=[0;0;0;0];
    [kk,ff]=FEASMBL2(kk,ff,k,f,index);   % assemble element
                                         matrices and vectors
end
%------------------------------------------------------------
% apply the natural boundary condition at the last node
%------------------------------------------------------------
for ifb=1:nfb
    ff(fbdof(ifb))=ff(fbdof(ifb))+fbval(ifb);
end
%----------------------------
% apply boundary conditions
%----------------------------
[kk,ff]=FEAPLYC2(kk,ff,bcdof,bcval);
%----------------------------
% solve the matrix equation
%----------------------------
uu=kk\ff;
%------------------------------------------------------------
% calculate the force at each node point
%------------------------------------------------------------
wt1 = wei(j)*(be-ay)/2;
for iel=1:nel+1            % loop for the total number of elements to
                          calculate the SSD at each point

    if (iel== nel+1)    % the last point
        kel= nel;
    else
        kel= iel;
    end

    nprop = matno(kel);  % material no. for the point
    nl=nodes(kel,1); nr=nodes(kel,2); % extract nodes for (iel)-
                                      th element
    xl=gcoord(nl); xr=gcoord(nr);% extract nodal coord values for
                                 the element
    eleng=xr-xl;          % element length
    index=FEELDOF1(kel,nnel,ndof);% extract system dofs
                                  associated with element
    [kd]= elematrx_freq(rho,dynamic_modulus,v,eleng);
    if (nprop==1)
        k=kd;
```

```
            end
            ii = index(1);
            jj = index(4);
            force(ii:jj) = k*uu(ii:jj);
            if (iel ~= nel+1)
                rsez = -force(index(1));
                rsz = force(index(2));
                ruz = uu(index(2));
                rue = -uu(index(1));
            else
                rsez= force(index(3));   % To get correct answer, "-" is
                                              needed for 'rsez' and
                rsz= -force(index(4));
                ruz= -uu(index(4));
                rue= uu(index(3));
            end
            for m=1:nlds
                xa=radius(m);
                tt3=rho*xa;
                fj0=besselj(0,tt3);
                fj1=besselj(1,tt3);
                con=fj1*xa*wt1;
                vv=con*vld(m);
                for n=1:nposition
                    xc=xp(n);
                    yc=yp(n);
                    xrel=xc-xpl(m);
                    yrel=yc-ypl(m);
                    radi=(xrel^2+yrel^2)^0.5;
                    thett=0.0;
                    if radi>0
                        thett=atan2(yrel,xrel);
                    end
                    sn=sin(thett);
                    cs=cos(thett);
                    tt1=rho*radi;
                    fj0=besselj(0,tt1);
                    fj1=besselj(1,tt1);
                    djr=rho/2;
                    if radi>0
                        djr=fj1/radi;
                    end
                    fj2=0.0;
                    fj3=0.0;
                    if tt1>0
                        fj2=2*fj1/tt1-fj0;
                        fj3=4*fj2/tt1-fj1;
                    end
                    np = iel;
%%%%%%%%%%%%%%%%%%%%%%%%%%%%%%%%%%%%%%%%%%%%%%%%%%%%%%%%%%%%%%%%%%%%%%%%%%%%%%%%%%%
                    if (np<nel+1)
                        er= vv*(rho*fj0-djr)*(-uu(index(1)));
                        etheta= vv*djr*(-uu(index(1)));
                        ez_freq_domain=(rsz-rho*c*(-uu(index(1))))/d;
                        ez = vv*fj0*ez_freq_domain;
                    else
                        er= vv*(rho*fj0-djr)*(-uu(index(3)));
                        etheta= vv*djr*(-uu(index(3)));
```

```
                        ez_freq_domain=(rsz-rho*c*(-uu(index(3))))/d;
                        ez = vv*fj0*ez_freq_domain;
                  end
                  sigmarr= a*er + b*etheta + c*ez;
                  sigmaoo= b*er + a*etheta + c*ez;
                  sigmazz= c*er + c*etheta + d*ez;
                  strainrr(np,n,iomega)= strainrr(np,n,iomega)+er;
                  strainoo(np,n,iomega)= strainoo(np,n,iomega)+etheta;
                  strainzz(np,n,iomega)= strainzz(np,n,iomega)+ez;
                  stressrr(np,n,iomega)= stressrr(np,n,iomega)+sigmarr;
                  stressoo(np,n,iomega)= stressoo(np,n,iomega)+sigmaoo;
                  stresszz_test(np,n,iomega)= stresszz_
                  test(np,n,iomega)+sigmazz;
%%%%%%%%%%%%%%%%%%%%%%%%%%%%%%%%%%%%%%%%%%%%%%%%%%%%%%%%%%%%%%%%%%%%%%%%%%%%%
                  sz=vv*rsz*fj0;
                  sro=0;
                  soz=0;
                  szr= - vv*rsez*fj1;
                  disr= - vv*(rue*fj1);
                  diso=0;
                  disz= vv*ruz*fj0;
                  %    rotate back -theta to x-y direction
                  sn=sin(-thett);
                  cs=cos(-thett);
                  c2t=cs*cs-sn*sn;
                  s2t=2.*sn*cs;
                  txz=cs*szr;
                  tzy=-sn*szr;
                  ux=disr*cs+diso*sn;
                  uy=-disr*sn+diso*cs;
                  stresszz(np,n,iomega)= stresszz(np,n,iomega)+sz;
                  stresszy(np,n,iomega)= stresszy(np,n,iomega)+tzy;
                  stressxz(np,n,iomega)= stressxz(np,n,iomega)+txz;
                  dispx(np,n,iomega)= dispx(np,n,iomega)+ux;
                  dispy(np,n,iomega)= dispy(np,n,iomega)+uy;
                  dispz(np,n,iomega)= dispz(np,n,iomega)+disz;

            end     % 'for n=1:nposition'
         end        % for 'm=1:nlds'

      end     % 'for iel=1:nel'

   end     % 'for j=1:20'

  end      % 'for i=1:NRHO'

end     % 'for it=1:mfft+1"
%------------------------------------------------
% Calculate stresses and strains
%------------------------------------------------
for n=1:nposition
    for np=1:nel+1
        for iomega=1:nfft
            if (iomega <= mfft+1)
                transfer_stresszz(iomega)= Fload(iomega)*stresszz(np,n,iome
                ga);
```

```
            transfer_stressrr(iomega)= Fload(iomega)*stressrr(np,n,iome
            ga);
            transfer_stressoo(iomega)= Fload(iomega)*stressoo(np,n,iome
            ga);
            transfer_stresszz_test(iomega)=Fload(iomega)*stresszz_
            test(np,n,iomega);
            transfer_strainzz(iomega)= Fload(iomega)*strainzz(np,n,iome
            ga);
            transfer_strainoo(iomega)= Fload(iomega)*strainoo(np,n,iome
            ga);
            transfer_strainrr(iomega)= Fload(iomega)*strainrr(np,n,iome
            ga);
            transfer_dispz(iomega)=Fload(iomega)*dispz(np,n,iomega);
        else
            transfer_stresszz(iomega)= conj(transfer_stresszz(nfft-
            iomega+2));
            transfer_stressrr(iomega)= conj(transfer_stressrr(nfft-
            iomega+2));
            transfer_stressoo(iomega)= conj(transfer_stressoo(nfft-
            iomega+2));
            transfer_stresszz_test(iomega)= conj(transfer_stresszz_
            test(nfft-iomega+2));
            transfer_strainzz(iomega)= conj(transfer_strainzz(nfft-
            iomega+2));
            transfer_strainoo(iomega)= conj(transfer_strainoo(nfft-
            iomega+2));
            transfer_strainrr(iomega)= conj(transfer_strainrr(nfft-
            iomega+2));
            transfer_dispz(iomega)=conj(transfer_dispz(nfft-iomega+2));
        end
    end
    response_stresszz = ifft(transfer_stresszz,nfft);
    response_stressrr = ifft(transfer_stressrr,nfft);
    response_stressoo = ifft(transfer_stressoo,nfft);
    response_stresszz_test = ifft(transfer_stresszz_test,nfft);
    response_strainzz = ifft(transfer_strainzz,nfft);
    response_strainoo = ifft(transfer_strainoo,nfft);
    response_strainrr = ifft(transfer_strainrr,nfft);
    response_dispz= ifft(transfer_dispz,nfft);
    for it=1:nfft
        stresszz(np,n,it)= real(response_stresszz(it));
        stressoo(np,n,it)= real(response_stressoo(it));
        stressrr(np,n,it)= real(response_stressrr(it));
        stresszz_test(np,n,it)= real(response_stresszz_test(it));
        strainzz(np,n,it)= real(response_strainzz(it));
        strainoo(np,n,it)= real(response_strainoo(it));
        strainrr(np,n,it)= real(response_strainrr(it));
        dispz(np,n,it)= real(response_dispz(it));
    end
    end
end
%------------------------------------------------------------
% print and plot the results
% ------------------------------------------------------------
figure(1)
plot(stresszz(1:36,1,1),-gcoord(1:36))
grid on
```

```
xlabel('stress/q')
ylabel('z/a')

figure(2)
plot(strainzz(1:36,1,1),-gcoord(1:36),strainzz(1:36,1,56),-gcoord(1:36),strai
nzz(1:36,1,156),...
            -gcoord(1:36),strainzz(1:36,1,556),-
            gcoord(1:36),strainrr(1:36,1,1),-gcoord(1:36),...
            strainrr(1:36,1,56),-gcoord(1:36),strainrr(1:36,1,156),-gcoord(1:
            36),strainrr(1:36,1,556),...
            -gcoord(1:36))
grid on
title('Fourier transform with respect to t')
xlabel('strain')
ylabel('z/a')

figure(3)
plot(stresszz(1:36,1,1),-gcoord(1:36),stresszz(1:36,1,56),-gcoord(1:36),stres
szz(1:36,1,156),...
            -gcoord(1:36),stresszz(1:36,1,556),-
            gcoord(1:36),stressrr(1:36,1,1),-gcoord(1:36),...
            stressrr(1:36,1,56),-gcoord(1:36),stressrr(1:36,1,156),-gcoord(1:
            36),stressrr(1:36,1,556),...
            -gcoord(1:36))
grid on
xlabel('stress/q')
ylabel('z/a')
title('Fourier transform with respect to t')
fid= fopen('strains.dat', 'w+')

for it=1:nfft
    time=[dt*(it-1)];
    fprintf(fid, 'Vertical section strain distribution at time
    %10.5f\n',time);
    fprintf(fid, 'depth_z  strainzz  strainoo  strainrr\n');
    for j=1:nnode
        temp = strainzz(j,1,it);
        epsiz(j)=temp;
        temp= strainoo(j,1,it);
        epsio(j)=temp;
        temp= strainrr(j,1,it);
        epsir(j)= temp;
    end
    out= [gcoord(1:nnode)'; epsiz(1:nnode); epsio(1:nnode); epsir(1:nnode)];
    fprintf(fid, '%6.4f    %20.14f   %20.14f    %20.14f\n',out);
end
fclose(fid);

fid=fopen('deflection.dat','w+');
uz(1:nfft)=dispz(1,1,1:nfft);
y=[t(1:nfft);uz(1:nfft)];
fprintf(fid, '%6.4f    %20.14f\n',y);
fclose(fid);

fid= fopen('stresses.dat', 'w+');
for it=1:nfft
    time=[dt*(it-1)];
    fprintf(fid, 'Vertical section stresses distribution at time
```

```
%10.5f\n',time);
    fprintf(fid, 'depth_z        stresszz        stressoo      stressrr\n');
    for j=1:nnode
        temp = stresszz(j,1,it);
        sigz(j)=temp;
        temp= stressoo(j,1,it);
        sigo(j)=temp;
        temp= stressrr(j,1,it);
        sigr(j)= temp;
    end
    out= [gcoord(1:nnode)'; sigz(1:nnode); sigo(1:nnode); sigr(1:nnode)];
    fprintf(fid, '%6.4f    %20.14f    %20.14f    %20.14f\n',out);
end
fclose(fid);
%-------------------------------------------------------------------------
```

Main program "Viscoelastic_1D_FE_Time.m"

```
%-------------------------------------------------------------------------
% Time Domain Viscoelastic Analysis of 1-D Soil Layer
% "Viscoelastic_1D_FE_Time.m"
% Variable descriptions
%   k = element matrix
%   f = element vector
%   force = node force vector
%   kk = system matrix
%   ff = system vector
%   index = a vector containing system dofs associated with each element
%   bcdof = a vector containing dofs associated with boundary conditions
%   bcval = a vector containing boundary condition values associated with
%           the dofs in 'bcdof'
%-------------------------------------------------------------------------
clear all
clf
format long
%-----------------------------------
% input data for control parameters
%-----------------------------------
nel=35              % number of elements
nnel=2;             % number of nodes per element
ndof=2;             % number of dofs per node
nnode=36;           % total number of nodes in system
sdof=nnode*ndof;    % total system dofs
%---------------------------------------
% input data for nodal coordinate values
%---------------------------------------
gcoord(1)=0;
N=10;
M=15;
P=19;
Q=35;
for i=1:N
    gcoord(i+1)=gcoord(i)+0.01;
end
for i=N+1:M
    gcoord(i+1)=gcoord(i)+0.02;
end
for i=M+1:P
```

```
        gcoord(i+1)=gcoord(i)+0.05;
end
for i=P+1:Q
    gcoord(i+1)=gcoord(i)+0.1;
end
%----------------------------------------------
% input data for nodal connectivity for each element
%----------------------------------------------
for iel=1:nel
    nodes(iel,1)=iel;
    nodes(iel,2)=iel+1;
end
%----------------------------------------
% input data for material properties
%----------------------------------------
nmat = 1      % total number of materials
              % Generalized Maxwell model- viscoelastic parameters
E =[1.0*0.7   1.0*0.7/7    1.0*0.7/7    1.0*0.7/7 ];
nE= length(E);
sumE=0;
for jj=1:nE
    sumE= sumE+E(jj);
end
poisson= 0.3;
lamda=[ 0.0    0.1    1.0    10.0];    %Each element of the array is the ratio
                                       %between
                                       %the elastic modulus of the spring and
                                       %the viscosity of the dashpot
%----------------------------------------------
% assign material property to each element
%----------------------------------------------
matno=ones(35,1);
%------------------------------------------------
% call function gauss to get gauss weights and points
%------------------------------------------------
[cod,wei]= GAUSS;
%---------------------------------------------------------
%   set time step parameters
%---------------------------------------------------------
deltat= 0.025;
t= 0.0:deltat:50;
nt= length(t);
%---------------------------------------------------------
% input vertical load and output position
%---------------------------------------------------------
nlds=1;                          % number of loads
vld= [1.0];                      % vertical load value
xpl= [0.0];                      % x-coord of the load center
ypl= [0.0];                      % y-coord of the load center
radius= [1.0];                   %radius of the load;
nposition=1;                     % number of output positions
xp= [0.0];                       % x-coords of output positions
yp= [0.0];                       % y-coords of output positions
%---------------------------------------------------------
% initialize stress and displacement
%---------------------------------------------------------
stressxx= zeros(nnode,nposition,nt);
stressyy= zeros(nnode,nposition,nt);
```

```
stresszz= zeros(nnode,nposition,nt);
stressxy= zeros(nnode,nposition,nt);
stresszy= zeros(nnode,nposition,nt);
stressxz= zeros(nnode,nposition,nt);
dispx= zeros(nnode,nposition,nt);
dispy= zeros(nnode,nposition,nt);
dispz= zeros(nnode,nposition,nt);
stressrr= zeros(nnode,nposition,nt);
stressoo= zeros(nnode,nposition,nt);
stresszz= zeros(nnode,nposition,nt);
stresszz_test= zeros(nnode,nposition,nt);
strainrr= zeros(nnode,nposition,nt);
strainoo= zeros(nnode,nposition,nt);
strainzz= zeros(nnode,nposition,nt);
% -------------------------------------------------------------
% set integration scheme parameter
%-------------------------------------------------------------
nrho= 10;                     % number of integration sections
drho= 1.495/radius(1);        % length of each section
u0= zeros(sdof,nrho*20);      % store the displacement in spacial frequency
                              % domain at previous time step.
%--------------------------------------------------------
% march forward in time
%--------------------------------------------------------
for it=1:nt
    counter= 0;               % the order of the current frequency
    be= 0.0;
    for i=1:nrho
        counter= (i-1)*20;
        ay=be;
        be=be+drho;
        for j=1:20
            counter= counter+1;
            rho=0.5*(be-ay)*cod(j)+0.5*(be+ay);
            %-----------------------------------
            % input data for boundary conditions
            %-----------------------------------
            ncb=2;                % number of d.o.f with specified values of 'u'
            bcdof= [71  72];      % d.o.f numbers of the d.o.f with specified 'u'
                                    values
            bcval= [0.0 0.0];     % the corresponding values of 'u'
            nfb=1;                % number of nodes with specified node force
            fbdof= [2];           % d.o.f numbers of the d.o.f with specified
                                    node force
            fbval=[1.0];          % the corresponding values of specified node
                                    force
            %-----------------------------------------
            % initialization of matrices and vectors
            %-----------------------------------------
            force = zeros(sdof,1);    % initialization of node force vector
            ff=zeros(sdof,1);         % initialization of system force
                                        vector
            kk=zeros(sdof,sdof);      % initialization of system matrix
            index=zeros(nnel*ndof,1); % initialization of index vector
            %------------------------------------------------------------------
            %  computation of element matrices and vectors and their assembly
```

```
%----------------------------------------------------------------
for iel=1:nel          % loop for the total number of elements
    nprop = matno(iel); % material no. for the element
    nl=nodes(iel,1); nr=nodes(iel,2); % extract nodes for (iel)-
                                      th element
    xl=gcoord(nl); xr=gcoord(nr);   % extract nodal coord values
                                      for the element
    eleng=xr-xl;                    % element length
    index=feeldof1(iel,nnel,ndof); % extract system dofs
                                      associated with element
    k_parameter=elematrx(rho,poisson,eleng); % compute element
                                               matrix parameters
    f=[0;0;0;0];
    %  if material is visco-elastic
    if nprop== 1
        k0=k_parameter*sumE;
        k=k0;
        if t(it)== t(2)
            kd1= k_parameter*E(2)*lamda(2);
            kd2= k_parameter*E(3)*lamda(3);
            kd3= k_parameter*E(4)*lamda(4);
            k=k0-(kd1+kd2+kd3)*deltat/2;
            f= deltat/2*kd1*exp(-lamda(2)*deltat)*u0(index(1):ind
            ex(4),counter)+...
                deltat/2*kd2*exp(-lamda(3)*deltat)*u0(index(1):ind
                ex(4),counter)+...
                deltat/2*kd3*exp(-lamda(4)*deltat)*u0(index(1):ind
                ex(4),counter);
        end
        if t(it)> t(2)
            kd1= k_parameter*E(2)*lamda(2);
            kd2= k_parameter*E(3)*lamda(3);
            kd3= k_parameter*E(4)*lamda(4);
            k= k0-(kd1+kd2+kd3)*deltat/2;
            f= kd1*exp(-lamda(2)*deltat)*(deltat/2*u0(index(1):in
            dex(4),counter)+ r1(index(1):index(4),counter))...
                +kd2*exp(-lamda(3)*deltat)*(deltat/2*u0(index(1):i
                ndex(4),counter)+r2(index(1):index(4),counter))...
                +kd3*exp(-lamda(4)*deltat)*(deltat/2*u0(index(1):i
                ndex(4),counter)+r3(index(1):index(4),counter));
        end

    end  % corresponding to ' if(nprop== 1)'
    [kk,ff]=feasmbl2(kk,ff,k,f,index);  % assemble element
                                          matrices and vectors
end
%-----------------------------------------------------------
% apply the natural boundary condition at the last node
%-----------------------------------------------------------
for ifb=1:nfb
    ff(fbdof(ifb))=ff(fbdof(ifb))+fbval(ifb);
end
%---------------------------
% apply boundary conditions
%---------------------------
[kk,ff]=feaplyc2(kk,ff,bcdof,bcval);
%---------------------------
%  solve the stiffness equation
```

```
%---------------------------
uu=kk\ff;
%-----------------------------------------------------------------
------
% calculate force at each node point
%-----------------------------------------------------------------
------
wt1 = wei(j)*(be-ay)/2;

for iel=1:nel+1          % loop for the total number of elements
    if (iel== nel+1)     % the last point
        kel= nel;
    else
        kel= iel;
    end
    nprop = matno(kel); % material no. for the element
    nl=nodes(kel,1); nr=nodes(kel,2); % extract nodes for (iel)-
                                    th element
    xl=gcoord(nl); xr=gcoord(nr);  % extract nodal coord values
                                    for the element
    eleng=xr-xl;                   % element length
    index=feeldof1(kel,nnel,ndof); % extract system dofs
                                    associated with element
    k_parameter=elematrx(rho,poisson,eleng); % compute element
                                    matrix
    ii = index(1);
    jj = index(4);
        if nprop==1
            k0=k_parameter*sumE;
            k=k0;
        if t(it)==t(1)
            force(ii:jj) =k*uu(ii:jj);
        end

        if t(it)== t(2)
            kd1= k_parameter*E(2)*lamda(2);
            kd2= k_parameter*E(3)*lamda(3);
            kd3= k_parameter*E(4)*lamda(4);
            k=k0-(kd1+kd2+kd3)*deltat/2;
            f= deltat/2*kd1*exp(-lamda(2)*deltat)*u0(index(1):ind
            ex(4),counter)+...
                deltat/2*kd2*exp(-lamda(3)*deltat)*u0(index(1):ind
                ex(4),counter)+...
                deltat/2*kd3*exp(-lamda(4)*deltat)*u0(index(1):ind
                ex(4),counter);
            force(ii:jj) =k*uu(ii:jj)-f;
        end

        if t(it)> t(2)
            kd1= k_parameter*E(2)*lamda(2);
            kd2= k_parameter*E(3)*lamda(3);
            kd3= k_parameter*E(4)*lamda(4);
            k= k0-(kd1+kd2+kd3)*deltat/2;
            f= kd1*exp(-lamda(2)*deltat)*(deltat/2*u0(index(1):in
            dex(4),counter)+ r1(index(1):index(4),counter))+...
                kd2*exp(-lamda(3)*deltat)*(deltat/2*u0(index(1)
                :index(4),counter)+ r2(index(1):index(4),count
                er))+...
```

```
                            kd3*exp(-lamda(4)*deltat)*(deltat/2*u0(index(1):in
                            dex(4),counter)+ r3(index(1):index(4),counter));
                 force(ii:jj) =k*uu(ii:jj)-f;
          end
    end                        % corresponding to 'if(nprop== 1)'
    if (iel~= nel+1)
         rsez= -force(index(1));   % To get correct answer, "-" is
                                        needed for 'rsez' and
         rsz= force(index(2));     % 'rue', why?
         ruz= uu(index(2));
         rue= -uu(index(1));
    else
         rsez= force(index(3));
         rsz= -force(index(4));
         ruz= -uu(index(4));
         rue= uu(index(3));
    end
    for m=1:nlds
         xa= radius(m);
         tt3= rho*xa;
         fj0= besselj(0,tt3);
         fj1= besselj(1,tt3);
         con= fj1*xa*wt1;
         vv= con*vld(m);
         for n=1:nposition
              xc= xp(n);
              yc= yp(n);
              xrel= xc-xpl(m);
              yrel= yc-ypl(m);
              radi= (xrel^2+yrel^2)^0.5;
              thett= 0.0;
              if radi>0
                   thett= atan2(yrel,xrel);
              end
              sn= sin(thett);
              cs= cos(thett);
              tt1= rho*radi;
              fj0= besselj(0,tt1);
              fj1= besselj(1,tt1);
              djr= rho/2;
              if radi>0
                   djr= fj1/radi;
              end
              fj2= 0.0;
              fj3= 0.0;
              if tt1>0
                   fj2= 2*fj1/tt1-fj0;
                   fj3= 4*fj2/tt1-fj1;
              end
              np = iel;   % the point number to be analyzed
              %%%%%%%%%%%%%%%%%%%%%%%%%%%%%%%%%%%%%%%%%%%%%%%%%%%%%%%
              if (np < nel+1)
                   er= vv*(rho*fj0-djr)*(-uu(index(1)));
                   etheta= vv*djr*(-uu(index(1)));
```

```
        else
            er= vv*(rho*fj0-djr)*(-uu(index(3)));
            etheta= vv*djr*(-uu(index(3)));
        end

        if (t(it)==t(1) & nprop==1)
            a= (1-poisson)/(1+poisson)/(1-2*poisson)*sumE;
            d= a;
            b= poisson/(1+poisson)/(1-2*poisson)*sumE;
            c=b;
            ez_freq_domain= (rsz-rho*c*(-uu(index(1))))/d;
        if (np==nel+1)
            ez_freq_domain= (rsz-rho*c*(-uu(index(3))))/d;
        end
            ez = vv*fj0*ez_freq_domain;
            sigmarr= a*er+ b*etheta+ c*ez;
            sigmaoo= b*er+ a*etheta+ c*ez;
            sigmazz= c*er+ c*etheta+ d*ez;
        end

        if (t(it)==t(2) && nprop==1)

            if (np==nel+1)
                index_no= index(3);
            else
                index_no= index(1);
            end
            c0= poisson/(1+poisson)/(1-2*poisson)*sumE;
            d0= (1-poisson)/(1+poisson)/(1-2*poisson)*sumE;
            c1_prime= poisson/(1+poisson)/
            (1-2*poisson)*E(2)*lamda(2);
            d1_prime= (1-poisson)/(1+poisson)/
            (1-2*poisson)*E(2)*lamda(2);
            rr_1(np,counter)= deltat/2*(exp(-
            lamda(2)*deltat)*(-u0(index_no,counter))-uu(index_
            no));
            c2_prime= poisson/(1+poisson)/
            (1-2*poisson)*E(3)*lamda(3);
            d2_prime= (1-poisson)/(1+poisson)/
            (1-2*poisson)*E(3)*lamda(3);
            rr_2(np,counter)= deltat/2*(exp(-
            lamda(3)*deltat)*(-u0(index_no,counter))-uu(index_
            no));

            c3_prime= poisson/(1+poisson)/
            (1-2*poisson)*E(4)*lamda(4);
            d3_prime= (1-poisson)/(1+poisson)/
            (1-2*poisson)*E(4)*lamda(4);
            rr_3(np,counter)= deltat/2*(exp(-
            lamda(4)*deltat)*(-u0(index_no,counter))-uu(index_
            no));
```

```
numerator= rsz- rho*c0*(-uu(index_no))+ rho*c1_
prime*rr_1(np,counter)+...
    d1_prime*deltat/2*exp(-lamda(2)*deltat)*ez0_
    freq_domain(np,counter)+...
    rho*c2_prime*rr_2(np,counter)+...
    d2_prime*deltat/2*exp(-lamda(3)*deltat)*ez0_
    freq_domain(np,counter)+...
    rho*c3_prime*rr_3(np,counter)+...
    d3_prime*deltat/2*exp(-lamda(4)*deltat)*ez0_
    freq_domain(np,counter);
denominator= d0-d1_prime*deltat/2-d2_
prime*deltat/2-d3_prime*deltat/2;
ez_freq_domain= numerator/denominator;
% the bottom point is added later on
ez= vv*fj0*ez_freq_domain;
rz_1(np,counter)= deltat/2*(exp(-
lamda(2)*deltat)*ez0_freq_domain(np,counter)+ez_
freq_domain);
rz_2(np,counter)= deltat/2*(exp(-
lamda(3)*deltat)*ez0_freq_domain(np,counter)+ez_
freq_domain);
rz_3(np,counter)= deltat/2*(exp(-
lamda(4)*deltat)*ez0_freq_domain(np,counter)+ez_
freq_domain);
convolution_er_1(np,counter)= deltat/2*(exp(-
lamda(2)*deltat)*er_previous_time(np,counter)+er);
convolution_etheta_1(np,counter)=
deltat/2*(exp(-lamda(2)*deltat)*etheta_previous_
time(np,counter)+etheta);
convolution_ez_1(np,counter)= deltat/2*(exp(-
lamda(2)*deltat)*ez_previous_time(np,counter)+ez);

convolution_er_2(np,counter)= deltat/2*(exp(-
lamda(3)*deltat)*er_previous_time(np,counter)+er);
convolution_etheta_2(np,counter)=
deltat/2*(exp(-lamda(3)*deltat)*etheta_previous_
time(np,counter)+etheta);
convolution_ez_2(np,counter)= deltat/2*(exp(-
lamda(3)*deltat)*ez_previous_time(np,counter)+ez);

convolution_er_3(np,counter)= deltat/2*(exp(-
lamda(4)*deltat)*er_previous_time(np,counter)+er);
convolution_etheta_3(np,counter)=
deltat/2*(exp(-lamda(4)*deltat)*etheta_previous_
time(np,counter)+etheta);
convolution_ez_3(np,counter)= deltat/2*(exp(-
lamda(4)*deltat)*ez_previous_time(np,counter)+ez);

sigmarr= d0*er+ c0*etheta+ c0*ez - d1_
prime*convolution_er_1(np,counter)-c1_prime...
    *convolution_etheta_1(np,counter)- c1_
    prime*convolution_ez_1(np,counter)-...
    d2_prime*convolution_er_2(np,counter)-c2_
    prime*convolution_etheta_2(np,counter)-...
    c2_prime*convolution_ez_2(np,counter)-...
    3_prime*convolution_er_3(np,counter)-c3_
    prime*convolution_etheta_3(np,counter)-...
```

```
                        c3_prime*convolution_ez_3(np,counter);
              sigmaoo= c0*er+ d0*etheta+ c0*ez - c1_
              prime*convolution_er_1(np,counter)-d1_prime...
                   *convolution_etheta_1(np,counter)- c1_
                   prime*convolution_ez_1(np,counter)-...
                   c2_prime*convolution_er_2(np,counter)-d2_
                   prime*convolution_etheta_2(np,counter)-...
                   c2_prime*convolution_ez_2(np,counter)-...
                   c3_prime*convolution_er_3(np,counter)-d3_
                   prime*convolution_etheta_3(np,counter)-...
                   c3_prime*convolution_ez_3(np,counter);
              sigmazz= c0*er+ c0*etheta+ d0*ez - c1_
              prime*convolution_er_1(np,counter)-c1_prime...
                   *convolution_etheta_1(np,counter)- d1_
                   prime*convolution_ez_1(np,counter)-...
                   c2_prime*convolution_er_2(np,counter)-c2_
                   prime*convolution_etheta_2(np,counter)-...
                   d2_prime*convolution_ez_2(np,counter)-...
                   c3_prime*convolution_er_3(np,counter)-c3_
                   prime*convolution_etheta_3(np,counter)-...
                   d3_prime*convolution_ez_3(np,counter);
         end
    if (t(it)>t(2) & nprop==1)
         if (np==nel+1)
              index_no= index(3);
         else
              index_no= index(1);
         end
         c0= poisson/(1+poisson)/(1-2*poisson)*sumE;
         d0= (1-poisson)/(1+poisson)/(1-2*poisson)*sumE;
         c1_prime= poisson/(1+poisson)/
         (1-2*poisson)*E(2)*lamda(2);
         d1_prime= (1-poisson)/(1+poisson)/
         (1-2*poisson)*E(2)*lamda(2);
         c2_prime= poisson/(1+poisson)/
         (1-2*poisson)*E(3)*lamda(3);
         d2_prime= (1-poisson)/(1+poisson)/
         (1-2*poisson)*E(3)*lamda(3);
         c3_prime= poisson/(1+poisson)/
         (1-2*poisson)*E(4)*lamda(4);
         d3_prime= (1-poisson)/(1+poisson)/
         (1-2*poisson)*E(4)*lamda(4);
         numerator= rsz-rho*c0*(-uu(index_no)) + rho*c1_
         prime*(exp(-lamda(2)*deltat)*(-r1(index_
         no,counter))+...
              deltat/2*(exp(-lamda(2)*deltat)*(-u0(index_
              no,counter))-uu(index_no)))+ d1_prime*...
              (exp(-lamda(2)*deltat)*rz_1(np,counter)+
              deltat/2*exp(-lamda(2)*deltat)*...
              ez0_freq_domain(np,counter))+rho*c2_
              prime*(exp(-lamda(3)*deltat)*(-r2(index_
              no,counter))+...
              deltat/2*(exp(-lamda(3)*deltat)*(-u0(index_
              no,counter))-uu(index_no)))+...
              d2_prime*(exp(-
              lamda(3)*deltat)*rz_2(np,counter)+
              deltat/2*exp(-lamda(3)*deltat)*...
              ez0_freq_domain(np,counter))+rho*c3_
```

```
            prime*(exp(-lamda(4)*deltat)*(-r3(index_
            no,counter))+...
            deltat/2*(exp(-lamda(4)*deltat)*(-u0(index_
            no,counter))-uu(index no)))+...
            d3_prime*(exp(-
            lamda(4)*deltat)*rz_3(np,counter)+
            deltat/2*exp(-lamda(4)*deltat)*...
            ez0_freq_domain(np,counter));
denominator= d0 - d1_prime*deltat/2-d2_
prime*deltat/2-d3_prime*deltat/2;
ez_freq_domain= numerator/denominator;
ez= vv*fj0*ez_freq_domain;
rz_1(np,counter)= deltat/2*(exp(-
lamda(2)*deltat)*ez0_freq_domain(np,counter)+ez_
freq_domain)...
            + exp(-lamda(2)*deltat)* rz_1(np,counter);
rz_2(np,counter)= deltat/2*(exp(-
lamda(3)*deltat)*ez0_freq_domain(np,counter)+ez_
freq_domain)...
            + exp(-lamda(3)*deltat)* rz_2(np,counter);
rz_3(np,counter)= deltat/2*(exp(-
lamda(4)*deltat)*ez0_freq_domain(np,counter)+ez_
freq_domain)...
            + exp(-lamda(4)*deltat)* rz_3(np,counter);
convolution_er_1(np,counter)= deltat/2*(exp(-
lamda(2)*deltat)*er_previous_
time(np,counter)+er)...
            + exp(-lamda(2)*deltat)*convolution_
               er_1(np,counter);
convolution_etheta_1(np,counter)=
deltat/2*(exp(-lamda(2)*deltat)*etheta_previous_
time(np,counter)+etheta)...
            + exp(-lamda(2)*deltat)*convolution_
               etheta_1(np,counter);
convolution_ez_1(np,counter)= deltat/2*(exp(-
lamda(2)*deltat)*ez_previous_
time(np,counter)+ez)...
            + exp(-lamda(2)*deltat)*convolution_
               ez_1(np,counter);
convolution_er_2(np,counter)= deltat/2*(exp(-
lamda(3)*deltat)*er_previous_
time(np,counter)+er)...
            + exp(-lamda(3)*deltat)*convolution_
               er_2(np,counter);
convolution_etheta_2(np,counter)=
deltat/2*(exp(-lamda(3)*deltat)*etheta_previous_
time(np,counter)+etheta)...
            + exp(-lamda(3)*deltat)*convolution_
               etheta_2(np,counter);
convolution_ez_2(np,counter)= deltat/2*(exp(-
lamda(3)*deltat)*ez_previous_
time(np,counter)+ez)...
            + exp(-lamda(3)*deltat)*convolution_
               ez_2(np,counter);
convolution_er_3(np,counter)= deltat/2*(exp(-
lamda(4)*deltat)*er_previous_
```

```
                time(np,counter)+er)...
                    + exp(-lamda(4)*deltat)*convolution_
                      er_3(np,counter);
                convolution_etheta_3(np,counter)=
                deltat/2*(exp(-lamda(4)*deltat)*etheta_previous_
                time(np,counter)+etheta)...
                    + exp(-lamda(4)*deltat)*convolution_
                      etheta_3(np,counter);
                convolution_ez_3(np,counter)= deltat/2*(exp(-
                lamda(4)*deltat)*ez_previous_
                time(np,counter)+ez)...
                    + exp(-lamda(4)*deltat)*convolution_
                      ez_3(np,counter);
                sigmarr= d0*er+ c0*etheta+ c0*ez - d1_
                prime*convolution_er_1(np,counter)-c1_prime...
                    *convolution_etheta_1(np,counter)- c1_
                    prime*convolution_ez_1(np,counter)-...
                    d2_prime*convolution_er_2(np,counter)-c2_
                    prime*convolution_etheta_2(np,counter)-...
                    c2_prime*convolution_ez_2(np,counter)-...
                    d3_prime*convolution_er_3(np,counter)-c3_
                    prime*convolution_etheta_3(np,counter)-...
                    c3_prime*convolution_ez_3(np,counter);
                sigmaoo= c0*er+ d0*etheta+ c0*ez - c1_
                prime*convolution_er_1(np,counter)-d1_prime...
                    *convolution_etheta_1(np,counter)- c1_
                    prime*convolution_ez_1(np,counter)-...
                    c2_prime*convolution_er_2(np,counter)-d2_
                    prime*convolution_etheta_2(np,counter)-...
                    c2_prime*convolution_ez_2(np,counter)-...
                    c3_prime*convolution_er_3(np,counter)-d3_
                    prime*convolution_etheta_3(np,counter)-...
                    c3_prime*convolution_ez_3(np,counter);
                sigmazz= c0*er+ c0*etheta+ d0*ez - c1_
                prime*convolution_er_1(np,counter)-c1_prime...
                    *convolution_etheta_1(np,counter)- d1_
                    prime*convolution_ez_1(np,counter)-...
                    c2_prime*convolution_er_2(np,counter)-c2_
                    prime*convolution_etheta_2(np,counter)-...
                    d2_prime*convolution_ez_2(np,counter)-...
                    c3_prime*convolution_er_3(np,counter)-c3_
                    prime*convolution_etheta_3(np,counter)-...
                    d3_prime*convolution_ez_3(np,counter);
            end
            % store the current strains for next time step
              calculation.
            ez0_freq_domain(np,counter)= ez_freq_domain;
            ez_previous_time(np,counter)= ez;
            er_previous_time(np,counter)= er;
            etheta_previous_time(np,counter)= etheta;

            strainrr(np,n,it)= strainrr(np,n,it)+er;
            strainoo(np,n,it)= strainoo(np,n,it)+etheta;
            strainzz(np,n,it)= strainzz(np,n,it)+ez;
            stressrr(np,n,it)= stressrr(np,n,it)+ sigmarr;
```

```
            stressoo(np,n,it)= stressoo(np,n,it)+ sigmaoo;
            stresszz_test(np,n,it)= stresszz_
            test(np,n,it)+sigmazz;
%%%%%%%%%%%%%%%%%%%%%%%%%%%%%%%%%%%%%%%%%%%%%%%%%%%%%%%%%%%%%%
            sr= vv*((rho*rue+bb*rsz)*fj0/cc-xm1*rue*djr);
            so= vv*((rho*rue*xm2+bb/cc*rsz)*fj0+xm1*rue*djr);
            sz= vv*rsz*fj0;
            sro=0;
            soz=0;
            szr= -vv*rsez*fj1;
            disr= -vv*(rue*fj1);
            diso=0;
            disz= vv*ruz*fj0;
            % rotate back -theta to x-y direction
            sn= sin(-thett);
            cs= cos(-thett);
            c2t= cs*cs-sn*sn;
            s2t= 2.*sn*cs;
            sxx= (sr+so)/2+(sr-so)*c2t/2+sro*s2t;
            syy= (sr+so)/2-(sr-so)*c2t/2-sro*s2t;
            txz= cs*szr;
            tzy= -sn*szr;
            txy= -(sr-so)*s2t/2;
            ux= disr*cs+diso*sn;
            uy= -disr*sn+diso*cs;
            stressxx(np,n,it)= stressxx(np,n,it)+sxx;
            stressyy(np,n,it)= stressyy(np,n,it)+syy;
            stresszz(np,n,it)= stresszz(np,n,it)+sz;
            stressxy(np,n,it)= stressxy(np,n,it)+txy;
            stresszy(np,n,it)= stresszy(np,n,it)+tzy;
            stressxz(np,n,it)= stressxz(np,n,it)+txz;
            dispx(np,n,it)= dispx(np,n,it)+ux;
            dispy(np,n,it)= dispy(np,n,it)+uy;
            dispz(np,n,it)= dispz(np,n,it)+disz;
            end         % corresponding to 'for n=1:nposition'
        end             % corresponding to 'm=1:nlds'
    end                 % corresponding to 'for iel=1:nel'

if t(it)==t(1)
    u0( :,counter)= uu;
elseif t(it)==t(2)
    r1(:,counter)= deltat/2*(exp(-lamda(2)*deltat)*u0(
      :,counter)+uu);
    r2(:,counter)= deltat/2*(exp(-lamda(3)*deltat)*u0(
      :,counter)+uu);
    r3(:,counter)= deltat/2*(exp(-lamda(4)*deltat)*u0(
      :,counter)+uu);
    u0( :,counter)=uu;
else
    r1(:,counter)= exp(-lamda(2)*deltat)*r1(:,counter)+
      deltat/2*(exp(-lamda(2)*deltat)*u0(:,counter)+uu);
    r2(:,counter)= exp(-lamda(3)*deltat)*r2(:,counter)+
      deltat/2*(exp(-lamda(3)*deltat)*u0(:,counter)+uu);
```

```
                    r3(:,counter)= exp(-lamda(4)*deltat)*r3(:,counter)+
                       deltat/2*(exp(-lamda(4)*deltat)*u0(:,counter)+uu);
                    u0(:,counter)=uu;
                end
            end                             %corresponding to 'for j=1:20'
          end                               %corresponding to 'for i=1:NRHO'
       end                                  % corresponding to 'for it=1:nt"
%--------------------------------------------------------------
% print and plot the results
% --------------------------------------------------------------
figure(1)
plot(stresszz(1:35,1,1),-gcoord(1:35))
title('Stress Variation')
xlabel('Strain')
ylabel('z/a')

fid= fopen('deflection', 'w+');
uz(1:nt)=dispz(1,1,1:nt);
y=[t(1:nt);uz(1:nt)];
fprintf(fid,'%6.4f %20.14f\n',y);
fclose(fid);

figure(2)
plot(strainzz(1:36,1,1),-gcoord(1:36),strainzz(1:36,1,56),-gcoord(1:36),strai
nzz(1:36,1,156),...
                   -gcoord(1:36),strainzz(1:36,1,556),-
                   gcoord(1:36),strainrr(1:36,1,1),-gcoord(1:36),...
                   strainrr(1:36,1,56),-gcoord(1:36),strainrr(1:36,1,156),-
                   gcoord(1:36),strainrr(1:36,1,556),...
                   -gcoord(1:36))
grid on
title('Direct Time Domain Analyses - One Layer')
xlabel('strain')
ylabel('z/a')

figure(3)
plot(stresszz(1:36,1,1),-gcoord(1:36),stresszz(1:36,1,56),-gcoord(1:36),stres
szz(1:36,1,156),...
                   -gcoord(1:36),stresszz(1:36,1,556),-
                   gcoord(1:36),stressrr(1:36,1,1),-gcoord(1:36),...
                   stressrr(1:36,1,56),-gcoord(1:36),stressrr(1:36,1,156),-
                   gcoord(1:36),stressrr(1:36,1,556),...
                   -gcoord(1:36))
grid on
xlabel('stress/q')
ylabel('z/a')
title('Direct Time Domain Analyses - One Layer')

%--------------------------------------------------------------
fid= fopen('strains.dat', 'w+')
for it=1:nt
    time=[deltat*(it-1)];
    fprintf(fid, 'Vertical section strain distribution at time
    %10.5f\n',time);
    fprintf(fid, 'depth_z          strainzz           strainoo
    strainrr\n');
```

```
    for j=1:nnode
        temp = strainzz(j,1,it);
        epsiz(j)=temp;
        temp= strainoo(j,1,it);
        epsio(j)=temp;
        temp= strainrr(j,1,it);
        epsir(j)= temp;
    end
    out= [gcoord(1:nnode)'; epsiz(1:nnode); epsio(1:nnode); epsir(1:nnode)];
    fprintf(fid, '%6.4f    %20.14f  %20.14f    %20.14f\n',out);
end
fclose(fid);

fid= fopen('stresses.dat', 'w+');
for it=1:nt
    time=[deltat*(it-1)];
    fprintf(fid, 'Vertical section stresses distribution at time
    %10.5f\n',time);
    fprintf(fid, 'depth_z        stresszz         stressoo
    stressrr\n');
    for j=1:nnode
        temp = stresszz(j,1,it);
        sigz(j)=temp;
        temp= stressoo(j,1,it);
        sigo(j)=temp;
        temp= stressrr(j,1,it);
        sigr(j)= temp;
    end
    out= [gcoord(1:nnode)'; sigz(1:nnode); sigo(1:nnode); sigr(1:nnode)];
    fprintf(fid, '%6.4f    %20.14f  %20.14f    %20.14f\n',out);
end
fclose(fid);
```

```
%--------------------------------------------------------------
```

Subprogram "FEAPLYC2"

```
function [kk,ff]=FEAPLYC2(kk,ff,bcdof,bcval)

%--------------------------------------------------------
% Purpose:
% Apply constraints to matrix equation [kk]{x}={ff}
%
% Synopsis:
% [kk,ff]=feaplybc(kk,ff,bcdof,bcval)
%
% Variable Description:
% kk - system matrix before applying constraints
% ff - system vector before applying constraints
% bcdof - a vector containging constrained d.o.f
% bcval - a vector containing contained value
%--------------------------------------------------------

    n=length(bcdof);
    sdof=size(kk);
```

```
    for i=1:n
        c=bcdof(i);
        for j=1:sdof
            kk(c,j)=0;
        end

        kk(c,c)=1;
        ff(c)=bcval(i);
    end

end
```

%--

Subprogram "FEASMBL2.m"

```
function [kk,ff]=FEASMBL2(kk,ff,k,f,index)
%------------------------------------------------------------
% Purpose:
% Assembly of element matrices into the system matrix &
% Assembly of element vectors into the system vector
%
% Synopsis:
% [kk,ff]=feasmbl2(kk,ff,k,f,index)
%
% Variable Description:
% kk - system matrix
% ff - system vector
% k  - element matrix
% f  - element vector
% index - d.o.f. vector associated with an element
%------------------------------------------------------------
    edof = length(index);
    for i=1:edof
        ii=index(i);
            ff(ii)=ff(ii)+f(i);
            for j=1:edof
                jj=index(j);
                    kk(ii,jj)=kk(ii,jj)+k(i,j);
            end
    end

end
```

%--

Subprogram "FEELDOF1.m"

```
function [index]=FEELDOF1(iel,nnel,ndof)
%------------------------------------------------------------
% Purpose:
% Compute system dofs associated with each element in one-
% dimensional problem
%
% Synopsis:
% [index]=feeldof1(iel,nnel,ndof)
%
% Variable Description:
```

```
% index - system dof vector associated with element "iel"
% iel - element number whose system dofs are to be determined
% nnel - number of nodes per element
% ndof - number of dofs per node

%-----------------------------------------------------------
edof = nnel*ndof;
start = (iel-1)*(nnel-1)*ndof;
for i=1:edof
    index(i)=start+i;
end

end

%-----------------------------------------------------------------
```

Subprogram "GAUSS.m"

```
function [COD,WEI]=GAUSS
        COD(11) = 0.0765265211;
        COD(12) = 0.2277858511;
        COD(13) = 0.3737060887;
        COD(14) = 0.5108670019;
        COD(15) = 0.6360536807;
        COD(16) = 0.7463319064;
        COD(17) = 0.8391169718;
        COD(18) = 0.9122344282;
        COD(19) = 0.9639719272;
        COD(20) = 0.9931285991;

        WEI(11) = 0.1527533871;
        WEI(12) = 0.1491729864;
        WEI(13) = 0.1420961093;
        WEI(14) = 0.1316886384;
        WEI(15) = 0.1181945319;
        WEI(16) = 0.1019301198;
        WEI(17) = 0.0832767415;
        WEI(18) = 0.0626720483;
        WEI(19) = 0.0406014298;
        WEI(20) = 0.0176140071;

for i= 1:10
    j= 20 - i + 1;
    WEI(i) = WEI(j);
    COD(i) = -COD(j);
end

end

%-------------------------------------------------------------------------
```

Subprogram "LAPIN.m"

```
function [H,E]=LAPIN(T1, TN, N, IMAN, ILAPIN, IKONV, NS1, NS2, ICON, IKOR,
CON, IER)
% Laplace transform calculation of the time dependent variables
global TA  TB  T0  CONOPT  ABSF  LVAL  HMONO
H= zeros(6,N);
E= zeros(3,NS1);
```

```
IER=0;
if (TN < T1)
    IER =1;
end

if (N < 1)
    IER = IER+10;
end
if (IMAN==0)
    if (NS1~=60)
        IER = IER+100000;
    end
    if (IER~=0)
        ERROR(IER);
        return;
    end
    ILAPIN =1;
    IKONV =2;
    ICON =1;
    IKOR = 0;
    NS2= 0;
elseif (IMAN== 1)
    if (ILAPIN<1 | ILAPIN > 2)
        IER= IER+1000;
    end
    if (IKONV<1 | IKONV>2)
        IER= IER+10000;
    end
    if (NS1<1 | (NS2<1 & (IKOR ==1)))
        IER=IER+100000;
    end
    if (ICON<0 |ICON>2|(ICON==2 & ILAPIN==2))
        IER= IER+1000000;
    end
    if (IKOR<0 | IKOR>1)
        IER = IER+10000000;
    end
    if (ICON==0 & CON< 0.0)
        IER= IER+100000000 ;
    end
    if (IER~=0)
        ERROR(IER);
        return;
    end
else
    IER= IER+100;
    ERROR(IER);
    return;
end

CON1= 20.0;
CON2= CON1-2.0;
ABSF = 0.0;
J3= (3-ICON)/2- mod((3-ICON)/2,1)+ N*((ICON/2)-mod(ICON/2,1));
TA = T1;
TB= TN;
if ( (ICON-1)< 0 )
    for  L3=1:J3
```

```
            LVAL =L3;
            HMONO =0;
            KOR1= IKOR;
            JUMP=0;
            [H,E]= LAPIN2(T1,TN,N,ILAPIN,IKONV,NS1,NS2,ICON,IKOR,CON,H,E,JUMP);
      end
end

if ( (ICON-1)>=0 )
      for L3=1:J3
            LVAL =L3;
            HMONO =0;
            KOR1= IKOR;
            % COMPUTATION OF THE OPTIMAL PARAMETERS
            if ((ICON-1)==1)
                  TA = T1+ L3*(TN-T1)/(N+1);
                  TB= TA;
            end
            NH= N/2-mod(N/2,1);
            T = TA+(TB-TA)*NH/(N+1);
            TK = (2-ILAPIN)*T + (ILAPIN-1)*TB;
            % COMPUTATION OF THE TRUNCTION ERROR (RNSUM)
            T0= T;
            CON= CON1;
            JUMP= 1;
            [H,E]= LAPIN2(T1,TN,N,ILAPIN,IKONV,NS1,NS2,ICON,IKOR,CON,H,E,JUMP);
            FN = H(1,L3);
            FNS1= E(1,NS1);
            CON= CON2;
            JUMP=2;
            [H,E]= LAPIN2(T1,TN,N,ILAPIN,IKONV,NS1,NS2,ICON,IKOR,CON,H,E,JUMP);
            if (~(FN~= H(1,L3) & FNS1~= E(1,NS1)))
                  CONOPT = CON;
                  ABSF =0.0;
                  if (CONOPT <= 0.0)
                        CONOPT=1.0;
                  end
                  JUMP= 6;
                  [H,E]= LAPIN2(T1,TN,N,ILAPIN,IKONV,NS1,NS2,ICON,IKOR,CON,H,E,JU
                  MP);
            else
                  RNSUM= TK*(FN-H(1,L3))/(exp(CON1)-exp(CON2));
                  if (ILAPIN~=2)
                        % COMPUTAION OF THE ACCELERATION FACTOR (DEL)
                        RACC = T*(FNS1-E(1,NS1))/(exp(CON1)-exp(CON2));
                        DEL = RNSUM/RACC;
                  end

                  if (IKOR~=1)
                  % OPTIMAL PARAMETERS (METHOD A)
                        T0= 2.0*TK +T;
                        CON = CON1/4.0;
                        JUMP=3;
                        [H,E]= LAPIN2(T1,TN,N,ILAPIN,IKONV,NS1,NS2,ICON,IKOR,CON,H,E,
                        JUMP);
                        FN = H(1,L3);
                        CONOPT= -TK/(2.0*TK+T)*log(abs(RNSUM/(TK*FN)));
```

```
        else
% OPTIMAL PARAMETERS FOR THE KORREKTUR METHOD (METHOD A)
            T0= 4.0*TK+T;
            CON = CON1/4.0;
            JUMP=4;
            [H,E]= LAPIN2(T1,TN,N,ILAPIN,IKONV,NS1,NS2,ICON,IKOR,CON,H,E,
            JUMP);
            FN = H(1,L3);
            T0= 8.0*TK+T;
            JUMP = 5;
            [H,E]= LAPIN2(T1,TN,N,ILAPIN,IKONV,NS1,NS2,ICON,IKOR,CON,H,E,
            JUMP);
            FN = FN-H(1,L3);
            CONOPT= -TK/(4.0*TK+T)*log(abs(RNSUM/(TK*FN)));
        end

% OPTIMAL PARAMETERS (METHOD B)
        if (ILAPIN~=1)
            ABSF = abs(exp(CONOPT)*RNSUM*2.0/TK);
            if (CONOPT <= 0.0)
                CONOPT=1.0;
            end
            JUMP= 6;
            [H,E]= LAPIN2(T1,TN,N,ILAPIN,IKONV,NS1,NS2,ICON,IKOR,CON,H,E,
            JUMP);
        else
            V1= CON1/T;
            V2= CONOPT/T;
            W= pi*NS1/T;
            [FREAL, FIMAG]= FEM1D(V2,W);
            RNSUMK = RNSUM*FREAL;
            [FREAL, FIMAG]= FEM1D(V1,W);
            RNSUMK = RNSUMK/FREAL;
            ABRN= (RNSUMK- RNSUM)/(V2-V1);
            CONOPT= -log(abs((ABRN/T+RNSUMK)/(T*(IKOR*2+2)*FN)))/
            (3+2*IKOR);
            V1=V2;
            V2= CONOPT/T;
            [FREAL,FIMAG]= FEM1D(V2,W);
            RNSUMK= RNSUMK*FREAL;
            [FREAL,FIMAG]= FEM1D(V1,W);
            RNSUMK= RNSUMK/FREAL;
            ABSF= exp(CONOPT)/T*abs(RNSUMK)+ abs(exp(-
            2.0*CONOPT)*FN*(IKOR-1)+ exp(-4.0*CONOPT)*FN*IKOR);
            if (CONOPT <= 0.0)
                CONOPT=1.0;
            end
            JUMP= 6;
            [H,E]= LAPIN2(T1,TN,N,ILAPIN,IKONV,NS1,NS2,ICON,IKOR,CON,H,E,
            JUMP);
        end            % corresponding to if (ILAPIN~=1)
    end                % corresponding to if (~(FN~= H(1,L3) & FNS1~=
                       %   E(1,NS1)))
  end                  % for  L3=1, J3
end                    % if ( (ICON-1)>=0 )

end

%-----------------------------------------------------------------------
```

Subprogram "LAPIN2.m"

```
function [H,E]= LAPIN2(T1,TN,N,ILAPIN,IKONV,NS1,NS2,ICON,IKOR,CON,H,E,JUMP)
%
%   LAPLACE - INVERSION WITH OPTICAL PARAMETERS
%
global TA TB T0 CONOPT ABSF LVAL HMONO

if (JUMP~=0 & JUMP>=6)
    T0=TA;
    TT=TB;
    I1=LVAL;
    J1=(2-ICON)*N+(ICON-1)*LVAL;
    CON=CONOPT;
    KOR1=IKOR;
elseif (JUMP==0)
    T0=T1;
    TT=TN;
    I1=1;
    J1=N;
    KOR1=IKOR;
    JUMP=6;
elseif (JUMP<6)
    KOR1=0;
    TT=T0;
    I1=LVAL;
    J1=LVAL;
end

DELTA=(TT-T0)/(N+1);
NSUM= NS1;
% COMPUTATION OF THE T-VALUE FROM T0,TT
for   K2=1:2
    if (ILAPIN~=1)
        TE= K2*TT;
        V=CON/TE;
        [FREAL, FIMAG]= FEM1D(V,0.0);
        RAL= -0.50*FREAL;
        PITE=pi/TE;
        for L=1:NSUM
            W = (L-1)*PITE;
            [FREAL,FIMAG]= FEM1D(V,W);
            E(2,L)=FREAL;
            E(3,L) = FIMAG;
        end
    end

    for K1=I1:J1
        TL= T0+K1*DELTA;
        if (ILAPIN~=2)
            if (K2==2)
                TL=3.0*TL;
            end
            V=CON/TL;
            FAKTOR=exp(V*TL)/TL;
            [FREAL,FIMAG]= FEM1D(V,0.0);
            RAL=-0.50*FREAL;
            PIT=pi/TL;
```

```
        EINS=1.0;
        SURE=0.0;

        % METHOD OF DURBIN
        for L=1:NSUM
            W=(L-1)*PIT;
            [FREAL,FIMAG]= FEM1D(V,W);
            SURE= SURE+FREAL*EINS;
            EINS=-EINS;
            E(1,L)= FAKTOR*(RAL+SURE);
        end
elseif (ILAPTN==2)
    if(K2==2)
        TL= TL+TE;
    end
    FAKTOR= exp(V*TL)/TE;
    SURE= 0.0;
    SUIM= 0.0;
    for L=1: NSUM
        W=(L-1)*PITE;
        SURE= SURE+E(2,L)*cos(W*TL);
        SUIM= SUIM+ E(3,L)*sin(W*TL);
        E(1,L)= FAKTOR*(RAL+SURE-SUIM);
    end
end
% SEARCH FOR STATIONARY VALUES
NMAX= NSUM*2/3-mod(NSUM*2/3,1);
MONOTO =0;
K=0;
if ((E(1,NSUM)-E(1,NSUM-1))<0)
    RICHTA= -1;
elseif ((E(1,NSUM)-E(1,NSUM-1))>0)
    RICHTA= 1;
end
FLAGK=0;
for L=1:NMAX
    J=NSUM-L;
    if ((E(1,J)-E(1,J-1))<0)
        RICHT=-1;
    elseif ((E(1,J)-E(1,J-1))>0)
        RICHT= 1;
    end
    if (RICHT~=RICHTA)
        K= K+1;
        E3(K) = E(1,J);
        RICHTA= RICHT;
        if (K==3)
            FLAGK=1;
            break;
        end
    end
end
if (K==0)
%C CURVE FITTING (CFM)
    X1= NSUM-2.0;
    X2= NSUM-1.0;
    X3= NSUM-0.0;
    Y1= E(1,NSUM-2);
```

```
        Y2= E(1,NSUM-1);
        Y3= E(1,NSUM);
        B= ((Y3-Y1)*X3*X3*(X1+X2)/(X1-X3)-(Y2-Y1)*X2*X2*(X1+X3)/(X1-X2))/
        (X3-X2);
        A= ((Y2-Y1)-B*(X1-X2)/(X1+X2))*(X2*X2*X1*X1)/(X1*X1-X2*X2);
        H(K2,K1)= Y1-(A/X1+B)/X1;
        MONOTO=1;
        if (K2==1)
            H(4,K1)=3;
        end
    elseif (K~=3)
        H(K2,K1) = E(1, NSUM);
        if(K2==1)
            H(4,K1)=0;
        end
    elseif (K==3 & FLAGK==1)
        KE=2;
        if ((E3(KE)-E(1,J))*RICHTA >0.0)
            H(K2,K1)=(E3(1)+E3(3))/4.0+E3(2)/2.0;
            if(K2==1)
                H(4,K1)=1;
            end
        else
            JMIN= NSUM/3- mod(NSUM/3,1);
            JMAX=J-1;
            FLAG=0;   % set a flag to tell going out of loop
            for JJ=JMIN:JMAX
                J=J-1;
                if ((E(1,J)-E(1,J-1))<0)
                    RICHT=-1;
                elseif ((E(1,J)-E(1,J-1))>0)
                    RICHT= 1;
                end
                if(RICHT~=RICHTA)
                    RICHTA= RICHT;
                    KE=3-KE;
                    if ((E3(KE)-E(1,J))*RICHTA > 0.0)
                        FLAG =1;
                        break;
                    end
                end
            end
            if ((E3(KE)-E(1,J))* RICHTA > 0.0 & FLAG==1 )
            %C MINIMUM-MAXIMUM METHOD (MINIMAX)
                H(K2,K1)=(E3(1)+E3(3))/4.0+E3(2)/2.0;
                if(K2==1)
                    H(4,K1)=1;
                end
            else
                MONOTO=1;
                if (IKONV~=2)
                    H(K2,K1)=(E3(1)+E3(3))/4.0+E3(2)/2.0;
                    if(K2==1)
                        H(4,K1)=1;
```

```
                    end
                end
                if (IKONV==2) % EPSILON-ALGORITHM (EPAL)
                    K=0;
                    NSUMM1= NSUM-1;
                    E2= E(1,1);
                    for L=1:NSUMM1
                        E1=E(1,1);
                        TM=0.0;
                        LP1= L+1;
                        for M=1:L
                            MM= LP1-M;
                            TM1= E(1,MM);
                            DIVI= E(1,MM+1)-E(1,MM);
                            if (abs(DIVI)> 1.0e-20)
                                E(1,MM)= TM+1.0/DIVI;
                                TM=TM1;
                            else
                                K=L;
                                break;
                            end
                        end
                        if (K==L)
                            break;
                         end
                        E2=E1;
                    end

                    if (abs(E1)> abs(E2))
                        E1=E2;
                    end
                    if (abs(E(1,1))>abs(E1))
                        E(1,1)=E1;
                    end
                    H(K2,K1)=E(1,1);
                    if (K2==1)
                        H(4,K1)=K+2;
                    end
                end                    % corresponding to if (IKONV==2)
            end                        % corresponding to if ((E3(KE)-E(1,J))*
                                       %   RICHTA > 0.0 & FLAG==1 )
        end                            % corresponding to if ((E3(KE)-
                                       %   E(1,J))*RICHTA >0.0)
    end                                % corresponding to if (K==0)

    H(3,K1)= CON;
    H(5,K1)= ABSF;
    HMONO= HMONO+K2*MONOTO*10^(6-JUMP);
    if (JUMP>=6)
        H(6,K1)= (2-K2)*HMONO+ (K2-1)*(2*MONOTO)+H(6,K1);
        HMONO= (HMONO/10-mod(HMONO/10,1))*10;
    end
end  % corresponding to for K1=I1:J1

if (KOR1==0)
    return;
end
if (K2==2)
```

```
            break;
        end
        NSUM = NS2;
end     % corresponding to for K2=1:2

%C  CORRECTOR METHOD
FAKTOR= -exp(-2.0*CON);
for K=I1:J1
    H(1,K)= H(1,K)+FAKTOR*H(2,K);
end

end
```

%--

Subprogram "FEM1D.m"

```
function [freal, fimag]= FEM1D(sr, si)
%-----------------------------------------------------------------------------
% Tailored 1-D FEM Code for Viscoelastic Analysis in Laplace Domain
%-----------------------------------------------------------------------------
% Variable descriptions
% k = element matrix
% f = element vector
% force = node force vector
% kk = system matrix
% ff = system vector
% index = a vector containing system dofs associated with each element
% bcdof = a vector containing dofs associated with boundary conditions
% bcval = a vector containing boundary condition values associated with
%           the dofs in 'bcdof'
%-----------------------------------------------------------------------------
s= complex(sr, si);
%-------------------------------------
% input data for control parameters
%-------------------------------------
nel=35;             % number of elements
nnel=2;             % number of nodes per element
ndof=2;             % number of dofs per node
nnode=36;           % total number of nodes in system
sdof=nnode*ndof;    % total system dofs
%---------------------------------------
%  input data for nodal coordinate values
%---------------------------------------
gcoord(1)=0;
N=10;
M=15;
P=19;
Q=35;
for i=1:N
    gcoord(i+1)=gcoord(i)+0.01;
end
for i=N+1:M
    gcoord(i+1)=gcoord(i)+0.02;
```

```
end
for i=M+1:P
    gcoord(i+1)=gcoord(i)+0.05;
end
for i=P+1:Q
    gcoord(i+1)=gcoord(i)+0.1;
end
%----------------------------------------------------
% input data for nodal connectivity for each element
%----------------------------------------------------
for iel=1:nel
    nodes(iel,1)=iel;
    nodes(iel,2)=iel+1;
end
%---------------------------------------
% input data for material properties
%---------------------------------------
nmat = 1;              % total number of materials
E =[969000*0.7  969000*0.7/7  969000*0.7/7  969000*0.7/7];
n_component= length(E);
yita= [0  969000  96900  9690];
poisson= 0.3;
%-----------------------------------------------
% assign material property to each element
%-----------------------------------------------
matno=ones(35,1);
%--------------------------------------------------------
%  call function gauss to get gauss weights and points
%--------------------------------------------------------
[cod,wei]= GAUSS;
%--------------------------------------------------------------
% input vertical load and output position
%--------------------------------------------------------------
nlds=1;          % number of loads
vld= [1];        % vertical load value
xpl= [0.0];      % x-coord of the load center
ypl= [0.0];      % y-coord of the load center
radius= [1.0];   %radius of the load;
nposition=1;     % number of output positions
xp= [0.0];       % x-coords of output positions
yp= [0.0];       % y-coords of output positions
%--------------------------------------------------------------
% Initialize stresses and displacements in Laplace domain for a certain s
%--------------------------------------------------------------
stressxx= zeros(nnode,nposition);
stressyy= zeros(nnode,nposition);
stresszz= zeros(nnode,nposition);
stressxy= zeros(nnode,nposition);
stresszy= zeros(nnode,nposition);
stressxz= zeros(nnode,nposition);
dispx= zeros(nnode,nposition);
dispy= zeros(nnode,nposition);
dispz= zeros(nnode,nposition);

stressrr= zeros(nnode,nposition);
stressoo= zeros(nnode,nposition);
stresszz= zeros(nnode,nposition);
strainrr= zeros(nnode,nposition);
strainoo= zeros(nnode,nposition);
```

```
strainzz= zeros(nnode,nposition);
stresszz_test= zeros(nnode,nposition);
%---------------==    --==-------------------------------------------
% set integration scheme
%-----------------------------------------------------------------
nrho= 10;   % number of integration sections
drho= 1.495/radius(1);   % length of each section
% Associated viscoelastic parameters for certain s
a = (1-poisson)/(1+ poisson)/(1-2*poisson)*E(1)*(1+1/7*s/(0.1+s)+1/7*s/
(1+s)+1/7*s/(10+s));
d = a;
b = poisson/(1+poisson)/(1-2*poisson)*E(1)*(1+1/7*s/(0.1+s)+1/7*s/
(1+s)+1/7*s/(10+s));
c = b;
f = 1/2/(1+poisson)*E(1)*(1+1/7*s/(0.1+s)+1/7*s/(1+s)+1/7*s/(10+s));
%%%%%%%%%%%%%%%%%%%%%%%%%%%%%%%%%%%%%%%%%%%%%%%%%%%%%%%%%%%%%%%%%%%%%%%%
counter= 0; % the order of the current frequency for spatial coordinate
be= 0.0;
for i=1:nrho
    counter= (i-1)*20;
    ay=be;
    be=be+drho;
    for j=1:20
        counter= counter+1;
        rho=0.5*(be-ay)*cod(j)+0.5*(be+ay);
        %----------------------------------
        % input data for boundary conditions
        %----------------------------------
        ncb=2;          % number of d.o.f with specified values of 'u'
        bcdof=[71 72]; % d.o.f numbers of the d.o.f with specified 'u' values
        bcval=[0.0 0.0];  % the corresponding values of 'u'

        nfb=1;          % number of nodes with specified node force
        fbdof= [2];     % d.o.f numbers of the d.o.f with specified node force
        fbval=[1.0];    % the corresponding values of specified node force
        %---------------------------------------
        % initialization of matrices and vectors
        %---------------------------------------
        force = zeros(sdof,1);  % initialization of node force vector
        ff=zeros(sdof,1);       % initialization of system force vector
        kk=zeros(sdof,sdof);    % initialization of system matrix
        index=zeros(nnel*ndof,1);  % initialization of index vector
        %-----------------------------------------------------------------
        % computation of element matrices and vectors and their assembly
        %-----------------------------------------------------------------
        for iel=1:nel            % loop for the total number of elements
            nprop = matno(iel);     % material no. for the element
            nl=nodes(iel,1); nr=nodes(iel,2);  % extract nodes for (iel)-th
            xl=gcoord(nl); xr=gcoord(nr);       % extract nodal coord values
            eleng=xr-xl;                        % element length
            index=FEELDOF1(iel,nnel,ndof);      % extract system dofs
            [kd]= elematrx_laplace(rho,poisson,E(1),eleng, s);
                                            % compute element matrix
            k=kd;
            f=[0;0;0;0];
            [kk,ff]=FEASMBL2(kk,ff,k,f,index); % assemble element matrices
```

```
end
%------------------------------------------------------------
% apply the natural boundary condition at the last node
%------------------------------------------------------------
for ifb=1:nfb
    ff(fbdof(ifb))=ff(fbdof(ifb))+fbval(ifb);
end
%----------------------------
% apply boundary conditions
%----------------------------
[kk,ff]=FEAPLYC2(kk,ff,bcdof,bcval);
%----------------------------
%  solve the matrix equation
%----------------------------
uu=kk\ff;
%------------------------------------------------------------
% calculate force at each node point
%------------------------------------------------------------
wt1 = wei(j)*(be-ay)/2;
for iel=1:nel+1      % loop for the total number of elements to
                         calculate the SSD at each point
    if (iel== nel+1)       % the last point
        kel= nel;
    else
        kel= iel;
    end
    nprop = matno(kel);   % material no. for the point
    nl=nodes(kel,1); nr=nodes(kel,2); % extract nodes for (iel)-th
    xl=gcoord(nl); xr=gcoord(nr);% extract nodal coord values
    eleng=xr-xl;               % element length
    index=FEELDOF1(kel,nnel,ndof);% extract system dofs
    [kd]= elematrx_laplace(rho,poisson,E(1),eleng, s);
                              % compute element matrix
    k=kd;
    ii = index(1);
    jj = index(4);
    force(ii:jj) =k*uu(ii:jj);

    if (iel ~= nel+1)
        rsez= -force(index(1));   % To get correct answer, "-" is
                                       needed for 'rsez' and
        rsz= force(index(2));     % 'rue', why?
        ruz= uu(index(2));
    else
        rsez= force(index(3));    % To get correct answer, "-" is
                                       needed for 'rsez' and
        rsz= -force(index(4));     % 'rue', why?
        ruz= -uu(index(4));
        rue= uu(index(3));
    end
    for m=1:nlds
        xa= radius(m);
        tt3= rho*xa;
        fj0= besselj(0,tt3);
        fj1= besselj(1,tt3);
        con= fj1*xa*wt1;
        vv= con*vld(m);
```

```
for n=1:nposition
    xc= xp(n);
    yc= yp(n);
    xrel= xc-xp1(m);
    yrel= yc-yp1(m);
    radi= (xrel^2+yrel^2)^0.5;
    thett= 0.0;
    if radi>0
        thett= atan2(yrel,xrel);
    end
    sn= sin(thett);
    cs= cos(thett);
    tt1= rho*radi;
    fj0= besselj(0,tt1);
    fj1= besselj(1,tt1);
    djr= rho/2;
    if radi>0
        djr= fj1/radi;
    end
    fj2= 0.0;
    fj3= 0.0;
    if tt1>0
        fj2= 2*fj1/tt1-fj0;
        fj3= 4*fj2/tt1-fj1;
    end
    np = iel;    % the point number to be analyzed
    %%%%%%%%%%%%%%%%%%%%%%%%%%%%%%%%%%%%%%%%%%%%%%%%%%%%%%%

    if (np<nel+1)
        er= vv/s*(rho*fj0-djr)*(-uu(index(1)));
        etheta= vv/s*djr*(-uu(index(1)));
        ez_freq_domain=(rsz-rho*c*(-uu(index(1))))/d;
        ez = vv/s*fj0*ez_freq_domain;
    else
        er= vv/s*(rho*fj0-djr)*(-uu(index(3)));
        etheta= vv/s*djr*(-uu(index(3)));
        ez_freq_domain=(rsz-rho*c*(-uu(index(3))))/d;
        ez = vv/s*fj0*ez_freq_domain;
    end
    sigmarr= a*er + b*etheta + c*ez;
    sigmaoo= b*er + a*etheta + c*ez;
    sigmazz= c*er + c*etheta + d*ez;
    strainrr(np,n)= strainrr(np,n)+er;
    strainoo(np,n)= strainoo(np,n)+etheta;
    strainzz(np,n)= strainzz(np,n)+ez;
    stressrr(np,n)= stressrr(np,n)+sigmarr;
    stressoo(np,n)= stressoo(np,n)+sigmaoo;
    stresszz_test(np,n)= stresszz_test(np,n)+sigmazz;
%%%%%%%%%%%%%%%%%%%%%%%%%%%%%%%%%%%%%%%%%%%%%%%%%%%%%%%%%%%%%%%%%%%%%%%%%%%
    sr= vv*((rho*rue+bb*rsz)*fj0/cc-xm1*rue*djr);
    so= vv*((rho*rue*xm2+bb/cc*rsz)*fj0+xm1*rue*djr);
    sz= vv/s*rsz*fj0;
    sro=0;
    soz=0;
    szr= -vv/s*rsez*fj1;
    disr= -vv/s*(rue*fj1);
```

```
                                diso=0;
                                disz= vv/s*ruz*fj0;
                                %  rotate back -theta to x-y direction
                                sn= sin(-thett);
                                cs= cos(-thett);
                                c2t= cs*cs-sn*sn;
                                s2t= 2.*sn*cs;
                                sxx= (sr+so)/2+(sr-so)*c2t/2+sro*s2t;
                                syy= (sr+so)/2-(sr-so)*c2t/2-sro*s2t;
                                txz= cs*szr;
                                tzy= -sn*szr;
                                txy= -(sr-so)*s2t/2;
                                ux= disr*cs+diso*sn;
                                uy= -disr*sn+diso*cs;
                                stressxx(np,n,it)= stressxx(np,n,it)+sxx;
                                stressyy(np,n,it)= stressyy(np,n,it)+syy;
                                stresszz(np,n )= stresszz(np,n)+sz;
                                stressxy(np,n,it)= stressxy(np,n,it)+txy;
                                stresszy(np,n)= stresszy(np,n)+tzy;
                                stressxz(np,n)= stressxz(np,n)+txz;
                                dispx(np,n)= dispx(np,n)+ux;
                                dispy(np,n )= dispy(np,n)+uy;
                                dispz(np,n)= dispz(np,n)+disz;

                    end       % corresponding to 'for n=1:nposition'

                 end          % corresponding to 'm=1:nlds'

             end              % corresponding to 'for iel=1:nel+1'

         end                  %corresponding to 'for j=1:20'

     end                      %corresponding to 'for i=1:NRHO'

freal= real(strainzz(1,1));
fimag= imag(strainzz(1,1));

end

%--------------------------------------------------------------------
```

Subprogram "elematrx_laplace.m"

```
function [kd]=elematrx_laplace(rho,poisson,E0,eleng, s)
%----------------------------------------------------------------
% Synopsis:
% [kd]=elematrx_laplace(rho,poisson,E0,eleng,s)
% Variable Description:
% kd - derivative of element matrix which is time dependent
% rho, E0, poisson - viscoelastic material properties
% eleng - element length
% s - Laplace variable
%----------------------------------------------------------------
h= eleng;
rh2= rho*rho;

a = (1-poisson)/(1+ poisson)/(1-2*poisson)*E0*(1+1/7*s/(0.1+s)+1/7*s/
(1+s)+1/7*s/(10+s));
```

```
d = a;
b = poisson/(1+poisson)/(1-2*poisson)*E0*(1+1/7*s/(0.1+s)+1/7*s/(1+s)+1/7*s/
(10¦¤));
c = b;
f = 1/2/(1+poisson)*E0*(1+1/7*s/(0.1+s)+1/7*s/(1+s)+1/7*s/(10+s));

kd(1,1)= rh2*h*a/3+ f/h;
kd(1,2)= -rho*c/2 + f*rho/2;
kd(1,3)= rh2*a*h/6-f/h;
kd(1,4)= rho*c/2+f*rho/2;
kd(2,1)= kd(1,2);
kd(2,2)= d/h+f*rh2*h/3;
kd(2,3)= -rho/2*c-rho/2*f;
kd(2,4)= -d/h+rh2*h*f/6;
kd(3,1)= kd(1,3);
kd(3,2)= kd(2,3);
kd(3,3)= rh2*h*a/3 + f/h;
kd(3,4)= rho*c/2-rho*f/2;
kd(4,1)= kd(1,4);
kd(4,2)= kd(2,4);
kd(4,3)= kd(3,4);
kd(4,4)= d/h + rh2/3*h*f;
end


%-------------------------------------------------------------------------
```

Subprogram "elematrx.m"

```
function [k_parameter]=elematrx(rho,poisson,eleng)
%-----------------------------------------------------------------
% Purpose:
% Get the element matrix for (a u'' + b u' + c u)
% using linear bar elements
%
% Synopsis:
% [k_parameter]=elematrx(rho,poisson,eleng)
%
% Variable Description:
% k_parameter - element matrix (size of 2x2)
% rho, poisson - viscoelastic material properties
% eleng - element length
%-----------------------------------------------------------------
% element matrix
a= (1-poisson)/(1+poisson)/(1-2*poisson);
c= poisson/(1+poisson)/(1-2*poisson);
f= 1/2/(1+poisson);
h= eleng;
rh2= rho*rho;
k_parameter(1,1)= rh2*a*h/3+f/h;
k_parameter(1,2)= -rho*c/2+f*rho/2;
k_parameter(1,3)= rh2*a*h/6-f/h;
k_parameter(1,4)= rho*c/2+f*rho/2;
k_parameter(2,1)= k_parameter(1,2);
k_parameter(2,2)= a/h+f/3*rh2*h;
k_parameter(2,3)= -c*rho/2-f*rho/2;
k_parameter(2,4)= -a/h+f/6*rh2*h;
k_parameter(3,1)= k_parameter(1,3);
k_parameter(3,2)= k_parameter(2,3);
```

```
k_parameter(3,3)= rh2*a*h/3+f/h;
k_parameter(3,4)= rho*c/2-rho*f/2;
k_parameter(4,1)= k_parameter(1,4);
k_parameter(4,2)= k_parameter(2,4);
k_parameter(4,3)= k_parameter(3,4);
k_parameter(4,4)= a/h+rh2*f/3*h;
end
```

%--

Subprogram "elematrx_freq.m"

```
function [ kd ]=elematrx_freq(rho,dynamic_modulus,v,eleng)
%----------------------------------------------------------------
% Synopsis:
% [kd]=elematrx_freq(rho,dynamic_modulus,v,eleng)
%
% Variable Description:
% kd - element matrix (size of 2x2)
% rho, v, dynamic_modulus - viscoelastic material properties
% eleng - element length
%----------------------------------------------------------------
% element matrix
h= eleng;
rh2= rho*rho;
a = (1-v)*dynamic_modulus/((1+v)*(1-2*v));
d = a;
b = v*dynamic_modulus/((1+v)*(1-2*v));
c = b;
f = dynamic_modulus/2/(1+v);

kd(1,1)= rh2*h*a/3+ f/h;
kd(1,2)= -rho*c/2 + f*rho/2;
kd(1,3)= rh2*a*h/6-f/h;
kd(1,4)= rho*c/2+f*rho/2;
kd(2,1)= kd(1,2);
kd(2,2)= d/h+f*rh2*h/3;
kd(2,3)= -rho/2*c-rho/2*f;
kd(2,4)= -d/h+rh2*h*f/6;
kd(3,1)= kd(1,3);
kd(3,2)= kd(2,3);
kd(3,3)= rh2*h*a/3 + f/h;
kd(3,4)= rho*c/2-rho*f/2;
kd(4,1)= kd(1,4);
kd(4,2)= kd(2,4);
kd(4,3)= kd(3,4);
kd(4,4)= d/h + rh2/3*h*f;

end
```

PART III
System Response: Methods of Analyses

Analytical Methods

7.1 Introduction

For some physical problems, the idealization of the problem may lead to the formulation of a simple mathematical model represented in terms of governing field equations (ordinary or partial differential equations) along with prescribed boundary conditions. In these situations, closed form solutions for the response variables can be obtained by well-established analytical methods either directly by doing the mathematics or using readily available mathematical packages such as MAPLE®, MATLAB®etc.

Whenever possible, we always first try to obtain analytical solution as it is 'exact' and also provides more insight into the nature of a problem. Additionally, analytical solutions become useful for the verification of the solutions obtained by approximate numerical methods. In this chapter some basic analytical solutions for some simple and useful problems in geomechanics are presented. Both the standard direct solutions and the computer aided indirect analytical solutions (developed using MAPLE) are presented. At a fundamental level, a wide range of geotechnical problems reduce to that of deformation and flow response of a saturated deformable porous media subjected to a variety of external loading (hydraulic and mechanical). The problems chosen belong to the following classes:

- Flow of water (seepage) through non-deformable porous medium (no deformation)
- Deformation of a column of deformable porous medium (no flow)
- Consolidation of a saturated deformable porous medium (decoupled flow and deformation)
- Consolidation of a saturated and nearly saturated deformable porous medium (coupled flow and deformation)

7.2 1-D Flow through a Land Mass: Island Recharge Problem

7.2.1 Problem Definition

Consider an island shown below that is surrounded by water mass which covers all its lateral boundaries and is underlain by impervious rock layer. The area close to the water is being considered for some development project. The concern here is that the rainfall to the island towards the unconfined soil mass causes the ground water table to get too close to the surface unless it is artificially drained. This situation requires the evaluation of depth, z_w above the sea level.

The actual problem is first idealized both in terms of the geometry and related physical process. The lateral boundaries are considered vertical and only a cross-section of the actual three dimensional island in *x-z* plane of dimensions *L×b* is taken (Figure 7.2-1). The soil is considered non-deformable and the flow is 1-D taking place only in the *x*-direction.

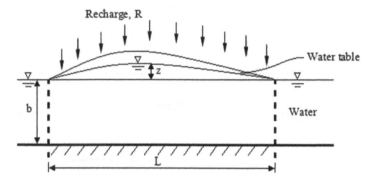

Figure 7.2-1 Island mass under recharge

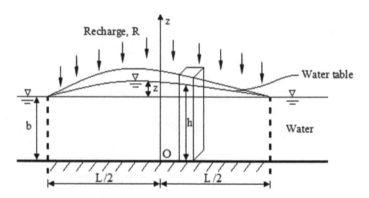

Figure 7.2-2 Idealized problem

7.2.2 Mathematical Model

A certain number of assumptions need to be made in the problem prior to the analytical solution. We assume that the D'Arcy's law governs the flow of water within the porous soil. Here, assuming that total head, *h* is constant along the *x*-direction, the flux along that direction becomes,

$$f_x = -k \frac{dh}{dx} \tag{7.2.1}$$

Considering a 1-D element with a unit length in the direction normal to the page (Figure 7.2-2), the rate of flow is:

$$Q_x = \rho_w f_x h \tag{7.2.2}$$

Now the mass balance of flow is considered through an elementary soil column as can be seen in Figure 7.2-3 as,

$$Q_{x+\frac{\Delta t}{1}} - Q_{x-\frac{\Delta t}{1}} - \rho_w R \Delta x = 0 \tag{7.2.3}$$

leading to,

$$\left(h+\frac{\partial h}{\partial x}\frac{\Delta x}{2}\right)\left(f_x+\frac{\partial f}{\partial x}\frac{\Delta x}{2}\right)-\left(h-\frac{\partial h}{\partial x}\frac{\Delta x}{2}\right)\left(f_x-\frac{\partial f_x}{\partial x}\frac{\Delta x}{2}\right)-R\Delta x=0 \qquad (7.2.4)$$

Neglecting the second order terms yields,

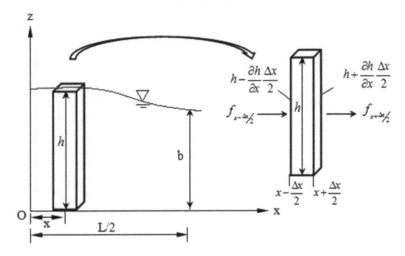

Figure 7.2-3 Solution domain and a soil column element

$$\frac{\partial h}{\partial x}f_x+h\frac{\partial f_x}{\partial x}-R=0 \qquad (7.2.5)$$

$$\frac{\partial}{\partial x}(hf_x)-R=0 \qquad (7.2.6)$$

Assuming the term in the parenthesis to be linearly varying,

$$hf_x=Rx \qquad (7.2.7)$$

and substituting (7.2.1) into this equation gives,

$$-hk\frac{\partial h}{\partial x}=Rx \qquad (7.2.8)$$

We get the solution in the form below,

$$h=\pm\sqrt{c-\frac{R}{k}x^2} \qquad (7.2.9)$$

c being a constant. The boundary condition that needs to be satisfied is,

$$h=b \ @ \ x=\frac{L}{2}$$

which readily yields the value of constant c as,

$$c=b^2+\frac{R}{k}\frac{L^2}{4} \qquad (7.2.10)$$

The particular solution can now be obtained as,

$$h = \sqrt{b^2 + \frac{R}{k}\left(\frac{L^2}{4} - x^2\right)}$$

(7.2.11)

Thus, the variation of the elevation of ground water table becomes,

$$z_w = h - b = \sqrt{b^2 + \frac{R}{k}\left(\frac{L^2}{4} - x^2\right)} - b$$

(7.2.12)

Computer Implementation

The above recharge problem is now solved in MAPLE as presented in Figure 7.2-4 and in a worksheet at the end of the chapter called "**Worksheet 1.mw**".

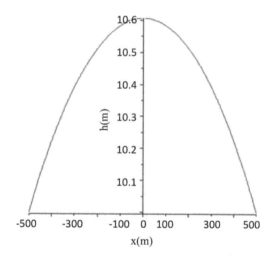

Figure 7.2-4 Variation of total head, $R = 5 \times 10^{-6}$, $b = 10$, $L = 1000$, $k = 0.1$

7.3 Regional Groundwater Flow: Steady State Seepage

7.3.1 Problem Definition

Consider the following situation in which over a large region the topography of the ground surface and water table are as shown in Figure 7.3-1. Because of the sloping water table, there may be a flow set up in the domain Ω^* (E_1-E_2-E_4-E_3^*). Here the objective is to construct a mathematical model for such a regional groundwater flow and develop an analytical solution for it.

Considering the region Ω^* of ground water flow to be of a large extent in the lateral direction, the slope of the water table will be c. Furthermore actual domain Ω^* is replaced by a rectangular domain Ω of (E_1-E_2-E_4-E_3). It is also assumed that there is no variation in the z-direction (normal to x-y plane). Therefore, we are considering only a two-dimensional (2-D) unconfined flow and the flow to be steady state and independent of time.

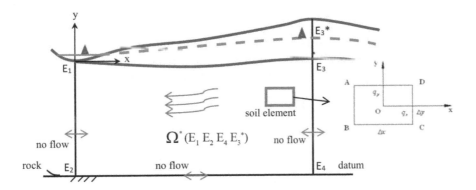

Figure 7.3-1 Physical model for 2-D steady state seepage

7.3.2 Mathematical Model

We should now consider a rectangular element of soil in the domain and related water flow in the x-y directions (Figure 7.3-1). The following physical laws govern the flow of fluid through the porous medium.

D'Arcy's Law

D'Arcy's law represents a constitutive law for the flow of pore fluid according to which the flux (i.e. volume of fluid flowing per unit time and per unit total area of the medium) is given by:

$$f_x = -k_x \frac{\partial h}{\partial x} \tag{7.3.1}$$

$$f_y = -k_y \frac{\partial h}{\partial y} \tag{7.3.2}$$

where mass flow through AB and DC are:

$$q_{x-\Delta x} = \left[\rho_w f_x - \frac{\partial}{\partial x}(\rho_w f_x)\frac{\Delta x}{2} \right]\Delta y \tag{7.3.3}$$

$$q_{x+\Delta x} = \left[\rho_w f_x - \frac{\partial}{\partial x}(\rho_w f_x)\frac{\Delta x}{2} \right]\Delta y \tag{7.3.4}$$

Net mass flux in the x-direction is:

$$Q_x = q_{x-\Delta x} - q_{x+\Delta x} = -\frac{\partial}{\partial x}(\rho_w f_x)\Delta x \Delta y \tag{7.3.5}$$

Similarly, the net mass flux in the y-direction can be written as,

$$Q_y = q_{y-\Delta y} - q_{y+\Delta y} = -\frac{\partial}{\partial y}(\rho_w f_y)\Delta y \Delta x \tag{7.3.6}$$

Law of Conservation of Mass

Considering the soil skeleton to be non-deformable and the pore fluid to be incompressible, the net mass flux due to flow is zero. This simply means the mass of fluid entering into the elementary volume must be same as the mass of fluid leaving it such that:

$$Q = Q_x + Q_y = 0 \tag{7.3.7}$$

leading to

$$\frac{\partial}{\partial x}\left(\rho_w k_x \frac{\partial h}{\partial x}\right) + \frac{\partial}{\partial y}\left(\rho_w k_y \frac{\partial h}{\partial y}\right) = 0 \tag{7.3.8}$$

If the water is homogeneous (i.e. ρ_w does not vary with the coordinate system), then the above equation reduces to,

$$\frac{\partial}{\partial x}\left(k_x \frac{\partial h}{\partial x}\right) + \frac{\partial}{\partial y}\left(k_y \frac{\partial h}{\partial y}\right) = 0 \tag{7.3.9}$$

If the soil is homogeneous (i.e. k_x and k_y do not vary with the coordinate system), then the above equation reduces to,

$$k_x \frac{\partial^2 h}{\partial x^2} + k_y \frac{\partial^2 h}{\partial y^2} = 0 \tag{7.3.10}$$

Furthermore, if we consider the soil within the flow domain to be isotropic (i.e. $k_x = k_y = k$) then the above form further reduces to the Laplace equation,

$$\frac{\partial^2 h}{\partial x^2} + \frac{\partial^2 h}{\partial y^2} = 0 \tag{7.3.11}$$

which is also written in the following form:

$$\nabla^2 h = 0 \tag{7.3.12}$$

The mathematical model can now be seen in Figure 7.3-2.

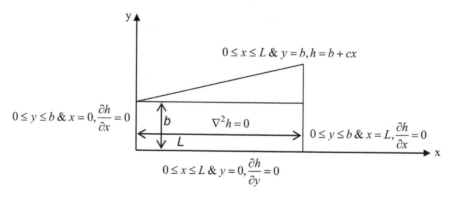

Figure 7.3-2 Mathematical model

7.3.3 Analytical Solution

The governing equation is now solved by the method of "separation of variables". Let us assume the solution to be of the form,

$$h(x, y) = F(x)G(y) \tag{7.3.13}$$

where $F(x)$ and $G(y)$ are independent functions which let the total head be written in terms of separated variables to be functions of x and y. We can then write,

$$\frac{\partial^2 h}{\partial x^2} = \frac{d^2 F}{dx^2} \cdot G \tag{7.3.14}$$

$$\frac{\partial^2 h}{\partial y^2} = \frac{d^2 F}{dy^2} \cdot F \tag{7.3.15}$$

Substituting the above into the governing equation of (7.3.12),

$$\frac{d^2 F}{dx^2} \cdot G + \frac{d^2 G}{dy^2} \cdot F = 0 \tag{7.3.16}$$

$$\frac{1}{F} \frac{d^2 F}{dx^2} + \frac{1}{G} \frac{d^2 G}{dy^2} = -\mu^2 \tag{7.3.17}$$

Hence we get two ordinary differential equations (ODE) to be of the form,

$$\frac{d^2 F}{dx^2} + \mu^2 F = 0 \tag{7.3.18}$$

$$\frac{d^2 G}{dy^2} - \mu^2 G = 0 \tag{7.3.19}$$

which have the solutions:

$$F(x) = A^* \cos \mu x + B^* \sin \mu x \tag{7.3.20}$$

$$G(y) = C^* e^{\mu y} + D^* e^{-\mu y} \tag{7.3.21}$$

respectively. The constants, A^*, B^*, C^* and D^* need to be evaluated using the boundary conditions. Now we combine the forms using (7.3.13) and get,

$$h(x, y) = (A^* \cos \mu x + B^* \sin \mu x)(C^* e^{\mu y} + D^* e^{-\mu y}) \tag{7.3.22}$$

The boundary conditions are such that at the left boundary: $x = 0$ & $0 \le y \le b \rightarrow \dfrac{\partial h}{\partial x} = 0$

resulting in,

$$\left.\frac{\partial h}{\partial x}\right|_{x=0} = (-\mu A^* \sin \mu x + \mu B^* \cos \mu x)G(y) = 0$$

yielding $B^* = 0$. Therefore equation (7.3.22) now becomes,

$$h(x, y) = A^* \cos \mu x (C^* e^{\mu y} + D^* e^{-\mu y})$$

or

$$h(x, y) = \cos \mu x (A e^{\mu y} + B e^{-\mu y}) \tag{7.3.23}$$

At the bottom boundary: $y = 0$ & $0 \le x \le L \to \dfrac{\partial h}{\partial y} = 0$

resulting in

$$\left.\frac{\partial h}{\partial y}\right|_{y=0} = \cos \mu x(\mu A e^{\mu y} - \mu B e^{-\mu y}) = 0$$

yielding $A = B$. Therefore equation (7.3.23) becomes,

$$h(x, y) = A \cos \mu x(e^{\mu y} + e^{-\mu y}) \tag{7.3.24}$$

At the right boundary: $x = L$ & $0 \le y \le b \to \dfrac{\partial h}{\partial x} = 0$

resulting in,

$$\left.\frac{\partial h}{\partial x}\right|_{x=L} = -\mu A \sin \mu x(e^{\mu y} + e^{-\mu y}) = 0$$

The above gives,

$$\sin(\mu L) = 0 \to \mu = \frac{n\pi}{L}, \ n = 0, 1, 2 \dots . \text{ We can then write,}$$

$$h(x, y) = 2 A_n \cos \frac{n\pi}{L} x \left(\frac{e^{\mu y} + e^{-\mu y}}{2} \right) \tag{7.3.25}$$

leading to

$$h(x, y) = B_n \left(\cos \frac{n\pi}{L} x \right) \left(\cosh \frac{n\pi}{L} y \right) \tag{7.3.26}$$

A general solution can therefore be obtained as,

$$h_n(x, y) = \sum_{n=0}^{\infty} B_n \left(\cos \frac{n\pi}{L} x \right) \left(\cosh \frac{n\pi}{L} y \right) \tag{7.3.27}$$

It should be noted here that for the Fourier series expansion, we have the following forms,

$$f(x) = \sum_{n=0}^{\infty} C_n \left(\cos \frac{n\pi}{L} x \right) \to C_0 = \frac{1}{L} \int_0^L f(x)dx \to C_n$$

$$= \frac{2}{L} \int_0^L f(x) \left(\cos \frac{n\pi}{L} x \right) dx, \ \ n = 1, 2 \dots \infty$$

$$\tag{7.3.28}$$

Finally at the top boundary,

$$y = b \ \& \ 0 \le x \le L \to h(x, b) = f(x) = b + cx \tag{7.3.29}$$

yielding

$$\rightarrow \sum_{n=0}^{\infty} B_n \left(\cosh \frac{n\pi}{L} b \right) \left(\cos \frac{n\pi}{L} x \right) \rightarrow \sum_{n=0}^{\infty} C_n \left(\cos \frac{n\pi}{L} x \right) = f(x) \tag{7.3.30}$$

where $C_n = B_n \left(\cosh \frac{n\pi}{L} b \right)$ and which is the Fourier series representation of $f(x)$. Thus, the constant C_n can be evaluated as:

$C_0 = b + \frac{cL}{2}$ and thus, $C_n = \frac{2}{L} \int_0^L (b + cx) \left(\cos \frac{n\pi}{L} x \right) dx$, $n = 1, 2 ... \infty$. Using "integration by parts", this leads to,

$$C_n = \frac{2cL}{n^2 \pi^2} (\cos(n\pi) - 1) \tag{7.3.31}$$

which makes B_n,

$$B_n = \frac{2cL}{n^2 \pi^2} \frac{(\cos(n\pi) - 1)}{\cosh\left(\frac{n\pi}{L} b \right)}, \quad n = 1, 2 ... \infty \tag{7.3.32}$$

The final form of the total head now becomes,

$$h_n(x, y) = b + \frac{cL}{2} + \sum_{n=0}^{\infty} \frac{2cL}{n^2 \pi^2} \frac{(\cos(n\pi) - 1)}{\cosh\left(\frac{n\pi}{L} b \right)} \left(\cos \frac{n\pi}{L} x \right) \left(\cosh \frac{n\pi}{L} y \right) \tag{7.3.33}$$

Computer Implementation

The above solution is implemented in MAPLE ("**Worksheet 2.mw**"). The results are presented below and showing the resulting total head, gradient and flux at specified numbers of grid points. Equipotential lines and flow lines that are orthogonal to them are plotted as well (Figures 7.3-3 and 7.3-4). The outcomes are compared with the solutions as referenced in Toth (1962) where the total head reduces from 10387 to 10010 in 20,000 ft (Figure 7.3-5).

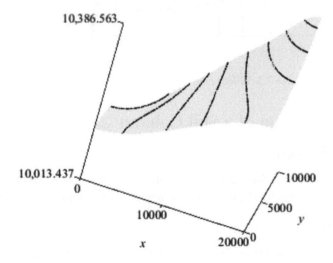

Figure 7.3-3 The variation of total head, $h(x, y)$, $L = 20000$, $b = 10000$, t $c = 0.02$, $f = b + cx$

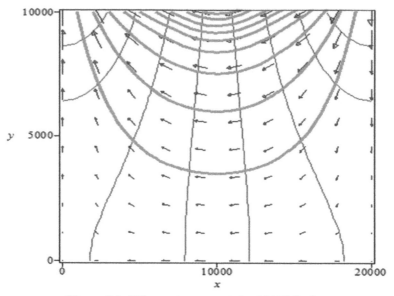

Figure 7.3-4 Flow net representation MAPLE plot

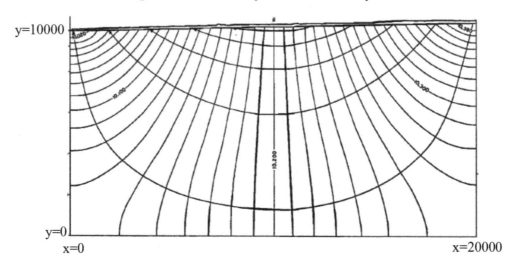

Figure 7.3-5 Flow net representation (after Toth, 1962)

7.4 1-D Deformation of a Soil Column

7.4.1 Physical Problem

Now consider a simple 1-D deformation of a soil mass below (Figure 7.4-1). Our objective is to develop analytical solutions for the deformation and stresses within this column.

7.4.2 Idealization

Firstly, we idealize the geometry of the soil column as being trapezoidal in shape (Figure 7.4-2). Additionally with respect to the constitutive behavior, soil is considered linear elastic, and the column remains in the state of equilibrium.

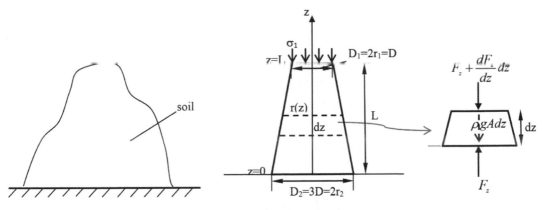

Figure 7.4-1 Soil mass   **Figure 7.4-2** Soil column and a selected element

7.4.3 Mathematical Model

For a truncated element seen below, we can write the equilibrium as:

$$\sum F_z = ma_z \tag{7.4.1}$$

$$F_z - \left(F_z + \frac{dF_z}{dz} dz \right) - \gamma\, A dz = (\gamma\, A dz)a_z \tag{7.4.2}$$

Let $r_1 = \dfrac{D_1}{2}$ and r_2 be the upper and lower radii of the column, respectively. Thus the variation of radius along the column is,

$$r(z) = r_2 + \frac{(r_1 - r_2)}{L} z \tag{7.4.3}$$

For the given problem it is,

$$r(z) = 3r_1 - \frac{2r_1 z}{L} \tag{7.4.4}$$

Ignoring the self-weight and the accelerations leads to,

$$\frac{dF}{dz} = \frac{d}{dz}(\sigma(z)A(z)) = 0 \tag{7.4.5}$$

$$\frac{d\sigma(z)}{dz} A + \frac{dA(z)}{dz}\sigma = 0 \tag{7.4.6}$$

$$\frac{d\sigma(z)}{dz}\pi r(z)^2 + 2\pi r(z)\sigma \frac{dr(z)}{dz} = 0 \tag{7.4.7}$$

$$\frac{d\sigma(z)}{dz} r(z) + 2\sigma \frac{dr(z)}{dz} = 0 \tag{7.4.8}$$

Integrating the above equation (with the boundary condition $\sigma(L) = \sigma_1$) the following solution for stress is readily obtained as,

$$\sigma(z) = \sigma_1 \left(\frac{r_1}{a+bz} \right)^2 \tag{7.4.9}$$

where a and b are determined from the linear variation of the cross section of the bar. Using the stress-strain relationship for linear elastic material, the above equation can be written as,

$$E \frac{\partial u(z)}{\partial z} = \sigma_1 \left(\frac{r_1}{a+bz} \right)^2 \tag{7.4.10}$$

Now this equation can be integrated with the boundary condition ($u(0) = 0$) to obtain the displacement variation as,

$$u(z) = \frac{\sigma_1 r_1^2}{Eb} \left(\frac{z}{a+bz} \right) \tag{7.4.11}$$

For the problem in Figure 7.4-2, cross section area varies as,

$$A(z) = \pi r(z)^2 = \pi \left(r_1 \left(3 - \frac{2z}{L} \right) \right)^2 \tag{7.4.12}$$

The analytical solution to the above form can also be obtained by integrating the governing equation for strain and using the Hooke's law as,

$$\int_0^z \varepsilon(z)dz = \int_0^z \frac{F}{EA(z)}dz = \frac{F}{EA_1}\int_0^z \frac{1}{\left(3 - \frac{2z}{L} \right)^2}dz \tag{7.4.13}$$

where $A_1 = \pi r_1^2$. So the displacement and stress variation in the pier are,

$$u = -\frac{FL}{EA_1}\left(\frac{z}{9L-6z} \right) = -\frac{\sigma_1 L}{E}\left(\frac{z}{9L-6z} \right) \tag{7.4.14}$$

$$\sigma(z) = -\frac{FL}{A_1}\left(\frac{9L}{(9L-6z)^2} \right) = -\sigma_1 L \left(\frac{9L}{(9L-6z)^2} \right) \tag{7.4.15}$$

Computer Implementation

The MAPLE implementation of the problem is given at the end of the chapter along with related plots below in Figure 7.4-3 resulting from the worksheet "**Worksheet 3.mw**".

7.5 1-D Consolidation of a Soil Column: Decoupled Flow and Deformation

After discussing the problems of flow and deformation separately, we are now considering the problem of 'consolidation' of saturated soils. For some situations (as discussed in section 7.7) the problem of coupled flow and deformation can be idealized as that of decoupled flow and deformation for which Terzaghi developed his 1-D consolidation theory (Terzaghi 1923, 1925).

7.5.1 Physical Problem

A structure supported on a soil site with a layer of saturated clay (Figure 7.5-1) will experience a time dependent settlement. The prediction of the amount and the rate of settlement is an important task.

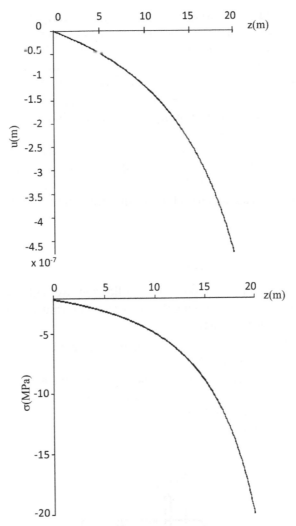

Figure 7.4-3 Variation of stress and displacement with depth, $E = 28$ MPa, $L = 20$, $s_1 = -20$, $r_1 = 0.25$, $r_2 = 0.75$

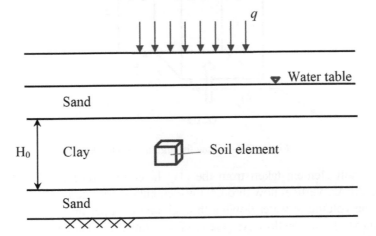

Figure 7.5-1 1-D consolidation of clay layer

Prior to the placement of the structure, underlying soils carry geostatic stresses due to the weight of the layers above and the portion of this stress increment in the pores is carried by the pore water resulting in an increase in pore water pressure referred to as excess (over hydrostatic portion) pore pressure. After the structural load is applied, the soil-water system is subjected to additional total stresses. Initially total head created in pore pressure field, the water will start to flow causing the pore water pressure to decrease and consequently part of the stress increment will be transformed to soil skeleton resulting in an increase in the effective stress which will subsequently cause compression of the soil.

For coarse grained soils with high permeability, this process (i.e. dissipation of pore water pressure and consequent increase in effective stress) is almost instantaneous. However, for fine-grained soils with very low permeability, this process is gradual. It takes time for the excess pore water pressure to dissipate and therefore for the stress increment to become effective and to cause compression. This gradual and time-dependent process of compression of fine-grained soils is called '*consolidation*'.

7.5.2 Mathematical Formulation

In order to develop a mathematical model of consolidation, we will first idealize the situation and then using some basic laws underlying the process, will formulate the governing equations. The following assumptions are made:

1. Only the consolidation of the clay layer which is assumed to be saturated, isotropic and homogeneous, will be considered.
2. The stress increase caused by the surface load 'q' within the entire clay layer is uniform.
3. Flow of pore water and compression of soil skeleton occurs only in the vertical direction.
4. Flow of pore water is governed by D'Arcy's law.
5. Pore water is incompressible.
6. The strains are small.

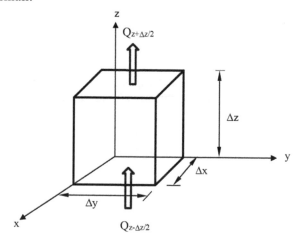

Figure 7.5-2 Flow through a soil element

Consider a soil element taken from the clay layer in Figure 7.5-2 with an elementary volume $V = \Delta x \, \Delta y \, \Delta z$. Vertical flow through the elementary volume is described in terms of the flux which is the volume of water flowing through unit area per unit time as described before within the context of seepage flow. As given in Equations (7.3.5) and (7.3.6), the net mass flux in the z-direction can be written similarly as:

$$Q_z = -\frac{\partial}{\partial z}(\rho_w f_z)V \tag{7.5.1}$$

which becomes after including the D'Arcy's law,

$$Q_z = -\frac{\partial}{\partial z}\left[\rho_w \frac{k_z}{\rho_w g}\left(\frac{\partial p_w}{\partial z}+1\right)\right]V \tag{7.5.2}$$

resulting in,

$$Q_z = -\frac{k_z}{g}\left(\frac{\partial^2 p_w}{\partial z^2}\right) \tag{7.5.3}$$

According to the *principle of conservation of mass*, net mass flux of pore water through the soil must be equal to the change in fluid mass due to the change in volume of the soil skeleton. Thus we have,

$$Q_z = \rho_w \frac{\partial \varepsilon_v}{\partial t} \tag{7.5.4}$$

where ε_v is the volumetric strain. For 1-D compression of the soil,

$$\frac{\partial \varepsilon_v}{\partial t} = m_v \frac{\partial \sigma_z'}{\partial t} \tag{7.5.5}$$

where m_v is the volumetric compressibility coefficient and $m_v = 1/D$ and $D = \dfrac{E(1-v)}{(1+v)(1-2v)}$

with E being the Young's modulus and v, Poisson's ratio. $\partial \sigma_z'$ is the change in effective stress which, in accordance with our assumption, is the same as the change in pore water pressure, ∂p_w. Therefore,

$$\frac{\partial \varepsilon_v}{\partial t} = m_v \frac{\partial p_w}{\partial t} \tag{7.5.6}$$

Combining Eqs. (7.5.3), (7.5.4) and (7.5.6) we get,

$$C_v \frac{\partial^2 p_w}{\partial z^2} = \frac{\partial p_w}{\partial t} \tag{7.5.7}$$

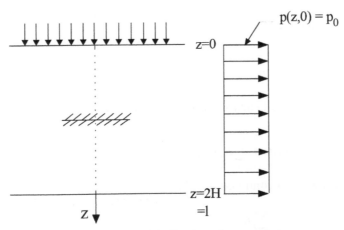

Figure 7.5-3 1-D idealization of consolidation

with $C_v = \dfrac{k_z}{m_v \rho_w g}$ being the coefficient of consolidation and p_w excess pore pressure. See Figure 7.5-3.

7.5.3 Analytical Solution

Eq. (7.5.7) can be solved for p_w using again the '*separation of variables*'. We write the field variable as:

$$p_w(z,t) = F(z)G(t) \tag{7.5.8}$$

Substitution of this definition to (7.5.7) yields,

$$F(z)\left(\frac{d}{dt}G(t)\right) = c_v\left(\frac{d^2}{dz^2}F(z)\right)G(t) \tag{7.5.9}$$

$$\frac{\dfrac{d^2}{dz^2}F(z)}{F(z)} = \frac{\dfrac{d}{dt}G(t)}{C_v G(t)} = -\lambda^2 \tag{7.5.10}$$

$$\frac{d^2}{dz^2}F(z) + \lambda^2 F(z) = 0 \tag{7.5.11a}$$

$$\frac{d}{dt}G(t) + \lambda^2 C_v G(t) = 0 \tag{7.5.11b}$$

Equations (7.5.11) are the two ordinary differential equations (ODE) obtained from the original partial differential equation (PDE) (7.5.7). The first one has a solution of:

$$F(z) = B_n \sin\left(\frac{n\pi z}{H}\right) \tag{7.5.12}$$

under the condition $F(0) = F(H) = 0$ where $\lambda = \dfrac{n\pi}{L}$, $n = 0, 1, \ldots N$ for a total of N numbers. The solution to the second ODE is:

$$G(t) = e^{\left(-Cv\left(\frac{n\pi}{H}\right)^2 t\right)} \tag{7.5.13}$$

From here, the general solution is obtained as:

$$p_w(z,t) = B_n \sin\left(\frac{n\pi z}{H}\right) e^{\left(-Cv\left(\frac{n\pi}{H}\right)^2 t\right)} \tag{7.5.14}$$

A linear combination of all such solutions is written as,

$$p_w(z,t) = \sum_{n=0}^{\infty} B_n \sin\left(\frac{n\pi z}{H}\right) e^{\left(-Cv\left(\frac{n\pi}{H}\right)^2 t\right)} \tag{7.5.15}$$

which must satisfy the initial condition of $p_w(z, 0) = f(z) = (p_w)_0$ which, if taken as 1.0 yields,

$$\sum_{n=0}^{\infty} B_n \sin\left(\frac{n\pi z}{H}\right) = 1 \tag{7.5.16}$$

From here B_n can be evaluated as,

$$B_n = \int_0^H \frac{2}{H} p_0 \sin\left(\frac{n\pi z}{H}\right) dz = \frac{2 - 2\cos(n\pi)}{n\pi} \tag{7.5.17}$$

So (7.5.15) becomes,

$$p_w(z,t) = \sum_{n=0}^{\infty} \frac{2 - 2\cos(n\pi)}{n\pi} (p_w)_0 \sin\left(\frac{n\pi z}{H}\right) e^{\left(-C_v\left(\frac{n\pi}{H}\right)^2 t\right)} \tag{7.5.18}$$

In terms of non-dimensional parameters where $P = \dfrac{p_w}{(p_w)_0}$, $T = \dfrac{t}{\tau}$ and $Z = \dfrac{z}{H}$ are chosen,

we can write $T = \dfrac{C_v t}{H^2}$ and $\dfrac{C_v}{H^2} = \dfrac{1}{\tau}$; now the above form becomes:

$$P(Z,T) = \sum_{n=0}^{\infty} \frac{2 - 2\cos(n\pi)}{n\pi} \sin(n\pi Z) e^{(-(n\pi)^2 T)} \tag{7.5.19}$$

The amount of consolidation taking place is evaluated using 'average degree of consolidation', U. Once the average degree of consolidation is known, actual consolidation at any time can be readily calculated:

$$U = 1 - \frac{\int_0^{2H} P(Z,T)dZ}{\int_0^{2H} P(Z,0)dZ} \tag{7.5.20}$$

Computer Implementation

For the purpose of verification, MAPLE worksheet (**Worksheet 4.mw**) is verified on the following problem with a consolidating soft clay layer between two sand layers (Figure 7.5-4) and the results (Table 7.5-1) are compared with the solution given below. The MAPLE program to solve this problem can be found in the Appendix. The variation of pore pressure is plotted in MAPLE (Figures 7.5-5, 7.5-6). Figure 7.5-7 presents the average degree and % consolidation.

Table 7.5-1 Results of Figure 7.5-4

z (m)	U %	p_w (kPa)
3	61	39
6	46	54
9	61	39
12	100	0

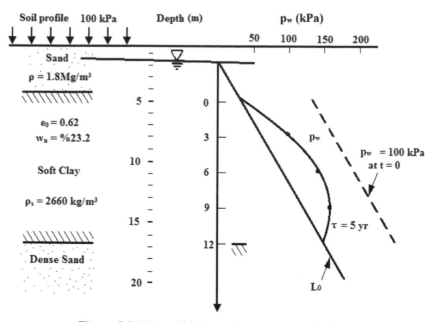

Figure 7.5-4 Consolidation of a clay layer in the field

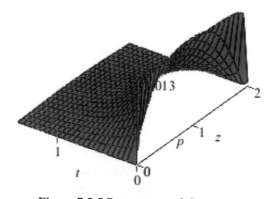

Figure 7.5-5 Pore pressure 3-D contours

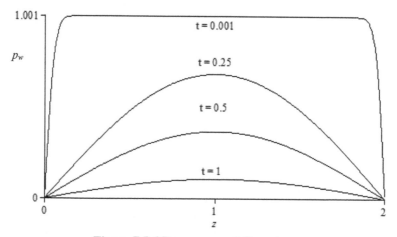

Figure 7.5-6 Pore pressure 2-D contours

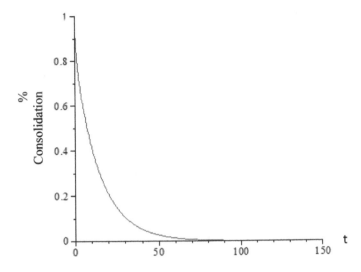

Figure 7.5-7 Average degree of consolidation

7.6 Contaminant Transport

The protection and management of groundwater resources require us to identify the sources and mechanism by which contaminants can enter groundwater flow system and to develop reliable predictions of the transport of contaminants within the flow system. In this section, the basic elements of modeling and computing for contaminant transport are presented.

7.6.1 Physical Problem

Consider a landfill on a site with soil profile shown below in Figure 7.6-1. The concern here is that contaminated 'leachate' from the fill may reach the aquifer below. The clay layer presented between the bottom of the landfill and the top of the aquifer may be considered as a natural barrier which, due to its low permeability, will try to impede the leachate flow and also the migration of contaminants. To evaluate such a system, it is required to analyze the transport of contaminants from the source (landfill) to the water bearing layer (aquifer).

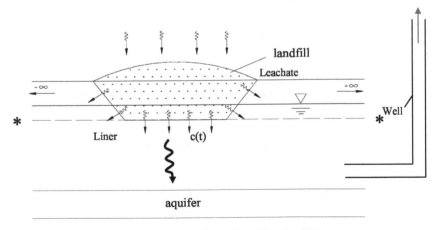

Figure 7.6-1 Soil profile with a landfill

7.6.2 Idealized Problem: 1-D

We consider the landfill to be of a very wide extent, and an inexhaustible source of contamination with a constant concentration of only a single contaminant. We can, therefore, idealize the flow and transport to be considered only in 1-D, in the z-direction. The problem is shown in Figure 7.6-2.

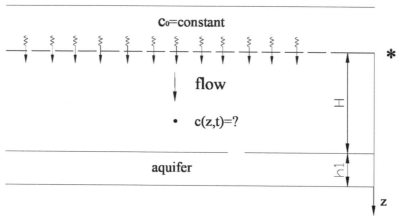

Figure 7.6-2 Idealized problem of contaminant transport

7.6.3 Mathematical Model: Formulation

Advection

Advection is the transport process that involves the movement of contaminant with flowing water. Bulk displacement of the solute due to flowing groundwater is the plug flow along the streamlines at an average velocity of,

$$q = nv \tag{7.6.1}$$

where q is superficial velocity, v is seepage velocity and n is porosity. Here q is also called the movement of contaminants at a speed corresponding to the groundwater velocity. The total mass transported from a contaminant source into a barrier during time, t is evaluated as,

$$m = A \int_0^t qc\,dt \tag{7.6.2}$$

where A is the cross sectional area of the landfill and c is mass concentration of the solute per unit volume of the fluid. Hence the total mass of solute moving through per unit time is,

$$m_t = qAc \tag{7.6.3}$$

The amount of solute moving per unit time per unit area is then,

$$f_z = nvc \tag{7.6.4}$$

Figure 7.6-3 presents the 1-D flow process.

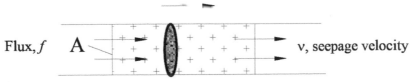

Figure 7.6-3 Transport and flow

Mechanical Dispersion

Hydraulic mixing is processed due to variation of velocity at the pore scale. There are three main sources for this as presented in Figure 7.6-4. From statistical consideration of pore velocity variation by averaging at a macro representative elementary volume (REV) level:

$$f_z^{md} = -nD_{md}\frac{\partial c}{\partial z} \tag{7.6.5}$$

where

$$D_{md} = \alpha v \tag{7.6.6}$$

with α being the *dispersivity* of the medium.

Molecular Diffusion

A solute will move from an area of its higher concentration to an area of lower concentration in response to the concentration gradient. This is known as *molecular diffusion* (Figure 7.6-5). The process will take place even in the absence of flow. The mass of solute diffusing is proportional to the concentration gradient, as described by Fick's first law. The mass flux due to this is written as,

$$f_z^e = -nD_e\frac{\partial c}{\partial z} \tag{7.6.7}$$

Now combining mechanical dispersion and the diffusion the flux due to 'hydrodynamic dispersion' is written as,

$$f_z^{hd} = -n(D_{md} + D_e)\frac{\partial c}{\partial z} = -nD\frac{\partial c}{\partial z} \tag{7.6.8}$$

Total flux is then,

$$f_z = nvc - nD\frac{\partial c}{\partial z} \tag{7.6.9}$$

Sorption

Removal of solutes from water by the process of chemical reaction (i.e. cation exchange) is called *sorption*. It is defined as,

$$q(c,t) = \rho k_d\frac{\partial c}{\partial t} \tag{7.6.10}$$

where $q(c, t)$ is the mass of the solute removed from solution per unit volume per unit time.

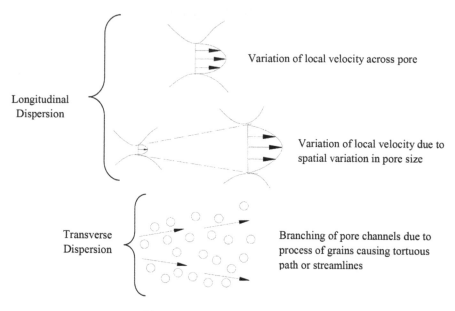

Figure 7.6-4 Dispersion types

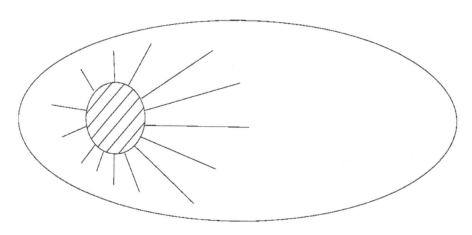

Figure 7.6-5 Movement from high concentration to low concentration

Radioactive and Biological Decay

The radioactive and biological decay is defined as,

$$p(c,t) = n\lambda c \tag{7.6.11}$$

where $p(c, t)$ is the mass of the solute reduced (lost) per unit volume and unit time due to biological and radioactive decay and λ is the decaying factor.

The processes controlling flux can now be summarized as: (i) Advection as the movement of solute by flowing water, (ii) Mechanical mixing (Figure 7.6-6) as the hydrodynamic dispersion and (iii) Diffusion.

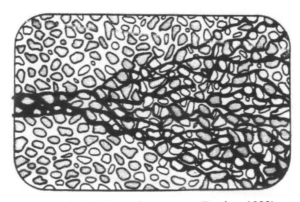

Figure 7.6-6 Dispersion process (Booker, 1989)

7.6.4 Mathematical Model: Governing Equation

Using the *principle of conservation of mass* in the transport process (Figure 7.6-7), we can write,

$$\Delta m = (\Delta m)_m - (\Delta m)_{out} - q(c, t) \tag{7.6.12}$$

$$n\Delta z\Delta A \frac{\partial c}{\partial t} = f_z \Delta A - (f_z + \frac{\partial f_z}{\partial z}\Delta z)\Delta A - q\Delta z\Delta A - p\Delta z\Delta A \tag{7.6.13}$$

Using equations (7.6.9) through (7.6.11) in the above leads to,

$$n\frac{\partial c}{\partial t} = \left(nD\frac{\partial^2 c}{\partial z^2} - nv\frac{\partial c}{\partial z} \right) - \rho k_d \frac{\partial c}{\partial t} - n\lambda c \tag{7.6.14}$$

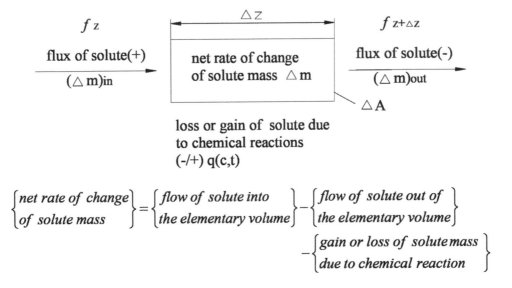

Figure 7.6-7 Conservation of mass

If we ignore the biological and radioactive decay term ($\lambda = 0$), then 1-D transport equation (7.6.14) can be written as,

$$nD\frac{\partial^2 c}{\partial z^2} - nv\frac{\partial c}{\partial z} = (n + \rho k_d)\frac{\partial c}{\partial t} \tag{7.6.15}$$

Rearranging the terms, we get,

$$\frac{D}{(1 + \rho k_d / n)}\frac{\partial^2 c}{\partial z^2} - \frac{v}{(1 + \rho k_d / n)}\frac{\partial c}{\partial z} = \frac{\partial c}{\partial t} \tag{7.6.16}$$

and finally reach,

$$D^*\frac{\partial^2 c}{\partial z^2} - v^*\frac{\partial c}{\partial z} = \frac{\partial c}{\partial t} \tag{7.6.17}$$

where $D^* = \dfrac{D}{(1 + \rho k_d / n)}$, and $v^* = \dfrac{v}{(1 + \rho k_d / n)}$. This partial differential equation is called the "Advective-Dispersive Transport" equation.

7.6.5 Analytical Solution

For the analytical solution of Eq. (7.6.17), a surface depository over a deep layer with no advection ($v = 0$) is considered (Figure 7.6-8). If we re-write the governing equation (7.6.17) by solely considering hydrodynamic dispersion,

$$D^*\frac{\partial^2 c}{\partial z^2} = \frac{\partial c}{\partial t} \tag{7.6.18}$$

and prescribe the boundary conditions as: $c(0, t) = c_0$ and $c(H = \infty, t) = 0$ along with the initial condition, $c(z, 0) = 0$, taking the Laplace transform (L[f]) of both the governing equation and the boundary conditions results in,

$$D^*\frac{d^2 c}{dz^2} = s\bar{c} - c(z, 0) \tag{7.6.19a}$$

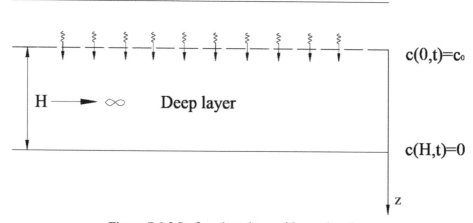

Figure 7.6-8 Surface depository with no advection

or

$$\frac{d^2c}{dz^2} = \frac{s}{D^*}\bar{c} = \alpha^2\bar{c} \tag{7.6.19b}$$

with

$$\bar{c}(0) = \frac{c_0}{s} \tag{7.6.20}$$

$$\bar{c}(\infty) = 0 \tag{7.6.21}$$

where $\alpha = \sqrt{\dfrac{s}{D^*}}$. Solution to (7.6.19) can be written as,

$$\bar{c} = Ae^{\alpha z} + Be^{-\alpha z} \tag{7.6.22}$$

Applying the boundary conditions we get $B = \dfrac{c_0}{s}$ from (7.6.20) and $A = 0$ from (7.6.21). Therefore,

$$\bar{c} = \frac{c_0}{s}e^{-\alpha z} \tag{7.6.23}$$

can be obtained. Now let $z = k\sqrt{D^*}$ and plug it in the above form to get,

$$\bar{c}(z) = c_0 \frac{e^{-Ks^{1/2}}}{s} \tag{7.6.24}$$

Taking the inverse transform, we obtain,

$$c(z,t) = c_0\left[1 - erf(\frac{z}{2(D^*t)^{1/2}})\right] \tag{7.6.25}$$

where $erf(\ )$ is the so called "error function". In the case of surface repository over a deep layer with advection and diffusion ($\neq$) (Figure 7.6-9), we recall equation (7.6.17),

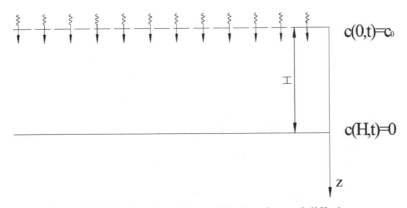

Figure 7.6-9 Surface depository with advection and diffusion

$$D^* \frac{\partial^2 c}{\partial z^2} = -v^* \frac{\partial c}{\partial z} = \frac{\partial c}{\partial t} \qquad (7.6.26)$$

Under the same boundary and initial conditions, with a similar approach, the solution to above equation can be readily obtained as,

$$c(z,t) = \frac{c_0}{2}\left[erfc(\frac{z - v^*t}{2\sqrt{D^*t}}) + \exp(\frac{v^*}{D^*})erfc(\frac{z - v^*t}{2\sqrt{D^*t}}) \right] \qquad (7.6.27)$$

where $erfc( )$ is the "complementary error function".

As a note, we remind the reader the Laplace transform and the error functions used in the solutions below,

$$L[f(t)] = \int_0^\infty e^{-st} f(t)dt = f(s) \qquad (7.6.28a)$$

$$f(t) = L^{-1}[f(s)] = \frac{1}{2\pi i} \int_{c-i\infty}^{c+i\infty} e^{st} f(s)ds \qquad (7.6.28b)$$

$$erf(z) = \frac{2}{\sqrt{\pi}} \int_0^z e^{-u^2} du \qquad (7.6.29a)$$

$$erfc(z) = 1 - erf(z) = \frac{2}{\sqrt{\pi}} \int_z^\infty e^{-u^2} du \qquad (7.6.29b)$$

along with typical Laplace transforms given in the table below.

Table 7.6-1 Typical Laplace transforms

$f(t) = inv(L[f(s)])$	$f(s) = L[f(t)]$
1	$\dfrac{1}{s}$
e^{-bt}	$\dfrac{1}{s+b}$
$\dfrac{\partial f}{\partial t}$	$sf(s) - f(0)$
$erf\left(\dfrac{k}{2\sqrt{t}}\right)$	$\dfrac{e^{-ks}}{s}\ (k > 0)$
$e^{a(k+ak)}erfc\left(\dfrac{a\sqrt{t}+k}{2\sqrt{t}}\right)$	$\dfrac{e^{-k\sqrt{s}}}{\sqrt{s}(a+\sqrt{s})}\ (k > 0)$

Computer Implementation

Implementation of the above-presented analytical solution into MAPLE is carried out (**Worksheet 5.mw**) and the results are presented below in Figures 7.6-10 and 7 6-11. The solution is given in terms of $erf(z)$ defined as in 7.6.29a. The shaded area in the figure below for $z > 0$ is $f(u) = \dfrac{2e^{-u^2}}{\sqrt{\pi}}$. The 3-D plot is obtained if we take $D^* = 1$ and $c_0 = 2$ in the equation 7.6.25 getting,

$$c(z,t) = 2 - 2erf\left(\frac{z}{2\sqrt{t}}\right) \qquad (7.6.30)$$

7.7 1-D Coupled Flow and Deformation

7.7.1 Introduction

The details of the equations governing the response of a saturated porous medium to external loading were presented in Chapter 3. In general, the underlying process is that of coupled flow and deformation of the porous medium which can, only under some conditions such as the assumptions made in section 7.5.2, be treated as a decoupled process. An example of this is

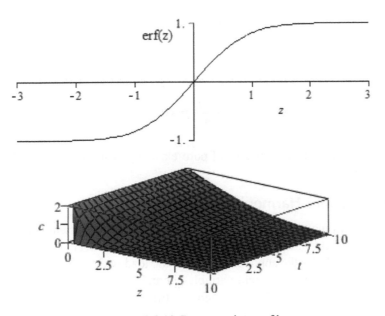

Figure 7.6-10 Concentration profile

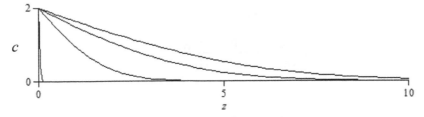

Figure 7.6-11 Concentration variation

Terzaghi's 1-D theory of consolidation. In this section, coupling of the related equations in the context of 1-D analysis of flow and deformation is presented.

7.7.2 Mathematical Formulation

Considering the saturated (or nearly saturated) porous medium composed of solid grains with different size and shape and of a viscous fluid (generally water), it can be viewed as a two-phase material where the particles of both phases are assumed to behave collectively as a continuous medium. For developing a general mathematical formulation governing the response of a saturated soil, the following two assumptions are made:

1. The water and the gas phases within the porous medium are considered as a single compressible fluid.
2. The effects of gas diffusing through water and movement of water vapor are ignored.

The action of either monotonic or cyclic loading on the saturated soil causes a stress field and associated pore water pressure in the domain. The response and stability of the soil elements collectively (i.e. integrated over the whole domain) govern the response and stability of the entire domain. For the case where all the inertial terms are neglected, the equations governing the "quasi-static" (QS) response (discussed briefly in Chapter 3) are obtained:

$$\frac{\partial \sigma}{\partial z} + \rho g = 0 \tag{7.7.1a}$$

$$-\frac{\partial p}{\partial z} + \rho_f g = \frac{\rho_f g}{k_z}\dot{\bar{w}} \tag{7.7.1b}$$

$$\frac{\partial}{\partial z}\dot{u} + \frac{\partial}{\partial z}\dot{\bar{w}} + \frac{n}{K_f}\dot{p} = 0 \tag{7.7.1c}$$

which are nothing but the Biot equations of poro-elasticity (with the Biot coefficient, $\alpha = 1$ as valid for most soils).

7.7.3 Response under Harmonic Loading

It is possible to manipulate the above set such that the resulting system consists of only two equations in terms of field variables u, solid displacement and $\bar{w}$, average relative pore fluid (i.e. water) displacement. Under the application of periodic load with $q_0 e^{i\omega t}$, q_0 being the amplitude, and ω the angular frequency, the field variables, u and $\bar{w}$ are replaced with their harmonic equivalents of $Ue^{i\omega t}$ and $\bar{W}e^{i\omega t}$ which are then substituted into the governing equations giving the following system,

$$\frac{\partial^2 U}{\partial \bar{z}^2} + \kappa\frac{\partial^2 \bar{W}}{\partial \bar{z}^2} = 0 \tag{7.7.2a}$$

$$\kappa\frac{\partial^2 U}{\partial \bar{z}^2} + \kappa\frac{\partial^2 \bar{W}}{\partial \bar{z}^2} = \frac{i}{\Pi_1}\bar{W} \tag{7.7.2b}$$

where

$$V_c^2 = \frac{D + K_f/n}{\rho} \tag{7.7.3a}$$

is the speed of compression wave, k is the permeability of the medium and

$$\kappa = \frac{K_f/n}{D + K_f/n} \tag{7.7.3b}$$

is the ratio of pore fluid bulk modulus to the total soil bulk modulus,

$$\Pi_1 = \frac{kV_c^2}{g\beta\omega H^2} \tag{7.7.3c}$$

controls how fast the compression wave travels through the porous medium in relation to the induced wave,

$$\beta = \frac{\rho_f}{\rho} \tag{7.7.3d}$$

as the density ratio and

$$\bar{z} = \frac{z}{H} \tag{7.7.3e}$$

as the normalized depth ratio are the non-dimensional parameters defined in the physical process. The resulting form is equivalent to the quasi-static consolidation problem including the compressibility of water. In matrix form the above equations become,

$$\begin{bmatrix} DD^2\kappa & \left(\dfrac{i}{\Pi_1} + DD^2\kappa\right) \\ (DD^2) & DD^2\kappa \end{bmatrix} \begin{bmatrix} U \\ \bar{W} \end{bmatrix} = 0 \tag{7.7.4}$$

The closed form solution of these equations is as follows:

$$\kappa\frac{\partial^2 U}{\partial\bar{z}^2} - \frac{\partial^2 U}{\partial\bar{z}^2} = \frac{i}{\Pi_1}\bar{W} \tag{7.7.5a}$$

$$\frac{\partial^2 U}{\partial\bar{z}^2} = \frac{i}{(\kappa-1)\Pi_1}\bar{W} \tag{7.7.5b}$$

which, if substituted into (7.7.2) becomes:

$$\frac{\partial^2 \bar{W}}{\partial\bar{z}^2} = \frac{-i}{\kappa(\kappa-1)\Pi_1}\bar{W} \tag{7.7.6}$$

with a solution of,

$$\bar{W} = C_1 e^{+\sqrt{A}\bar{z}} + C_2 e^{-\sqrt{A}\bar{z}} \tag{7.7.7}$$

where A is $\dfrac{-i}{\kappa(\kappa-1)\Pi_1}$ and C_1 and C_2 are the constants. Substitution of (7.7.7) into (7.7.5) and the following successive integration with respect to $\bar{z}$ yields:

$$U = -\kappa(C_1 e^{\sqrt{A}\bar{z}} + C_2 e^{-\sqrt{A}\bar{z}}) + C_3\bar{z} + C_4 \tag{7.7.8}$$

Using the boundary conditions from Figure 7.7-1, we have,

$$\left[\begin{array}{l} z = 0 \rightarrow \bar{z} = 0 \rightarrow \partial U\big/\partial\bar{z} = \dfrac{\bar{q}}{D} \rightarrow \dfrac{\bar{q}}{D} = -\kappa\sqrt{A}(C_1 - C_2) + C_3 \\[3mm] \rightarrow C_3 = \dfrac{\bar{q}}{D} + \kappa\sqrt{A}(C_1 - C_2) \end{array}\right. \tag{7.7.9a}$$

$$\left[\begin{array}{l} z = H \rightarrow \bar{z} = 1 \rightarrow U = 0 \rightarrow \kappa(C_1 e^{+\sqrt{A}} + C_2 e^{-\sqrt{A}}) = C_3 + C_4 \\[3mm] \rightarrow C_4 = \kappa(C_1 e^{+\sqrt{A}} + C_2 e^{-\sqrt{A}}) - \dfrac{\bar{q}}{D} - \kappa\sqrt{A}(C_1 - C_2) \end{array}\right. \tag{7.7.9b}$$

$$\left[z = 0 \rightarrow \bar{z} = 0 \rightarrow \partial\bar{W}\big/\partial\bar{z} = -\dfrac{\bar{q}}{D} \rightarrow \dfrac{\bar{q}}{\sqrt{AD}} = (C_1 - C_2) \right. \tag{7.7.9c}$$

$$\left[z = H \rightarrow \bar{z} = 1 \rightarrow \bar{W} = 0 \rightarrow (C_1 e^{+\sqrt{A}} + C_2 e^{-\sqrt{A}}) = 0 \rightarrow C_2 = (-C_1 e^{+2\sqrt{A}}) \right. \tag{7.7.9d}$$

resulting in $C_1 = \dfrac{-\bar{q}}{\sqrt{AD}(1 + e^{+2\sqrt{A}})}$, $C_2 = \dfrac{\bar{q}e^{+2\sqrt{A}}}{\sqrt{AD}(1 + e^{+2\sqrt{A}})}$, $C_3 = \dfrac{\bar{q}}{D}(1 - \kappa)$ and $C_4 = -\dfrac{\bar{q}}{D}(1 - \kappa)$

So, (7.7.7) and (7.7.8) become:

$$\bar{W} = \dfrac{-\bar{q}}{\sqrt{AD}(1 + e^{+2\sqrt{A}})} e^{\sqrt{A}\bar{z}} + \dfrac{\bar{q}}{\sqrt{AD}(1 + e^{+2\sqrt{A}})} e^{\sqrt{A}(2-\bar{z})} \tag{7.7.10a}$$

$$U = -\kappa\left(\dfrac{-\bar{q}e^{\sqrt{A}\bar{z}}}{\sqrt{AD}(1 + e^{2\sqrt{A}})} + \dfrac{\bar{q}e^{\sqrt{A}(2-\bar{z})}}{\sqrt{AD}(1 + e^{2\sqrt{A}})} \right) + \dfrac{\bar{q}}{D}(1 - \kappa)(\bar{z} - 1) \tag{7.7.10b}$$

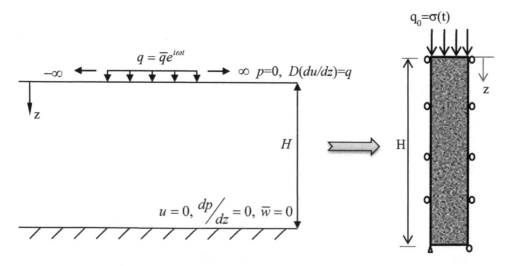

Figure 7.7-1 A layer of saturated porous medium under constant harmonic load

The actual displacements are $u = Ue^{i\omega t}, w = \bar{W}e^{i\omega t}$, thus finally,

$$\bar{w}(\bar{z},t) = \frac{\bar{q}}{D\sqrt{A}(1+e^{+2\sqrt{A}})}(e^{\sqrt{A}(2-\bar{z})} - e^{\sqrt{A}\bar{z}})e^{i\omega t} \tag{7.7.11a}$$

$$u(\bar{z},t) = \left[-\kappa\left(\frac{-\bar{q}e^{\sqrt{A}\bar{z}}}{\sqrt{A}D(1+e^{2\sqrt{A}})} + \frac{\bar{q}e^{\sqrt{A}(2-\bar{z})}}{\sqrt{A}D(1+e^{2\sqrt{A}})}\right) + \frac{\bar{q}}{D}(1-\kappa)(\bar{z}-1)\right]e^{i\omega t} \tag{7.7.11b}$$

$$\rightarrow u(\bar{z},t) = -\frac{\bar{q}}{D}\left[\kappa\left(\frac{e^{\sqrt{A}(2-\bar{z})} - e^{\sqrt{A}\bar{z}}}{\sqrt{A}(1+e^{2\sqrt{A}})}\right) + (1-\kappa)(\bar{z}-1)\right] \tag{7.7.12}$$

By using (7.7.11) and recalling D'Arcy's law, pore pressure can be calculated as,

$$p(z,t) = i\omega\frac{\rho_f g}{k}\frac{\bar{q}H}{AD(1+e^{2\sqrt{A}})}e^{i\omega t}\left[e^{\sqrt{A}\frac{z}{H}} + e^{\sqrt{A}\left(2-\frac{z}{H}\right)}\right] + C_5 \tag{7.7.13}$$

where C_5 can be found from the boundary conditions on the pore pressure as,

$$z = 0, \ p = 0, \ \rightarrow C_5 = -i\omega\frac{\bar{q}\rho_f g H}{DAk}e^{i\omega t} \tag{7.7.14}$$

$$p(z,t) = i\omega\frac{\rho_f g\bar{q}H}{ADk}\left[\frac{e^{\sqrt{A}\frac{z}{H}} + e^{\sqrt{A}\left(2-\frac{z}{H}\right)}}{(1+e^{2\sqrt{A}})} - 1\right]e^{i\omega t} \tag{7.7.15}$$

Computer Implementation

The preceding solution has been implemented in MATLAB code "**Coupled_1D_QS.m**" and following results have been obtained for a column of soil (Figure 7.7-2) for various soil and wave parameters. Figure 7.7-3 shows the variation of pore pressure profile for various soil permeability and Young's moduli.

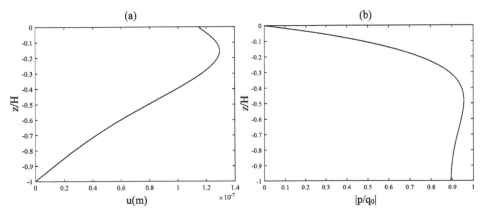

Figure 7.7-2 QS solution for (a) Displacement and (b) Pore pressure
for $H = 10$ m, $k_z = 10^{-4}$ m/s, $\omega = 3.4$ Hz

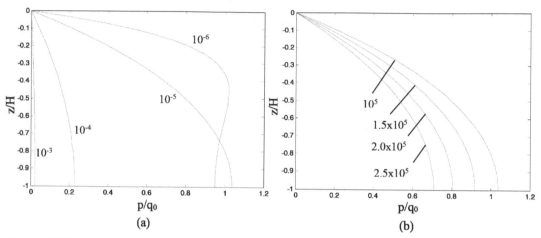

Figure 7.7-3 Pore pressure response for various (a) Permeability (m/s) (E = 100 MPa), (b) Young's modulus (kPa) (k = 10^{-5} m/s), T = 10 s

7.8 2-D Coupled Flow and Deformation

In this section, the solution of the 2-D response under progressive wave loading is presented. Writing the governing equations of coupled-flow and deformation (7.7-1) in 3-D form yields,

$$\sigma_{ij,j} + \rho g_i = 0 \tag{7.8.1a}$$

$$-p_{w,i} + \rho_f g_i - \frac{\rho_f g_i}{k_i}\dot{\bar{w}}_i = 0 \tag{7.8.1b}$$

$$\dot{u}_{i,i} + \dot{\bar{w}}_{i,i} + \frac{n}{K_f}\dot{p}_w = 0 \tag{7.8.1c}$$

7.8.1 Response of Soil Layer under Progressive Wave

Another basic problem of interest is a layer of saturated porous medium under a simple harmonic wave motion propagating over the surface of, for example, an ocean wave generated in the deep water progressing towards the shore. Analytical solutions are readily developed (Ulker, 2009; Ulker and Rahman, 2009) assuming that the response variables will also be harmonic in time and vary also in the horizontal direction (Figure 7.8-1). So we first write the scalar form of (7.7.1) with $Q = \dfrac{K_f}{n}$ and get,

$$K\frac{\partial^2(u_x)}{\partial x^2} + \lambda\frac{\partial^2(u_z)}{\partial x\partial z} + G\left(\frac{\partial^2(u_x)}{\partial z^2} + \frac{\partial^2(u_z)}{\partial x\partial z}\right) + Q\left(\frac{\partial^2(u_x)}{\partial x^2} + \frac{\partial^2(u_z)}{\partial x\partial z}\right) + Q\left(\frac{\partial^2(\bar{w}_x)}{\partial x^2} + \frac{\partial^2(\bar{w}_z)}{\partial x\partial z}\right) = 0 \tag{7.8.2a}$$

$$K\frac{\partial^2(u_z)}{\partial z^2} + \lambda\frac{\partial^2(u_x)}{\partial x\partial z} + G\left(\frac{\partial^2(u_z)}{\partial x^2} + \frac{\partial^2(u_x)}{\partial x\partial z}\right) + Q\left(\frac{\partial^2(u_z)}{\partial z^2} + \frac{\partial^2(u_x)}{\partial x\partial z}\right) + Q\left(\frac{\partial^2(\bar{w}_z)}{\partial z^2} + \frac{\partial^2(\bar{w}_x)}{\partial x\partial z}\right) = 0 \tag{7.8.2b}$$

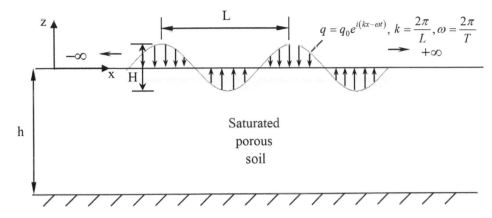

Figure 7.8-1 Two dimensional saturated porous medium under progressive harmonic load

$$Q\left(\frac{\partial^2(u_x)}{\partial x^2} + \frac{\partial^2(u_z)}{\partial x\partial z}\right) + Q\left(\frac{\partial^2(\bar{w}_x)}{\partial x^2} + \frac{\partial^2(\bar{w}_z)}{\partial x\partial z}\right) = \frac{\rho_f g}{k_x}(\dot{\bar{w}}_x) \qquad (7.8.2c)$$

$$Q\left(\frac{\partial^2(u_z)}{\partial z^2} + \frac{\partial^2(u_x)}{\partial x\partial z}\right) + Q\left(\frac{\partial^2(w_z)}{\partial z^2} + \frac{\partial^2(w_x)}{\partial x\partial z}\right) = \frac{\rho_f g}{k_z}(\dot{\bar{w}}_z) \qquad (7.8.2d)$$

Considering all the response variables to be of the form again, $u(x, z, t) = U(z)e^{i(kx-\omega t)}$, we can obtain a harmonic complex form of the governing equations where $U(z)$ represents the amplitudes of the response variables which in this case are the solid and relative fluid displacements, $U_x, U_z, \bar{W}_x, \bar{W}_z$. If we divide equations (7.8-1) by $K + \dfrac{K_f}{n}$ and introduce the normalized depth ratio, $\bar{z} = \dfrac{z}{h}$, then the resulting system becomes,

$$h^2\left(-k^2\right)U_x + ikh\left(\frac{\dfrac{K_f}{n} + \lambda + G}{K + \dfrac{K_f}{n}}\right)\frac{dU_z}{d\bar{z}} + h^2\left(\frac{-k^2\dfrac{K_f}{n}}{K + \dfrac{K_f}{n}}\right)\bar{W}_x + \frac{ikh\dfrac{K_f}{n}}{K + \dfrac{K_f}{n}}\frac{d\bar{W}_z}{d\bar{z}} + \frac{G}{K + \dfrac{K_f}{n}}\frac{d^2U_x}{d\bar{z}^2} = 0 \quad (7.8.3a)$$

$$\frac{d^2U_z}{d\bar{z}^2} + \frac{ikh\left(\dfrac{K_f}{n} + \lambda + G\right)}{K + \dfrac{K_f}{n}}\frac{dU_x}{d\bar{z}} + \frac{\dfrac{K_f}{n}}{K + \dfrac{K_f}{n}}\frac{d^2\bar{W}_z}{d\bar{z}^2} + \frac{ikh\dfrac{K_f}{n}}{K + \dfrac{K_f}{n}}\frac{d\bar{W}_x}{d\bar{z}} + h^2\left(\frac{-k^2G}{K + \dfrac{K_f}{n}}\right)U_z = 0 \quad (7.8.3b)$$

$$\left(\frac{-k^2h^2\dfrac{K_f}{n}}{K + \dfrac{K_f}{n}}\right)U_x + h\left(\frac{\dfrac{i\omega\rho_f hg}{k_x} - k^2h\dfrac{K_f}{n}}{K + \dfrac{K_f}{n}}\right)\bar{W}_x + ikh\frac{\dfrac{K_f}{n}}{K + \dfrac{K_f}{n}}\left(\frac{dU_z}{d\bar{z}} + \frac{d\bar{W}_z}{d\bar{z}}\right) = 0 \qquad (7.8.3c)$$

$$\frac{\frac{K_f}{n}}{K+\frac{K_f}{n}}\left(\frac{d^2U_z}{d\bar{z}^2}+\frac{d^2\bar{W}_z}{d\bar{z}^2}\right)+\frac{ikh\frac{K_f}{n}}{K+\frac{K_f}{n}}\frac{dU_x}{d\bar{z}}+\frac{ikh\frac{K_f}{n}}{K+\frac{K_f}{n}}\frac{d\bar{W}_x}{d\bar{z}}+h^2\left(\frac{i\omega\rho_f g\Big/k_z}{K+\frac{K_f}{n}}\right)\bar{W}_z=0 \qquad (7.8.3\text{d})$$

In addition to the ones in the previous 1-D set, following non-dimensional parameters are now introduced:

$$\kappa_1=\frac{\lambda}{K+\frac{K_f}{n}} \qquad (7.8.4)$$

as the ratio of the volumetric part of Lame's constants to the bulk modulus of the system;

$$\kappa_2=\frac{G}{K+\frac{K_f}{n}} \qquad (7.8.5)$$

as the ratio of the shear modulus to the bulk modulus of the system;

$$m=kh \qquad (7.8.6)$$

as the spatial variation of the loading where $k=2\pi/L$ is the wave number, h is the depth of the medium and L is the wavelength;

$$\Pi_{1x}=\frac{k_x V_c^2}{g\beta\omega h^2} \qquad (7.8.7)$$

as the ratio of time for pore fluid flow in the x-direction to time for compression wave to travel;

$$\Pi_{1z}=\frac{k_z V_c^2}{g\beta\omega h^2} \qquad (7.8.8)$$

same as above, now accounting for the flow in z-direction. Here due to second dimension, Π_1 has now two components (Π_{1x} and Π_{1z}). An increase in Π_{1x} or Π_{1z} means a more drained behavior or a slower loading; however, higher frequency of the loading causes a decrease in Π_{1x} and Π_{1z}. An increase in m corresponds to a smaller wavelength implying faster spatial fluctuation of the loading. As m goes to zero, the response of porous medium will approach to one corresponding to 1-D analysis. The governing equations, in terms of these parameters, can be written in the following matrix form below where DD denotes $\partial/d\bar{z}$ and DD^2, $\partial^2/\partial\bar{z}^2$.

$$\begin{bmatrix} -m^2\kappa & im\kappa DD & \left(\frac{i}{\Pi_{1x}}-m^2\kappa\right) & im\kappa DD \\ im\kappa DD & \kappa DD^2 & im\kappa DD & \left(\frac{i}{\Pi_{1z}}+\kappa DD^2\right) \\ (-m^2+\kappa_2 DD^2) & im(\kappa+\kappa_1+\kappa_2)DD & -m^2\kappa & im\kappa DD \\ im(\kappa+\kappa_1+\kappa_2)DD & (-m^2\kappa_2+DD^2) & im\kappa DD & \kappa DD^2 \end{bmatrix}\begin{bmatrix} U_x \\ U_z \\ \bar{W}_x \\ \bar{W}_z \end{bmatrix}=0 \quad (7.8.9)$$

If in the above matrix form, we neglect the wave number, k, and take $m = 0$, then it becomes the 1-D solution provided that $\Pi_{1x} = 0$ and $\Pi_{1z} = \Pi_1$.

Boundary Conditions

The boundary conditions that are used for the problem in Figure 7.8-1 are: at $z = -h$, the three displacements go to zero, $U_x = U_z = \bar{W}_z = 0$ and at $z = 0$, vertical effective stress and shear stress vanish ($\sigma'_{zz} = \tau_{xz} = 0$) and pore pressure becomes $p_w = q_0 e^{i(kx-\omega t)}$ where q_0 is the amplitude of the load.

Closed Form Solution

Using these boundary conditions, a 6×6 matrix with elements $\xi_{ij}(i, j = 1, 2, 3, 4, 5, 6)$ is obtained in terms of eigenvalues η_i ($i = 1, 2, 3, 4, 5, 6$) that are obtained from the characteristic equation, $\det [M]_j = 0$ defined as,

$$(\alpha_1 DD^6 + \alpha_2 DD^4 + \alpha_3 DD^2 + \alpha_4)_j = 0. \tag{7.8.10}$$

Here, depending on the formulation, α coefficients will be functions of the non-dimensional parameters defined before. The roots of the characteristic equation are evaluated and the desired wave-induced displacements, stresses and pore pressure can be calculated as,

$$u_x = \left[\sum_{j=1}^{6} a_j e^{\eta_j \frac{z}{h}} \right] e^{i(kx-\omega t)} \tag{7.8.11a}$$

$$u_z = \left[\sum_{j=1}^{6} a_j b_j e^{\eta_j \frac{z}{h}} \right] e^{i(kx-\omega t)} \tag{7.8.11b}$$

$$\bar{w}_x = \left[\sum_{j=1}^{6} a_j c_j e^{\eta_j \frac{z}{h}} \right] e^{i(kx-\omega t)} \tag{7.8.11c}$$

$$\bar{w}_z = \left[\sum_{j=1}^{6} a_j d_j e^{\eta_j \frac{z}{h}} \right] e^{i(kx-\omega t)} \tag{7.8.11d}$$

$$\sigma'_{xx} = \left[\sum_{j=1}^{6} \left(ikK + b_j \lambda \frac{\eta_j}{h} \right) a_j e^{\eta_j \frac{z}{h}} \right] e^{i(kx-\omega t)} \tag{7.8.11e}$$

$$\sigma'_{zz} = \left[\sum_{j=1}^{6} \left(ik\lambda + b_j K \frac{\eta_j}{h} \right) a_j e^{\eta_j \frac{z}{h}} \right] e^{i(kx-\omega t)} \tag{7.8.11f}$$

$$\tau_{xz} = G \left[\sum_{j=1}^{6} \left(\frac{\eta_j}{h} + ikb_j \right) a_j e^{\eta_j \frac{z}{h}} \right] e^{i(kx-\omega t)} \tag{7.8.11g}$$

$$p_w = -\frac{K_f}{n} \left[\sum_{j=1}^{6} \left(ik(1 + c_j) + \frac{\eta_j}{h}(b_j + d_j) \right) a_j e^{\eta_j \frac{z}{h}} \right] e^{i(kx-\omega t)} \tag{7.8.11h}$$

The details of the solution can be found in the Appendix and other sources as Ulker and Rahman (2009). Here, the response obtained using the solutions developed is compared with the one obtained by Madsen (1978) and Rahman et al. (1994). Firstly, the pore pressure response for two saturations and bulk moduli are compared with Madsen (1978) (Figure 7.8-2).

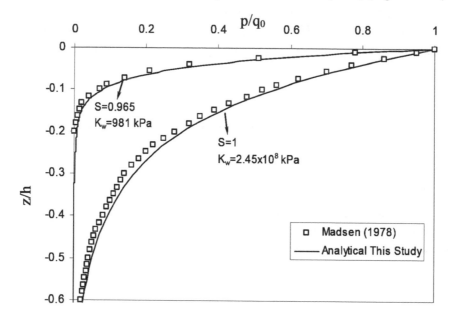

Figure 7.8-2 Pore pressure variation normalized with wavelength from Madsen (1978)

Another verification attempt was made with a comparison using the analytical solutions developed by Rahman et al. (1994). In that study, the inertial terms associated with both the soil skeleton motion and the fluid motion are also neglected. Here the normalized pore pressure (p/q_0), normalized vertical effective (σ'_{zz}/q_0) and shear stress (σ_{xz}/q_0) responses are compared with each other and the results are found to be in well agreement (Figure 7.8-3). A parametric study is also carried out and the resulting response plots are presented in a set of two figures below (Figure 7.8-4) for various representative soil and wave parameters followed by another set of results including stress and solid displacement variations in Figure 7.8-5. The soil parameters affecting the response are the permeability (k_z), degree of saturation (S), and the depth of the soil (h/L), whereas for the wave loading, wave steepness (H/L) and the wave period (T) are the dominant factors. Soil is assumed primarily to be fully saturated ($S = 1$); however the effect of slight unsaturation on the response is also investigated. Additionally, the soil is assumed to be isotropic in terms of permeability which is taken as $k_z = 0.002$ m/s.

In summary, 1-D and 2-D analytical solutions to basic problems are presented. Physical models are described with related math models through prescribed boundary conditions in terms of governing partial differential equations. Exact solutions are derived for simplified domains and material parameters. The results are presented in terms of variation of field variables and stresses in time and space. Specifically, in the solution of 2-D coupled problem, non-dimensional parameters play a key role to define the characteristics of loading and the porous medium (i.e. soil).

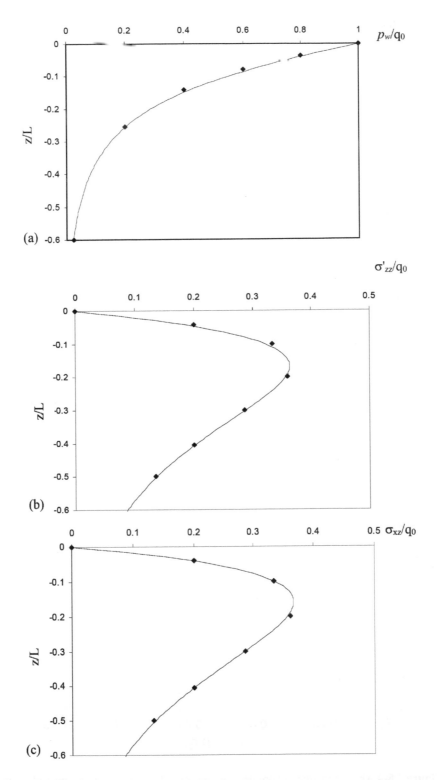

Figure 7.8-3 Comparison between Ulker and Rahman (2009) (solid lines) and Rahman et al. (1994) (markers) for: (a) Pore pressure, (b) Effective vertical stress, (c) Shear stress response

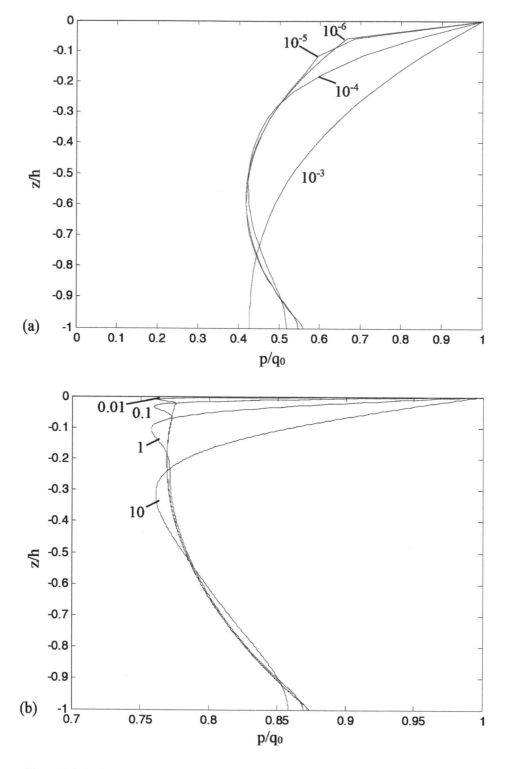

Figure 7.8-4 Absolute value of pore pressure response for various: (a) Permeability (m/s) with $T = 10$ s, (b) Wave period (s) with $k_x = k_z = 10^{-5}$ m/s

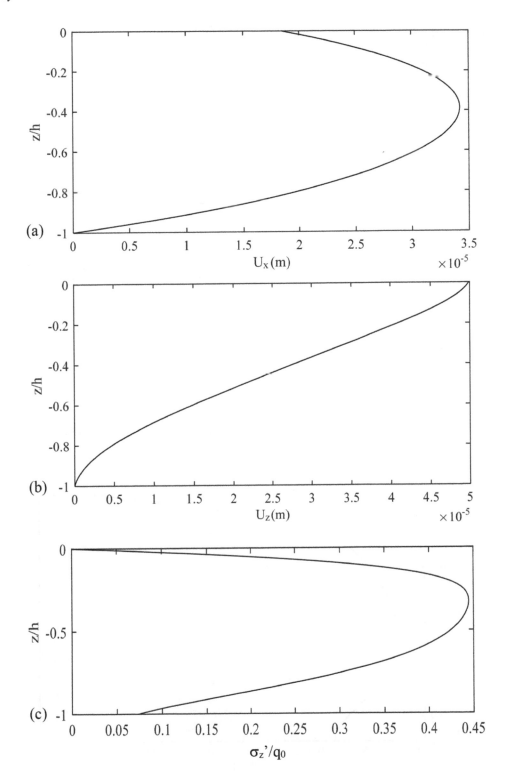

Figure 7.8-5 Absolute value of response variations: (a) Horizontal displacement, (b) Vertical displacement (s) with $k_x = k_z = 10^{-4}$ m/s, (c) Vertical effective stress, cont'd.

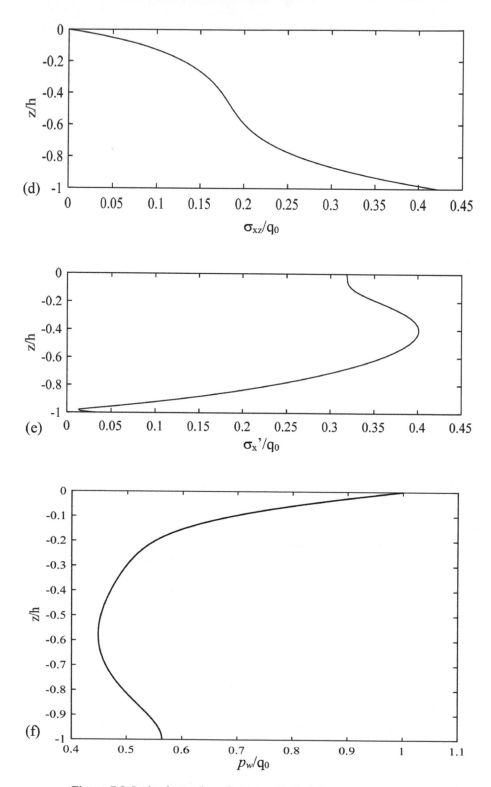

Figure 7.8-5 Absolute value of response variations: (d) Shear stress,
(e) Horizontal effective stress, (f) Pore pressure

7.9 Exercise Problems

7.9.1   A stream flows through an alluvial valley bounded by impermeable rocks as shown in the figure below. The lands adjacent to the stream are being considered for a development. Recharge to the unconfined alluvial aquifer is expected to increase which may cause the water table to become too close to the land surface unless artificial drainage is used. Treating this as a recharge problem, develop a MAPLE program to calculate and plot the water table elevation as a function of distance from the stream relative to the water level in the stream, for any geometry, recharge rate and the coefficient of permeability.

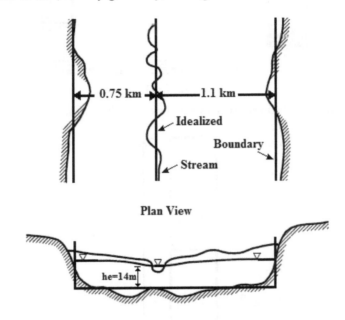

Figure 7.9-1 Recharge problem (McWhorter and Sunada, 1981)

7.9.2   For the regional ground water flow to take place across the following region in Figure 7.9-2, solve the 2-D steady state seepage problem for total head and also sketch the flow lines and equipotential lines for various head differences and geometrical dimensions.

7.9.3   Consider a column of soil given in Figure 7.9-3. Under its own weight, calculate the displacement and stress profile and the top displacement as well as bottom stress for E = 10 MPa and v = 0.3.

7.9.4   See the soil profile below (Figure 7.9-4) under a constant load q. The soil sample recovered from the middle of the clay layer (see the figure) has been tested in an oedometer for 1-D consolidation and later another sample from the same location has been tested in a triaxial device in the lab. These tests led to the following soil properties: $k_z = 10^{-7}$ m/s, $v = 0.45$, $E = 50$ MPa, $\gamma_{clay} = 20$ kN/m³, $\gamma_w = 9.81$ kN/m³. Obtain an analytical solution for the consolidation of the clay layer in terms of pore water pressure, p considering a column of that layer and all the necessary physical parameters such as H, k_z, γ_w etc. Hint: You should consider the stress distribution in the sand layer (reduction in the surface load, q with depth) to evaluate the effective stress increase in the mid-point of clay layer. Hydrostatic pore pressure at any depth below GWT is evaluated as $p = \gamma_w z_w$. What is the total consolidation settlement of the clay layer? How much time is needed for 90% (t_{90}) of that settlement to complete? Also develop a

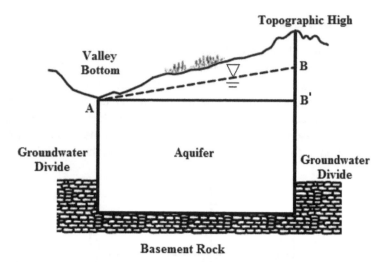

Figure 7.9-2 Regional ground water flow (McWhorter and Sunada, 1981)

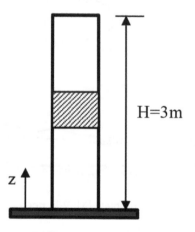

Figure 7.9-3 Regional ground water flow

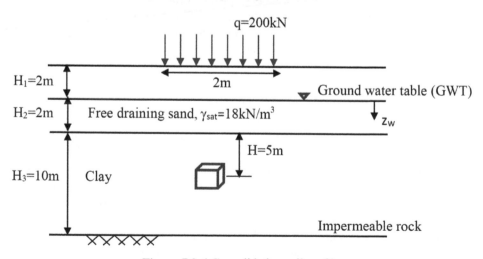

Figure 7.9-4 Consolidating soil profile

MATLAB program and investigate the effect of C_v (coefficient of consolidation) on the pore pressure variation with depth.

7.9.5   For an infinitely deep deposit with a constant surface concentration, c_0, develop an analytical solution to the advection-dispersion equation. Your results should match with those from Ogata and Banks (1961) for following conditions, $c(z, 0) = 0 \rightarrow z > 0$, $c(0, t) = c_0 \rightarrow t \geq 0$ and $c(\infty, 0) = 0 \rightarrow t \geq 0$ (see Figure 7.9-5).

7.9.6   For a column of two-phase soil, develop a MAPLE worksheet and plot the variation of pore pressure and soil displacement in depth and time for various soil and wave parameters. Compare your results with the ones you got from problem 7.9.4 for the same input and soil properties.

7.9.7   Resolve the same problem in Figure 7.9-4 now considering a soil column under constant harmonic wave loading with 10 sec. period acting on the surface of the clay layer.

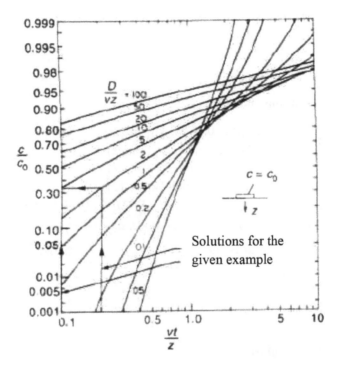

Figure 7.9-5 Variation of contaminant concentration for an infinitely deep soil deposit (Ogata and Banks, 1961); graphical values are obtained for $v_a = 0.002$ m/a, diffusion coefficient, $D = 0.01\text{-}0.02$ m²/a, porosity $n = 0.4$, $\rho K_d = 0\text{-}1.2$, source concentration $c_0 = 1500$ mg/l, time = 80-160 years, $z = 2$ m.

MAPLE® Worksheets

Worksheet 1: 1-D Flow: Island Recharge Problem

restart:
with(*DEtools*) :
with(*LinearAlgebra*) :
with(*Vector Calculus*) :
Interface (*warnlevel* = 0) :
Interface (*rtablesize* = 15) :
with(*plots*) :

Governing Equation:

> *ode* : = -*k*.*h*(*x*). *diff* (*h*(*x*),*x*) = *Rx*,

Boundary Conditions:

> *ics* : = $h\left(\dfrac{L}{2}\right) = b$;

Idealized Problem (Solution Domain):

pi : = *plot*([[[— 0.5, 0], [4, 0]], [[0, –0.5], [0, 3]]], *color* = *black*) :
p2 : = *plot*([[[0, 2], [3, 2]], [[3, 2], [3, 0]]], *color* = *black*) :
p3 := *textplot*({ [3.7, 1, "h(L/2,z) = b"], [1.5, 2.2, "h(x,z) = f(x)"],
    [3.1, 2.2, "(L/2,b)"]}, *font* = [*TIMES,ROMAN*, 12]) :
p4 : = *textplot* ({[4, 0.2, "x"], [0.2, 3, "z"]}, *font* = [*TIMES,ITALIC*, 12]):
display ([*p*|(1..4)], *scaling* = *constrained, axes* = *none, title* = "Figure 1. Solution Domain")

Top Boundary Condition: Total Head

R : = 0.000005
b : = 10
L : = 1000
k : = 0.1

Analytical Closed Form Solution:

> *s* : = *dsolve*(*ode*);

> *sol* : = $h(x) = \dfrac{\sqrt{-k(Rx^2 - C1\,k)}}{k}$

> eql : = *eval* $\left(rhs(sol), \ x = \dfrac{L}{2}\right) = b$;

> _*C1* := *solve*(*eql, _C1*);
> *h* := *eval*(*rhs*(*sol*));
plot(*h,x* =–500 ..500, *title* = "Figure 1. Variation of h", *labels* = [*x*,'*h*'])
> *z* := *h – b*
plot(*z,x* = –500 ..500, *title* = "Figure 2. Water table elevation", *labels*
    = [*x*,'z_w'])

Worksheet 2: Steady State Seepage

with(*plots*) :
with(*DEtools*) :
with(*LinearAlgebra*) :
with(*VectorCalculus*) :
Interface(*warnlevel* = 0) :
Interface(*rtablesize* = 15) :

Governing Equation:

$h_{xx} + h_{yy} = 0$

Boundary Conditions:

$h'(x, 0) = 0 \quad h'(0, y) = 0 \quad h'(L, y) = 0 \quad h(x, b) = f(x)$

Idealized Problem (Solution Domain):

p1 := *plot*([[[−0.5, 0], [4, 0]], [[0, −0.5], [0, 3]]], *color* = *black*) :
p2 := *plot*([[[0, 2], [3, 2]], [[3, 2], [3, 0]]], *color* = *black*) :
p3 := *textplot*({[− .7, 1, "*h*'(0, *y*) = 0"], [3.7, 1, "*h*'(*L*, *y*) = 0"], [1.5, 2.2,
    "*h*(*x*, *b*) = *f*(*x*)"], [1.5,−.3, "*h*'(*x*, 0) = 0"], [3.1, 2.5, "(*a*, *b*)"]}, *font* =
    [*TIMES,ROMAN,* 12]):
p4 := *textplot*({[4, 0.2, "*x*"], [0.2, 3, "*y*"]}, *font* =[*TIMES,ITALIC,* 12]):
display([*p*|(1 ..4)], *scaling* = *constrained*, *axes* = *none*, *title*
    = "Figure 1. Solution Domain")

Top Boundary Condition: Total Head

> $f := b + c.x$
$c := 1$
$b := 5$
$L := 10$
plot(*f*, *x* = 0..*L*, *scaling* = *constrained*, *color* = *red*, *labels* =[*x*, *h*],
    tickmarks = [10, 5], *title* = "Figure 2. Top Boundary Condition")

> $\dfrac{\partial^2}{\partial x^2} h + \dfrac{\partial^2}{\partial y^2} h = 0$

$h := F(x){\cdot}G(y) \xrightarrow{\text{differentiate w.r.t. x}} \left(\dfrac{d}{dx} F(x)\right) G(y) \xrightarrow{\text{differentiate w.r.t.x}} \left(\dfrac{d^2}{dx^2} F(x)\right) G(y)\, F(x)\, G(y)$

$h := \quad F(x){\cdot}G(y) \xrightarrow{\text{differentiate w.r.t. y}} F(x)\left(\dfrac{d}{dy} G(y)\right) \xrightarrow{\text{differentiate w.r.t.x}} F(x)\left(\dfrac{d^2}{dx^2} G(y)\right)$

$\Rightarrow \dfrac{\partial^2}{\partial x^2} h + \dfrac{\partial^2}{\partial y^2} h = 0$

$\left(\dfrac{d^2}{dx^2} F(x)\right) G(y) = - F(x)\left(\dfrac{d^2}{dy^2} G(y)\right)$

$$\xrightarrow{\text{manipulate equation}} \frac{\dfrac{d^2}{dx^2}F(x)}{F(x)} + \frac{\dfrac{d^2}{dy^2}G(y)}{G(y)} = \mu^2$$

$$\frac{d^2}{dx^2}F(x) + F(x)\mu^2 = 0$$

$$\frac{d^2}{dx^2}F(x) + F(x)\mu^2 = 0$$

$$> \frac{d^2}{dx^2}F(x) + F(x)\mu^2 = 0$$

$$\xrightarrow{\text{solve } DE} F(x)\_Cl\sin(\mu\,x) + \_C2\cos(\mu x) \quad \frac{d^2}{dy^2}G(y) - \mu^2 G(y) = 0 \xrightarrow{\text{solve } DE}$$

The constant coefficients C_1 and C_2 in both general solutions are obtained using the boundary conditions stated above.

$> h = (\_C_1\sin(\mu\,x) + \_C_2\cos(\mu x) \cdot (\_C_3\,e^{\mu y} + \_C4\,e^{-\mu y}$

Boundary Condition 1:

$$\frac{\partial}{\partial x}h = 0 \ @\ x = 0, 0 < y < b$$

$$(\_C_1\sin(\mu\,x) + - C_2\cos(\mu\,x))\,(\_C_3 e^{\mu y} + \_C_4^{-\mu y}) \xrightarrow{\text{differentiate w.r.t. x}}$$

$$(\_C_1\cos(\mu\,x)\mu - C_2\sin(\mu\,x)\mu)\,(\_C_3 e^{\mu y} + \_C_4 e^{-\mu y}) \xrightarrow{\text{evaluate at point}}$$

$$\_C_1\mu(\_C_3 e^{\mu y} + \_C_4 e^{-\mu y})$$

$$\_C_1 := 0$$

$$> h := \_C_2\cos(\mu x) \cdot (\_C_3 e^{\mu y} + \_C_4 e^{-\mu y})$$

Boundary Condition 2:

$$\frac{\partial}{\partial y}h = 0 \ @\ y = 0, 0 < x < L$$

$$> \ \_C_2\cos(\mu x) \cdot (\_C_3 e^{\mu y} + \_C_4 e^{-\mu y})$$

$$\xrightarrow{\text{differentiate w.r.t. y}} \_C_2\cos(\mu\,x)\,(\_C_3\,\mu e^{\mu y} - \_C_4\mu e^{-\mu y}) \xrightarrow{\text{evaluate at point}}$$

$$\_C_2\cos(\mu x)\,(\_C_3\mu - \_C_4\mu)\ \_C_3 = \_C_4$$

$$\_C3 = \_C4$$

$$h := \_C_2 \cdot C_3\cos(\mu x) \cdot (e^{\mu y} + e^{-\mu y})$$

Boundary Condition 3:

$$\frac{\partial}{\partial y}h = 0 @ x = L, 0 < y < b$$

$$\_C_2\_C_3\cos(\mu\,x)\,(e^{\mu y} + e^{-\mu y}) \xrightarrow{\text{differentiate w.r.t. x}} - \_C_2\_C_3\sin(\mu x)\mu\,(e^{\mu y} + e^{-\mu y})$$

$$\xrightarrow{\text{evalu at eat point}} - \_C_2 \_C_3 \sin(\mu L)\,\mu\,(e^{\mu y} + e^{-\mu y})$$

$$\sin(\mu L) = 0 \rightarrow \mu = \frac{n}{L}\pi, \, n = 0, 1, 2\ldots$$

Let $\_C2 \cdot \_C3 = A_n$

$$\mu := \frac{n}{L}\pi$$

$$h := \frac{2 \cdot A_n \cos(\mu x) \cdot (e^{\mu y} + e^{-\mu y})}{2}$$

$$h := A_n \cos\left(\frac{n\pi x}{L}\right) \cdot \cosh\left(\frac{n\pi y}{L}\right)$$

General Solution:

$$> h := \sum_{n=1}^{\text{infinity}} A_n \cos\left(\frac{n\,Pi\,x}{L}\right)\cosh\left(\frac{nPiy}{L}\right)$$

Boundary Condition 4:

$$f = b + cx @ y = b, \, 0 < x < L$$
$$> f := b + c \cdot x$$

$$A_0 = b + \frac{c \cdot L}{2}$$

$$A_n := \frac{2}{L} \cdot \int_0^L f \cdot \cos\left(\frac{n\pi x}{L}\right) dx$$

evaluate at point
$$\rightarrow n = 1$$

$$A_n := \frac{2 \cdot c \cdot L \cdot (\cos(n \cdot pi) - 1)}{n^2 \cdot pi^2 \cosh\left(\frac{b \cdot n \cdot pi}{L}\right)}$$

$$> h(x, y) = b + \frac{c \cdot L}{2} + \sum_{n=1}^{\text{infinity}} A_n \cos\left(\frac{n\,Pi\,x}{L}\right)\cosh\left(\frac{n\,Pi\,y}{L}\right)$$

$B := value\,(A_n)$ assuming $n :: integer$:
$pi := 3.142$:

$$h[5] := b + \frac{c \cdot L}{2} + sum\left(B \cdot \cos\left(\frac{n \cdot pi \cdot x}{L}\right) \cdot \cosh\left(\frac{n \cdot pi \cdot y}{L}\right), n = 1\ldots5\right)$$

>   *plot3d(h[5], x = 0 ..L, y = 0 ..b, color = yellow, axes = frame, style*
       = patchcontour, tickmarks = [4, 4, 2], scaling = constrained, title
       = "Figure 3. Potential Lines: 3D Plot")

>   *p5 := contourplot(h[5], x = 0 ..L, y = 0 ..b, color = red, contours = 15,*
       view = [0 ..L, 0 ..b], axes = bax, title

= "Figure 4. Equipotential Lines: 2-D Plot")

It is useful to see how the values of the coefficient decreases as we take more terms in the solution.

$s := seq(B, n = 1 ..5)$

We need to draw the flow lines in our domain. The flow lines are orthogonal to the equipotential lines, that is, they are along the negatives of the gradient vectors of $h(x,y)$. The vectors $-\nabla h$ are tangent vectors along the flow lines, which themselves are obtained by integrating the differential equations $x(m) = -h_x (x(m), y(m))$ and $y(m) = -h_y (x(m), y(m))$. So we need to evaluate the gradients of the total head first with respect to x and y. Here the minus sign indicates energy loss from lower to higher values whereas for example heat flows from higher to lower temperatures, but gradient vectors point from lower to higher function values.

$gh[5] : = - Gradient(h[5], [x, y])$:

The differential equations for the flow lines are:

$d1 := diff(x (m), m) = eval(gh[5][1], \{x = x (m), y = y (m)\})$:

$d2 := diff(y (m), m) = eval(gh[5][2], \{x = x (m), y = y (m)\})$:

Now we pick some initial points $[m_0, x(m_0), y(m_0)]$ from which to start the flow lines, designating them with letter G:

$$> initsG := seq\left(\left[0, \frac{k}{2}\right], k = 1..9\right)$$

Use the command "Fieldplot" for ploting the gradient vectors:

$p6 : = fieldplot(gh[5], x = 0 ..L, y = 0 .. b, grid = [10, 10], color = blue,$
  $view = [0 ..L, 0 ..b], axes = box, arrows = medium, title$
  $= $ "Figure 5. Flow Vectors")

Use the comman "Fieldplot" for plotting the gradient vectors:

$p6 := fieldplot(gh[5], x = 0 ..L, y = 0 ..b, grid = [10, 10], color = blue,$
  $view =[0 ..L, 0 ,.b], axes = box, arrows = medium, title$
  $= $ "Figure 5. Flow Vectors")

Use the command "DEplot" to solve the pair of differential equations for the flow lines.

$> p7 :=DEplot ([d1, d2], [x(m), y(m)], m = 0 ..20, x = 0 ..L, y = 0 ..b,$
  $[initsG], arrows = none, view = [0 ..L, 0 ..b], axes = box, linecolor$
  $= green, stepsize = 0.2, title = $ "Figure 6. Flow Lines") :

$> display([p_5, p_7], view =[0 .L, 0 ..b], axes = box)$

Now let us see all the figures combined:

$p5 := contourplot(h[5], x=0..L, y=0..b, color=red, view=[0..L,0..b], axes=box)$:
$gh[5] := Gradient(h[5],[x,y])$:
$d1 : = diff(x(m),m) = eval(gh[5][1], \{x=x(m),y=y(m)\})$:
$d2 := diff(y(m),m) = eval(gh[5][2], \{x=x(m),y=y(m)\})$:
$initsG := seq([0,k/2,5],k=1..9)$:
$p6 :=fieldplot((gh[5]),x=0..L,y=0..b,grid=[10,10], color=blue, arrows=medium)$:
$p7 := DEplot([d1,d2],[x(m),y(m)], m=0..20, x=0..L, y=0..b, [initsG], arrows=none,$
 $linecolor=green, stepsize=.1)$:
$display([p||(5..7)], tickmarks=[4,4], view=[0..L,0..b], title =$"*Figure 7.2-D Steady State*

Seepage");

Worksheet 3: 1-D Deformation of a Soil Column

Initializations:
restart;
interface(warnlevel=0):
interface(rtablesize=15):
with(plots):
with(DEtools):
with(VectorCalculus):
with(LinearAlgebra):
BasisFormat(false):

Varying Cross Section Area:

> a: =r2;
> b:=(r1–r2)/L;
> r(z) := (a + b·z);
> A(z) := π·(r(z)²);
> Eq1: =r(z) *diff(s(z),z) +2*s(z) *diff(r(z),z)=0;
> sol1:= simplify(dsolve({Eq1,s(L)=s1},s(z)));
> Eq2: =E*diff(u(z),z)-s1*(r1/(a+b*z))^2=0;
> sol 2:= simplify(dsolve [{Eq2, μ(0) = 0},u(z)));

Numerical Example:

L:=20;
s1:=20;
r1:=0.25;
>r2 = 0.75;
>E:=28000000
> stress := rhs(sol1);
> displacement := rhs(sol2);
> plot(stress, z = 0..20, color=black, labels = [z, sigma(z)], scaling
*   = constrained, tickmarks = [5, 4], thickness = 1, title*
*   = "Figure 1. Stress Variation")*
>plot (displacement, z=0 ..20, color = black, labels = [z, disp(z)],
*   tickmarks = [5, 8], thickness = 1, title*
*   = "Figure 2. Displacement Variation")*

Worksheet 4: 1-D Consolidation

with(plots) :
with(DEtools) :
with(Vector Calculus) :

Problem Statement:

Governing Equation	(1)	$p_t = c_v p_{zz}$	$t' > '0$
Boundary condition	(2)	$p(0, t) = 0$	top of the layer
Boundary condition	(3)	$p(2H, t) = 0$	bottom of the layer
Initial condition	(4)	$p(z, 0) = f(z)$	initial pore pressure profile

Closed Form Solution:

$P: = F(z)*G(t);$
$eval(P, z=0) = 0;$
$eval(P, z=L) = 0;$
$F(0) = 0$
$F(L) = 0$
$q: = diff(p(z,t),t) = kappa*diff(p(z,t),z,z);$
$q1: = eval(q, p(z,t) = P);$

$$\frac{G'(t)}{\kappa G(t)} = \frac{F'(z)}{F(z)} = \text{-lambda}^2$$

$\kappa Q_J)\ Hz)$
$q2: = q1/kappa/P;$
$q3:= rhs(q2) = -lambda\hat{}2;$
$q4:= lhs(q2) = -lambda\hat{}2;$

Clearing the denominator in each leads to;

$F''(z) = -\lambda^2 F(z)$
$G'(t) = -\lambda^2 \kappa G(t)$
$q5:= q3*F(z);$
$q6: = q4*G(t)*kappa;$
$q7:= lhs(q5) - rhs(q5) = 0;$
$q8:= lhs(q6) - rhs(q6) = 0;$
$q9:= eval(q8, lambda = n*Pi/L);$
$q10:= dsolve(q9,G(t));$
$P[n]:= D[n]*sin(n*Pi*z/L)*exp(-kappa*n\hat{}2*Pi\hat{}2/L\hat{}2*t);$

A linear combination of all such eigen solutions

$$p(z, t) = \sum_{n=1}^{\infty} D_n \sin\left(\frac{n\pi z}{L}\right) e^{-\frac{\kappa n^2 \pi^2 t}{L^2}}$$

$simplify(eval(Sum(P[n],n=0..infinity), t=0)) = f(z);$
means the coefficients D_n in the series for $p(z, t)$ must therefore be the Fourier sine series coefficients for representing $f(z)$ on the interval $[0, L]$, *we have:*

$$D_n = \frac{2}{L} \int_0^L f(z) \sin\left(\frac{n\pi x}{L}\right) dz$$

f := 1:
plot(f,z=0..2,color=cyan, discont=true, labels=[z,p], scaling=constrained, tickmarks=[3,2], thickness=3, title="Figure 1. Initial Pore Pressure");

qb:= (2/L)*Int(f*sin(n*pi*z/L),z=0..L):

While Maple writes that integral in a strange fashion, it will compute its value in a simplified form if we tell Maple that n is an integer. The value of the integral is then:

b:= combine(value(qb)) assuming n::integer:

The solution to the boundary value problem is therefore the infinite series,

$$p(z,t) = \frac{2}{\pi}$$

$$\sum_{n=1}^{\infty} \left(\frac{1- \cos(n \cdot \pi)}{n} \right) \sin\left(\frac{n \pi z}{L} \right) e^{-\frac{\kappa n^2 \pi^2 t}{L^2}}$$

At $t = 0$ the series is the Fourier sine series for $f(z)$, a series whose coefficients slowly decrease in magnitude since $b_n = O\left(\frac{1}{n}\right)$. However, for $t > 0$ the exponential term rapidly becomes small with increasing. Hence, for $t > 0$ only a few terms of this series are needed for an accurate approximation whereas for $t = 0$ a very large number of terms are needed to represent the discontinuity in $f(z)$. Approximating the exact solution with the finite sum,

$$p_{50}(x, t) = \frac{2}{\pi} \sum_{n=1}^{50} \left(\frac{1- \cos(n \cdot \pi)}{n} \right) \sin\left(\frac{n \cdot \pi \cdot z}{L} \right) e^{-\frac{\kappa n^2 \pi^2 t}{L^2}}$$

pp:= sum(b*sin(1/2*n*pi*z)*exp(-1/4*n^2*pi^2*t),n=1..50):
plot3d(pp, z=0..2,t=0..1.5, shading=Z, axes=normal, labels = [z,t,p], tickmarks=[3,3,2], scaling=constrained, orientation=[-145,60], title="Figure 2. Pore Pressure Distribution");
pf1 = pp |_{t= 0.001} :

pf2:= eval(pp, t=.25):
pf3:= eval(pp, t=.5):
pf4:= eval(pp, t=1):
p1:= plot([pf1,pf2,pf3,pf4], z = 0..2, color=black, tickmarks=[3,2], scaling=constrained, labels=[z,p]):
p2:= textplot({[1,0.95,"t = 0.001"],[1,0.75,"t = 0.25"],[1,.5,"t = 0.5 "], [1,.15,"t = 1"]}):
display([p1,p2], title="Figure 3. Pore Pressure Profile");
animate(pp, z=0..2, t=0..1, frames = 10, color=black, labels=[z,p], tickmarks=[3,2], scaling=constrained);

Finding the Average Degree of Consolidation:

pi := 3.14

$$ft := sum\left(An \cdot sin\left(\frac{n \cdot 3.14 \cdot Z}{12}\right) \cdot exp\left(\frac{-1 \cdot n^2 \cdot 3.14^2 \cdot T}{144}\right), n = 1..100\right)$$

q16 := int(ft, Z = 0 ..12):
Y := eval(q16, T = 0):

$$U := 1 - \frac{q16}{Y}:$$

plot(U, T = 0 ..150, labels = [Time, PC])

$$R := \frac{q16}{Y}$$

plot(R, T = 0 ..150, labels = [Time, ADC])

Worksheet 5: 1-D Contaminant Transport

restart;
with(inttrans):
with(plots):
The boundary value problem consisting of the one-dimensional transport equation
$c_t(z, t) = Ds\ c_{zz}(z, t)$
that is,

q1:= diff(c(z,t),t) = Ds*diff(c(z,t),z,z);

The boundary condition $uc(0, t), = c_0$, that is,

B[1]:= c(0,t) = c0;

and the initial condition $c(z, 0) = 0$, that is,

B[2]:= c(z,0) = 0;

The partial Laplace transform with respect to t, with the alias

alias(C(z,s)=laplace(c(z,t),t,s));

gives

$sC(z, s) - c(z, 0) = Ds\ C_{zz}(z, s)$

q2:= laplace(q1,t,s);

which the initial condition $c(z, 0) = 0$ simplifies to,

$s\ C(z, s) = Ds\ C_{zz}(z, s)$
q3:= eval(q2, B[2]);

We have the ODE,

$s\ F(z) = Ds F''(z)$
q4:= eval(q3, C(z,s)=F(z));

with general solution

$$F(z) = c_1 e^{z\sqrt{\frac{s}{Ds}}} + c_2 e^{-z\sqrt{\frac{s}{Ds}}}$$

that is,
q5:= dsolve(q4,F(z));
q6:= eval(q5,_C1=0);

The boundary condition $uc(0, t) = c0$ transforms to,

$$C(0, s) = \frac{C0}{s} = F(0)$$

that is,
laplace(B[1],t,s);
q7 := eval(rhs(q6), z=0) = c0/s;
q8 := eval(rhs(q6),q7);

Recovery of $u(x, t)$ is by Maple's "invlaplace" command which gives,

'c(z,t)':= convert(invlaplace(q8,s,t),erf) assuming z>0,Ds>0;

The solution is given in terms of erf(z), the error function defined as

$$erf(z) = \frac{2}{\sqrt{Pi}} \int_0^z e^{-\sigma^2} d\sigma$$

Since erf $(-z) = -erf(z)$, the error function is odd, and
Limit(erf(z), z=infinity) = limit(erf(z), z=infinity);
p1:= plot(erf(z), z= -3..3, color=black, scaling=constrained, tickmarks=[6,2]):
p2:= textplot([-.6,.8, 'erf(z)'], font=[TIMES,ROMAN,12]):
display([p1,p2], labels=[z,''], title= "Figure 1");
*f:= 2*exp(-sigma^2)/sqrt(Pi);*
p3:= plot(f,sigma=-2..2, color=black, scaling=constrained, tickmarks=[4,2], labels=[" "," "]):
p4:= plot(f,sigma=0.. .8, filled=true, color=yellow):
p5:= textplot([1.5,-.2, 's'], font=[SYMBOL,10]):
p6:= textplot([.8,-.15, 'z'], font=[COURIER,10]):
display([p||(3..6)], title="Figure 2");

Finally, if in $c(z, t)$ we set $c_0 = 2$ and Ds = 1, we get the function

$$c(z, t) = 2\left(1 - erf\left(\frac{x}{2\sqrt{1}}\right)\right)$$

c:= unapply(eval('c(z,t)', {c0=2,Ds=1}),z,t):
'c'(z,t) = c(z,t);
plot3d(c(z,t), z=0..10, t=0..10, axes=box, scaling=constrained, shading=Z, tickmarks=[5,5,3],

labels=[z,t,c], orientation=[-50,75], title="Figure 3");

$$pf1 := c(z,t)\Big|_{t=0.001}; -1$$

pf2:= eval(c(z,t), t=1.):
pf3:= eval(c(z,t), t=5):
pf4:= eval(c(z,t), t=10):
p1:= plot([pf1,pf2,pf3,pf4], z=0..10, color=black, tickmarks=[3,2], scaling=constrained, labels=[z,p]): display([p1], title="Figure 4. Concentration Profile");
ct:= eval(c(z,t), z=5):
p1:= plot([ct], t = 0..10, color=black, tickmarks=[3,2], scaling=constrained, labels=[z,p]): display([p1], title="Figure 5. Concentration Variation");
animate(c(z,t), z=0..10, t=0..30, frames=5, color=red, scaling=constrained, labels=['z', 'c'], tickmarks=[10,2]);

MATLAB® Scripts

1-D Coupled Flow and Deformation: "Coupled_1D_QS.m"

```
%--------------------------------------------------------------
% Analytical Solution of 1-D Quasi-Static Coupled Flow-Deformation
% Biot Equations
% 1-D Solution of quasi static u-p form under periodic (e^iwt) loading
% Compressibility of the fluid, Kf, is included
%--------------------------------------------------------------
clear all
clc
H=10;          % Depth of layer
q0=1;          % Amplitude of the load
om=3.379;      % Frequency of the load
Kf=3330000;    % Bulk modulus of the fluid
n=0.3333;      % Porosity
E=749200;      % Elasticity Modulus
mu=0.2;        % Poisson's Ratio
k=0.0001;      % Permeability (m/s)
g=9.81;        % Acc. of Grav.
rho=3;         % Total Unit Weight
rhof=1;        % Fluid Unit Weight
D=E*(1-mu)/((1+mu)*(1-2*mu));    % Constrained Modulus
kappa=(Kf/n)/(D+Kf/n);
beta=rhof/rho;
Vc=sqrt((D+Kf/n)/rho);     % Velocity of sound in water
% Spatial Coordinate Discretization
N=100;
dz=H/N;
z=0:dz:H;
Z=z/H;
% Time Discretization
t0=0.;
dt=0.01;       % Time increment
nt=5000;
tf=t0+(nt-1)*dt;
t=t0:dt:tf;
% Dimensionless Constants
P1=k*Vc^2/(g*beta*om*H^2);
P2=(om*H/Vc)^2;
A=-sqrt(-1)/(kappa*(kappa-1)*P1);
% Periodic Load
q=q0*exp(sqrt(-1)*om*t);
% Pore Pressure
pp=H*(sqrt(-1)*om*rhof*g*(1/k)*q'*H/
(A*D))*((exp(sqrt(A)*Z)+exp(sqrt(A)*(2-Z)))/(1+exp(2*sqrt(A)))-1);
ppi=imag(pp);
ppr=real(pp);
Pabs=abs(pp);
% Displacement
uu=-(q'/D)*(kappa*(exp(sqrt(A)*(2-Z))-exp(sqrt(A)*Z))/
(sqrt(A)*(1+exp(2*sqrt(A))))+(1-kappa)*(Z-1));
uui=imag(uu);
uur=real(uu);
uabs=abs(uu);
%% Plots
% Pore Pressure
```

```
subplot(211)
plot(t,ppr/q0)
title('Pore pressure-Time history')
ylabel('p/q')
xlabel('t(s)')
subplot(212)
plot(Pabs(1,:)/q0,-Z)
title('Pore pressure distribution')
ylabel('z/H')
xlabel('p/q')
% axis([0 1.5 -1 0])
figure
subplot(211)
plot(ppi(1:nt/50:nt,1:N+1)'/q0,-Z)
title('Pore pressure distribution')
ylabel('z/H')
xlabel('p(Im)/q0')
subplot(212)
plot(ppr(1:nt/50:nt,1:N+1)'/q0,-Z)
title('pore pressure distribution')
ylabel('z/H')
xlabel('p(Re)/q0')
%% Displacement
figure
subplot(211)
plot(t,uabs)
title('Displacement-time history')
ylabel('u(m)')
xlabel('t(s)')
subplot(212)
plot(uabs,-Z)
title('Displacement variation')
ylabel('z/H')
xlabel('u(m)')
```

2-D Quasi-Static Solution: "Coupled_Flow_Deformation_2D_QS.m"

```
%------------------------------------------------------------
% Analytical Solution of 2-D Coupled Flow and Deformation
% Reduced Biot Equations
% Quasi-Static Form
% z=>-inf
%------------------------------------------------------------
% disp ('Enter the # of elements in z direction')
% nz=input ('prompt')
nz=100;
%----------------------------------------
% input data for seabed and material
%----------------------------------------
H=30;                   % Depth of soil (m)
T=8;                    % Wave Period (sec)
E=250000;               % Elasticity Modulus (kPa)
mu=0.3;                 % Poisson's Ratio
kz=0.0001;              % Vertical Permeability (m/s)
kx=kz;                  % Horizontal Permeability (m/s)
kw=2.4525*10^6;         % Bulk Modulus of water (kPa)
S=1;                    % Saturation
n=0.4;                  % Porosity
```

```
rho=2;                      %Unit weight of soil+water (kN/m^3/g)
rhof=1;                     % Fluid Unit Weight
g=9.81;                     % Acc. of Gravity (m/s^2)
bt=rhof*g;                  % body force of water
%-------------------------------------------
% input data for wave
%-------------------------------------------
wL=100;                     % Wavelength (m)
q0=1;                       % Load amplitude (kN)
p_ref=100;                  % Reference pressure (kPa)
w=2*pi/T;                   % Frequency
k=2*pi/wL;                  % Wave number
kf=p_ref*kw/(p_ref+kw*(1-S));      % Bulk modulus of pore fluid
M=E*(1-mu)/((1+mu)*(1-2*mu));      % Constrained Modulus (kPa)
G=E/(2*(1+mu));                    % Shear Modulus (kPa)
lambda=2*G*mu/(1-2*mu);            % Lame's Constant (kPa)
%------------------------------------------------------------
% input data for the loading history and discretization
%------------------------------------------------------------
dz=-H/nz;
z=0:dz:-H;
ZH=z/H;
%------------------------------------------------------------
% Non-dimensional Parameters
%------------------------------------------------------------
Q=kf/n;
kappa=Q/(M+Q);
kappa1=lambda/(M+Q);
kappa2=G/(M+Q);
beta=rhof/rho;
m=k*H;
Vc=sqrt((M+Q)/rho); % Velocity of sound in water
P1=(kz*Vc^2)/(g*beta*w*H^2);
P11=(kx*Vc^2)/(g*beta*w*H^2);
%------------------------------------------------------------
% Aij Parameters
%------------------------------------------------------------
AA=m^2*kappa;
AA13=(sqrt(-1)/P11)-m^2*kappa;
A51=(1-kappa)*kappa;
A52=m^2*kappa;
A53=AA13+A52;
A61=kappa+kappa1+kappa2;
A75=sqrt(-1)/P1;
A76=A61-kappa;%dyn
A77=1+kappa2^2;
A83=AA13*(A61^2-A77);
A84=kappa2*A75;
%------------------------------------------------------------
% Coefficients of Characteristic Eqn.
%------------------------------------------------------------
% det(M)=0
%------------------------------------------------------------
a1=A51*A84;
a21=-A51*(A52^2);
a22=A52*(A52*(A76^2+kappa-A77)+2*A51*A52+2*AA13*(A51-kappa*A76)+...
```

```
      A83+kappa*A84);
a23=AA13*A84;
a2=a21+a22+a23;
a31=(A52^2)*(A51*kappa2*(A53/(kappa^2))-A75*(1-2*A61));
a32=A52*A75*(A83/kappa-A52);
a3=a31+a32;
a4=A84*m^2*(m^2*AA13+A52^2);
%-------------------------------------------------------------------
% Roots of Characteristic Eqn.
%-------------------------------------------------------------------
poly=[a1 0 a2 0 a3 0 a4];
L=roots(poly);
%-------------------------------------------------------------------
% Entries of 4x4 differential matrix
%-------------------------------------------------------------------
for i=1:6
    B(i)=sqrt(-1)*m*kappa*L(i);
    C(i)=L(i)^2*kappa;
    A24(i)=sqrt(-1)/P1+kappa*(L(i)^2);
    A31(i)=-m^2+kappa2*(L(i)^2);
    D(i)=sqrt(-1)*m*L(i)*(kappa+kappa1+kappa2);
    A42(i)=-m^2*kappa2+L(i)^2;
%-------------------------------------------------------------------
% Coefficients of soln. entries of eigenvectors (bi, ci, di)
%-------------------------------------------------------------------
    detm = (B(i)^2)*(A42(i)-C(i))-B(i)*D(i)*(C(i)-A24(i))+AA*(C(i)^2-
    A24(i)*A42(i));
    b(i)= (B(i)^3+B(i)*(C(i)*(A31(i)-AA)-D(i)*B(i)-
    A31(i)*A24(i))+AA*D(i)*A24(i))/detm;
    c(i)= (-B(i)^2*A42(i)+2*B(i)*C(i)*D(i)-A31(i)*(C(i)^2-A24(i)*A42(i))-
    D(i)^2*A24(i))/detm;
    d(i)= (B(i)*D(i)*(D(i)-B(i))+B(i)*A42(i)*(AA-A31(i))+C(i)*(A31(i)*B(i)-
    AA*D(i)))/detm;
%-------------------------------------------------------------------
% 6x6 Matrix to find ai
%-------------------------------------------------------------------
    a5(i)= (M*b(i)*L(i))/H+sqrt(-1)*k*lambda;
    a6(i)= G*(L(i)/H+sqrt(-1)*k*b(i));
    e(i)= sqrt(-1)*k*(1+c(i))+L(i)*(b(i)+d(i))/H;
end
CM=[ exp(-L(1)) exp(-L(2)) exp(-L(3)) exp(-L(4)) exp(-L(5)) exp(-L(6));...
    b(1)*exp(-L(1)) b(2)*exp(-L(2)) b(3)*exp(-L(3)) b(4)*exp(-L(4))
    b(5)*exp(-L(5)) b(6)*exp(-L(6));...
    d(1)*exp(-L(1)) d(2)*exp(-L(2)) d(3)*exp(-L(3)) d(4)*exp(-L(4))
    d(5)*exp(-L(5)) d(6)*exp(-L(6));...
    a5(1)   a5(2)   a5(3)   a5(4)   a5(5)   a5(6);...
    a6(1)   a6(2)   a6(3)   a6(4)   a6(5)   a6(6);...
    e(1)    e(2)    e(3)    e(4)    e(5)    e(6)];
% RHS Vector (known)
Fv=[0 0 0 0 0 -q0/Q];
Fv=Fv';
% Solve for the Coefficient Vector of Analytical Solution (unknown) XX=[a1 a2
a3 a4 a5 a6];
XX=CM\Fv;

Ux=XX(1)*exp(L(1)*ZH)+XX(2)*exp(L(2)*ZH)+XX(3)*exp(L(3)*ZH)+...
    XX(4)*exp(L(4)*ZH)+XX(5)*exp(L(5)*ZH)+XX(6)*exp(L(6)*ZH);
```

```
Uz=XX(1)*b(1)*exp(L(1)*ZH)+XX(2)*b(2)*exp(L(2)*ZH)+XX(3)*b(3)*exp(L(3)*
    ZH)+...
    XX(4)*b(4)*exp(L(4)*ZH)+XX(5)*b(5)*exp(L(5)*ZH)+XX(6)*b(6)*exp(L(6)*ZH);

P=-Q*( (sqrt(-1)*k*(1+c(1))+L(1)*(b(1)+d(1))/H)*XX(1)*exp(L(1)*ZZ)+...
       (sqrt(-1)*k*(1+c(2))+L(2)*(b(2)+d(2))/H)*XX(2)*exp(L(2)*ZZ)+...
       (sqrt(-1)*k*(1+c(3))+L(3)*(b(3)+d(3))/H)*XX(3)*exp(L(3)*ZZ)+...
       (sqrt(-1)*k*(1+c(4))+L(4)*(b(4)+d(4))/H)*XX(4)*exp(L(4)*ZZ)+...
       (sqrt(-1)*k*(1+c(5))+L(5)*(b(5)+d(5))/H)*XX(5)*exp(L(5)*ZZ)+...
       (sqrt(-1)*k*(1+c(6))+L(6)*(b(6)+d(6))/H)*XX(6)*exp(L(6)*ZZ)  );

Szz=((M*b(1)*L(1))/H+sqrt(-1)*k*lambda)*XX(1)*exp(L(1)*ZZ)+...
    ((M*b(2)*L(2))/H+sqrt(-1)*k*lambda)*XX(2)*exp(L(2)*ZZ)+...
    ((M*b(3)*L(3))/H+sqrt(-1)*k*lambda)*XX(3)*exp(L(3)*ZZ)+...
    ((M*b(4)*L(4))/H+sqrt(-1)*k*lambda)*XX(4)*exp(L(4)*ZZ)+...
    ((M*b(5)*L(5))/H+sqrt(-1)*k*lambda)*XX(5)*exp(L(5)*ZZ)+...
    ((M*b(6)*L(6))/H+sqrt(-1)*k*lambda)*XX(6)*exp(L(6)*ZZ);

Sxx=(M*sqrt(-1)*k+(b(1)*L(1)*lambda)/H)*XX(1)*exp(L(1)*ZZ)+...
    (M*sqrt(-1)*k+(b(2)*L(2)*lambda)/H)*XX(2)*exp(L(2)*ZZ)+...
    (M*sqrt(-1)*k+(b(3)*L(3)*lambda)/H)*XX(3)*exp(L(3)*ZZ)+...
    (M*sqrt(-1)*k+(b(4)*L(4)*lambda)/H)*XX(4)*exp(L(4)*ZZ)+...
    (M*sqrt(-1)*k+(b(5)*L(5)*lambda)/H)*XX(5)*exp(L(5)*ZZ)+...
    (M*sqrt(-1)*k+(b(6)*L(6)*lambda)/H)*XX(6)*exp(L(6)*ZZ);

Txz=G*((L(1)/H+sqrt(-1)*k*b(1))*XX(1)*exp(L(1)*ZZ)+...
    (L(2)/H+sqrt(-1)*k*b(2))*XX(2)*exp(L(2)*ZZ)+...
    (L(3)/H+sqrt(-1)*k*b(3))*XX(3)*exp(L(3)*ZZ)+...
    (L(4)/H+sqrt(-1)*k*b(4))*XX(4)*exp(L(4)*ZZ)+...
    (L(5)/H+sqrt(-1)*k*b(5))*XX(5)*exp(L(5)*ZZ)+...
    (L(6)/H+sqrt(-1)*k*b(6))*XX(6)*exp(L(6)*ZZ));

Uxabs=abs(Ux');
Uzabs=abs(Uz');
Pr=real(P');
ppabs=abs(P')/q0;
Szzr=real(Szz');
Szzabs=abs(Szz')/q0;
Sxxr=real(Sxx');
Sxxabs=abs(Sxx')/q0;
Txzr=real(Txz');
Txzabs=abs(Txz')/q0;
ZH=ZH';
% ------------------------------------------------------------
% Plots
% ------------------------------------------------------------
% SOLID DISPLACEMENTS
% ------------------------------------------------------------
subplot 211
plot(Uxabs,ZH,'black')
title('Absolute value of Horizontal Solid Displacement')
ylabel('z/h')
xlabel('Ux(m)')
subplot 212
plot(Uzabs,ZH,'black')
title('Absolute value of Vertical Solid Displacement')
ylabel('z/h')
xlabel('Uz(m)')
```

```
%------------------------------------------------------------
% PORE PRESSURE
%------------------------------------------------------------
figure
plot(ppabs,ZH)
title('pore pressure distribution')
ylabel('z/H')
xlabel('p/q0')
%------------------------------------------------------------
% STRESSES
%------------------------------------------------------------
figure
subplot 311
plot(Szzabs,ZH,'black')
title('Vertical eff. stress distribution')
ylabel('z/H')
xlabel('Szz/q0')
subplot 312
plot(Txzabs,ZH,'black')
title('Shear Stress distribution')
ylabel('z/H')
xlabel('Sxz/q0')
subplot 313
plot(Sxxabs,ZH,'black')
title('Horizontal Effective Stress distribution')
ylabel('z/H')
xlabel('Sxx/q0')
```

Semi-Analytical Methods

8.1 Introduction

Having developed the formulation of a mathematical model (in terms of governing field equations and the boundary conditions), wherever possible we seek its solution by Analytical Methods (Chapter 7). When we are unable to develop analytical solutions, we resort to numerical methods such as the Finite Difference Method (Chapter 9), Finite Element Method (Chapter 10), or several other alternatives. These numerical methods are now accepted as valuable tools for the analysis of geotechnical engineering problems. Their strength is due to their ability to capture complex geometries, a variety of different material properties and associated complex constitutive behaviors as well as complex boundary conditions.

For several problems in geomechanics and geotechnical engineering, soil deposits may be idealized as horizontally layered systems with their properties not varying in the horizontal direction. For these cases, it is unnecessary nor is it efficient to use the full power of the numerical methods (FDM or FEM). The recognition of this has led to the development of a variety of less general but more efficient Semi-Analytical Methods (SAM) for particular classes of problems. Among these, Finite Layer Method (FLM) is quite useful. The general approach in a SAM is to use analytical solutions for the basics of the problem in developing the solution while making use of numerical transform methods for handling the complexity of loading and/or the geometry.

In this chapter, we introduce the basic elements of the FLM. We are using the problems of 'stress analysis' and 'QS analysis of multi-layer porous media under waves' as a context. We limit our discussion to the plain strain situations and consider the layer properties to be homogeneous, isotropic and linearly elastic. The complexity of the surface loading is handled by discrete Fourier transform. All the details of the method is first presented for the stress analysis of an elastic half-space under a general surface loading. These are implemented in a simple MATLAB® program. We then extend the formulation to the general case of multi-layer system and also subsequently include the porous medium properties.

It should be noted that the general approach of the SAM has been further extended to a wide range of geotechnical problems including layered elastic system under strip, circular and rectangular loadings, consolidation of layered systems, viscoelastic layered systems, wave-induced seabed response, contaminant migration (Booker and Small, 1985), and response of saturated porous media.

8.2 Stress Analysis

Let us consider the problem of load-deformation of a foundation soil (considered as linear elastic material) subjected to a general surface loading shown in Figure 8.2-1. For such problems, the main focus is firstly the stress analysis of the soil in the profile. In that case analytical solutions are possible only with severe idealizations of the type of loading, soil properties and geometry. For general situations, we will have to resort to numerical approaches as mentioned before. However, in the case of a horizontally layered soil profile, which is ideally the case in practice, with each layer idealized as linearly elastic (or viscoelastic for that matter), SAM may be conveniently used. The reason for that is, soil response can be thought of having an isotropic linear elastic behavior under loading while the loading itself can still sustain its arbitrary character. Such a combination will give more accurate results in the sense that the complex nature of the surface load which, often times, is what engineers need to deal with in projects, will at least be captured in the problem. Further complexities associated with other components of the problem such as material nonlinearities, heterogeneities or boundary conditions (even if there are no structural elements in the model such as the free field case) will certainly require more sophisticated approaches leading to a set of nonlinear algebraic equations to solve. Such a method then is surely no longer a 'semi-analytical method' as it is only possible to solve nonlinear equations by means of a numerical method which consists of iterative schemes.

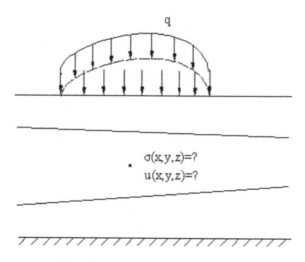

Figure 8.2-1 Stress analysis: General problem

In this section, the fundamentals of SAM are presented for the analysis of stresses in a layered elastic soil profile in plane strain subjected to a general surface loading.

8.2.1 Elastic Response: Plane Strain Condition

Let us recall the equations (given in Ch. 3) governing the problem in Figure 8.2-1. The equation of equilibrium needs to be written first:

$$\frac{\partial \sigma_{xx}}{\partial x} + \frac{\partial \sigma_{xz}}{\partial z} = 0 \tag{8.2.1a}$$

$$\frac{\partial \sigma_{zx}}{\partial x} + \frac{\partial \sigma_{zz}}{\partial z} = 0 \tag{8.2.1b}$$

Constitutive equation in terms of the stress-strain relationship (through linear elasticity) which, in this case is the Hooke's law, comes next:

$$\sigma_{xx} = \lambda\varepsilon_v + 2G\varepsilon_{xx} \tag{8.2.2a}$$

$$\sigma_{zz} = \lambda\varepsilon_v + 2G\varepsilon_{zz} \tag{8.2.2b}$$

$$\sigma_{xz} = G\gamma_{xz} \tag{8.2.2c}$$

where $\lambda = \dfrac{vE}{(1+v)(1-2v)} = K - \dfrac{2}{3}G$ with K being the bulk modulus and G, the shear modulus. Volumetric strain is calculated as $\varepsilon_v = \varepsilon_{xx} + \varepsilon_{zz}$. Combining equations (8.2.1) and (8.2.2) yields:

$$G\nabla^2 u_x + (\lambda + G)\frac{\partial\varepsilon_v}{\partial x} = 0 \tag{8.2.3a}$$

$$G\nabla^2 u_z + (\lambda + G)\frac{\partial\varepsilon_v}{\partial z} = 0 \tag{8.2.3b}$$

Differentiating (8.2.3) with respect to x and z, and adding together we get:

$$\nabla^2\varepsilon_v = 0 \tag{8.2.4}$$

Equation (8.2.4) can also be written as:

$$\nabla^2(\sigma_{xx} + \sigma_{zz}) = 0 \tag{8.2.5}$$

If we choose a stress function φ such that,

$$\sigma_{xx} = \frac{\partial^2\varphi}{\partial z^2} \tag{8.2.6a}$$

$$\sigma_{zz} = \frac{\partial^2\varphi}{\partial x^2} \tag{8.2.6b}$$

$$\sigma_{xz} = -\frac{\partial^2\varphi}{\partial x\partial z} \tag{8.2.6c}$$

Then, combining equations (8.2.4) and (8.2.5), we get:

$$\nabla^4\varphi = 0 \tag{8.2.7}$$

where $\nabla^4 = \dfrac{\partial^4}{\partial x^4} + 2\dfrac{\partial^4}{\partial x^2\partial y^2} + \dfrac{\partial^4}{\partial z^4}$. In the following sections, the solutions to the above equation are presented for various stress states.

8.2.2 Stresses in a Semi-Infinite Elastic Soil under Harmonic Pressure

Let us now consider the problem in Figure 8.2-2. The solution to the governing equation (8.2.7) is assumed to be of the form below:

$$\varphi = (Ae^{-\alpha z} + Bze^{-\alpha z})\sin\alpha x \tag{8.2.8}$$

We need to choose φ so that it satisfies the governing equation (8.2.7) as well as the following boundary conditions:

$$\sigma_{xz}\big|_{z=0} = 0 \tag{8.2.9a}$$

$$\sigma_{zz}\big|_{z=0} = -p_0 \sin \alpha x \tag{8.2.9b}$$

Using the above boundary conditions and equation (8.2.8), we find:

$$\varphi = \frac{p_0}{\alpha^2}(1 + \alpha z)e^{-\alpha z}\sin \alpha x \tag{8.2.10}$$

From equations (8.2.6) and (8.2.10), we can get:

$$\sigma_{xx} = -p_0(1 - \alpha z)e^{-\alpha z}\sin \alpha x \tag{8.2.11a}$$

$$\sigma_{zz} = -p_0(1 + \alpha z)e^{-\alpha z}\sin \alpha x \tag{8.2.11b}$$

$$\sigma_{xz} = p_0 \alpha z e^{-\alpha z}\cos \alpha x \tag{8.2.11c}$$

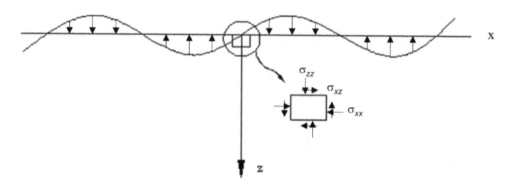

Figure 8.2-2 Stresses in elastic half space under a harmonic surface load

For details, see Box 8.1 below.

Box 8.1: Details for the solution in (8.2.11)

$$\phi = (Ae^{-\alpha z} + Bze^{-\alpha z})\sin \alpha x$$

$$\sigma_{zz} = \frac{\partial^2 \varphi}{\partial z^2} = -\alpha^2(Ae^{-\alpha z} + Bze^{-\alpha z})\sin \alpha x$$

$$\sigma_{zz}\big|_{z=0} = -P_0 \sin \alpha x \Rightarrow A = \frac{P_0}{\alpha^2}$$

$$\sigma_{xz} = -\frac{\partial^2 \varphi}{\partial x \partial z} = -\alpha(-\alpha Ae^{-\alpha z} + Bze^{-\alpha z} - \alpha Bze^{-\alpha z})\cos \alpha x$$

$$\sigma_{xz}\big|_{z=0} = 0 \Rightarrow B = \frac{P_0}{\alpha}$$

$$\varphi = \frac{P_0}{\alpha}(1+\alpha z)e^{-\alpha z}\sin\alpha x$$

$$\sigma_{xx} = \frac{\partial^2\varphi}{\partial z^2} = -P_0(1-\alpha z)e^{-\alpha z}\sin\alpha x$$

$$\sigma_{zz} = \frac{\partial^2\varphi}{\partial x^2} = -\frac{P_0}{\alpha}(1+\alpha z)e^{-\alpha z}\sin\alpha x$$

$$\sigma_{xz} = -\frac{\partial^2\varphi}{\partial x\partial z} = P_0\alpha z e^{-\alpha z}\cos\alpha x$$

8.2.3 Stresses in a Semi-Infinite Elastic Soil under a Surface Load $e^{i\alpha x}$

The solution presented in the previous section can alternatively be obtained in exponential form with load $e^{i\alpha x}$ and the following boundary conditions:

$$\sigma_{xz}\big|_{z=0} = 0 \tag{8.2.12a}$$

$$\sigma_{zz}\big|_{z=0} = -e^{i\alpha x} \tag{8.2.12b}$$

Now let $\phi = (Ae^{-\alpha z} + Bze^{-\alpha z})e^{i\alpha x}$ be the solution to (8.2.7). For the boundary conditions (8.2.12), we have:

$$\phi = \frac{1}{\alpha^2}(1+\alpha z)e^{-\alpha z}e^{i\alpha x} \tag{8.2.13}$$

From equations (8.2.6) and (8.2.13), we obtain:

$$\sigma_{zz} = -(1+\alpha z)e^{-\alpha z}e^{i\alpha x} \tag{8.2.14a}$$

$$\sigma_{xx} = -(1-\alpha z)e^{-\alpha z}e^{i\alpha x} \tag{8.2.14b}$$

$$\sigma_{xz} = i\alpha z e^{-\alpha z}e^{i\alpha x} \tag{8.2.14c}$$

The details for the above solution are presented in Box 8.2 below. Alternatively, we write the above as:

$$\sigma_{zz} = HS_{zz}e^{i\alpha x} \tag{8.2.15a}$$

$$\sigma_{xx} = HS_{xx}e^{i\alpha x} \tag{8.2.15b}$$

$$\sigma_{xz} = HS_{xz}e^{i\alpha x} \tag{8.2.15c}$$

where the transfer functions *HSs* are derived as:

$$HS_{zz} = -(1+\alpha z)e^{-\alpha z} \tag{8.2.16a}$$

$$HS_{xx} = -(1-\alpha z)e^{-\alpha z} \tag{8.2.16b}$$

$$HS_{xz} = i\alpha z e^{-\alpha z} \qquad\qquad (8.2.16c)$$

The details are in Box 8.2.

8.2.4 Stresses in a Semi-Infinite Elastic Soil Subjected to a Strip Surface Load

If the surface load is of an arbitrary type varying along x as shown below then the solution can be developed using Fourier transformation. The basic idea here is, to decompose the load into its Fourier components, find the response to each component separately in frequency domain, and sum up all the solutions to get the final solution back in time domain. The process is carried out in the following steps:

Step 1. Discretize the load function $p(x_m)$.

Figure 8.2-3 Discretized load function

Box 8.2: Details for the solution in (8.2.14)

$$\phi = (Ae^{-\alpha z} + Bze^{-\alpha z})\sin \alpha x$$

$$\sigma_{zz} = \frac{\partial^2 \varphi}{\partial z^2} = -\alpha^2(Ae^{-\alpha z} + Bze^{-\alpha z})\sin \alpha x$$

$$\sigma_{zz}\big|_{z=0} = -e^{i\alpha x} \Rightarrow A = \frac{1}{\alpha^2}$$

$$\sigma_{xz} = -\frac{\partial^2 \varphi}{\partial x \partial z} = -\alpha(-\alpha Ae^{-\alpha z} + Bze^{-\alpha z} - \alpha Bze^{-\alpha z})\cos \alpha x$$

$$\sigma_{xz}\big|_{z=0} = 0 \Rightarrow B = \frac{1}{\alpha}$$

$$\varphi = \frac{1}{\alpha^2}(1 + \alpha z)e^{-\alpha z}\sin \alpha x$$

$$\sigma_{xx} = \frac{\partial^2 \varphi}{\partial z^2} = -(1 - \alpha z)e^{-\alpha z}\sin \alpha x$$

$$\sigma_{zz} = \frac{\partial^2 \varphi}{\partial x^2} = -(1 + \alpha z)e^{-\alpha z}\sin \alpha x$$

$$\sigma_{xz} = -\frac{\partial^2 \varphi}{\partial x \partial z} = i\alpha z e^{-\alpha z}$$

Step 2. Using the Fast Fourier Transform (FFT) evaluate the Fourier components $PP(\alpha_n)$ of $p(x_n)$.

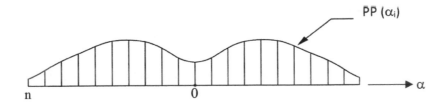

Figure 8.2-4 Fourier components

In FFT, the following relationships exist:

$$\alpha_n = 2\pi f_n$$

$$f_n = \frac{(n-1)}{nfft} \cdot \frac{1}{\Delta x} \quad \text{for } n = 1, \ldots, \frac{nfft}{2} + 1$$

$$f_n = -\frac{(nfft - n + 1)}{nfft} \cdot \frac{1}{\Delta x} \quad \text{for } n = \frac{nfft}{2} + 2, \ldots, nfft \text{ (negative frequency)}$$

For continuous load function, we have,

$$PP(\alpha) = \frac{1}{2\pi} \int_{-\infty}^{\infty} p(x)e^{-i\alpha x} dx$$

For discrete load function, we have,

$$PP(\alpha_n) = \Delta x \sum_{m=1}^{nfft} p(x_m)e^{-i\frac{2\pi nm}{nfft}}$$

This step simply breaks the original problem down to many problems of evaluating the stresses due to harmonic loadings.

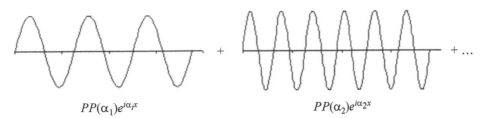

Figure 8.2-5 Fourier series expansion

Step 3. Evaluate the response (i.e. stresses) for each harmonic component with unit amplitude (i.e. $+1e^{i\alpha_n x}$).

$$\underbrace{e^{i\alpha_n x}}_{\text{Input}} \longrightarrow \begin{cases} HS_{zz}.e^{i\alpha_n x} \\ HS_{xx}.e^{i\alpha_n x} \\ HS_{xz}.e^{i\alpha_n x} \end{cases}$$

$$\underbrace{\phantom{HS_{zz}.e^{i\alpha_n x}HS_{xx}}}_{\text{Transfer functions}}$$

Step 4. Evaluate the response for each harmonic component with $PP(\alpha_n)$ (actual amplitudes associated with corresponding frequencies).

$$\underbrace{PP(\alpha_n)e^{i\alpha_n x}}_{\text{Input}} \longrightarrow \begin{cases} PP(\alpha_n)HS_{zz}.e^{i\alpha_n x} \\ PP(\alpha_n)HS_{xx}.e^{i\alpha_n x} \\ PP(\alpha_n)HS_{xz}.e^{i\alpha_n x} \end{cases}$$

$$\underbrace{\phantom{PP(\alpha_n)HS_{zz}.e^{i\alpha_n x}PP(\alpha_n)HS}}_{\text{Response in frequency domain}}$$

Step 5. Using the inverse Fourier transform (IFFT), sum the responses due to all of the harmonic terms to obtain the actual response (stresses) to the original loading.

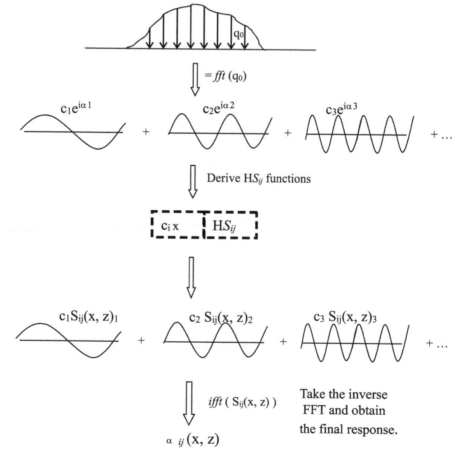

Figure 8.2-6 Summary of the Fourier transform process of solving the boundary value problem

$$\sigma_{zz} = ifft(S_{zz})$$

$$\sigma_{xx} - ifft(S_{xx})$$

$$\sigma_{xz} = ifft(S_{xz})$$

Figure 8.2-6 summarizes the process in a more elaborate fashion.

An exact form of Fourier transform can be obtained below for, $f(t) = \begin{cases} 0 & t < 0 \\ e^{-t} & t \geq 0 \end{cases}$ as

$$F[f(t)] = \int_{-\infty}^{\infty} e^{-t} e^{-i\omega t}\, dt = \int_{0}^{\infty} e^{-(1+i\omega)t}\, dt \qquad (8.2.17)$$

with the property that $F(i\omega) = \dfrac{1}{1+i\omega}$ and $|F(i\omega)| = \dfrac{1}{\sqrt{1+\omega^2}}$.

Computer Implementation

Figure 8.2-7 presents the FFT of an exponential input function whose MATLAB® code **"fft_example.m"** is given at the end.

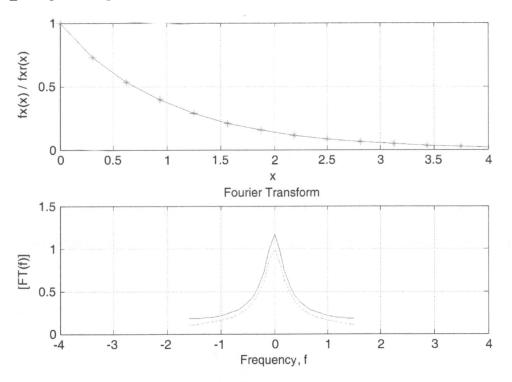

Figure 8.2-7 Fourier transform example, dotted line (...) is exact, solid line is FFT

8.2.5 Stresses in a System of Elastic Soil Layers

Let us now consider the analysis of horizontally layered elastic soil shown in the figure below.

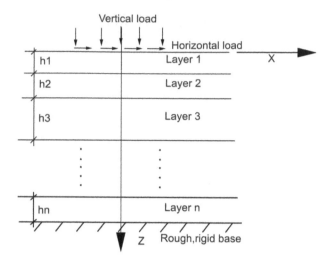

Figure 8.2-8 Elastic soil layers

On the surface of the layered elastic system, general shear and normal loading represented by the following functions are applied:

$$\sigma_{xx} = f_x(x) \tag{8.2.18a}$$

$$\sigma_{zz} = f_z(x) \tag{8.2.18b}$$

These loading functions can be Fourier transformed as:

$$F_x(\alpha) = -\frac{1}{2\pi i} \int_{-\infty}^{\infty} f_x e^{-i\alpha x} dx \tag{8.2.19a}$$

$$F_z(\alpha) = \frac{1}{2\pi} \int_{-\infty}^{\infty} f_z e^{-i\alpha x} dx \tag{8.2.19b}$$

In this section, all the elements of the analysis of this problem are presented.

8.2.6 Elastic Response of a Single Layer to the Fourier Component $1e^{i\alpha_j x}$

Let us recall the governing equations developed earlier,

$$G\nabla^2 u_x + (\lambda + G)\frac{\partial \varepsilon_v}{\partial x} = 0 \tag{8.2.20a}$$

$$G\nabla^2 u_z + (\lambda + G)\frac{\partial \varepsilon_v}{\partial z} = 0 \tag{8.2.20b}$$

$$\nabla^2 \varepsilon_v = 0 \tag{8.2.20c}$$

Let the solution be:

$$u_x = -iUe^{i\alpha x} \tag{8.2.21a}$$

$$u_z = We^{i\alpha x} \tag{8.2.21b}$$

$$\epsilon_{y} = E_y e^{i\alpha x} \tag{8.2.21c}$$

where,

$$E_v = \left(\frac{dW}{dz} + \alpha U \right) \tag{8.2.22}$$

Combining equations (8.2.20) with the constitutive relations of (8.2.2), stresses can be obtained in the following form:

$$\sigma_{zz} = Ne^{i\alpha x} \tag{8.2.23a}$$

$$\sigma_{xx} = He^{i\alpha x} \tag{8.2.23b}$$

$$\sigma_{xz} = -iTe^{i\alpha x} \tag{8.2.23c}$$

with

$$N = \frac{Ev}{(1+v)(1-2v)}\alpha U + \frac{E(1-v)}{(1+v)(1-2v)}\frac{dW}{dz} \tag{8.2.24a}$$

$$H = \frac{E(1-v)}{(1+v)(1-2v)}\alpha U + \frac{Ev}{(1+v)(1-2v)}\frac{dW}{dz} \tag{8.2.24b}$$

$$T = \frac{E}{2(1+v)}\left(\frac{dU}{dz} - \alpha W \right) \tag{8.2.24c}$$

If we replace λ and G with the Young's modulus E and Poisson's ratio v, the following set of equations are arrived by substituting equation set (8.2.24) into the governing equations (8.2.20):

$$\frac{d^2U}{dz^2} - \alpha^2 U - \frac{\alpha}{1-2v}E_v = 0 \tag{8.2.25a}$$

$$\frac{d^2W}{dz^2} - \alpha^2 W + \frac{1}{1-2v}\frac{dE_v}{dz} = 0 \tag{8.2.25b}$$

$$E_v = \left(\frac{dW}{dz} + \alpha U \right) \tag{8.2.25c}$$

These equations provide a set of ordinary differential equations with constant coefficients and can be solved in a straightforward manner. Using equation (8.2.20) with the above expression for E_v, we get a second order ordinary differential equation for E_v:

$$\frac{d^2W}{dz^2} - \alpha^2 Z = 0 \tag{8.2.26}$$

The solution to this equation can be written as:

$$E_v = c_1 \underbrace{\cosh(\alpha z)}_{C} + c_2 \underbrace{\sinh(\alpha z)}_{S} \tag{8.2.27}$$

with E_v evaluated, U and W can be solved from the first two equations of (8.2.25). Finally, the stress components can be evaluated from (8.2.24),

$$\begin{bmatrix} N \\ T \\ U \\ W \end{bmatrix} = \begin{bmatrix} 2GC & 2GZS & 2GS & 2GZC \\ -2GS & -2G(ZC+S) & -2GC & -2G(ZS+C) \\ \dfrac{C}{\alpha} & \dfrac{-ZS-2(1-v)C}{\alpha} & -\dfrac{S}{\alpha} & \dfrac{-ZC-2(1-v)S}{\alpha} \\ \dfrac{S}{\alpha} & \dfrac{ZC-(1-2v)S}{\alpha} & \dfrac{C}{\alpha} & \dfrac{ZS-(1-2v)C}{\alpha} \end{bmatrix} \begin{bmatrix} C_1 \\ C_2 \\ C_3 \\ C_4 \end{bmatrix} \tag{8.2.28}$$

where,

$C = \cosh Z$, $S = \sinh Z$ and $Z = \alpha z$.

A Single Soil Layer Underlain by a Rigid Layer

From the above equation (8.2.28), the constants C_1, C_2, C_3 and C_4 can be evaluated from the boundary conditions. For example, for a single layer with unit vertical load and a horizontal load applied at the surface and with a rough and rigid base, we can use the boundary conditions seen in Figure 8.2-9.

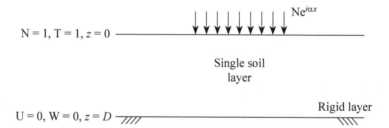

Figure 8.2-9 A single soil layer underlain by a rigid layer

Using these boundary conditions, (8.2.28) can be written in the form of:

$$1 = C_1 N_1(z{=}0) + C_2 N_2(z{=}0) + C_3 N_3(z{=}0) + C_4 N_4(z{=}0) \tag{8.2.29a}$$

$$1 = C_1 T_1(z{=}0) + C_2 T_2(z{=}0) + C_3 T_3(z{=}0) + C_4 T_4(z{=}0) \tag{8.2.29b}$$

$$0 = C_1 U_1(z{=}D) + C_2 U_2(z{=}D) + C_3 U_3(z{=}D) + C_4 U_4(z{=}D) \tag{8.2.29c}$$

$$0 = C_1 W_1(z{=}D) + C_2 W_2(z{=}D) + C_3 W_3(z{=}D) + C_4 W_4(z{=}D) \tag{8.2.29d}$$

where

$$N_i = \sum_{j=1}^{4} C_j N_j (z_i) ; \, i = m, \, p \tag{8.2.30a}$$

$$T_i = \sum_{j=1}^{4} C_j T_j (z_i) \, ; \, i = m, \, p \tag{8.2.30b}$$

$$U_i = \sum_{j=1}^{4} C_j U_j (z_i) \, ; \, i = m, \, p \tag{8.2.30c}$$

$$W_i = \sum_{j=1}^{4} C_j W_j (z_i) \, ; \, i = m, \, p \tag{8.2.30d}$$

From these equations, the constants C_1–C_4 can be readily calculated and the solution for other response variables is obtained.

Single Layer as a Finite Element: Element Equations

For a single layer, we use subscripts 'p, m' to denote upper surface and lower surface respectively so that for example $U = U_p$ when $z = z_p$. Next, we need to find out the relationship between displacement and traction. To do this, let:

$$\bar{A} = (U_m, W_m, U_p, W_p)^T \tag{8.2.31a}$$

$$\bar{F} = (T_m, N_m, -T_p, -N_p)^T \tag{8.2.31b}$$

$$\bar{C} = (C_1, C_2, C_3, C_4) \tag{8.2.31c}$$

Then we find that:

$$\bar{A} = \bar{\bar{X}}(\alpha)\bar{C} \tag{8.2.32a}$$

$$\bar{F} = \bar{\bar{Y}}(\alpha)\bar{C} \tag{8.2.32b}$$

where,

$$\bar{\bar{X}}(\alpha) = \begin{bmatrix} U_1(z_m) & U_2(z_m) & U_3(z_m) & U_4(z_m) \\ W_1(z_m) & W_2(z_m) & W_3(z_m) & W_4(z_m) \\ U_1(z_p) & U_2(z_p) & U_3(z_p) & U_4(z_p) \\ W_1(z_p) & W_2(z_p) & W_3(z_p) & W_4(z_p) \end{bmatrix} \tag{8.2.33a}$$

$$\bar{\bar{Y}}(\alpha) = \begin{bmatrix} T_1(z_m) & T_2(z_m) & T_3(z_m) & T_4(z_m) \\ N_1(z_m) & N_2(z_m) & N_3(z_m) & N_4(z_m) \\ -T_1(z_p) & -T_2(z_p) & -T_3(z_p) & -T_4(z_p) \\ -N_1(z_p) & -N_2(z_p) & -N_3(z_p) & -N_4(z_p) \end{bmatrix} \tag{8.2.33b}$$

We can also write:

$$\bar{C} = \bar{\bar{X}}^{-1}(\alpha)\bar{A} \tag{8.2.34a}$$

$$\bar{F} = \bar{\bar{Y}}(\alpha)\bar{\bar{X}}^{-1}(\alpha)\bar{A} \qquad (8.2.34b)$$

where $K = \bar{\bar{Y}}\bar{\bar{X}}^{-1}$ is the layer stiffness matrix. The stiffness layer relationship can be established as:

$$\begin{bmatrix} T_m \\ N_m \\ -T_p \\ -N_p \end{bmatrix} = \begin{bmatrix} k_{11} & k_{12} & k_{13} & k_{14} \\ k_{21} & k_{22} & k_{23} & k_{24} \\ k_{31} & k_{32} & k_{33} & k_{34} \\ k_{41} & k_{42} & k_{43} & k_{44} \end{bmatrix} \begin{bmatrix} U_m \\ W_m \\ U_p \\ W_p \end{bmatrix} \qquad (8.2.35)$$

Elastic Response for Multiple Layers

To illustrate the way in which the layer stiffness matrix may be used to solve multiple layered elastic system, let us consider a three layered system as an example. If we adopt the notation that the subscript '*j*' denotes the value of a particular quantity on the node plane $z = z_j$, the stiffness relationships have the form

$$\begin{bmatrix} T_1 \\ N_1 \\ -T_2 \\ -N_2 \end{bmatrix} = \begin{bmatrix} a_{11} & a_{12} & a_{13} & a_{14} \\ a_{21} & a_{22} & a_{23} & a_{24} \\ a_{31} & a_{32} & a_{33} & a_{34} \\ a_{41} & a_{42} & a_{43} & a_{44} \end{bmatrix} \begin{bmatrix} U_1 \\ W_1 \\ U_2 \\ W_2 \end{bmatrix} \qquad (8.2.36a)$$

$$\begin{bmatrix} T_2 \\ N_2 \\ -T_3 \\ -N_3 \end{bmatrix} = \begin{bmatrix} b_{11} & b_{12} & b_{13} & b_{14} \\ b_{21} & b_{22} & b_{23} & b_{24} \\ b_{31} & b_{32} & b_{33} & b_{34} \\ b_{41} & b_{42} & b_{43} & b_{44} \end{bmatrix} \begin{bmatrix} U_2 \\ W_2 \\ U_3 \\ W_3 \end{bmatrix} \qquad (8.2.36b)$$

$$\begin{bmatrix} T_3 \\ N_3 \\ -T_4 \\ -N_4 \end{bmatrix} = \begin{bmatrix} c_{11} & c_{12} & c_{13} & c_{14} \\ c_{21} & c_{22} & c_{23} & c_{24} \\ c_{31} & c_{32} & c_{33} & c_{34} \\ c_{41} & c_{42} & c_{43} & c_{44} \end{bmatrix} \begin{bmatrix} U_3 \\ W_3 \\ U_4 \\ W_4 \end{bmatrix} \qquad (8.2.36c)$$

These three layer matrices may be assembled to give the final layer stiffness matrix:

$$K\bar{A} = \bar{F} \qquad (8.2.37)$$

where $\bar{A} = (U_1, W_1, U_2, W_2, U_3, W_3, 0, 0)^T$, $\bar{F} = (R, S, 0, 0, 0, 0, T_4$, and R, S are the coefficients of the surface loadings. For the particular value of α, the global stiffness matrix K is as follows:

$$K = \begin{bmatrix} a_{11} & a_{12} & a_{13} & a_{14} & 0 & 0 & 0 & 0 \\ a_{21} & a_{22} & a_{23} & a_{24} & 0 & 0 & 0 & 0 \\ a_{31} & a_{32} & a_{33}+b_{11} & a_{34}+b_{12} & b_{13} & b_{14} & 0 & 0 \\ a_{41} & a_{42} & a_{43}+b_{21} & a_{44}+b_{22} & b_{23} & b_{24} & 0 & 0 \\ 0 & 0 & b_{31} & b_{32} & b_{33}+c_{11} & b_{34}+c_{12} & c_{13} & c_{14} \\ 0 & 0 & b_{41} & b_{42} & b_{43}+c_{21} & b_{44}+c_{22} & c_{23} & c_{24} \\ 0 & 0 & 0 & 0 & c_{31} & c_{32} & c_{33} & c_{34} \\ 0 & 0 & 0 & 0 & c_{41} & c_{42} & c_{43} & c_{44} \end{bmatrix}$$

(8.2.38)

For N number of layers, Figure 8.2-10, the stiffness matrix is symmetric and has a bandwidth of 6 with 2N degrees of freedom.

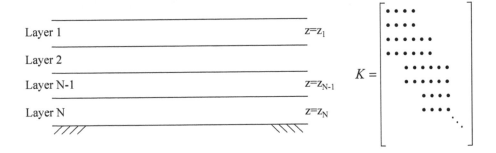

Figure 8.2-10 Horizontally layered soil and stiffness matrix

Summary

$$\sigma_{xz} = \int_{-\infty}^{\infty} S(\alpha)e^{i\alpha x} d\alpha$$

Approximate the surface traction (*fft*)

$$\sigma_{zz} = \int_{-\infty}^{\infty} P(\alpha)e^{i\alpha x} d\alpha$$

$$\Downarrow \quad \begin{matrix} S_1 \; S_2 \; .....S_N \\ P_1 \; P_2 \; .....P_N \\ \alpha_1 \; \alpha_2 \; .....\alpha_N \end{matrix}$$

$$\bar{\bar{K}}\bar{A} = \bar{F}$$

$$\alpha = \alpha_n$$
$$P = P_n$$
$$S = S_n$$

Form stiffness matrix and load vectors for each harmonics

$$\Downarrow \quad \text{Do for each } \alpha$$

$$\overline{A} = \overline{\overline{K}}^{-1}\overline{F}$$

$$\overline{A} = [u_1 \ w_1 \ ...]^T$$

Solve stiffness equation

⇩

$$r(x,z) = \int\limits_{-\infty}^{\infty} Ae^{i\alpha x} d\alpha$$

Calculate response variables at desired
location through *ifft*

Computer Implementation

The formulation presented in the previous section is implemented into a simple MATLAB
program **"Elast_Stress.m"**. This program is used to analyze stresses within an elastic half
space subjected to a rectangular strip load. All the data (both input and the results) are clearly
identified in the program.

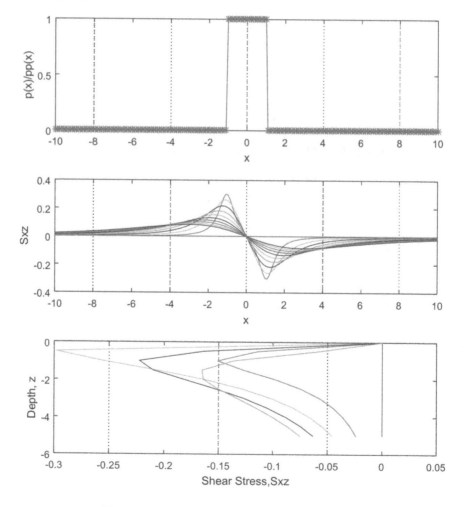

Figure 8.2-11 Computed stresses from MATLAB®

The results from the program are presented in Figure 8.2-11. The top figure simply reproduces the geometry of the input loading. The middle figure shows the variation of shear stress along the x-direction at various depths and the bottom figure shows the variation of shear stresses along the z-direction at various distances in the x-direction.

8.3 Quasi-Static Analysis of Multi-Layer Porous Media under Waves

In this section, we take into account the coupled solution of flow and deformation problem given in section 7.8 in Chapter 7 considering a multi-layer porous media under waves as in Figure 8.3-1. A semi-analytical solution developed by Ulker (2015) to the quasi-static response of saturated-layered porous media under harmonic waves is presented. The equations of (7.8-1) governing the response of porous media are used and the necessary continuity conditions are defined in addition to the boundary conditions prescribed earlier in section 7.8 (Figure 7.8-1).

8.3.1 Semi-Analytical Solution

A semi-analytical solution to the dynamic response of a saturated multi-layer porous medium under harmonic waves is presented as illustrated in Figure 8.3-1. Equation (7.8.1c) is integrated in time and rewritten in tensor form as,

$$Q(\mathbf{u}_{i,i} + \bar{\mathbf{w}}_{i,i})_i = -\mathbf{p}_{,i} \tag{8.3.1}$$

If we substitute (8.3.1) into (7.8.1b) and use the effective stress relation, while neglecting the body forces in the system, we get two final equations that are valid for the layers which are numbered with subscript ($l = 1,2...N$) as,

$$(\sigma'_{ij,j})_l + Q_l\left[(\mathbf{u}_{i,i} + \bar{\mathbf{w}}_{i,i})_i\right]_l = \rho_l(\ddot{\mathbf{u}}_i)_l + (\rho_f)_l(\ddot{\bar{\mathbf{w}}}_i)_l \tag{8.3.2a}$$

$$Q_l\left[(\mathbf{u}_{i,i} + \bar{\mathbf{w}}_{i,i})_i\right]_l = (\rho_f)_l(\ddot{\mathbf{u}}_i)_l + \left(\frac{\rho_f}{n}\right)_l(\ddot{\bar{\mathbf{w}}}_i)_l + \left(\frac{\rho_f g_i}{k_i}\right)_l(\dot{\bar{\mathbf{w}}}_i)_l \tag{8.3.2b}$$

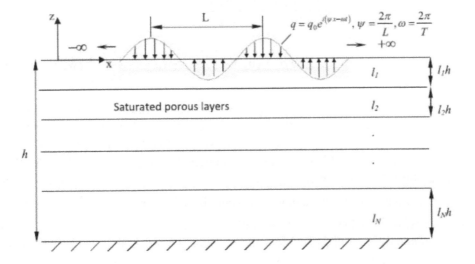

Figure 8.3-1 Multi-layer porous media under progressive wave load

Considering the elastic stress-strain relationship using the Lame's parameters λ and G, we have,

$$(\sigma_{ij}{}') = \lambda_l\,(\varepsilon_{kk})_l\,\delta_{ij} + 2Gl\,(\varepsilon_{ij}) \tag{8.3.3}$$

where ε_{kk} is volumetric strain. Thus, the scalar form of (8.3.2) incorporating (8.3.3) is written as,

$$K_l\,\frac{\partial^2\,(u_x)_l}{\partial x^2} + \lambda_l\,\frac{\partial^2\,(u_z)_l}{\partial x\partial z} + G_l\left(\frac{\partial^2\,(u_x)_l}{\partial z^2} + \frac{\partial^2\,(u_z)_l}{\partial x\partial z}\right) + Q_l\left(\frac{\partial^2\,(u_x)_l}{\partial x^2} + \frac{\partial^2\,(u_z)_l}{\partial x\partial z}\right)$$

$$+ Q_l\left(\frac{\partial^2\,(\bar{w}_x)_l}{\partial x^2} + \frac{\partial^2\,(\bar{w}_z)_l}{\partial x\partial z}\right) = \rho_l\ddot{u}_x + (\rho_f)_l\,\ddot{\bar{w}}_x \tag{8.3.4a}$$

$$K_l\,\frac{\partial^2\,(u_z)_l}{\partial z^2} + \lambda_l\,\frac{\partial^2\,(u_x)_l}{\partial x\partial z} + G_l\left(\frac{\partial^2\,(u_z)_l}{\partial x^2} + \frac{\partial^2\,(u_x)_l}{\partial x\partial z}\right) + Q_l\left(\frac{\partial^2\,(u_z)_l}{\partial z^2} + \frac{\partial^2\,(u_x)_l}{\partial x\partial z}\right)$$

$$+ Q_l\left(\frac{\partial^2\,(\bar{w}_z)_l}{\partial z^2} + \frac{\partial^2\,(\bar{w}_x)_l}{\partial x\partial z}\right) = \rho_l\,(\ddot{u}_z)_l + (\rho_f)_l\,(\ddot{\bar{w}}_z)_l \tag{8.3.4b}$$

$$Q_l\left(\frac{\partial^2\,(u_x)_l}{\partial x^2} + \frac{\partial^2\,(u_z)_l}{\partial x\partial z}\right) + Q_l\left(\frac{\partial^2\,(\bar{w}_x)_l}{\partial x^2} + \frac{\partial^2\,(\bar{w}_z)_l}{\partial x\partial z}\right) = (\rho_f)_l\,(\ddot{u}_x)_l$$

$$(\rho_f)_l\,(\ddot{u}_x)_l + \left(\frac{\rho_f}{n}\right)_l\,(\ddot{\bar{w}}_x)_l + \left(\frac{\rho_f g}{k_x}\right)_l\,(\dot{\bar{w}}_x)_l \tag{8.3.4c}$$

$$Q_l\left(\frac{\partial^2\,(u_z)_l}{\partial z^2} + \frac{\partial^2\,(u_x)_l}{\partial x\partial z}\right) + Q_l\left(\frac{\partial^2\,(w_z)_l}{\partial z^2} + \frac{\partial^2\,(w_x)_l}{\partial x\partial z}\right) =$$

$$(\rho_f)_l\,(\ddot{u}_z)_l + \left(\frac{\rho_f}{n}\right)_l\,(\ddot{\bar{w}}_z)_l + \left(\frac{\rho_f g}{k_z}\right)_l\,(\dot{\bar{w}}_z)_l \tag{8.3.4d}$$

Therefore, for each layer we have $(l, k_x, k_z, n, \lambda, G, K, K_f, \rho, \rho_f)_l$. Considering all the response variables to be of the form, $r(x, z, t) = \mathbf{R}(z)e^{i(\psi x - \omega t)}$, we can obtain a harmonic complex form of the equations where $r(x, z, t)$ are the ultimate response variables and $\mathbf{R}(z)$ represents the amplitudes of these response variables which are the solid and fluid displacements, $\mathbf{U}_x, \mathbf{U}_z, \bar{\mathbf{W}}_x, \bar{\mathbf{W}}_z$, pore pressure, $\mathbf{P}_w$ and the stresses, $\sigma'_{xx}, \sigma'_{zz}, \sigma'_{xz}$, (e.g. $p_w\,(x, z, t) = \mathbf{P}_w(z)e^{i(\psi x - \omega t)}$). However, now that there are N number of layers, a modified normalized depth ratio, $\mathbf{Z} = \dfrac{z}{lh}$ is introduced into the linear set (with 'l' being the thickness ratio of each layer and h the total depth) which is now correspondingly multiplied by l or l^2 terms. Here, it is again necessary to define a set of non-dimensional parameters as was also stated in (7.8.4)-(7.8.8). So, $\kappa, \kappa_1, \kappa_2$ and β stay the same but $m = \psi l h$ is now modified to account for the effect of layering on the spatial variation of loading with $\psi = 2\pi/L$ being the wave number and L, the wavelength. The main parameters,

$$\Pi_{1x} = \frac{k_x V_c^2}{g\beta\omega l^2 h^2} \tag{8.3.5}$$

$$\Pi_{1z} = \frac{k_z V_c^2}{g\beta\omega l^2 h^2} \tag{8.3.6}$$

are now also modified due to layering. In addition, it is now necessary to account for the ratios of the non-dimensional parameters of each layer which will be of more use in the case of two distinct layers with distinct parameters. Thus for ($l = 2, 3...N$), we define,

$$(R_{\Pi_{1x}})_{l-1} = \frac{(\Pi_{1x})_{l-1}}{(\Pi_{1x})_l} \tag{8.3.7a}$$

$$(R_{\Pi_{1z}})_{l-1} = \frac{(\Pi_{1z})_{l-1}}{(\Pi_{1z})_l} \tag{8.3.7b}$$

$$(R_m)_{l-1} = \frac{m_{l-1}}{m_l} \tag{8.3.7c}$$

It should be noted here that these ratios are used mainly for two distinct layers in a few layer system.

Boundary Conditions

The boundary conditions used for the problem in Figure 8.3-1 are:
At $z = -h$, displacements,

$$(\mathbf{U_x})_N = (\mathbf{U_z})_N = (\mathbf{W_z})_N = 0 \tag{8.3.8}$$

and at $z = 0$, vertical effective stress and shear stress of the first layer ($l = 1$) vanish,

$$(\sigma'_{zz})_1 = (\sigma_{xz})_1 = 0 \tag{8.3.9}$$

while pore pressure takes the form,

$$\mathbf{P}_{w1} = q_0 e^{i(\psi x - \omega t)} \tag{8.3.10}$$

where

$$q_0 = \frac{\gamma_w H}{2\cosh(\psi d)} \tag{8.3.11}$$

and H is the wave height. Using these boundary conditions, a 6×6 matrix with entries $(\xi_{ij})_l$ ($i, j = 1...6$) is obtained in terms of eigenvalues $(\eta_i)_l$ that are calculated from the characteristic equations. Interested readers can refer to Ulker and Rahman (2009) for more details including the necessary differences in obtaining the response for various formulations.

Continuity Conditions

The semi-analytical solution for the dynamic response of layered porous media requires the use of continuity conditions at layer interfaces as defined below with 'j' now indicating the layer number for the system. For $z = -l_j h \rightarrow Z = -l_j$:

$$(\mathbf{U_x})_j = (\mathbf{U_x})_{j+1} \tag{8.3.12a}$$

Similarly, at a depth of $z = -(l_j + l_{j+1})\, h \rightarrow Z = -(l_j + l_{j+1})$, $(\mathbf{U_x})_{j+1} = (\mathbf{U_x})_{j+2}$ is satisfied and for $z = -(l_j + l_{j+1} + \ldots l_{N-1})\, h \rightarrow Z = -(l_j + l_{j+1} + \ldots l_{N-1}) \rightarrow (\mathbf{U_x})_{N-1} = (\mathbf{U_x})_N$ can be written. We can write the rest of the continuity conditions as:

$$(\mathbf{U_z})_j = (\mathbf{U_z})_{j+1} \tag{8.3.12b}$$

$$(\mathbf{f_z}) = (\mathbf{f_z})_{j+1} \rightarrow \left(\frac{k_z}{\rho_w g} \frac{\partial \mathbf{P}}{\partial \mathbf{z}} \right)_j = \left(\frac{k_z}{\rho_w g} \frac{\partial \mathbf{P}}{\partial \mathbf{z}} \right)_{j+1} \tag{8.3.12c}$$

$$(\sigma_{zz})_j = (\sigma_{zz})_{j+1} \rightarrow \left(D\frac{\partial \mathbf{U_z}}{\partial \mathbf{z}} + \lambda \frac{\partial \mathbf{U_x}}{\partial \mathbf{x}} \right)_j = \left(D\frac{\partial \mathbf{U_z}}{\partial \mathbf{z}} + \lambda \frac{\partial \mathbf{U_x}}{\partial \mathbf{x}} \right)_{j+1} \tag{8.3.12d}$$

$$(\tau_{xz})_j = (\tau_{xz})_{j+1} \rightarrow G_j\left(\frac{\partial \mathbf{U_x}}{\partial \mathbf{z}} + \frac{\partial \mathbf{U_z}}{\partial \mathbf{x}} \right)_j = G_{j+1}\left(\frac{\partial \mathbf{U_x}}{\partial \mathbf{z}} + \frac{\partial \mathbf{U_z}}{\partial \mathbf{x}} \right)_{j+1} \tag{8.3.12e}$$

$$(\mathbf{Pw})_j = (\mathbf{Pw})_{j+1} \tag{8.3.12f}$$

It should be noted that fluid flux from the D'Arcy's law in the vertical direction (f_z) must be continuous at layer interfaces as indicated in (8.3.12c). See the Appendix for related details. These equations result in the linear system,

$$[\xi]_{M \times M}\, \{\mathbf{a}\}_{M \times 1} = \{\mathbf{F}\}_{M \times 1} \tag{8.3.13}$$

where M = $6N$, $[\xi]$ is the matrix of equations obtained from both boundary and continuity equations (or $(\xi_{ij})_l$ equations) that is referred to as '*layer stiffness matrix*' by some (Rahman et al. 1994), $\{\mathbf{a}\}$ is the vector of coefficients a_i and $\{\mathbf{F}\}$ is the forcing vector of the form, $\{0, 0, F, \ldots 0\}$ with F being the only non-zero term coming from the force boundary condition at the surface written as $F = \dfrac{q_0}{Q_1}$ with $Q_1 = \left(\dfrac{K_f}{n} \right)_1$ for the first layer. The equation system of (8.3.13) should be separately written for all formulations. Then the wave-induced stresses and pore pressures for a particular layer, l, can be evaluated as,

$$(\sigma'_{zz})_l = \left[\sum_{j=1}^{6} \left(i\psi\lambda + b_j K \frac{\eta_j}{lh} \right) a_j e^{\eta_j \frac{\mathbf{z}}{lh}} \right]_l e^{i(\psi \mathbf{x} - \acute{E}t)} \tag{8.3.14}$$

$$(\tau_{xz})_l = G_l \left[\sum_{j=1}^{6} \left(\frac{\eta_j}{lh} + i\psi\, b_j \right) a_j e^{\eta_j \frac{\mathbf{z}}{lh}} \right]_l e^{i(\psi \mathbf{x} - \omega t)} \tag{8.3.15}$$

$$\mathbf{p}_{wl} = -\left(\frac{K_f}{n} \right)_l \left[\sum_{j=1}^{6} \left(i\psi\, (1 + c_j) + \frac{\eta_j}{lh}(b_j + d_j) \right) a_j e^{\eta_j \frac{\mathbf{z}}{lh}} \right]_l e^{i(\psi \mathbf{x} - \omega t)} \tag{8.3.16}$$

8.3.2 Quasi-Static Response of Multi-Layer System

In this section, the quasi-static response of a multi-layer system is presented in terms of absolute values of pore pressure and shear stress variations in the medium. Here non-dimensional parameters for each layer and their ratios $(R_{\Pi_{1x}}, R_{\Pi_{1z}}, R_m)$ play a key role in defining the continuous behavior of multiple layers. It is determined that in case there are only a few layers with distinct physical properties, depending on their respective thicknesses, the first set of distinct consecutive layers closest to the surface are important in identifying the necessary formulation governing the response. If there are more than two layers with thicknesses being relatively much larger or that there is a larger difference between some of the physical properties (i.e. hydraulic conductivity, degree of saturation), a weighted average of those properties using the relation below will be sufficient,

$$A_{eq} = \frac{\sum\limits_{l=1}^{N} h_l A_l}{\sum\limits_{l=1}^{N} h_l} \tag{8.3.17}$$

Here, A_{eq} is the equivalent response parameter to be weighted.

Computer Implementation

The presented solution above requires stress and pore pressure results to be plotted by implementing into a MATLAB program. That is called "**Semi_Analytical_Layered_Porous.m**". The semi-analytical solution is first validated using the published data considering a single (Figure 8.3-2) and a multiple layer solution (Figure 8.3-3). The multi-layer solution converges to the single-layer one for zero layer thickness. Response results match well with other solutions as presented. The code currently considers a two-layer system.

Figures 8.3-4 through 8.3-7 present the dynamic response of a layered system for the quasi-static formulation. The other formulations, namely the dynamic ones where inertial terms associated with the motion of solids and that of pore fluid are out of the scope of this book. The effect of the hydraulic conductivity in terms of R_{Π_1} and the layer thickness ratio in terms of R_m on the solid displacement, pore pressure and stress distributions are illustrated.

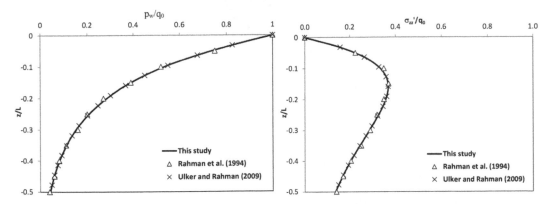

Figure 8.3-2 Validation of the presented solution for a single-layer case, (a) Normalized pore pressure variation with depth, (b) Normalized vertical effective stress with depth

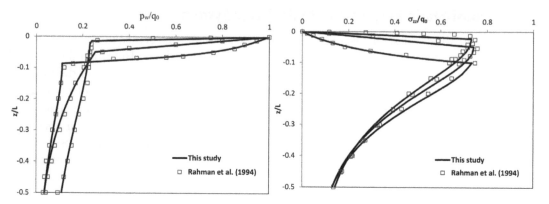

Figure 8.3-3 Validation of the presented solution for a multi-layer case, (a) Normalized pore pressure variation with depth, (b) Normalized vertical effective stress with depth

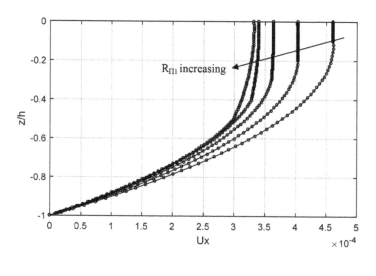

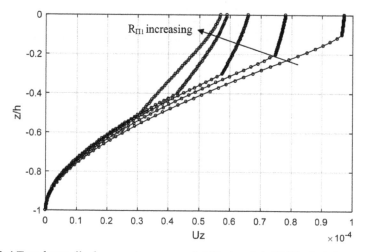

Figure 8.3-4 Two-layer displacement response, (a) Horizontal, (b) Vertical for various ratios of $R_{\Pi_1} = 100, 25, 11.1, 6.25, 4$, $R_m = 0.1$-0.5 (values in m)

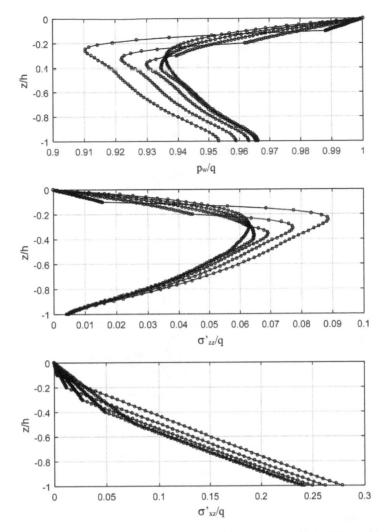

Figure 8.3-5 Two-layered response, (a) Pore pressure, (b) Vertical stress, (c) Shear stress for various ratios of R_{Π_1} = 100, 25, 11.1, 6.25, 4, R_m = 0.1-0.5 (values in m)

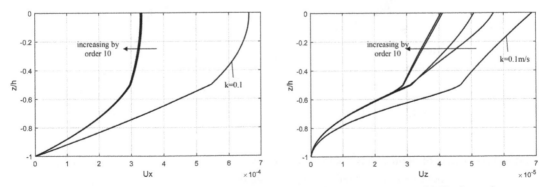

Figure 8.3-6 Effect of permeability coefficient on displacement response, (a) Horizontal, (b) Vertical for various ratios of R_{Π_1} = 4-40,000, R_m = 0.1-0.5 (values in m)

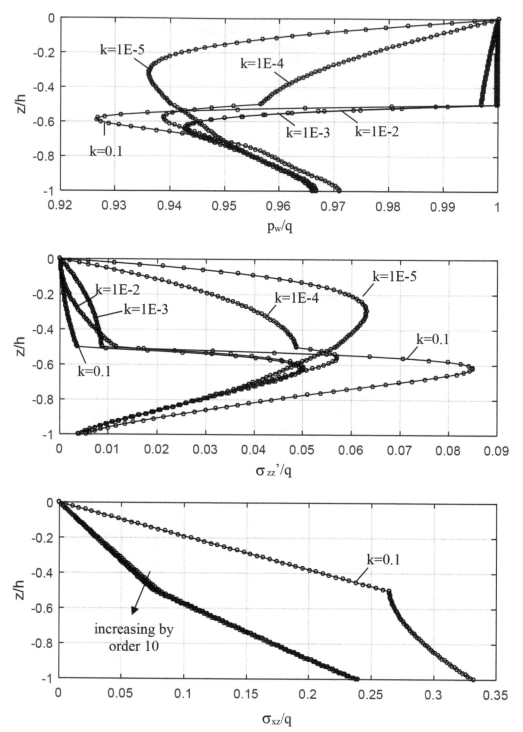

Figure 8.3-7 Effect of permeability coefficient on (a) Pore pressure, (b) Vertical stress, (c) Shear stress response for various ratios of R_{Π_1} = 4-40,000, R_m = 0.1-0.5 (values in m)

8.4 Exercise Problems

8.4.1 Develop a MATLAB program to evaluate and plot the elastic stress distribution within a homogeneous and infinitely deep soil profile subjected to an arbitrarily varying surface load.

8.4.2 Using the program, generate and plot the influence values for vertical stress distribution under a very long embankment and compare your results with those given in the figure below.

8.4.3 Your program should be able to evaluate all the stress components at any specified location. Evaluate the vertical stress, horizontal stress and shear stress at points A, B, C and D.

8.4.4 Compare the vertical stresses obtained in part (c) with those you calculate using the available solutions using the Osterberg (1957)'s chart given below.

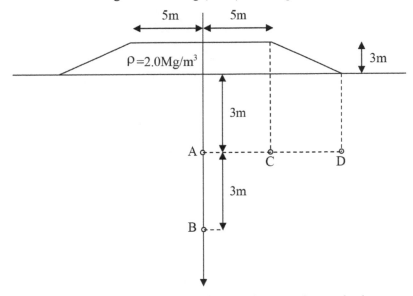

Figure 8.4-1 Stress distribution problem under a very long embankment

Solution for Comparison

Let us calculate the vertical stress under the centerline at depths of 3 and 6 m:

First we need to calculate the applied surface stress q_0 and the dimensions of the embankment in terms of a and b. $q_0 = \rho g h = 2.0$ Mg/m^3 × 9.81 m/s^2 × 3 m = **59 kPa**. From Figure 8.4-2, b = 5 m, $a = 2 \times 3$ m = 6 m.

Next, we calculate the vertical stress for $z = 3$ m. $a/z = 6/3 = 2$, $b/z = 5/3 = 1.67$

From Figure 8.4-2, $I = 0.488$, thus $\sigma_z = q_0 I = 59 \times 0.488 = $ **29 kPa** for one half of the embankment or 58 kPa for the whole. Next for $z = 6$ m, $a/z = 6/6 = 1$, $b/z = 5/6 = 0.83$, thus $I = 0.44$ and $\sigma_z = q_0 I = 59$ kPa × 0.44 × 2 = **52 kPa**.

8.4.3 For a two-layer soil system, investigate the effect of layer rigidity and the wave period on the quasi-static coupled response in terms of displacement, pore pressure and stress variations with the depth. Run the computer program "**Semi_Analytical_Layered_Porous.m**" given in this chapter for a parametric study. Interpret the results.

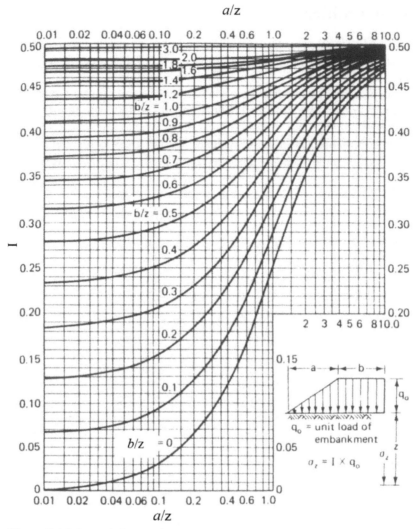

Figure 8.4-2 Stress influence factor under embankment load (Osterberg, 1957)

MATLAB® Scripts

Main program "Elast_Stress.m"

```
%*******************************************************************
% "Elast_Stress.m"
% Description: This program evaluates the elastic stress distribution within
% homogenous and infinitely deep soil profile under plain strain condition
% subjected to an arbitrarily varying surface load
%*******************************************************************
clear all
clf
im=sqrt(-1);
%FFT parameters defined
N=512;
M=N/2;
% [Step-1]Discrete 'x'are defined for the definition of 'p'
x1=-10;
dx=0.1;
nx=201;
xn=x1+(nx-1)*dx;
x=x1:dx:xn;
% load function, p(x) defined
for i=1:nx
    p(i)=0.;
end
for i=91:111
    p(i)=1.;
end

% [Step-2]Fourier Transform 'pfft(f) of 'p(x)' is evaluated
pfft=fft(p',N);
% [Verification]Inverse Fourier Transform pifft(x) of 'pfft(f)' evaluated
pifft=ifft(pfft,N);
% Real parts of 'pifft' (pp) should be the same as theoriginal load function
  verified below
pp=real(pifft);

% Plot the input load function
subplot(3,1,1)
plot (x(1:nx),p(1:nx),x(1:nx),pp(1:nx),'*')
title('input load function')
xlabel('x')
ylabel('p(x)/pp(x)')
grid
axis([-10 10 0 1])
%*******************************************************************
% Points for stress analysis ( z spatial co-ordinates) are defined
nz=11;
z1=0.0;
dz=0.5;
zn=z1+(nz-1)*dz;
z=z1:dz:zn;
%*******************************************************************
% frequencies associated with Fourier Terms
fs=1./(N*dx);
```

```
f=fs*linspace(-M,M-1,N);
f=fftshift(f');
alpha=2*pi.*f;
%*****************************************************************
%[Step-3]Evaluate the transfer function for each Fourier term (Frequency
 Domain)
for iz=1:nz
    alphaz=alpha.*z(iz);
    HSzz(1:N,iz)=(1.+alphaz).*exp(-alphaz);
    HSxx(1:N,iz)=(1.-alphaz).*exp(-alphaz);
    HSxz(1:M+1,iz)=(im.*alphaz(1:M+1).*exp(-alphaz(1:M+1)));
    HSxz(M+2:N,iz)=conj(HSxz(M:-1:2,iz));
end
%*****************************************************************
% [Step-4]Evaluate the response for each Fourier term (Response in Frequency
  Domain)
for iz=1:nz
    Szzf(1:N,iz)=HSzz(1:N,iz).*(pfft);
    Sxxf(1:N,iz)=HSxx(1:N,iz).*(pfft);
    Sxzf(1:N,iz)=HSxz(1:N,iz).*(pfft);
end
%*****************************************************************
% [Step-5]Evaluate the actual response (Response in spatial domain)
for iz=1:nz
    Szz(1:N,iz)=ifft(Szzf(1:N,iz),N);
    Sxx(1:N,iz)=ifft(Sxxf(1:N,iz),N);
    Sxz(1:N,iz)=ifft(Sxzf(1:N,iz),N);
end
Szz=real(Szz);
Sxx=real(Sxx);
Sxz=real(Sxz);

% Plot the stresses
subplot(3,1,2)
plot(x(1:nx), Sxz(1:nx,1:nz))
xlabel('x')
ylabel('Sxz')
grid
subplot(3,1,3)
plot(Sxz(101:5:121,1:nz),-z(1:nz))
xlabel('Shear Stress,Sxz')
ylabel('Depth, z')
grid
```

Main program "fft_example.m"

```
%*****************************************************************
% This program illustrates the use of 'fft' & 'ifft' commands in MATLAB
%*****************************************************************
clear all
clf

%FFT parameters defined
N= 32;        % Number of Points (=nfft)
M= N/2;
L=10.;                          % Length/Period
fs=1/L;                         % Frequency Spacing
```

```
% independent variables (x, z spatial co-ordinates) are defined
x1=0.,
dx=L/N;                          % Interval
xn=(N-1)*dx;
x=x1:dx:xn;                      % Discrete Values of x

% input function, f(x) defined
fx=exp(-x);
% Fourier Transform 'pfft(f) of 'p(x)' is evaluated
fftfx=(fft(fx,N));
% Inverse Fourier Transform pifft(x) of 'pfft(f)' evaluated
ifftfx=ifft(fftfx,N);
% Real parts of 'ifftfx' should be the same as the original function
fxr=real(ifftfx);

subplot(2,1,1)
plot (x(1:N),fx(1:N),x(1:N),fxr(1:N),'*')
title('input  function')
xlabel('x')
ylabel('fx(x) / fxr(x)')
axis([0 4 0 1.])
grid

% frequencies associated with Fourier Terms
f=fs*linspace(-M,M-1,N);
w=2*pi*f;
%Modulus of the Fourier Transform
FT1 =dx*abs(fftfx);              % computed by MATLAB'
FT=fftshift(FT1);               % The FT values rearranged
FTexact= 1./sqrt(1+w.^2);       % from analytical solution
[f', FT', FTexact']

subplot(2,1,2)
plot (f,FT, f,FTexact,'--')
axis([-4 4 0 1.5])
title('Fourier Transform')
xlabel(' Frequence, f')
ylabel('[FT(f)] ')
```

Main program "Semi_Analytical_Layered_Porous.m"

```
%*********************************************************************
% Analytical Solution for Multi-Layer Quasi-Static Response
% "Semi_Analytical_Porous.m"
% Note: Currently 2 layers are considered, code needs modification for nl>2
% Seabed soil system is considered with necessary parameters for wave load
%*********************************************************************
clear all
clc
nz=50;
nl=2;                           % Constant for 2 layers for now!
%----------------------------------------
%  input data for seabed and material
%----------------------------------------
H=3.51642;                      % Depth of seabed (m)
g=9.80665;
kw=245250000;                   % Bulk Modulus of water (kPa)
```

```
%------------------------------------------
%   input data for wave
%------------------------------------------
T=10;                              % Wave period (s)
wL=62.8319;                        % Wave length (m)
h=4.604;                           % Wave height (m)
wD=20;                             % Water depth (d)(m)
k=2*pi/wL;                         % Wave number
w=2*pi/T;                          % Freq.
%------------------------------------------
% Material Properties of Layers, [L1, L2]
%------------------------------------------
l_acc=[0.5 0.5];                   % Thickness ratio
E=[14300 14300];                   % Elasticity Modulus (kPa)
mu=[0.3 0.3];                      % Poisson's Ratio
kz=[0.00001 0.00001];              % Vertical Permeability (m/s)
S=[1 1];                           % Saturation
n=0.35*[1 1];                      % Porosity
rhof=[1 1];                        % Fluid Density
rhos=1.67*[1 1];                   % Solid Density

if sum(l_acc)== 1.0
    % do nothing
else
    error ('sum of the layer thickness ratios must be 1')
end
%----------------------------------------------------------------
%   input data for the load
%----------------------------------------------------------------
dz=-H/nz;
z=0:dz:-H;
ZL=z/wL;
ZH=z/H;
ti=0;
tf=101;
dt=0.01;
t=ti:dt:tf;
nt=tf/dt+1;
q0=rhof(1)*g*h/(2*cosh(k*wD));
ZHl=l_acc'*ZH;
ll=l_acc;
for i=2:nl
    l_acc(i)=l_acc(i)+l_acc(i-1);
    for j=1:length(ZHl)
        ZHl(i,j)=ZHl(i,j)-l_acc(i-1);
    end
end
ZHl=ZHl';
N=6*nl;
ZLl=l_acc'*ZL;
ll=l_acc;
for i=2:nl
    l_acc(i)=l_acc(i)+l_acc(i-1);
    for j=1:length(ZLl)
        ZLl(i,j)=ZLl(i,j)-l_acc(i-1);
    end
end
end
```

```
ZLl=ZLl';
%-------------     -----------------------------------------------
% Non-Dimensional Parameters
%------------------------------------------------------------  --
for i=1:nl
    m(i)=k*H*ll(i);
    rho(i)=rhof(i)*n(i)+rhos(i)*(1-n(i));      % Unit weight of soil+water
                                                 (kN/m^3/g);
    bf(i)=rhof(i)*g;                            % Body force of water
    kx(i)=kz(i);                                % Isotropic Permeability (m/s)
    kf(i)=bf(i)*wD*kw/(bf(i)*wD+kw*(1-S(i)));   % Bulk modulus of pore fluid
    M(i)=E(i)*(1-mu(i))/((1+mu(i))*(1-2*mu(i))); % Constrained Modulus (kPa)
    G(i)=E(i)/(2*(1+mu(i)));                    % Shear Modulus (kPa)
    lambda(i)=2*G(i)*mu(i)/(1-2*mu(i));         % Lame's Constant (kPa)
    Q(i)=kf(i)/n(i);
    kappa(i)=Q(i)/(M(i)+Q(i));
    kappa1(i)=lambda(i)/(M(i)+Q(i));
    kappa2(i)=G(i)/(M(i)+Q(i));
    beta(i)=rhof(i)/rho(i);
    Vc(i)=sqrt((M(i)+Q(i))/rho(i));             % Velocity of sound in water
    P1(i)=(kz(i)*Vc(i)^2)/(g*beta(i)*w*H^2*ll(i)^2);
    P11(i)=(kx(i)*Vc(i)^2)/(g*beta(i)*w*H^2*ll(i)^2);
%----------------------------------------------------------------
% Aij Parameters
%-----------------------------------------    -------------------
    AA(i)=m(i)^2*kappa(i);
    AA13(i)=(sqrt(-1)/P11(i))-m(i)^2*kappa(i);
    A51(i)=(1-kappa(i))*kappa(i);
    A52(i)=m(i)^2*kappa(i);
    A53(i)=AA13(i)+A52(i);
    A61(i)=kappa(i)+kappa1(i)+kappa2(i);
    A75(i)=sqrt(-1)/P1(i);
    A76(i)=A61(i)-kappa(i);
    A77(i)=1+kappa2(i)^2;
    A83(i)=AA13(i)*(A61(i)^2-A77(i));
    A84(i)=kappa2(i)*A75(i);
%----------------------------------------------------------------
% Coefficients of Characteristic Eqn. det(M)=0
%----------------------------------------------------------------
    a_static_1(i)=A51(i)*A84(i);
    a_static_21(i)=-A51(i)*(A52(i)^2);
    a_static_22(i)=A52(i)*(A52(i)*(A76(i)^2+kappa(i)-A77(i))+2*A51(i)*A52(i)+
    2*AA13(i)*(A51(i)-kappa(i)*A76(i))+A83(i)+kappa(i)*A84(i));
    a_static_23(i)=AA13(i)*A84(i);
    a_static_2(i)=a_static_21(i)+a_static_22(i)+a_static_23(i);
    a_static_31(i)=A52(i)^2*(A51(i)*kappa2(i)*(A53(i)/(kappa(i)^2))-
    A75(i)*(1-2*A61(i)));
    a_static_32(i)=A52(i)*A75(i)*(A83(i)/kappa(i)-A52(i));
    a_static_3(i)=a_static_31(i)+a_static_32(i);
    a_static_4(i)=A84(i)*m(i)^2*(m(i)^2*AA13(i)+A52(i)^2);
%----------------------------------------------------------------
% Roots of Characteristic Eqn. Matlab roots
%----------------------------------------------------------------
    poly_static(i,1)=a_static_1(i);
    poly_static(i,2)=0;
    poly_static(i,3)=a_static_2(i);
```

```
    poly_static(i,4)=0;
    poly_static(i,5)=a_static_3(i);
    poly_static(i,6)=0;
    poly_static(i,7)=a_static_4(i);
    L_static(i,:)=roots(poly_static(i,:));
%--------------------------------------------------------------
    for j=1:6
    %--------------------------------------------------------------
        B_static(i,j)=sqrt(-1)*m(i)*kappa(i)*L_static(i,j);
        C_static(i,j)=(L_static(i,j)^2)*kappa(i);
        A24_static(i,j)=sqrt(-1)/P1(i)+kappa(i)*(L_static(i,j)^2);
        A31_static(i,j)=-m(i)^2+kappa2(i)*(L_static(i,j)^2);
        D_static(i,j)=sqrt(-1)*m(i)*L_static(i,j)*(kappa(i)+kappa1(i)+kappa2
        (i));
        A42_static(i,j)=-m(i)^2*kappa2(i)+L_static(i,j)^2;
    %--------------------------------------------------------------
    % Coefficients of soln. entries of eigenvectors (bi, ci, di)
    %--------------------------------------------------------------
        detm_static(i,j) = B_static(i,j)^2*(A42_static(i,j)-C_static(i,j))-B_
        static(i,j)*D_static(i,j)*(C_static(i,j)-A24_static(i,j))+AA(i)*(C_
        static(i,j)^2-A24_static(i,j)*A42_static(i,j)); % AA'nin isareti
        degisti
        b_static(i,j)= (B_static(i,j)^3+B_static(i,j)*(C_static(i,j)*(A31_
          static(i,j)-AA(i))-D_static(i,j)*B_static(i,j)-A31_static(i,j)*A24_
          static(i,j))+AA(i)*D_static(i,j)*A24_static(i,j))/detm_static(i,j);
          % AA'nin isareti degisti
        c_static(i,j)= (-B_static(i,j)^2*A42_static(i,j)+2*B_static(i,j)*C_
          static(i,j)*D_static(i,j)-A31_static(i,j)*(C_static(i,j)^2-A24_
          static(i,j)*A42_static(i,j))-D_static(i,j)^2*A24_static(i,j))/detm_
          static(i,j);
        d_static(i,j)= (B_static(i,j)*D_static(i,j)*(D_static(i,j)-B_
          static(i,j))+B_static(i,j)*A42_static(i,j)*(AA(i)-A31_
          static(i,j))+C_static(i,j)*(A31_static(i,j)*B_static(i,j)-AA(i)*D_
          static(i,j)))/detm_static(i,j);% AA'nin isareti degisti
    end
end
%--------------------------------------------------------------
% Matrices after applying c.c. N = 6*(nl)
%--------------------------------------------------------------
N=6*nl;
for i=1:nl
for j=1+6*(i-1):6*i   % i=j and j=s in the text
    % Boundary Conditions
    if i==1
        a_static(1,j)= (M(1)*b_static(1,j)*L_static(1,j))/(H*ll(1))+sqrt(-
          1)*k*lambda(1);
        a_static(2,j)= G(1)*(L_static(1,j)/(H*ll(1))+sqrt(-1)*k*b_
          static(1,j));
        a_static(3,j)= sqrt(-1)*k*(1+c_static(1,j))+L_static(1,j)*(b_
          static(1,j)+d_static(1,j))/(H*ll(1));
    elseif i==nl
        a_static(N-2,j)= exp(-L_static(nl,j-(nl-1)*6));
        a_static(N-1,j)= b_static(nl,j-(nl-1)*6)*exp(-L_static(nl,j-
          (nl-1)*6));
        a_static(N,j)= d_static(nl,j-(nl-1)*6)*exp(-L_static(nl,j-(nl-1)*6));
```

```
else
    % do nothing
end
% Continuity Conditions
if i < nl
    a_static(4+6*(i-1),j)= exp(-L_static(i,j-(i-1)*6)*l_acc(i));
    a_static(5+6*(i-1),j)= b_static(i,j-(i-1)*6)*exp(-L_static(i,j-(i-
        1)*6)*l_acc(i));
    a_static(6+6*(i-1),j)= (kz(i)/bf(i))*Q(i)*L_static(i,j-(i-
        1)*6)*exp(-L_static(i,j-(i-1)*6)*l_acc(i))*(sqrt(-1)*k*(1+c_
        static(i,j-(i-1)*6))+L_static(i,j-(i-1)*6)*(b_static(i,j-(i-
        1)*6)+d_static(i,j-(i-1)*6))/(H*ll(i)))/(H*ll(i));
    a_static(7+6*(i-1),j)= exp(-L_static(i,j-(i-1)*6)*l_acc(i))*(sqrt(-
        1)*k*lambda(i)+M(i)*L_static(i,j-(i-1)*6)*b_static(i,j-(i-1)*6)/
        (H*ll(i)));
    a_static(8+6*(i-1),j)= G(i)*exp(-L_static(i,j-(i-1)*6)*l_
        acc(i))*(sqrt(-1)*k*b_static(i,j-(i-1)*6)+L_static(i,j-(i-1)*6)/
        (H*ll(i)));
    a_static(9+6*(i-1),j)= Q(i)*exp(-L_static(i,j-(i-1)*6)*l_
        acc(i))*(sqrt(-1)*k*(1+c_static(i,j-(i-1)*6))+L_static(i,j-(i-
        1)*6)*(b_static(i,j-(i-1)*6)+d_static(i,j-(i-1)*6))/(H*ll(i)));
else
    % do nothing
end

if i > 1
    a_static(4+6*(i-2),j)= -exp(-L_static(i,j-(i-1)*6)*l_acc(i-1));
    a_static(5+6*(i-2),j)= -b_static(i,j-(i-1)*6)*exp(-L_static(i,j-(i-
        1)*6)*l_acc(i-1));
    a_static(6+6*(i-2),j)= -(kz(i)/bf(i))*Q(i)*L_static(i,j-(i-
        1)*6)*exp(-L_static(i,j-(i-1)*6)*l_acc(i-1))*(sqrt(-1)*k*(1+c_
        static(i,j-(i-1)*6))+L_static(i,j-(i-1)*6)*(b_static(i,j-(i-1)*6)+d_
        static(i,j-(i-1)*6))/(H*ll(i)))/(H*ll(i));
    a_static(7+6*(i-2),j)= -exp(-L_static(i,j-(i-1)*6)*l_acc(i-
        1))*(sqrt(-1)*k*lambda(i)+M(i)*L_static(i,j-(i-1)*6)*b_static(i,j-
        (i-1)*6)/(H*ll(i)));
    a_static(8+6*(i-2),j)= -G(i)*exp(-L_static(i,j-(i-1)*6)*l_acc(i-
        1))*(sqrt(-1)*k*b_static(i,j-(i-1)*6)+L_static(i,j-(i-1)*6)/
        (H*ll(i)));
    a_static(9+6*(i-2),j)= -Q(i)*exp(-L_static(i,j-(i-1)*6)*l_acc(i-
        1))*(sqrt(-1)*k*(1+c_static(i,j-(i-1)*6))+L_static(i,j-(i-1)*6)*(b_
        static(i,j-(i-1)*6)+d_static(i,j-(i-1)*6))/(H*ll(i)));
else
    % do nothing
end
end
end
% RHS Vector (known)
NL=length(a_static);
RHS=zeros(1,NL);
RHS(3)=-q0/Q(1);
% Solve for the Coefficient Vector of Analytical Solution (unknown ai) XX=[a1
a2 a3 a4 a5 a6];
XX_static=a_static\RHS';
```

```
%-----------------------------------------------------------------
%  Stresses and Displacements
%-----------------------------------------------------------------
for i=1:nl
    Ux_static(i,:)=XX_static(6*i-5)*exp(L_static(i,1)*ZHl(:,i))+XX_
    static(6*i-4)*exp(L_static(i,2)*ZHl(:,i))+XX_static(6*i-3)*exp(L_
    static(i,3)*ZHl(:,i))+...
        XX_static(6*i-2)*exp(L_static(i,4)*ZHl(:,i))+XX_static(6*i-1)*exp(L_
        static(i,5)*ZHl(:,i))+XX_static(6*i)*exp(L_static(i,6)*ZHl(:,i));
    Uz_static(i,:)=XX_static(6*i-5)*b_static(i,1)*exp(L_
    static(i,1)*ZHl(:,i))+XX_static(6*i-4)*b_static(i,2)*exp(L_
    static(i,2)*ZHl(:,i))+XX_static(6*i-3)*b_static(i,3)*exp(L_
    static(i,3)*ZHl(:,i))+...
        XX_static(6*i-2)*b_static(i,4)*exp(L_static(i,4)*ZHl(:,i))+XX_
        static(6*i-1)*b_static(i,5)*exp(L_static(i,5)*ZHl(:,i))+XX_
        static(6*i)*b_static(i,6)*exp(L_static(i,6)*ZHl(:,i));
    P_static(i,:)=-Q(i)*((sqrt(-1)*k*(1+c_static(i,1))+L_static(i,1)*(b_
    static(i,1)+d_static(i,1))/(H*ll(i)))*XX_static(6*i-5)*exp(L_
    static(i,1)*ZHl(:,i))+...
            (sqrt(-1)*k*(1+c_static(i,2))+L_static(i,2)*(b_static(i,2)+d_
            static(i,2))/(H*ll(i)))*XX_static(6*i-4)*exp(L_
            static(i,2)*ZHl(:,i))+...
            (sqrt(-1)*k*(1+c_static(i,3))+L_static(i,3)*(b_static(i,3)+d_
            static(i,3))/(H*ll(i)))*XX_static(6*i-3)*exp(L_
            static(i,3)*ZHl(:,i))+...
            (sqrt(-1)*k*(1+c_static(i,4))+L_static(i,4)*(b_static(i,4)+d_
            static(i,4))/(H*ll(i)))*XX_static(6*i-2)*exp(L_
            static(i,4)*ZHl(:,i))+...
            (sqrt(-1)*k*(1+c_static(i,5))+L_static(i,5)*(b_static(i,5)+d_
            static(i,5))/(H*ll(i)))*XX_static(6*i-1)*exp(L_
            static(i,5)*ZHl(:,i))+...
            (sqrt(-1)*k*(1+c_static(i,6))+L_static(i,6)*(b_
            static(i,6)+d_static(i,6))/(H*ll(i)))*XX_static(6*i)*exp(L_
            static(i,6)*ZHl(:,i)));
    Szz_static(i,:)=((M(i)*b_static(i,1)*L_static(i,1))/(H*ll(i))+sqrt(-
1)*k*lambda(i))*XX_static(6*i-5)*exp(L_static(i,1)*ZHl(:,i))+...
        ((M(i)*b_static(i,2)*L_static(i,2))/(H*ll(i))+sqrt(-
1)*k*lambda(i))*XX_static(6*i-4)*exp(L_static(i,2)*ZHl(:,i))+...
        ((M(i)*b_static(i,3)*L_static(i,3))/(H*ll(i))+sqrt(-
1)*k*lambda(i))*XX_static(6*i-3)*exp(L_static(i,3)*ZHl(:,i))+...
        ((M(i)*b_static(i,4)*L_static(i,4))/(H*ll(i))+sqrt(-
1)*k*lambda(i))*XX_static(6*i-2)*exp(L_static(i,4)*ZHl(:,i))+...
        ((M(i)*b_static(i,5)*L_static(i,5))/(H*ll(i))+sqrt(-
1)*k*lambda(i))*XX_static(6*i-1)*exp(L_static(i,5)*ZHl(:,i))+...
        ((M(i)*b_static(i,6)*L_static(i,6))/(H*ll(i))+sqrt(-
1)*k*lambda(i))*XX_static(6*i)*exp(L_static(i,6)*ZHl(:,i));
    Txz_static(i,:)=G(i)*((L_static(i,1)/(H*ll(i))+sqrt(-1)*k*b_
static(i,1))*XX_static(6*i-5)*exp(L_static(i,1)*ZHl(:,i))+...
            (L_static(i,2)/(H*ll(i))+sqrt(-1)*k*b_static(i,2))*XX_static(6*i-
            4)*exp(L_static(i,2)*ZHl(:,i))+...
            (L_static(i,3)/(H*ll(i))+sqrt(-1)*k*b_static(i,3))*XX_static(6*i-
            3)*exp(L_static(i,3)*ZHl(:,i))+...
            (L_static(i,4)/(H*ll(i))+sqrt(-1)*k*b_static(i,4))*XX_static(6*i-
            2)*exp(L_static(i,4)*ZHl(:,i))+...
            (L_static(i,5)/(H*ll(i))+sqrt(-1)*k*b_static(i,5))*XX_static(6*i-
            1)*exp(L_static(i,5)*ZHl(:,i))+...
```

```
            (L_static(i,6)/(H*ll(i))+sqrt(-1)*k*b_static(i,6))*XX_
            static(6+i)*exp(L_static(i,6)*ZHl(:,i)));
end
Uxabs_static=abs(Ux_static');
Uzabs_static=abs(Uz_static');
Pabs_static=abs(P_static')/q0;
Szzabs_static=abs(Szz_static')/q0;
Txzabs_static=abs(Txz_static')/q0;
%-------------------------------------------------------------
% PLOTS
%-------------------------------------------------------------
RP1=P1(1)/P1(2);
Rm=m(1)/m(2);
%-------------------------------------------------------------
% Stresses and Pore Pressure
%-------------------------------------------------------------
% figure
for s=1:nl
    subplot 311
    plot(Pabs_static(:,s),ZHl(:,s),'-ko','MarkerSize',3)
                                %,'LineWidth',1.45,...
% 'MarkerEdgeColor','black',...
% 'MarkerFaceColor','w',...
% 'MarkerSize',4)
    ylabel('z/h')
    xlabel('p/q')
    grid on
    hold on
    subplot 312
    plot(Szzabs_static(:,s),ZHl(:,s),'-ko','MarkerSize',3)
    ylabel('z/h')
    xlabel('Szz/q')
    grid on
    hold on
    subplot 313
    plot(Txzabs_static(:,s),ZHl(:,s),'-ko','MarkerSize',3)%,'LineWidth',1.45)
    ylabel('z/h')
    xlabel('Sxz/q')
    grid on
    hold on
end
%-------------------------------------------------------------
% DISPLACEMENTS
%-------------------------------------------------------------
figure
for s=1:nl
    subplot 211
    hold on
    plot(Uxabs_static(:,s),ZHl(:,s),'-k','LineWidth',1.45)%'MarkerSize',3)%
% 'MarkerEdgeColor','red',...
% 'MarkerFaceColor','w',...
% 'MarkerSize',4)
    ylabel('z/h')
    xlabel('Ux')
    grid on
```

```
    hold on
    subplot 212
    hold on
    plot(Uzabs_static(:,s),ZHl(:,s),'-k','LineWidth',1.45)
                                %'MarkerSize',3)%,'LineWidth',1.45)
    ylabel('z/h')
    xlabel('Uz')
    grid on
    hold on
end
```

Finite Difference Method

9.1 Introduction

As mentioned earlier in Chapter 7, after a mathematical model is developed in the form of a set of governing equations, a solution for the field variables is sought. Analytical solutions for these equations are often not possible to develop for most of the real life practical problems. For these situations, we resort to numerical methods of solution. Among these methods, there is a direct and rather simple one known as the "Finite Difference Method (FDM)". In this method, the 'derivatives' in governing equations are replaced by their approximate "finite difference" relations defined at discrete nodal points of a grid or a mesh. Thus, the governing differential equations are transformed (discretized) into a set of algebraic equations, which are then solved easily with the help of a computer. FDM is often used in conjunction with FEM for the time-dependent problems to discretize the temporal derivatives. In this chapter, the fundamentals of FDM are presented along with applications to a few common geotechnical problems.

9.2 Finite Difference Approximation of Derivatives

9.2.1 Ordinary Derivatives for $\Phi = \Phi(x)$: First Order Derivative $\dfrac{\partial \Phi}{\partial x}$

Let us consider a function shown in Figure 9.2-1. For any location, within $x_1 \leq x \leq x_2$ (point C in the figure), $\dfrac{\partial \Phi}{\partial x}$ is approximated between the points A and B by:

$$\frac{d\Phi}{dx} = \frac{\Phi_2 - \Phi_1}{h} \tag{9.2.1}$$

where the approximation is defined using three different methods namely, "the forward, backward or central difference schemes" depending upon where the related approximation is being made. This is presented as below:

Forward finite difference (FFD) is stated as:

$$\left(\frac{d\Phi}{dx}\right)_{x=x_1} = \Phi_1' = \frac{\Phi_2 - \Phi_1}{h} \tag{9.2.2}$$

Backward finite difference (BFD):

$$\left(\frac{d\Phi}{dx}\right)_{x=x_2} = \Phi_2' = \frac{\Phi_2 - \Phi_1}{h} \tag{9.2.3}$$

and the central finite difference (CFD) is:

$$\left(\frac{d\Phi}{dx}\right)_{x=\frac{x_1+x_2}{2}} = \Phi_0' = \frac{\Phi_2 - \Phi_1}{h} \tag{9.2.4}$$

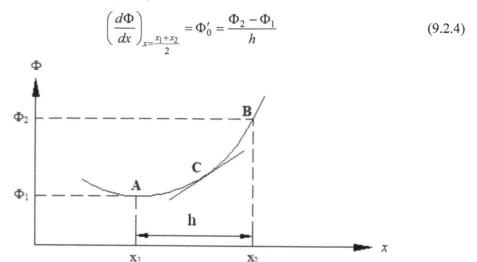

Figure 9.2-1 FD approximation for first order derivative

It is noted that the tangents at points, x_1, x_2 and $\frac{x_1+x_2}{2}$ are approximated by the slope of the chord |AB|. Obviously this approximation is the best for location, $x = \frac{x_1+x_2}{2}$. Here, h (or also called Δx) is finite and small but not necessarily infinitesimally small. Plus, the accuracy of the approximation relies heavily on the chosen increment h. That is, as we choose smaller h, the result of the calculation should theoretically get better. Also, these definitions are equivalent in the continuum but lead to different approximations in discrete sense. The question is then which one to choose, and whether there is a way to quantify the error committed. For that, we need to perform a Taylor series expansion around the point of interest.

9.2.2 Taylor Series Expansion

It is mostly possible to write the Taylor Series for approximating the derivatives at a given point on the curve. If we expand the function ϕ at x_{i+h} about the point x_i,

$$\Phi(x_i + \Delta x_i) = \Phi(x_i) + \Delta x_i \left(\frac{d\Phi}{dx}\right)_{x_i} + \frac{\Delta x_i^2}{2!}\left(\frac{d^2\Phi}{dx^2}\right)_{x_i} + \frac{\Delta x_i^3}{3!}\left(\frac{d^3\Phi}{dx^3}\right)_{x_i} + \dots \tag{9.2.5}$$

with $\Delta x_i = h$. The Taylor series can be rearranged to read as,

$$\frac{\Phi(x_i+h) - \Phi(x_i)}{h} = \left(\frac{d\Phi}{dx}\right)_{x_i} + \underbrace{\frac{h}{2!}\left(\frac{d^2\Phi}{dx^2}\right)_{x_i} + \frac{h^2}{3!}\left(\frac{d^3\Phi}{dx^3}\right)_{x_i} + \dots}_{\textit{Truncation Error}} \tag{9.2.6}$$

which clearly states the truncation error on the right hand side as the higher order terms. It should be noted here that the error is linearly dependent on the increment h which is maximum for single order. Therefore, it is safe to say that the order of error is first for the above approximation of first derivative, thus we write:

$$\left(\frac{d\Phi}{dx}\right)_{x_i} = \frac{\Phi(x_i + h) - \Phi(x_i)}{h} + O(h) \tag{9.2.7}$$

The same relation of (9.2.5) can be written to approximate the backward difference such that:

$$\Phi(x_i - \Delta x_{i-1}) = \Phi(x_i) - \Delta x_{i-1}\left(\frac{d\Phi}{dx}\right)_{x_i} + \frac{\Delta x_{i-1}^2}{2!}\left(\frac{d^2\Phi}{dx^2}\right)_{x_i} - \frac{\Delta x_{i-1}^3}{3!}\left(\frac{d^3\Phi}{dx^3}\right)_{x_i} + \dots \tag{9.2.8}$$

with $\Delta x_{i-1} = h$. If we write what we had in (9.2.6) again for (9.2.8), we get:

$$\frac{\Phi(x_i) - \Phi(x_i - h)}{h} = \left(\frac{d\Phi}{dx}\right)_{x_i} \underbrace{- \frac{h}{2!}\left(\frac{d^2\Phi}{dx^2}\right)_{x_i} + \frac{h^2}{3!}\left(\frac{d^3\Phi}{dx^3}\right)_{x_i} + \dots}_{Truncation\ Error} \tag{9.2.9}$$

which also shows that the order of magnitude of the analysis or in other words the order of error diminishing as the analysis marches in time is again linear. Thus,

$$\left(\frac{d\Phi}{dx}\right)_{x_i} = \frac{\Phi(x_i) - \Phi(x_i - h)}{h} + O(h) \tag{9.2.10}$$

Higher order approximation of the first derivative can be obtained by combining the two Taylor series equations (9.2.5) and (9.2.8). Multiplying the first by Δx_{i-1} and the second by Δx_i and adding both equations we get,

$$\frac{1}{\Delta x_i + \Delta x_{i-1}}\left[\Delta x_{i-1}\frac{\Phi(x_{i+1}) - \Phi(x_i)}{\Delta x_i} + \Delta x_i\frac{\Phi(x_i) - \Phi(x_{i-1})}{\Delta x_{i-1}}\right] = \left(\frac{d\Phi}{dx}\right)_{x_i} + \frac{\Delta x_{i-1}\Delta x_i}{3!}\left(\frac{d^3\Phi}{dx^3}\right)_{x_i} + \dots \tag{9.2.11}$$

where $x_{i-1} = x_i - \Delta x_{i-1}$ and $x_{i+1} = x_i + \Delta x_i$. Assuming $\Delta x_{i-1} = \Delta x_i = \Delta x = h$, we get,

$$\left[\frac{\Phi(x_{i+1}) - \Phi(x_{i-1})}{2\Delta x}\right] = \left(\frac{d\Phi}{dx}\right)_{x_i} + \frac{\Delta x^2}{3!}\left(\frac{d^3\Phi}{dx^3}\right)_{x_i} + \dots \tag{9.2.12}$$

finally giving:

$$\left(\frac{d\Phi}{dx}\right)_{x_i} = \left[\frac{\Phi(x_{i+1}) - \Phi(x_{i-1})}{2\Delta x}\right] + O(\Delta x^2) \tag{9.2.13}$$

which is also the *Central Finite Difference* method (CFD) as mentioned before.

9.2.3 Second Order Derivative $\dfrac{d^2\Phi}{dx^2}$

Let us now consider a function shown in Figure 9.2-2. For any location, within $x_1 \leq x \leq x_2$ a finite difference approximation for the second order derivative is defined as:

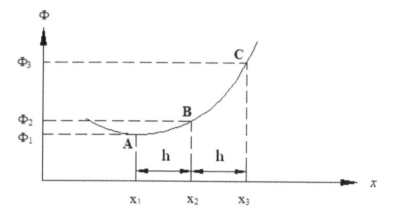

Figure 9.2-2 FD approximation for second order derivative

$$\Phi'' = \left(\frac{d^2\Phi}{dx^2}\right) = \frac{d}{dx}\left(\frac{d\Phi}{dx}\right) = \frac{\left(\frac{d\Phi}{dx}\right)_{x=x_2} - \left(\frac{d\Phi}{dx}\right)_{x=x_1}}{h} \qquad (9.2.14)$$

which leads to the relation below using the FFD scheme:

$$\Phi'' = \left(\frac{d^2\Phi}{dx^2}\right) = \frac{\dfrac{\Phi_3 - \Phi_2}{h} - \dfrac{\Phi_2 - \Phi_1}{h}}{h} = \frac{\Phi_1 - 2\Phi_2 + \Phi_3}{h^2} \qquad (9.2.15)$$

The above approximation can naturally be pursued using any of the forward, backward or central difference definitions depending on the point at which $\dfrac{d^2\Phi}{dx^2}$ is being defined. Below are the three finite difference formulae for the second derivative approximation of the arbitrary function in Figure 9.2-2 to be calculated at point $x = x_2$:

(i) Forward finite difference (FFD):

$$\left(\frac{d^2\Phi}{dx^2}\right)_{x=x_2} = \frac{\left(\frac{d\Phi}{dx}\right)_{x=x_3} - \left(\frac{d\Phi}{dx}\right)_{x=x_2}}{h} = \frac{\dfrac{\Phi_3 - \Phi_2}{h} - \dfrac{\Phi_2 - \Phi_1}{h}}{h} = \frac{\Phi_1 - 2\Phi_2 + \Phi_3}{h^2} \qquad (9.2.16)$$

(ii) Backward finite difference (BFD):

$$\left(\frac{d^2\Phi}{dx^2}\right)_{x=x_2} = \frac{\left(\frac{d\Phi}{dx}\right)_{x=x_2} - \left(\frac{d\Phi}{dx}\right)_{x=x_1}}{h} = \frac{\dfrac{\Phi_3 - \Phi_2}{h} - \dfrac{\Phi_2 - \Phi_1}{h}}{h} = \frac{\Phi_1 - 2\Phi_2 + \Phi_3}{h^2} \qquad (9.2.17)$$

(iii) Central finite difference (CFD):

$$\left(\frac{d^2\Phi}{dx^2}\right)_{x=x_2} = \frac{\left(\frac{d\Phi}{dx}\right)_{x=x_3} - \left(\frac{d\Phi}{dx}\right)_{x=x_1}}{2h} = \frac{\frac{\Phi_3 - \Psi_2}{h} - \frac{\Phi_2 - \Phi_1}{h}}{2h} = \frac{\Phi_1 - 2\Phi_2 + \Psi_3}{2h^2} \quad (9.2.18)$$

It should be noted that the above approximation is exact for the quadratic polynomial $\varphi(x)$ $= a_0 + a_1x + a_2x^2$.

9.2.4 Ordinary Derivatives: Summary

Considering Figure 9.2-3, we write the half step formulae for both the forward difference and backward difference methods and get:

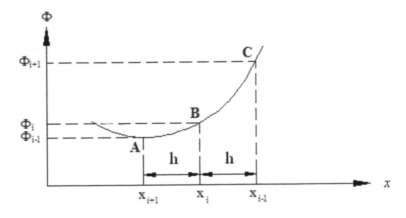

Figure 9.2-3 Forward difference approximation for ordinary derivatives

$$\Phi'_{i+\frac{1}{2}} = \frac{\Phi_{i+1} - \Phi_i}{h} \quad (9.2.19a)$$

$$\Phi'_{i-\frac{1}{2}} = \frac{\Phi_i - \Phi_{i-1}}{h} \quad (9.2.19b)$$

which when used in the CFD becomes

$$\Phi''_i = \frac{\Phi'_{i+\frac{1}{2}} - \Phi'_{i-\frac{1}{2}}}{h} \quad (9.2.20)$$

yielding the central difference approximations for derivatives at point i:

$$\Phi''_i = \frac{\Phi_{i-1} - 2\Phi_i + \Phi_{i+1}}{h^2} \quad (9.2.21)$$

or in a more general form as:

$$\Phi'_i = \frac{\Phi_{i+1} - \Phi_{i-1}}{2h} \quad (9.2.22)$$

9.2.5 Partial Derivatives for $\Phi = \Phi(x, y)$

Let us consider a 2-D function as shown below in Figure 9.2-4. Note the x = constant and y = constant lines on x-y plane and the values of the function along those lines. Also note the associated grid on x-y plane and the way grid lines and nodes are being named with (i, j) indices.

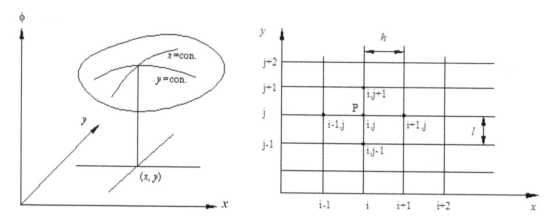

Figure 9.2-4 2-D function and its representation grid on x-y plane as (i, j) [x, y: spatial co-ordinates]

Following the definitions of partial derivatives we write:

$$\frac{\partial \Phi}{\partial x} = \frac{\partial \Phi(x, y_j)}{\partial x} = \left. \frac{d\Phi}{dx} \right|_{y=y_j} \tag{9.2.23}$$

$$\frac{\partial \Phi}{\partial y} = \frac{\partial \Phi(x_i, y)}{\partial y} = \left. \frac{d\Phi}{dy} \right|_{x=x_i} \tag{9.2.24}$$

$$\left(\frac{\partial^2 \Phi}{\partial x^2} \right)_{i,j} = \frac{\partial^2 \Phi(x, y_j)}{\partial x^2} = \left. \frac{d^2\Phi}{dx^2} \right|_{y=y_j} = \frac{\Phi_{i-1,j} - 2\Phi_{i,j} + \Phi_{i+1,j}}{h^2} \tag{9.2.25}$$

$$\left(\frac{\partial^2 \Phi}{\partial y^2} \right)_{i,j} = \frac{\partial^2 \Phi(x_i, y)}{\partial y^2} = \left. \frac{d^2\Phi}{dy^2} \right|_{x=x_i} = \frac{\Phi_{i,j-1} - 2\Phi_{i,j} + \Phi_{i,j+1}}{l^2} \tag{9.2.26}$$

9.2.6 Partial Derivatives for $\Phi = \Phi(x, t)$

Following the above steps, we can write the partial derivatives for this function as,

$$\left(\frac{\partial^2 \Phi}{\partial x^2} \right)_{i,k} = \frac{\Phi_{i-1,k} - 2\Phi_{i,k} + \Phi_{i+1,k}}{h^2} \tag{9.2.27}$$

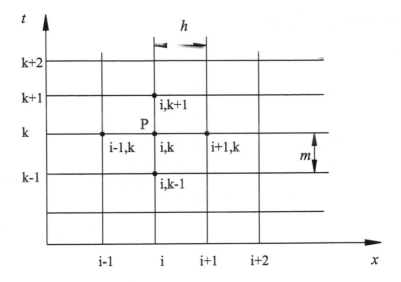

Figure 9.2-5 A 2-D function and grid on *x-t* plane [*x*: spatial co-ordinate, *t*: time]

Forward finite difference:

$$\left(\frac{\partial \Phi}{\partial t} \right)_{i,k} = \frac{\Phi_{i,k+1} - \Phi_{i,k}}{m} \tag{9.2.28}$$

Backward finite difference:

$$\left(\frac{\partial \Phi}{\partial t} \right)_{i,k} = \frac{\Phi_{i,k} - \Phi_{i,k-1}}{m} \tag{9.2.29}$$

Central finite difference:

$$\left(\frac{\partial \Phi}{\partial t} \right)_{i,k} = \frac{\Phi_{i,k+1} - \Phi_{i,k-1}}{2m} \tag{9.2.30}$$

9.2.7 Partial Derivatives for $\Phi = \Phi(x, y, t)$

Similarly we can write the finite difference approximations for these 3-D functions in indicial notation as:

$$\left(\frac{\partial^2 \Phi}{\partial x^2} \right)_{i,j,k} = \frac{\Phi_{i-1,j,k} - 2\Phi_{i,j,k} + \Phi_{i+1,\,j,k}}{h^2} \tag{9.2.31}$$

$$\left(\frac{\partial^2 \Phi}{\partial y^2} \right)_{i,j,k} = \frac{\Phi_{i,j-1,k} - 2\Phi_{i,j,k} + \Phi_{i,j+1,k}}{l^2} \tag{9.2.32}$$

$$\left(\frac{\partial \Phi}{\partial t} \right)_{i,j,k} = \frac{\Phi_{i,j,k+1} - \Phi_{i,j,k}}{m} \tag{9.2.33}$$

9.3 FDM for Consolidation (Parabolic) Equation

9.3.1 Problem Definition

Let us consider the following idealized problem of 1-D consolidation (diffusion). Here in a fully saturated layer of homogeneous soil of unit thickness, an initial pore water pressure $u(x, 0)$ is generated due to some loading. Notice that we now use 'u' as the pore pressure symbol to suit to the classical soil mechanics terminology. With the passage of time, the pore pressure will start dissipating due to the flow of water from the top and bottom free-draining boundaries, which results in a gradual compression (consolidation) of the soil layer. In order to solve such problems, our objective is to evaluate the time and space dependent pore water pressure $u(x, t)$.

The coupled flow and deformation phenomenon and its equations as were briefly presented in Chapter 7 and in details in Ulker (2009) are what we use to arrive at the uncoupled Terzaghi's theory of consolidation modeled by the following equation also given in Eq. (3.5.21):

$$C_v \frac{\partial^2 u}{\partial x^2} = \frac{\partial u}{\partial t} \quad 0 \le x \le 1 \tag{9.3.1}$$

with the boundary conditions:

$$u(z,t) = \begin{cases} 0 & z = 0 \\ 0 & z = 1 \end{cases} \tag{9.3.2}$$

and the initial conditions:

$$u(x,0) = \begin{cases} 2x & 0 \le x \le 1/2 \\ 2(1-x) & 1/2 \le x \le 1 \end{cases} \tag{9.3.3}$$

The analytical solution for the above problem was derived in Chapter 7 which is now modified to account for the boundary and initial conditions of Figure 9.3-1. Therefore we have:

$$u = \frac{8}{\pi^2} \sum_{n=1}^{\infty} \frac{1}{n} \left(\sin \frac{n\pi}{2} \right) (\sin n\pi x) \exp(-n^2 \pi^2 t) \tag{9.3.4}$$

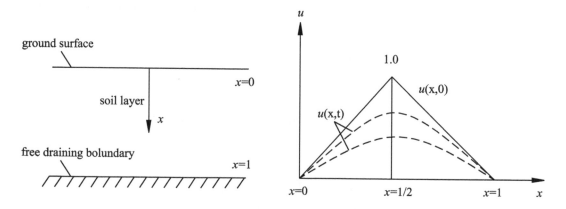

Figure 9.3-1 1-D consolidation of saturated soil layer

9.3.2 Finite Difference Schemes and Direct Method of Solution

In order to develop a numerical solution to the above equations, let us use the grid shown below in Figure 9.3-2. At any grid point $P(i, k)$, FD approximations for derivatives in Eq. (9.3.1) can be written as:

$$\left(\frac{\partial u}{\partial t}\right)_{i,k} = \frac{u_{i,k+1} - u_{i,k}}{\Delta t} \tag{9.3.5}$$

$$\left(\frac{\partial^2 u}{\partial x^2}\right)_{i,k} = \frac{1}{\Delta x^2}\left[\theta(u_{i-1,k+1} - 2u_{i,k+1} + u_{i+1,k+1}) + (1-\theta)(u_{i-1,k} - 2u_{i,k} + u_{i+1,k})\right] \tag{9.3.6}$$

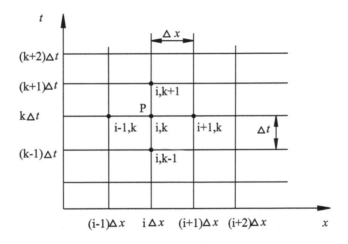

Figure 9.3-2 A finite difference grid for 1-D consolidation

with the following definition of θ:

$\theta = 0$ Explicit Scheme

$\theta = 0.5$ Crank-Nicolson Scheme - Implicit Scheme

$\theta = 1$ Fully Implicit Scheme - Implicit Scheme

Substituting equations (9.3.5) and (9.3.6) into (9.3.1), we get:

$$\frac{u_{i,k+1} - u_{i,k}}{\Delta t} = \frac{C_v}{\Delta x^2}\left[\theta(u_{i-1,k+1} - 2u_{i,k+1} + u_{i+1,k+1}) + (1-\theta)(u_{i-1,k} - 2u_{i,k} + u_{i+1,k})\right] \tag{9.3.7}$$

Here we have $r = \dfrac{C_v \Delta t}{(\Delta x)^2}$, the above equation can be written as:

$$u_{i,k+1} - u_{i,k} = r\left[\theta(u_{i-1,k+1} - 2u_{i,k+1} + u_{i+1,k+1}) + (1-\theta)(u_{i-1,k} - 2u_{i,k} + u_{i+1,k})\right] \tag{9.3.8}$$

The value of 'r' factor in above defines the nature of discretization with respect to the soil property C_v. The above equation provides us with different schemes for the solution with the following conditions:

$1/2 \leq \theta \leq 1$ Unconditionally stable and convergent numerical solution

$0 \le \theta < 1/2$ Stable and convergent numerical solution for

$$r \le \frac{1}{2(1-2\theta)} \quad (\theta = 0, r \le 1/2)$$

Explicit Scheme ($\theta = 0$)

For $\theta = 0$, Eq. (9.3.8) becomes:

$$u_{i,k+1} = r(u_{i-1,k} + u_{i+1,k}) + (1-2r)u_{i,k} \qquad (9.3.9)$$

With the help of few example calculations, we now study the effect of factor 'r' on the nature of numerical solution.

Case (a) $r = 0.1$:

Let $r = 0.1$, then Eq. (9.3.8) becomes:

$$u_{i,k+1} = 0.1(u_{i-1,k} + u_{i+1,k}) + 0.8u_{i,k} \qquad (9.3.10)$$

Case (b) $r = 0.5$:

Let $r = 0.5$, we have:

$$u_{i,k+1} = 0.5(u_{i-1,k} + u_{i+1,k}) \qquad (9.3.11)$$

Table 9.3-1 shows the results for $r = 0.1$.

Table 9.3-1 Numerical results for $r = 0.1$

$r = 0.1$		$i = 0$	1	2	3	4	5	6
		$x = 0$	0.1	0.2	0.3	0.4	0.5	0.6
$k = 0$	$t = 0.000$	0	0.2000	0.4000	0.6000	0.8000	1.0000	0.8000
1	0.001	0	0.2000	0.4000	0.6000	0.8000	0.9600	0.8000
2	0.002	0	0.2000	0.4000	0.6000	0.7960	0.9280	0.7960
3	0.003	0	0.2000	0.4000	0.5996	0.7896	0.9016	0.7896
4	0.004	0	0.2000	0.4000	0.5986	0.7818	0.8792	0.7818
5	0.005	0	0.2000	0.3999	0.5971	0.7732	0.8597	0.7732
:	:	:	:	:	:	:	:	:
10	0.01	0	0.1996	0.3968	0.5822	0.7218	0.7867	0.7281
:	:	:	:	:	:	:	:	:
20	0.02	0	0.1938	0.3781	0.5373	0.6484	0.6891	0.6486

Table 9.3-2 Comparison with analytical solution

$r = 0.1$	Finite-difference solution ($x = 0.3$)	Analytical solution ($x = 0.3$)	Absolute Error	% Error
$t = 0.005$	0.5971	0.5966	0.0005	0.08
$t = 0.01$	0.5822	0.5799	0.0023	0.4
$t = 0.02$	0.5373	0.5334	0.0039	0.7
$t = 0.10$	0.2472	0.2444	0.0028	0.1

Case (c) $r = 1$:

Let $r = 1$, we have,

$$u_{i,k+1} = u_{i-1,k} + u_{i+1,k} - u_{i,k} \qquad (9.3.12)$$

Tables 9.3-3–9.3-5 present related results for a number of time and space increments.

Table 9.3-3 Numerical results for $r = 0.5$

$r = 0.5$		$i = 0$	1	2	3	4	5	6
		$x = 0$	0.1	0.2	0.3	0.4	0.5	0.6
$k = 0$	$t = 0.00$	0	0.2000	0.4000	0.6000	0.8000	1.0000	0.8000
1	0.005	0	0.2000	0.4000	0.6000	0.8000	0.8000	0.8000
2	0.010	0	0.2000	0.4000	0.6000	0.7000	0.8000	0.7000
3	0.015	0	0.2000	0.4000	0.5500	0.7000	0.7000	0.7000
4	0.020	0	0.2000	0.3750	0.5500	0.6250	0.7000	0.6250
:	:	:	:	:	:	:	:	:
20	0.100	0	0.0949	0.1717	0.2484	0.2778	0.3071	0.2788

Table 9.3-4 Comparison with analytical solution

$r = 0.5$	Finite-difference solution $(x = 0.3)$	Analytical solution $(x = 0.3)$	Absolute Error	% Error
$t = 0.005$	0.6000	0.5966	0.0034	0.57
$t = 0.01$	0.6000	0.5799	0.0201	3.5
$t = 0.02$	0.5500	0.5334	0.0166	3.1
$t = 0.10$	0.2484	0.2444	0.0040	1.6

Table 9.3-5 Numerical results for $r = 1$

$r = 1$		$i = 0$	1	2	3	4	5	6
		$x = 0$	0.1	0.2	0.3	0.4	0.5	0.6
$k = 0$	$t = 0.00$	0	0.2	0.4	0.6	0.8	1.0	0.8
1	0.01	0	0.2	0.4	0.6	0.8	0.6	0.8
2	0.02	0	0.2	0.4	0.6	0.4	1.0	0.4
3	0.03	0	0.2	0.4	0.2	1.2	−0.2	1.2
4	0.04	0	0.2	0.0	0.0	−1.2	2.6	−1.2

In the following, we present the solutions from the explicit scheme for two values of the factor 'r' ($r = 0.48$ and $r = 0.52$). As it has been seen, the solution is stable as long as the value of 'r' remains less than or equal to 0.5, as noted previously. Figure 9.3-3 shows the comparison of results after Smith (1965).

Crank-Nicolson Scheme ($\theta = 0.5$):

For $\theta = 0.5$, Eq. (9.3.8) becomes,

$$-\frac{1}{2}ru_{i-1,k+1} + (1+r)u_{i,k+1} - \frac{1}{2}ru_{i+1,k+1} = \frac{1}{2}ru_{i-1,k} + (1-r)u_{i,k} + \frac{1}{2}ru_{i+1,k} \qquad (9.3.13)$$

Let $r = 1$, Eq. (9.3.8) becomes,

$$-u_{i-1,k+1} + 4u_{i,k+1} - u_{i+1,k+1} = u_{i-1,k} + u_{i+1,k} \qquad (9.3.14)$$

Using the above, for each time step we will get a set of algebraic equations which can be solved directly hence this Crank-Nicolson scheme is also called the "Direct Method of FD scheme". Table 9.3-6 shows the results for $r = 1$ and the comparison with analytical solution is given in Table 9.3-7.

Table 9.3-6 Numerical results for $r = 1$

$r = 1$		$i = 0$	1	2	3	4	5
		$x = 0$	0.1	0.2	0.3	0.4	0.5
$k = 0$	$t = 0.00$	0	0.2	0.4	0.6	0.8	1.0
1	0.01	0	0.1989	0.3956	0.5834	0.7381	0.7691
2	0.02	0	0.1936	0.3789	0.5400	0.6461	0.6921
$\vdots$	$\vdots$	$\vdots$	$\vdots$	$\vdots$	$\vdots$	$\vdots$	$\vdots$
10	0.10	0	0.0948	0.1803	0.2482	0.2918	0.3069
Analytical solution $t = 0.10$		0	0.0934	0.1776	0.2444	0.2873	0.3021

Table 9.3-7 Comparison with analytical solution

	Finite-difference solution ($x = 0.5$)	*Analytical solution ($x = 0.5$)*	*Absolute error*	*% Error*
$t = 0.01$	0.7691	0.7743	−0.0052	−0.7
$t = 0.02$	0.6921	0.6809	+0.0112	+1.6
$t = 0.10$	0.3069	0.3021	0.0048	1.6

9.3.3 Iterative Method of Solution

Implicit Method (Crank-Nicolson)

Substituting $\theta = 1/2$ into Eq. (9.3.8) we get:

$$u_{i,k+1} - u_{i,k} = \frac{r}{2}\left[(u_{i-1,k+1} - 2u_{i,k+1} + u_{i+1,k+1}) + (u_{i-1,k} - 2u_{i,k} + u_{i+1,k})\right] \qquad (9.3.15)$$

In the above equation, for the next time step $u_{i,k+1}$, $(u_{i-1,k+1} - 2u_{i,k+1} + u_{i+1,k+1})$ are unknown, while for the current time $u_{i,k}$, $(u_{i-1,k} - 2u_{i,k} + u_{i+1,k})$ are known. At every time step k, (9.3.15) is to be solved iteratively. The above equation is now written in the following form with m representing the iteration number while suppressing the second subscript $k + 1$.

$$u_i^{(m+1)} = \frac{r}{2}(u_{i-1}^{(m)} - 2u_i^{(m)} + u_{i+1}^{(m)}) + b_i \tag{9.3.16a}$$

$$b_i = u_{i,k} + \frac{r}{2}(u_{i-1,k} - 2u_{i,k} + u_{i+1,k}) \tag{9.3.16b}$$

The above scheme of solution converges only for $0 \leq r \leq 1/2$ which makes this inefficient. An iterative method is convergent if the differences between the exact solution and the successive approximations tend to zero as the number of iterations increases.

Jacobi's Method

With a little rearrangement of the terms, (9.3.16) can be rewritten as,

$$u_i^{(m+1)} = \frac{r}{2(1+r)}\left[u_{i-1}^{(m)} + u_{i+1}^{(m)}\right] + \frac{b_i}{1+r} \tag{9.3.17a}$$

$$b_i = u_{i,k} + \frac{r}{2}(u_{i-1,k} - 2u_{i,k} + u_{i+1,k}) \tag{9.3.17b}$$

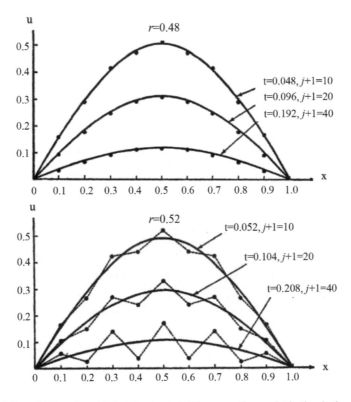

Figure 9.3-3 Instability of explicit scheme, solid lines are the analytical solutions, markers show the FD solution (Smith, 1965)

The above scheme converges for all values of r. For example, calculation, let $\Delta x = 1$, $\Delta t = 0.01$, $C_v = 1$ will yield $r = 1$. The above equation gets simplified to the following:

$$u_i^{(m+1)} = \frac{1}{4}(u_{i-1}^{(m)} + u_{i+1}^{(m)} + u_{i+1,j} + u_{i-1,j}) \tag{9.3.18}$$

Using the above iterative scheme, the numerical results for the first time step are obtained. After 10 iterations, the results are the same as those obtained by the direct method. Note that for the above,

$$u_1^{(k+1)} = \frac{1}{4}\{u_2^{(k)} + 0.4\}$$

(9.3.19a)

$$u_2^{(k+1)} = \frac{1}{4}\{u_1^{(k)} + u_3^{(k)} + 0.8\}$$

(9.3.19b)

$$u_3^{(k+1)} = \frac{1}{4}\{u_2^{(k)} + u_4^{(k)} + 1.2\}$$

(9.3.19c)

$$u_4^{(k+1)} = \frac{1}{4}\{u_3^{(k)} + u_5^{(k)} + 1.6\}$$

(9.3.19d)

$$u_5^{(k+1)} = \frac{1}{4}\{2u_4^{(k)} + 1.6\}$$

(9.3.19e)

Table 9.3-8 summarizes the Jacobi iterative results.

Table 9.3-8 Jacobi iterative results

$r = 1$	$i = 0$	1	2	3	4	5
	$u_{i,0} = 0$	0.2	0.4	0.6	0.8	1.0
$k = 0$	0	0.2	0.4	0.6	0.8	1.0
1	0	0.2	0.4	0.6	0.8	0.8
2	0	0.2	0.4	0.6	0.75	0.8
3	0	0.2	0.4	0.5875	0.75	0.7750
4	0	0.2	0.3969	0.5875	0.7460	0.7750
5	0	0.1992	0.3969	0.5844	0.7460	0.7703
6	0	0.1992	0.3959	0.5844	0.7387	0.7703
7	0	0.1990	0.3959	0.5836	0.7387	0.7693
:	:	:	:	:	:	:
11	0	0.1989	0.3956	0.5834	0.7381	0.7691

Gauss-Seidel Method

This method is similar to the Jacobi method with a difference that in this method, we use the most recent values available for a particular grid point. Thus, (9.3.16a) now becomes:

$$u_i^{(m+1)} = \frac{r}{2}\left[u_{i-1}^{(m+1)} - 2u_i^{(m+1)} + u_{i+1}^{(m)}\right] + b_i$$

(9.3.20)

which can also be written as,

$$u_i^{(m+1)} = \frac{r}{2(1+r)} \left[u_{i-1}^{(m+1)} + u_{i+1}^{(m)} \right] + \frac{b_i}{1+r} \tag{9.3.21}$$

$$b_i = u_{i,k} + \frac{r}{2}(u_{i-1,k} - 2u_{i,k} + u_{i+1,k}) \tag{9.3.22}$$

The above scheme also converges for all values of r. With $\Delta x = 1$, $\Delta t = 0.01$, $C_v = 1$, $r = 1$:

$$u_i^{(m+1)} = \frac{1}{4}(u_{i-1}^{(m+1)} + u_{i+1}^{(m)} + u_{i+1,J} + u_{i-1,J}) \tag{9.3.23}$$

Table 9.3-9 Gauss-Seidel iterative results

	$i = 0$	1	2	3	4	5
	$u_{i,0} = 0$	0.2	0.4	0.6	0.8	1.0
$k = 0$	0	0.2	0.4	0.6	0.8	1.0
1	0	0.2	0.4	0.6	0.8	0.8
2	0	0.2	0.4	0.6	0.75	0.775
3	0	0.2	0.4	0.5875	0.7406	0.7703
4	0	0.2	0.3969	0.5844	0.7387	0.7693
5	0	0.1992	0.3959	0.5836	0.7382	0.7691
6	0	0.1990	0.3957	0.5835	0.7381	0.7691
7	0	0.1989	0.3956	0.5834	0.7381	0.7693

For the first time step:

$$u_1^{(k+1)} = \frac{1}{4}\{0 + u_2^{(k)} + 0.4 + 0\} \tag{9.3.24a}$$

$$u_2^{(k+1)} = \frac{1}{4}\{u_1^{(k+1)} + u_3^{(k)} + 0.6 + 0.2\} \tag{9.3.24b}$$

$$u_3^{(k+1)} = \frac{1}{4}\{u_2^{(k+1)} + u_4^{(k)} + 0.8 + 0.4\} \tag{9.3.24c}$$

$$u_4^{(k+1)} = \frac{1}{4}\{u_3^{(k+1)} + u_5^{(k)} + 1.0 + 0.6\} \tag{9.3.24d}$$

$$u_5^{(k+1)} = \frac{1}{4}\{2u_4^{(k)} + 1.6\} \tag{9.3.24e}$$

We note that while Jacobi method requires 11 iterations, Gauss-Seidel requires seven iterations to get the same result. Table 9.3-9 summarizes the Gauss-Seidel iterative results.

Computer Implementation

MATLAB® programs "**FDM_I.m**" and "**FDM_II.m**" are developed to analyze the 1-D consolidation for which Figure 9.3-4 shows the pore pressure response under triangular initial load. Average degree consolidation with time and pore pressure isochrones are given in Figure 9.3-5.

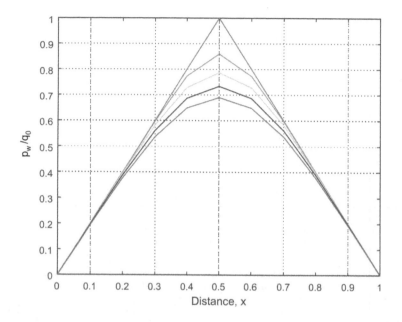

Figure 9.3-4 1-D Consolidation response under triangular initial load

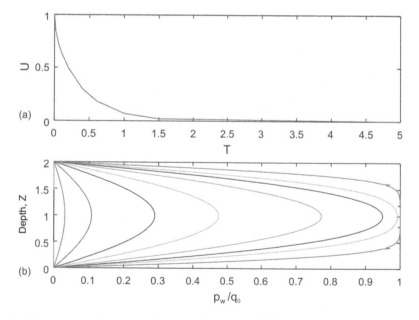

Figure 9.3-5 (a) Average degree % (U) consolidation and (b) Pore pressure ratio-depth relations

9.4 FDM for Seepage (Laplace) Equation: 2-D Steady State Flow

9.4.1 Problem Definition and Governing Equations

Let us consider the problem of seepage through the foundation soil underneath a concrete dam shown in the figure below. Here the dam holds a reservoir of water on the left side and a pool of water on the right, with water levels assumed to remain constant. We also assume that the foundation soil does not undergo any deformation, hence rigid soil skeleton. Under a difference of total head of ($\varphi_L - \varphi_R$), a steady state flow (seepage) of water will be set up through the porous soil. In order to solve this problem, we need to develop a solution for the distribution of total head $\varphi(x, y)$ within the soil. Only then we can obtain other variables of practical interests, viz. rate of flow (seepage loss) and uplift pressure on the base of the dam. Here, in this section an application of FDM is presented for such 2-D steady state flow problem.

Following what is presented in Chapter 2, the equation governing the steady state flow of water in the foundation soil is written as:

$$\frac{\partial}{\partial x}\left(k_x \frac{\partial \varphi}{\partial x}\right) + \frac{\partial}{\partial y}\left(k_y \frac{\partial \varphi}{\partial y}\right) = 0 \tag{9.4.1}$$

where φ is total head $\varphi(x, y)$ and k_x and k_y are the coefficients of permeability in the x and y directions. In Figure 9.4.1, we represent the infinitely extended foundation soil by a truncated domain. The lateral boundaries are at considerable distances from the dam, while the bottom boundary may represent an actual interface with an impervious rock layer or a truncated boundary at considerable distance below the base of the dam. On the grid spaces (elements) immediately next to the boundaries (which should be of small size), the following boundary conditions are imposed and identified with symbols. The interior grids are where $\varphi(x, y)$ needs to be evaluated. Therefore,

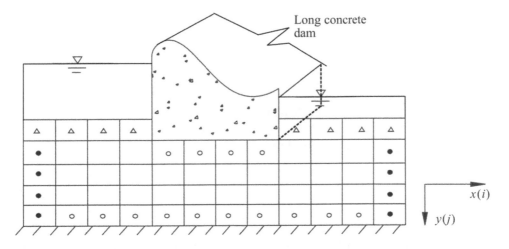

Figure 9.4-1 Seepage under a concrete dam

$$\circ: \frac{\partial \varphi}{\partial y} = 0 \text{ (No flow in } y\text{-direction)} \tag{9.4.2}$$

$$\bullet: \frac{\partial \varphi}{\partial x} = 0 \text{ (No flow in } x\text{-direction)} \tag{9.4.3}$$

$$\Delta : \varphi = \varphi^* \text{ (Known total heads)} \tag{9.4.4}$$

9.4.2 Finite Difference Approximation

In order to develop the finite difference equations, we use a grid like the one in Figure 9.4-1, a segment thereof is shown below in Figure 9.4-2. Note the way we use the indicial notations to identify the grids and their dimensions. The finite difference approximation of the two terms in the governing equation (9.4.1) can be written as:

$$\frac{\partial}{\partial x}\left(T_x \frac{\partial \varphi}{\partial x}\right)_{i,j} = \frac{1}{\Delta x_i}\left[\left(T_x \frac{\partial \varphi}{\partial x}\right)_{i+\frac{1}{2},j} - \left(T_x \frac{\partial \varphi}{\partial x}\right)_{i-\frac{1}{2},j}\right] \tag{9.4.5a}$$

$$\frac{\partial}{\partial y}\left(T_y \frac{\partial \varphi}{\partial y}\right)_{i,j} = \frac{1}{\Delta y_i}\left[\left(T_y \frac{\partial \varphi}{\partial y}\right)_{i,j+\frac{1}{2}} - \left(T_y \frac{\partial \varphi}{\partial y}\right)_{i,j-\frac{1}{2}}\right] \tag{9.4.5b}$$

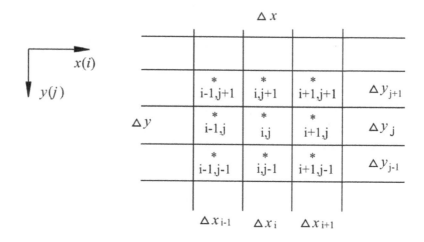

Figure 9.4-2 Finite difference grid

Now considering the continuity of flow,

$$\left(T_x \frac{\partial \varphi}{\partial x}\right)_{i+\frac{1}{2},j} = T_{x_i}\frac{\varphi_{i+\frac{1}{2},j} - \varphi_{i,j}}{\dfrac{\Delta x_i}{2}} = T_{x_{i+1}}\frac{\varphi_{i+1,j} - \varphi_{i+\frac{1}{2},j}}{\dfrac{\Delta x_{i+1}}{2}} \tag{9.4.6}$$

$\varphi_{i+\frac{1}{2},j}$ can be evaluated from Eq. (9.4.6), which can then be eliminated from the expression giving,

$$\left(T_x \frac{\partial \varphi}{\partial x}\right)_{i+\frac{1}{2},j} = \frac{2 T_{x_{i,j}} T_{x_{i+1,j}}}{T_{x_i}\Delta x_{i+1} + T_{x_{i+1}}\Delta x_i}(\varphi_{i+1,j} - \varphi_{i,j}) \tag{9.4.7}$$

Similarly, we can obtain:

$$\left(T_x \frac{\partial \varphi}{\partial x}\right)_{i-\frac{1}{2},j} = \frac{2T_{x_{i,j}} T_{x_{i-1,j}}}{T_{x_i} \Delta x_{i-1} + T_{x_{i-1}} \Delta x_i} (\varphi_{i,j} \quad \varphi_{i-1,j}) \tag{9.4.8}$$

Substitute Eqs. (9.4.7) and (9.4.8) into (9.4.6),

$$\frac{\partial}{\partial x}\left(T_x \frac{\partial \varphi}{\partial x}\right)_{i,j} = R_{i,j}(\varphi_{i+1,j} - \varphi_{i,j}) - L_{i,j}(\varphi_{i,j} - \varphi_{i-1,j}) \tag{9.4.9}$$

where

$$R_{i,j} = \frac{2T_{xi,j} T_{xi+1,j}}{(T_{xi,j} \Delta x_{i+1} + T_{xi+1,j} \Delta x_i) \Delta x_i} \tag{9.4.10}$$

$$L_{i,j} = \frac{2T_{xi,j} T_{xi+1,j}}{(T_{xi,j} \Delta x_{i+1} + T_{xi+1,j} \Delta x_i) \Delta x_i} \tag{9.4.11}$$

Similarly, considering the flow in the y-direction, we can now obtain the following finite difference equation approximation,

$$\frac{\partial}{\partial y}\left(T_y \frac{\partial \varphi}{\partial y}\right)_{i,j} = B_{i,j}(\varphi_{i,j+1} - \varphi_{i,j}) - A_{i,j}(\varphi_{i,j} - \varphi_{i,j-1}) \tag{9.4.12}$$

In the above,

$$B_{i,j} = \frac{2T_{yi,j} T_{yi,j+1}}{(T_{yi,j} \Delta y_{j+1} + T_{yi,j+1} \Delta y_j) \Delta y_j} \tag{9.4.13}$$

$$A_{i,j} = \frac{2T_{yi,j} T_{yi,j-1}}{(T_{yi,j} \Delta y_{j-1} + T_{yi,j-1} \Delta y_j) \Delta y_j} \tag{9.4.14}$$

Now substituting the finite difference approximations of (9.4.9) and (9.4.12) into the governing equation (9.4.5), following discrete equation returns,

$$\boxed{R_{i,j}(\varphi_{i+1,j} - \varphi_{i,j}) - L_{i,j}(\varphi_{i,j} - \varphi_{i-1,j}) + B_{i,j}(\varphi_{i,j+1} - \varphi_{i,j}) - A_{i,j}(\varphi_{i,j} - \varphi_{i,j-1}) = 0} \tag{9.4.15}$$

Writing such equations for each (i, j) grid point and incorporating the boundary conditions, a set of algebraic equations appear in the following form:

$$\boxed{[A]\{\varphi\} = \{b\}} \tag{9.4.16}$$

This equation is then solved for the total head, φ by some efficient numerical method.

Modified Equations for the Boundary Conditions

Equation (9.4.15) is for the interior points, which need to be modified for the no-flow boundary conditions. After rearrangement and arithmetic manipulation, equation (9.4.15) can be written as,

$$\left(\frac{R_{i,j}}{2}+\frac{R_{i,j}}{2}\right)\varphi_{i+1,j}+\left(\frac{L_{i,j}}{2}+\frac{L_{i,j}}{2}\right)\varphi_{i-1,j}+\left(\frac{B_{i,j}}{2}+\frac{B_{i,j}}{2}\right)\varphi_{i,j+1}+\left(\frac{A_{i,j}}{2}+\frac{A_{i,j}}{2}\right)\varphi_{i,j-1}$$

$$-\left(\frac{A_{i,j}}{2}+\frac{A_{i,j}}{2}+\frac{B_{i,j}}{2}+\frac{B_{i,j}}{2}+\frac{R_{i,j}}{2}+\frac{R_{i,j}}{2}+\frac{L_{i,j}}{2}+\frac{L_{i,j}}{2}\right)\varphi_{i,j}=0 \qquad (9.4.17)$$

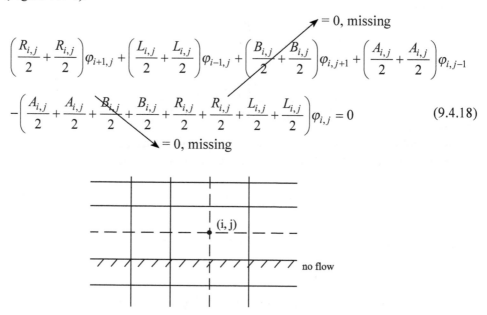

Figure 9.4-3 A representation of (9.4.15) for interior nodes

For the no flow boundary condition, Eq. (9.4.17) can be modified by eliminating the missing contribution from $A_{i,j}/2$ or $B_{i,j}/2$. For example, consider a grid right above a no flow boundary (Figure 9.4-4).

$= 0$, missing

$$\left(\frac{R_{i,j}}{2}+\frac{R_{i,j}}{2}\right)\varphi_{i+1,j}+\left(\frac{L_{i,j}}{2}+\frac{L_{i,j}}{2}\right)\varphi_{i-1,j}+\left(\frac{B_{i,j}}{2}+\frac{B_{i,j}}{2}\right)\varphi_{i,j+1}+\left(\frac{A_{i,j}}{2}+\frac{A_{i,j}}{2}\right)\varphi_{i,j-1}$$

$$-\left(\frac{A_{i,j}}{2}+\frac{A_{i,j}}{2}+\frac{B_{i,j}}{2}+\frac{B_{i,j}}{2}+\frac{R_{i,j}}{2}+\frac{R_{i,j}}{2}+\frac{L_{i,j}}{2}+\frac{L_{i,j}}{2}\right)\varphi_{i,j}=0 \qquad (9.4.18)$$

$= 0$, missing

Figure 9.4-4 Modification of boundary for boundary condition

Since there is no flow from below this grid, all the B coefficients will be zero where $\frac{B_{i,j}}{2} = 0$ and $\varphi_{i,j+1}$ is the missing node. Therefore, the finite difference equation for such a grid will be modified as:

$$R_{i,j}\varphi_{i+1,j} + L_{i,j}\varphi_{i-1,j} + A_{i,j}\varphi_{i,j-1} - (R_{i,j} + L_{i,j} + A_{i,j})\varphi_{i,j} = 0 \qquad (9.4.19)$$

This is the modified equation for the boundary grid point $(i, j+1)$ shown above. Similarly, modified equations for other no flow boundaries can readily be obtained.

9.4.3 Iterative Method of Solution

Eq. (9.4.15) or its modified form can be written in the following iterative form also:

$$\varphi_{i,j}^{k+1} = \frac{R_{i,j}\varphi_{i+1,j}^{(k)} + L_{i,j}\varphi_{i-1,j}^{(k+1)} + B_{i,j}\varphi_{i,j+1}^{(k)} + A_{i,j}\varphi_{i,j-1}^{(k+1)}}{(R_{i,j} + L_{i,j} + B_{i,j} + A_{i,j})} \qquad (9.4.20)$$

In order to solve such a system, we should first assume a set of values for all unknown $\varphi_{i,j}$. Then using the above equation we solve for each $\varphi_{i,j}$ in the next iteration. A number of iterations will have to be performed until a convergent solution is achieved.

Computer Implementation

The above FD formulation is implemented in a simple MATLAB program called "**FDM_III.m**" which gives a result of seepage flow as in Figure 9.4-5 for the seepage under a concrete dam problem.

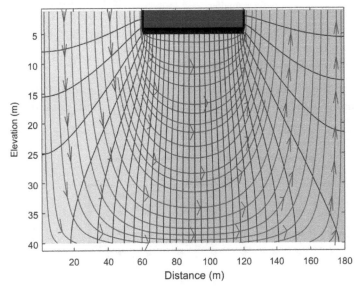

Figure 9.4-5 Flow and equipotential lines under a dam

9.5 FDM for Groundwater Flow: Aquifer Simulation

9.5.1 Problem Definition and Governing Equations

Consider the problem of groundwater flow in an aquifer. Figure below represents the problem of flow in confined and unconfined aquifers due to recharge or discharge (withdrawal) of water into them. In this case, the flow is considered to be a time-dependent 2-D flow through deformable porous soil. The basic problem requires a solution for the total head $\varphi(x, y, t)$. Considering all the relevant physics (of flow and deformation), as was presented earlier in

Chapter 2, for this problem the governing equation can be written as:

$$\frac{\partial}{\partial x}\left(k_x \frac{\partial \varphi}{\partial x}\right) + \frac{\partial}{\partial y}\left(k_y \frac{\partial \varphi}{\partial y}\right) = S\frac{\partial \varphi}{\partial t} - W(x,y,t) \qquad (9.5.1)$$

in which $k_x = K_x b$ and $k_y = K_y b$ are transmissibility values in x- and y-directions. S is the storativity of the aquifer, which represents the volumetric compressibility of the soil in aquifer. And, φ as mentioned before, is total head. For confined aquifer 'b' represents its thickness while for an unconfined aquifer this is the time-dependent height of the phreatic surface. W is (–) for external recharge and (+) for discharge.

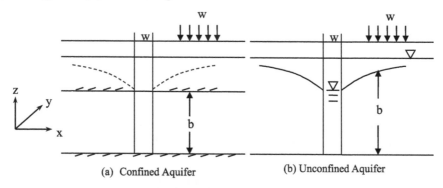

(a) Confined Aquifer   (b) Unconfined Aquifer

Figure 9.5-1 Groundwater flow in aquifers

9.5.2 Finite Difference Approximation

In order to develop the FD equations, we use a grid whose segment is shown in Figure 9.5-2. A node located in the center of each grid is identified with indicial notation as shown below,

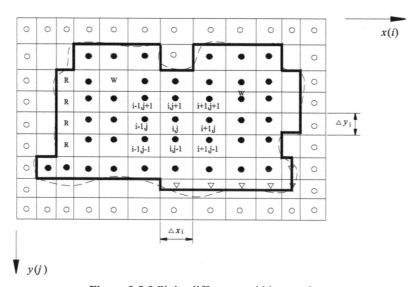

Figure 9.5-2 Finite difference grid in x-y plane

$\odot : = T_x = T_y = 0 \quad \dfrac{\partial \varphi}{\partial x} = 0, \dfrac{\partial \varphi}{\partial y} = 0$ •: internal nodes W: Discharge well   R: Recharge well

$\triangledown$: $\varphi = \varphi^*$ (known constant head) -------: Aquifer boundary ———: Mathematical model boundary

Equation (9.5.1) is rewritten for node (i, j) as,

$$\frac{\partial}{\partial x}\left(T_x \frac{\partial \psi}{\partial x}\right)_{i,j} + \frac{\partial}{\partial y}\left(T_y \frac{\partial \varphi}{\partial y}\right)_{i,j} - W_{i,j,k} = S_{i,j}\left(\frac{\partial \varphi}{\partial t}\right)_{i,j} \tag{9.5.2}$$

For the first two terms that have already been obtained in the previous section (9.4.9 and 9.4.12) the discrete version of equation (9.5.2) can be written as,

$$R_{i,j}(\varphi_{i+1,j,k} - \varphi_{i,j,k}) - L_{i,j}(\varphi_{i,j,k} - \varphi_{i,j-1,k}) + B_{i,j}(\varphi_{i,j+1,k} - \varphi_{i,j,k})$$

$$- A_{i,j}(\varphi_{i,j,k} - \varphi_{i,j-1,k}) - W_{i,j,k} = \frac{S_{i,j}}{\Delta t}(\varphi_{i,j,k} - \varphi_{i,j,k-1}) \tag{9.5.3}$$

In order to simplify the equation above, we drop the indices: i, j, k and retain the indices only when they are not i, j, k.

$$R\left(\varphi_{i+1} - \varphi\right) - L\left(\varphi - \varphi_{i-1}\right) + B\left(\varphi_{j+1} - \varphi\right) - A\left(\varphi - \varphi_{j-1}\right) - W = \frac{S}{\Delta t}(\varphi - \varphi_{k-1}) \tag{9.5.4}$$

Rearranging the terms yields,

$$R\varphi_{i+1} + L\varphi_{i-1} + B\varphi_{j+1} + A\varphi_{j-1} - \left(R + L + B + A + \frac{S}{\Delta t}\right) \times \varphi = -\frac{S}{\Delta t}\varphi_{k-1} + W \tag{9.5.5}$$

in which $\dfrac{S}{\Delta t}\varphi_{k-1} + W$ is known. In the matrix notation we have,

$$[M]_{(k-1)}\{\varphi\}_{(k)} = \{b\}_{(k-1)} \tag{9.5.6}$$

Note that, R, L, A and B have to be modified for the boundary nodes with no flow as explained in the previous section. The coefficient matrix, M in the above equation will be time dependent for an unconfined aquifer.

9.6 FDM for Consolidation of a Layered System

9.6.1 Problem Definition and Governing Equations

Figure 9.6-1 represents a soil site with layers of soil with different properties subjected to a time varying loading on the ground surface. The loading as well as the soil properties do not vary in the lateral directions reducing this problem to that of 1-D consolidation with decoupled flow (of pore water) and deformation (of soils). In order to solve this problem, we need to evaluate the distribution of time dependent pore water pressure $u(z, t)$. As presented before, the equation governing the field variable $u(z, t)$ can be written in a slightly modified form as:

$$C_v \frac{\partial^2 u}{\partial z^2} = \frac{\partial u}{\partial t} - \frac{\partial \sigma_z}{\partial t} \tag{9.6.1}$$

where $C_v = \dfrac{k_z}{m_v \gamma_w}$ is the coefficient of consolidation, k_z is the coefficient of permeability in z-direction, m_v is the coefficient of volume change and γ_w is unit weight of water. And the total vertical stress σ_z is given by the following,

$$\sigma_z = \sigma_{z0} + \Delta\sigma_z \tag{9.6.2}$$

$$\frac{\partial\sigma_z}{\partial t} = \frac{dq}{dt} \tag{9.6.3}$$

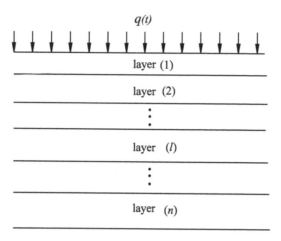

Figure 9.6-1 Consolidation of a layered system

For each layer ($l = 1, 2..., n$), the above equations, with layers being identified by superscript l can be written as:

$$C_v^l \frac{\partial^2 u^l}{\partial z^2} = \frac{\partial u^l}{\partial t} - \frac{dq}{dt} \tag{9.6.4}$$

In the above layered system, each layer is subdivided into a number of layers as shown in Figure 9.6-2. The conditions at the top and bottom of the layered site can be written in the following general form in which different choices for the coefficients a, b, and c yield different conditions as given in Table 9.6-1.

$$z = 0, \; a^1 \frac{\partial u^1}{\partial z}(0,t) - b^1 u^1(0,t) = -c^1 \tag{9.6.5}$$

$$z = H, \; a^n \frac{\partial u^n}{\partial z}(H,t) + b^n u^n(H,t) = c^n \tag{9.6.6}$$

In the above table, u_0, u_p are the specified excess pore pressures at upper and lower boundaries, respectively and v_0^1, v_p^n are the specified flow velocities. K^0, h^0 are coefficient of permeability and thickness of impedance layer above the upper boundary, respectively. K^{n+1}, h^{n+1} are coefficient of permeability and thickness of impedance layer below the lower boundary. At the interface of any two layers, the following equations of continuity are imposed:

$$u^l(z_r,t) = u^{l+1}(z_r,t) \tag{9.6.7}$$

$$K^l \frac{\partial u^l(z_r,t)}{\partial t} = K^{l+1} \frac{\partial u^{l+1}(z_r,t)}{\partial t} \tag{9.6.8}$$

Additionally, the load history, $q(t)$ at the surface is discretized as shown in Figure 9.6-3.

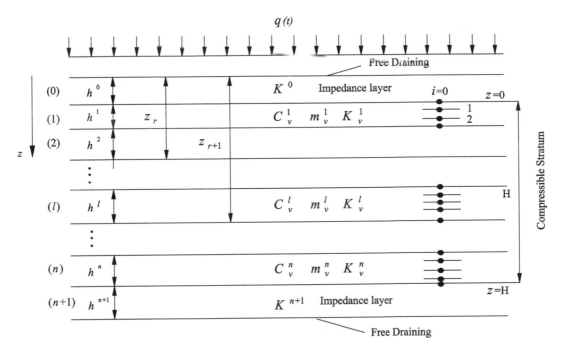

Figure 9.6-2 Finite difference discretization for consolidation of a layered system (reproduced from Desai and Christian, 1977)

Table 9.6-1 Drainage conditions at the boundaries

Boundary condition	Upper boundary			Lower boundary		
	a^l	b^l	c^l	a^n	b^n	c^n
Excess pore pressure	0	1	u_0	0	1	u_p
Free draining	0	−1	0	0	1	0
Velocity	1	0	$\dfrac{\gamma_w}{K^1 v_0^1}$	1	0	$\dfrac{\gamma_w}{K^1}\tau_p^n$
Impervious	1	0	0	1	0	0
Impeded	h^1	$\lambda^1 = \dfrac{K^0 h^1}{K^1 h^0}$	0	h^n	$\lambda^n = \dfrac{K^{n+1} h^n}{K^n h^{n+1}}$	0

9.6.2 Finite Difference Approximation

Using the mesh (grid) in z, t space (shown below in Figure 9.6-4), derivatives in the governing equation (9.6.4) can be written as:

$$\frac{\partial u}{\partial t} = \frac{u_{i,j+1} - u_{i,j}}{\Delta t} \tag{9.6.9}$$

$$\frac{\partial q}{\partial t} = \frac{q_{j+1} - q_j}{\Delta t} \tag{9.6.10}$$

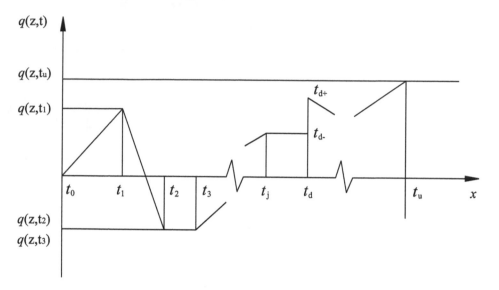

Figure 9.6-3 Discretized load history

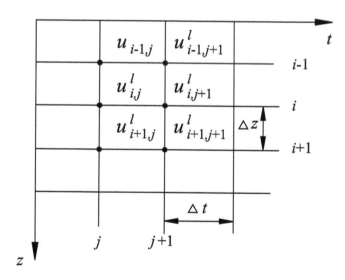

Figure 9.6-4 Finite difference mesh

$$\frac{\partial^2 u}{\partial z^2} = \frac{1}{2(\Delta z)^2}\Big[(u_{i-1,j+1} - 2u_{i,j+1} + u_{i+1,j+1}) + (u_{i-1,j} - 2u_{i,j} + u_{i+1,j})\Big] \qquad (9.6.11)$$

For Interior Nodes

Substituting the above finite difference approximations for derivatives into Eq. (9.6.1), we get:

$$(u_{i,j+1}^l - u_{i,j}^l) - (q_{j+1} - q_j) = \frac{C_v^l \Delta t}{2(\Delta z^l)^2}\Big[(u_{i-1,j+1}^l - 2u_{i,j+1}^l + u_{i+1,j+1}^l) + (u_{i-1,j}^l - 2u_{i,j}^l + u_{i+1,j}^l)\Big]$$

$$(9.6.12)$$

$$F^l u^l_{i-1,j+1} - Q^l u^l_{i,j+1} + F^l u^l_{i+1,j+1} = -E^l_{i,j} \qquad (9.6.13)$$

$$E^l_{i,j} = F^l u^l_{i-1,j} - P^l u^l_{i,j} + F^l u^l_{i+1,j} + \Delta q \qquad (9.6.14)$$

We know that $F^l = \dfrac{R^l}{2} = \dfrac{C^l_v \Delta t}{2(\Delta z^l)^2}$, $Q^l = 1 + R^l$, $P^l = 1 - R^l$, $\Delta q = q_{i+1} - q_i$. For details, see Box A-1.

For Layers' Interface Nodes

From Eqs. (9.6.7) and (9.6.8), considering the interface between layer l and layer $l+1$,

$$A^l u^l_{i-1,j+1} - B^l u^l_{i,j+1} + R^{l+1} u^{l+1}_{i+1,j+1} = -G^l_{i,j} \qquad (9.6.15)$$

$$G^l_{i,j} = A^l u^l_{i-1,j} - C^l u^l_{i,j} + R^{l+1} u^{l+1}_{i+1,j} + (1 + r^l \beta^l)\Delta q \qquad (9.6.16)$$

in which,

$$A^l = \frac{\alpha^l R^{l+1}}{\beta^l}$$

$$B^l = Q^{l+1} + r^l \beta^l Q^l$$

$$C^l = P^{l+1} + r^l \beta^l P^l$$

$$\alpha^l = \frac{K^l}{K^{l+1}}$$

$$\beta^l = \frac{\delta^l}{\delta^{l+1}}$$

$$r^l = \frac{m^l_v}{m^{l+1}_v}$$

For details, see Box A-2.

For the Boundary Nodes:

For $i = 0$, $z = 0$ we get:

$$-\left(\frac{\Delta z^1 b^1}{a^1} + Q^1\right) u^1_{0,j+1} + R^1 u^1_{1,j+1} + \frac{\Delta z^1 R^1 C^1}{a^1} = -E^1_{0,j} \qquad (9.6.17)$$

$$E^1_{0,j} = \left(P^1 - \frac{\Delta z^1 b^1 R^1}{a^1}\right) u^1_{0,j} + R^1 u^1_{1,j} + \frac{\Delta z^1 b^1 C^1}{a^1} + \Delta q \qquad (9.6.18)$$

For details see Box A-3. In the case $i = p$, $z = H$:

$$R^n u_{p-1,j+1}^n - \left(\frac{\Delta z^n b^n R^n}{a^n} + Q^n \right) u_{p,j+1}^n + \frac{\Delta z^n R^n C^n}{a^n} = -E_{p,j}^n \qquad (9.6.19)$$

$$E_{p,j}^n = R^n u_{p-1,j}^n + \left(P^n - \frac{\Delta z^n b^n R^n}{a^n} \right) u_{p,j}^n + \frac{\Delta z^n b^n C^n}{a^n} + \Delta q \qquad (9.6.20)$$

Using the appropriate difference equations, we write the algebraic equation for each node from top to bottom together and develop the complete set of difference equations in the following form,

$$[K]\{u\}_{t+\Delta t} = \{F\}_t \qquad (9.6.21)$$

The above equation can readily be solved in MATLAB® or MAPLE®.

Box A-1

FDM for interior nodes

For interior nodes,

$$(u_{i,j+1}^l - u_{i,j}^l) - (q_{j+1} - q_j) = \frac{C_v^l \tau}{2(\delta^l)^2} \left[(u_{i-1,j+1}^l - 2u_{i,j+1}^l + u_{i+1,j+1}^l) + (u_{i-1,j}^l - 2u_{i,j}^l + u_{i+1,j}^l) \right] \qquad (A.1.1)$$

Moving all the terms with $j+1$ on one side:

$$F^l u_{i-1,j+1}^l - (1 + 2F^l) u_{i,j+1}^l + F^l u_{i+1,j+1}^l$$
$$= -F^l u_{i-1,j}^l - (1 - 2F^l) u_{i,j}^l - F^l u_{i+1,j}^l - (q_{j+1} - q_j) \qquad (A.1.2)$$

In the above,

$$F^l = \frac{C_v^l \tau}{(\delta^l)^2} \qquad (A.1.3)$$

Now let

$$\begin{cases} R^l = 2F^l \\ Q^l = 1 + R^l \\ P^l = 1 - R^l \\ \Delta q = q_{i+1} - q_i \end{cases}$$

Then the above equation can be written as:

$$F^l u_{i-1,j+1}^l - Q^l u_{i,j+1}^l + F^l u_{i+1,j+1}^l = -E_{i,j}^l \qquad (A.1.4)$$

where

$$E_{i,j}^l = F^l u_{i-1,j}^l - P^l u_{i,j}^l + F^l u_{i+1,j}^l + \Delta q \qquad (A.1.5)$$

Box A-2

FDM for layer interface nodes

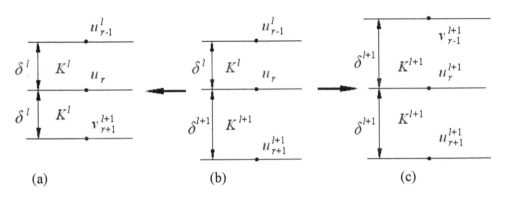

(a)   (b)   (c)

Figure A.2-1

For interface nodes:

   r – position

   j – time

   l – layer

Using continuity condition,

$$u^l(z_r,t) = u^{l+1}(z_r,t) \tag{A.2.1}$$

$$K^l \frac{\partial u^l}{\partial z}(z_r,t) = K^{l+1}\frac{\partial u^{l+1}}{\partial z}(z_r,t) \tag{A.2.2}$$

For Figure A.2-1,

$$u_r^l = u_r^{l+1} = u_r \tag{A.2.3}$$

$$K^l \frac{v_{r+1}^l - u_{r-1}^l}{2\delta^l} = K^{l+1}\frac{u_{r+1}^{l+1} - v_{r-1}^{l+1}}{2\delta^{l+1}} \tag{A.2.4}$$

Using:

$$\alpha^l = \frac{K^l}{\delta^l} \tag{A.2.5}$$

$$\alpha^{l+1} = \frac{K^{l+1}}{\delta^{l+1}} \tag{A.2.6}$$

$$\alpha^l v_{r+1}^l - \alpha^l u_{r-1}^l = \alpha^{l+1}u_{r+1}^{l+1} - \alpha^{l+1}v_{r-1}^{l+1} \tag{A.2.7}$$

Put v in one side and u in another side, then

$$\alpha^{l+1}v_{r-1}^{l+1} + \alpha^l v_{r+1}^l = \alpha^{l+1}u_{r+1}^{l+1} + \alpha^l u_{r-1}^l \tag{A.2.8}$$

Since continuity must hold for all the steps, so just write the equation for j and $j + 1$ time. For step j

$$\alpha^{l+1}v_{r-1,j}^{l+1} + \alpha^l v_{r+1,j}^l = \alpha^{l+1}u_{r+1,j}^{l+1} + \alpha^l u_{r-1,j}^l \tag{A.2.9}$$

For step $j+1$

$$\alpha^{l+1}v_{r-1,j+1}^{l+1} + \alpha^l v_{r+1,j+1}^l = \alpha^{l+1}u_{r+1,j+1}^{l+1} + \alpha^l u_{r-1,j+1}^l \tag{A.2.10}$$

Write the governing difference equation for Figure A.2-1(a) and A.2-1(b), for Figure A.2-1(a)

$$(u_{i,j+1}^l - u_{i,j}^l) - (q_{j+1} - q_j) = \frac{C_v^l \tau}{2(\delta^l)^2}\left[(u_{r-1,j+1}^l - 2u_{r,j+1}^l + v_{r+1,j+1}^l) + (u_{r-1,j}^l - 2u_{r,j}^l + v_{r+1,j}^l)\right] \tag{A.2.11}$$

for Figure A.2-1(b),

$$(u_{i,j+1}^l - u_{i,j}^l) - (q_{j+1} - q_j) = \frac{C_v^{l+1}\tau}{2(\delta^{l+1})^2}\left[(v_{r-1,j+1}^{l+1} - 2u_{r,j+1}^{l+1} + v_{r+1,j+1}^{l+1}) + (u_{r-1,j}^{l+1} - 2u_{r,j}^{l+1} + v_{r+1,j}^{l+1})\right] \tag{A.2.12}$$

With the continuity condition,

$$K^l \frac{v_{r+1}^l - u_{r-1}^l}{2\delta^l} = K^{l+1}\frac{u_{r+1}^{l+1} - v_{r-1}^{l-1}}{2\delta^{1+l}} \tag{A.2.13}$$

put it into equations (A.2.9) through (A.2.12) and plug v in (A.2.11) and (A.2.12) on one side, then,

$$\alpha^{l+1}v_{r-1,j}^{l+1} + \alpha^l v_{r+1,j}^l = \alpha^{l+1}u_{r+1,j}^{l+1} + \alpha^l u_{r-1,j}^l \tag{A.2.14}$$

$$\alpha^{l+1}v_{r-1,j+1}^{l+1} + \alpha^l v_{r+1,j+1}^l = \alpha^{l+1}u_{r+1,j+1}^{l+1} + \alpha^l u_{r-1,j+1}^l \tag{A.2.15}$$

$$F^l v_{r+1,j+1}^l + F^l v_{r+1,j}^l = (2F^l + 1)u_{r,j+1} + (2F^l - 1)u_{r,j}$$
$$- F^l u_{r-1,j+1} - F^l u_{r-1,j} - (q_{j+1} - q_j) \tag{A.2.16}$$

$$F^{l+1}v_{r-1,j+1}^{l+1} + F^{l+1}v_{r-1,j}^{l+1} = (2F^{l+1} + 1)u_{r,j+1} + (2F^{l+1} - 1)u_{r,j}$$
$$- F^{l+1}u_{r+1,j+1} - F^{l+1}u_{r+1,j} - (q_{j+1} - q_j) \tag{A.2.17}$$

Using Eqs. (A.2.14), (A.2.15), (A.2.16) and (A.2.17), solve $v_{r-1,j}^{l+1}$, $v_{r-1,j+1}^{l+1}$, $v_{r+1,j}^l$, $v_{r+1,j+1}^l$, put whichever $v_{i,j}^l$ from (A.2.14) through (A.2.17), and place u_{j+1} on one side and u_j on another side, we get the following equation:

$$A^l u_{i-1,j+1}^l - B^l u_{r,j+1}^l + R^{l+1}u_{r+1,j+1}^{l+1} = -G_{r,j}^l \tag{A.2.18}$$

where

$$G_{r,j}^l = A^l u_{r-1,j}^l - C^l u_{r,j}^l + R^{l+1}u_{r+1,j}^{l+1} + (1 + \gamma^l \beta^l)\Delta q \tag{A.2.19}$$

in which

$$R^l = 2F^l$$

$$F^l = \frac{C_v^l \tau}{2(\Delta \delta^l)^2}$$

$$C_v^l = \frac{K^l}{\gamma_w m_v^l}$$

$$A^l = \frac{\alpha^l R^{l+1}}{\beta^l}$$

$$B^l = Q^{l+1} + r^l \beta^l Q^l$$

$$C^l = P^{l+1} + r^l \beta^l P^l$$

$$\alpha^l = \frac{K^l}{K^{l+1}}$$

$$\beta^l = \frac{\delta^l}{\delta^{l+1}}$$

$$r^l = \frac{m_v^l}{m_v^{l+1}}$$

$$Q^l = 1 + R^l$$

$$P^l = 1 - R^l$$

Box A-3

FDM for layer boundary nodes:
Top boundary node, $z = 0$, $i = 0$

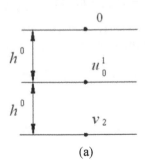 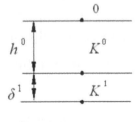 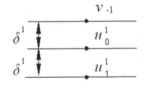

| (a) | (b) | (c) |

Figure A.3-1

Considering continuity at the boundary node,

$$u^0_{0,j} = u^1_{0,j} \tag{A.3.1}$$

$$u^0_{0,j+1} = u^1_{0,j+1} \tag{A.3.2}$$

$$K_0^0 \frac{v_1 - 0}{2h^0} = K^1 \frac{u_1^1 - v_{-1}}{2\delta^1} \tag{A.3.3}$$

$$\frac{K_0^0 \delta^1}{K^1 h^0} v_1 = u_1^1 - v_{-1} \tag{A.3.4}$$

$$\lambda^1 = \frac{K_0^0 \delta^1}{K^1 h^0} \tag{A.3.5}$$

For time j,

$$\lambda^1 v_{1,j} = u_{1,j}^1 - v_{-1,j} \tag{A.3.6a}$$

For time $j + 1$,

$$\lambda^1 v_{1,j+1} = u_{1,j+1}^1 - v_{-1,j+1} \tag{A.3.6b}$$

Note that layer is an impedance layer (0), therefore variation of pore pressure within this layer will be linear.

$$v_1 = 2u_0^1$$

Therefore above equations become:

$$2\lambda^1 u_{0,j} = u_{1,j}^1 - v_{-1,j} \tag{A.3.7a}$$

$$2\lambda^1 u_{0,j+1} = u_{1,j+1}^1 - v_{-1,j+1} \tag{A.3.7b}$$

Rearranging the above equations:

$$v_{-1,j} = u_{1,j}^1 - 2\lambda^1 u_{0,j} \tag{A.3.8a}$$

$$v_{-1,j+1} = u_{1,j+1}^1 - 2\lambda^1 u_{0,j+1} \tag{A.3.8b}$$

$$u_{0,j+1}^1 - u_{0,j}^1 - q_{j+1} + q_j = \frac{R^1}{2}\left[(v_{-1,j+1} - 2u_{0,j+1}^1 + u_{1,j+1}^1) + (v_{-1,j} - 2u_{0,j}^1 + u_{1,j}^1)\right] \tag{A.3.9}$$

where $R^1 = \frac{C_v \tau}{(\delta^1)^2}$. Now writing the finite difference approximation of the governing differential equation at the boundary node for Figure A.3-1. Now collecting all the $j + 1$ terms on one side, we get,

$$\frac{R^1}{2} v_{-1,j+1} - R^1 u_{0,j+1}^1 + \frac{R^1}{2} u_{1,j+1}^1 - u_{0,j+1}^1 = -u_{0,j}^1 - \Delta q - \frac{R^1}{2} v_{-1,j} + R^1 u_{0,j}^1 - \frac{R^1}{2} u_{1,j}^1 \tag{A.3.10}$$

Now substituting for $v_{-1,j+1}$ and $v_{-1,j}$ from Eq. (A.3.8) and rearranging the terms, we get

$$-(R^1 \lambda^1 + Q^1) u_{0,j+1}^1 + R^1 u_{1,j+1}^1 + \frac{\Delta z^1 R^1 C^1}{a^1} = -E_{0,j}^1 \tag{A.3.11}$$

where

$$E_{0,j}^1 = -\left[(P^1 - R^1 \lambda^1) u_{0,j}^1 + R^1 u_{1,j}^1 + \Delta q\right]$$

9.7 FDM for Laterally Loaded Piles: Soil-Structure Interaction

9.7.1 Problem Definition and Governing Equations

There are many geotechnical problems, which involve interaction of soil with some structural elements. Here, we consider the case of a laterally loaded pile which is shown in Figure 9.7-1 together with assumed variations of deflection, bending moment, and shear along the depth of the pile. For the analysis of this problem, the objective is to evaluate the primary field variable, the deflection, $y(z)$.

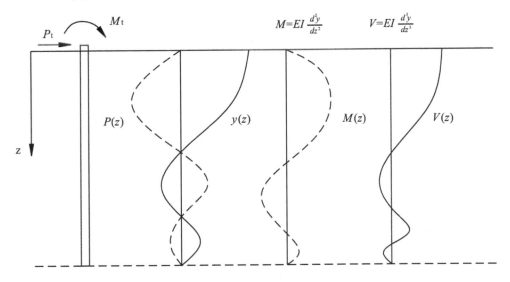

Figure 9.7-1 A laterally loaded pile

Recalling the basic Euler's 'deflection of beam' equations, the governing equation for the deflection of an embedded pile can be written as,

$$\frac{d^2 M}{dz^2} + E_s y = 0 \qquad (9.7.1)$$

where $M = EI \dfrac{d^2 y}{dz^2}$ is the moment. $E_s(= KB)$ is soil elastic modulus, and $R(= EI)$ is flexural stiffness of the pile. For convenience, some details of the above equation is presented in Box B-1 at the end of this section.

9.7.2 Finite Difference Model

In order to develop a finite difference model, the pile shown in Figure 9.7-2 is discretized as shown below.

$$\left(\frac{d^2 M}{dz^2}\right)_m = \frac{\left(\dfrac{d^2 M}{dz^2}\right)_{m-\frac{1}{2}} - \left(\dfrac{d^2 M}{dz^2}\right)_{m+\frac{1}{2}}}{h} = \frac{M_{m-1} - M_m - (M_m - M_{m+1})}{h^2} \qquad (9.7.2a)$$

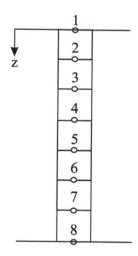

Figure 9.7-2 Finite difference discretization of pile

$$\left(\frac{d^2 M}{dz^2}\right)_m = \frac{R_{m-1}(y_{m-2}-2y_{m-1}+y_m)-2R_m(y_{m-1}-2y_m+y_{m+1})+R_{m+1}(y_m-2y_{m+1}+y_{m+2})}{h^4}$$

(9.7.2b)

Combining equations (9.7.1) and (9.7.2b):

$$R_{m-1}y_{m-2}+(-2R_{m-1}-R_m)y_{m-1}+(R_{m-1}+4R_m+R_{m+1})y_m$$
$$+(-2R_m-2R_{m+1})y_{m+1}+R_{m+1}y_{m+2}+h^4(E_s)_m y_m = 0$$

(9.7.3)

At the tip of the pile ($m = 0$) we get,

$$M_0 = \left(EI\frac{d^2 y}{dz^2}\right)_0$$

(9.7.4)

$$V_0 = \left(EI\frac{d^3 y}{dz^3}\right)_0$$

(9.7.5)

$$\frac{R_0}{h^2}(y_{-1}-2y_0+y_1) = 0$$

(9.7.6)

$$\frac{R_0}{h^3}(y_{-2}-2y_{-1}+2y_1-y_2) = 0$$

(9.7.7)

Also Eq. (9.7.3) should apply to pile top ($m = 0$) giving,

$$R_{-1}y_{-2}+(-2R_{-1}-2R_0)y_{-1}+y_0(R_{-1}+4R_0+R_1)$$
$$+(-2R_0-2R_1)y_1+R_1y_2+h^4 E_{s0}y_0 = 0$$

(9.7.8)

From Equations (9.7.6), (9.7.7) and (9.7.8), we can write,

$$\left(2+\frac{h^4 E_{s0}}{R_0}\right)y_0 - 4y_1 + 2y_2 = 0$$

(9.7.9)

At the pile tip ($m = t$),

$$M_t = EI \frac{d^2 y}{dz^2} \tag{9.7.10}$$

$$P_t = EI \frac{d^3 y}{dz^3} \tag{9.7.11}$$

$$y_{t-1} - 2y_t + y_{t+1} = \frac{M_t h^2}{R_t} \tag{9.7.12}$$

$$y_{t-2} - 2y_{t-1} + 2y_{t+1} - y_{t+2} = \frac{2h^3 P_t}{R_t} \tag{9.7.13}$$

$$2y_{t-2} - 4y_{t-1} + \left(2 + \frac{h^4 E_{st}}{R_t}\right) = \frac{2M_t h^2}{R_t} + \frac{2P_t h^2}{R_t} \tag{9.7.14}$$

Finally, we can develop a set of algebraic equations by using,

(a) The finite difference Eq. (9.7.3) for all the interior nodes ($m = 1$ to $m = t - 1$);
(b) The finite difference Eq. (9.7.9) for the pile top ($m = 0$);
(c) The finite difference Eq. (9.7.14) for the pile tip ($m = t$),

in the form of

$$[A]\{y\} = \{L\} \tag{9.7.15}$$

where $\{y\}$ is pile deflection at discrete points ($m = 1$ to $m = t - 1$) and can be solved readily.

Box B-1

Differential equation of the pile deflection
Let us recall the very basic equations for the deflection of a beam shown below,

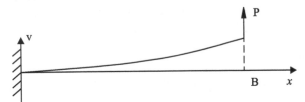

Figure B.1-1 Beam deflection

$$\kappa = \frac{1}{\rho} = \frac{d\theta}{ds} \qquad \frac{dv}{dx} = \tan\theta \tag{B.1.1}$$

Using $ds \approx dx$,

$$\kappa = \frac{1}{\rho} = \frac{d\theta}{dx} = \frac{d^2 v}{dx^2} \tag{B.1.2}$$

If the beam is linearly elastic and follows Hooke's law, then the curvature is,

$$\kappa = \frac{M}{EI} = \frac{d^2v}{dx^2} \tag{B.1.3}$$

For non-prismatic beams *EI* is not constant:

$$\frac{d}{dx}\left(EI\frac{d^2v}{dx^2}\right) = \frac{dM}{dx} = V \tag{B.1.4}$$

$$\frac{d^2}{dx^2}\left(EI\frac{d^2v}{dx^2}\right) = \frac{dV}{dx} = -q, \quad \frac{d^2M}{dx^2} = -q(x) \tag{B.1.5}$$

A pile embedded in soil is subjected to a lateral load due to restraints from the surrounding soil. Using the theory of subgrade reaction, this lateral load intensity can be written as:

$$q(x) = Bq_0(x) = K_sBy(x) = E_sy(x) \tag{B.1.6}$$

leading to the final equation,

$$EI\frac{d^2y}{dx^2} + E_sy(x) = 0, \quad \frac{d^2y}{dx^2} + 4\lambda^4 y = 0 \tag{B.1.7}$$

where $\lambda = \sqrt[4]{\dfrac{E_s}{4EI}}$ is relative stiffness of the pile.

9.8 Error, Convergence and Stability

9.8.1 Introduction

In this section, some very preliminary discussions of errors, convergence and stability associated with finite difference approximation and methods of solution are presented. Let us recall the basic steps we take in modeling and developing numerical solutions for a problem. In Figure 9.8-1, these steps and associated solutions are shown.

9.8.2 Definitions

In FDM, discretization is achieved by replacing various derivatives in the governing equation by their finite difference approximations.

Errors

The main sources of error in an FD analysis are,

Discretization error: $\qquad e_D = \varphi_E - \varphi_D$

Numerical error (Round off): $\qquad e_N = \varphi_D - \varphi_N$

Total error: $\qquad e = \varphi_E - \varphi_N$

Here, the discretization error, e_D, will depend on the time step and the size of the grid and is written as,

$$e_D = f(\Delta x, \Delta y, \Delta z, \Delta t) \tag{9.8.1}$$

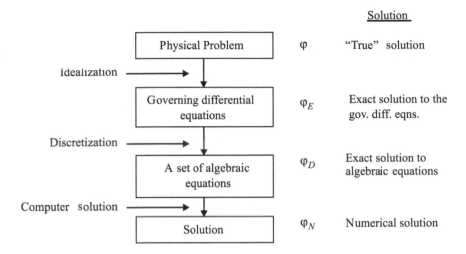

Figure 9.8-1 FDM steps and solutions

Convergence

In its simplest form, as is also the case for any numerical analysis carried out in time and space, a finite difference solution scheme is convergent, if $e_D \to 0$, as $\Delta z, \Delta x, \Delta y, \Delta t \to 0$.

Stability

As for the stability of a finite difference solution, it is satisfied if cumulative effect of all the numerical (round off) errors $|e_N|$ remains negligible.

9.8.3 Errors in Finite Difference Approximations

Error in φ' Central Scheme

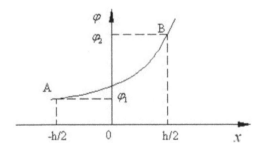

Figure 9.8-2 Error in central scheme for φ'

Recalling Taylor Series expansion,

$$\varphi(x+h) = \varphi(x) + h\varphi'(x) + \frac{h^2}{2!}\varphi''(x) + \frac{h^3}{3!}\varphi'''(x) + \cdots\cdots$$

$$= \varphi(x) + h\varphi'(x) + \frac{h^2}{2!}\varphi''(x) + \frac{h^3}{3!}\varphi'''(\xi) \qquad 0 \le \xi \le h \qquad (9.8.2)$$

in which

$$\frac{h^3}{3!}\varphi'''(x)+\cdots\cdots=\frac{h^3}{3!}\varphi'''(\xi) \tag{9.8.3}$$

$$\varphi_1 = \varphi_0 - \frac{1}{2}h\varphi_0' + \frac{1}{2!}\left(\frac{h}{2}\right)^2\varphi_0'' - \frac{1}{3!}\left(\frac{h}{2}\right)^3\varphi'''(\xi_1), \quad -\frac{h}{2} \le \xi_1 \le 0 \tag{9.8.4}$$

$$\varphi_2 = \varphi_0 + \frac{1}{2}h\varphi_0' + \frac{1}{2!}\left(\frac{h}{2}\right)^2\varphi_0'' + \frac{1}{3!}\left(\frac{h}{2}\right)^3\varphi'''(\xi_2), \quad 0 \le \xi_2 \le \frac{h}{2} \tag{9.8.5}$$

$$\frac{\varphi_2 - \varphi_1}{h} = \varphi' + \frac{1}{3!}\left(\frac{1}{2}\right)^3 h^2\left[\varphi'''(\xi_1) + \varphi'''(\xi_2)\right] \tag{9.8.6}$$

$$\left[\frac{\varphi_2 - \varphi_1}{h} - \varphi_0'\right] \le \frac{1}{3!}\left(\frac{1}{2}\right)^3 h^2 2\left|\varphi'''(x)\right|_{max}$$

$$\le \frac{1}{24}h^2\left|\varphi'''(x)\right|_{max} \tag{9.8.7}$$

$$Error = \left[\frac{\varphi_2 - \varphi_1}{h} - \varphi_0'\right] \tag{9.8.8}$$

If $\varphi''(x)$ is bounded between points (1) and (2), the error involved in the FD approximation is $e \le ch^2$ with a constant c and $e \approx O(h^2)$ error is of the order, h^2 where if h is halved, error decreases by a factor of 4. That is, $\varphi_0' = \frac{\varphi_2 - \varphi_1}{h}$ if $\varphi''(x) = 0$.

Example 9.8-1: Determine a bound for the error in an FD approximation if $\varphi = x^3$ and $\varphi_0' = 0$.

Solution:

$$\varphi = x^3 \rightarrow \varphi' = 3x^2 \rightarrow \varphi_0' = 0$$
$$\varphi' = 6x$$
$$\varphi'' = 6$$

Actual
$$\begin{cases} \varphi_1 = -\frac{h^3}{8} \\[2mm] \varphi_2 = \frac{h^3}{8} \\[2mm] \frac{\varphi_2 - \varphi_1}{h} - \varphi_0' = \frac{h^2}{4} \end{cases} \tag{9.8.9}$$

From error analysis we have:

$$\left\{\left|\frac{\varphi_2 - \varphi_1}{h} - \varphi_0'\right| \le \frac{1}{24}h^2\left|\varphi'''(x)\right|_{max} \le \frac{h^2}{4}\right. \tag{9.8.10}$$

Error in φ' of Backward and Forward Schemes

Recalling the first order approximation of both schemes (Figure 9.2-1) in the figure below,

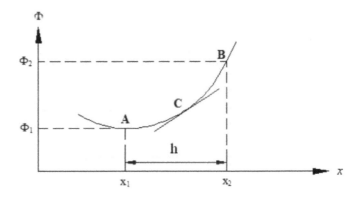

Figure 9.8-3 Function approximation in backward and forward schemes

for the forward scheme we write,

$$\varphi_2 = \varphi_1 + h\varphi_1' + \frac{h^2}{2!}\varphi_1'' + \frac{h^3}{3!}\varphi_1''' + \cdots\cdots \tag{9.8.11}$$

$$\varphi_2 = \varphi_1 + h\varphi_1' + \frac{h^2}{2!}\varphi''(\xi_1), \quad x_1 \le \xi_1 \le x_2 \tag{9.8.12}$$

in which

$$\frac{h^2}{2!}\varphi_1'' + \frac{h^3}{3!}\varphi_1''' + \cdots\cdots = \frac{h^2}{2!}\varphi''(\xi_1) \tag{9.8.13}$$

$$\left|\frac{\varphi_2 - \varphi_1}{h} - \varphi_1'\right| \le \frac{h}{2}\left|\varphi''(x)\right|_{\max} \tag{9.8.14}$$

$e = O(h)$ error is of the order h.

For the backward scheme, similarly we should have $e = O(h)$. We note then that the central difference scheme for φ' is more accurate than forward and backward finite difference schemes.

Error in $\varphi''(x)$

Figure 9.8-4 shows the use of mid-point existence in the definition of error for $\varphi''(x)$ through a central scheme. Therefore we write the Taylor series as,

$$\varphi_1 = \varphi_2 - h\varphi_2' + \frac{1}{2!}h^2\varphi_2'' - \frac{1}{3!}h^3\varphi_2''' + \frac{1}{4!}h^4\varphi''''(\xi_1), \quad -h \le \xi_1 \le 0 \tag{9.8.15a}$$

$$\varphi_1 = \varphi_2 + h\varphi_2' + \frac{1}{2!}h^2\varphi_2'' + \frac{1}{3!}h^3\varphi_2''' + \frac{1}{4!}h^4\varphi''''(\xi_2), \quad 0 \le \xi_2 \le h_2 \tag{9.8.15b}$$

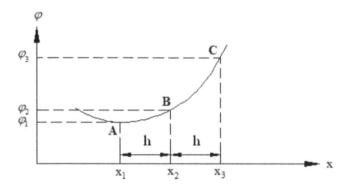

Figure 9.8-4 Error in central scheme for $\varphi''(x)$

With the second order approximation by using all the three point function values given as,

$$\frac{(\varphi_1 + \varphi_3 - 2\varphi_2)}{h^2} = \varphi_2'' + \frac{1}{4}h^2 \left[\varphi''''(\xi_1) + \varphi''''(\xi_2) \right] \qquad (9.8.16)$$

we arrive at,

$$\left| \frac{1}{h^2}(\varphi_1 - 2\varphi_2 + \varphi_3) - \varphi_2'' \right| \leq \frac{1}{12}h^2 \max \left| \varphi''''(x) \right| \qquad (9.8.17)$$

Here, if $|\varphi'''(x)|$ is finite $\approx O(h^2)$ but if $\dfrac{\varphi_1 - 2\varphi_2 + \varphi_3}{h^2}$ is taken as φ_1'' or φ_3'' then error becomes first order $O(h)$.

9.8.4 Error Propagation

Let us understand the propagation of error in the numerical solution for the consolidation equation as reminded below,

$$C_v \frac{\partial^2 u}{\partial x^2} = \frac{\partial u}{\partial t} \qquad 0 \leq x \leq 1 \qquad (9.8.18)$$

Boundary conditions:

$$\begin{cases} u(0,t) = 0 \\ u(1,t) = 0 \end{cases} \qquad (9.8.19)$$

Initial conditions:

$$u(x,0) = \begin{cases} 2x & 0 \leq x \leq 1/2 \\ 2(1-x) & 1/2 \leq x \leq 1 \end{cases} \qquad (9.8.20)$$

$$\varphi_{i,j+1} = \varphi_{i,j} + r\left(\varphi_{i-1,j} - 2\varphi_{i,j} + \varphi_{i+1,j}\right) \qquad (9.8.21)$$

where $r = \dfrac{C_v \Delta t}{\Delta z^2} = 0.5$ (assumed for this example)

$$\varphi_{i,j+1} = \frac{1}{2}\left(\varphi_{i-1,j} + \varphi_{i+1,j}\right) \qquad (9.8.22)$$

For the above FD scheme, we get the following values of error in magnitude in Table 9.8-1 for each time step at each grid point.

Table 9.8-1 Error accumulation

	i = 0	1	2	3	4	5	6	7	8	9	10
j = 0	0	0	0	0	0	1.0e	0	0	0	0	0
1	0	0	0	0	1/2e	0	1/2e	0	0	0	0
2	0	0	0	1/4e	0	1/2e	0	1/4e	0	0	0
3	0	0	1/8e	0	0	0	0	0	1/8e	0	0
4	0	1/16	0	1/4e	0	3/8e	0	1/4e	0	1/16	0
5	0	0	5/32	0	5/16	0	5/16	0	5/32	0	0
6	0	0	0	0	0	0	0	0	0	0	0

The maximum error along each row gradually decreases as j increases (i.e. as we march the solution in time). Therefore, this scheme is stable.

9.9 Exercise Problems

9.9.1 1-D FD consolidation equation (9.3.1) is required such that:

A. Use the finite difference formulation presented in section 9.3 and develop a MATLAB program to study the explicit scheme for solving 1-D consolidation equation. Your program should have the following features:
- Solution can be obtained for '$\theta = 0$' and any value of 'r'.
- Should be able to handle any initial condition.
- Should be able to plot your results for $u(x, t)$.
- Exercise your program on the problem shown below, and compare your results with the analytical solution. For the explicit method ($\theta = 0$), explore the stability of the numerical solution, using various values of 'r'.

B. Develop another MATLAB program to solve the 1-D consolidation problem using Crank-Nicolson scheme ($\theta = 0.5$) solving the final algebraic equations by the Direct Method. Exercise your program on the same problem shown below.

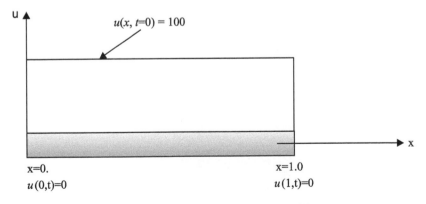

Figure 9.9-1 1-D consolidation problem

9.9.2 For the 2-D steady state seepage problem

A. Using the finite difference formulation, presented above, develop a MATLAB program to solve the 2-D seepage equation. Your program:
   - Should be able to handle anisotropy and inhomogeneity of soil.
   - Should use the iterative scheme of solving the algebraic equations.
   - Should be able to plot: the contour of total head (equi-potential lines) and also the flow lines.

B. Exercise your program on the problem shown below with the following three conditions:
   - Homogenous and isotropic foundation soil, with $k_x = k_y = 0.0005$ ft/s.
   - Homogeneous and anisotropic foundation soil, with $k_x = 0.0025$ ft/s, $k_y = 0.0005$ ft/s.
   - Layered and isotropic foundation soil, with $k_x = k_y = 0.0025$ ft/s in top 20 ft and $k_x = k_y = 0.0005$ ft/s at the bottom 20 ft of the soil.

Present your results in a nice way and write a brief summary and conclusions.

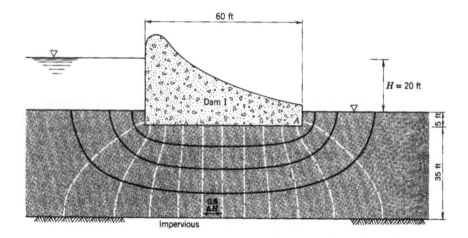

Figure 9.9-2 Seepage under a dam problem (Lambe and Whitman, 1969)

MATLAB® Scripts

Main Program "FDM_I.m"

```
% "FDM_I.m"
% One-Dimensional Consolidation: Basics
% This program evaluates the pore pressure, u(x,t), response using the
  explicit scheme
clear all

theta=0.;
dt=0.001;                    % Time increment
dx=0.1;                      % Length increment
cv=1.;                       % Coef. of consolidation
r=cv*dt/dx/dx                % Time factor
x=0.:0.1:1.;                 % Length vector
t=0.:0.001:0.02;             % Time vector
nx=length(x);                % Size of x
nt=length(t);                % Size of t
%INITIALIZATION
for j=1,nt
    for i=1:nx
        u(i,j)=0.;
    end
end

% INITIAL CONDITION
u(:,1)= [0. 0.2 0.4 0.6 0.8 1.0 0.8 0.6 0.4 0.2 0.];

% BOUNDARY CONDITION
for j=1:nt
    u(1,j)=0.;
    u(nx,j)=0.;
end

% Calculate the pore pressure values for all the interior nodes
for j=1:nt
    for i= 2:nx-1
        u(i,j+1)=r*(u(i-1,j)+u(i+1,j))+(1-2*r)*u(i,j);
    end
end
% Plot
for j=1:5:nt
    plot (x,u(:,j));
    hold on
end
title ('Explicit Scheme')
xlabel('Distance, x')
ylabel('Pore Pressure, u')
grid on
```

Main Program "FDM_II.m"

```
% "FDM_II.m"
% One-Dimensional Consolidation
% This program evaluates the pore pressure ratio, UU, and average degree
% of consolidation, W
clear all
```

```
H=2;                                            % Layer Thickness
dz=0.01;                                        % Depth increment
T=[0.0, .02, .06, .1, .2, .4, .6 , 1, 1.5, 5]; % Time vector
Z=0.:dz:H;                                      % Depth vector
NT=length(T);                                   % Size of time
NZ=length(Z);                                   % Size of depth
W=zeros(NT,NZ);
UU=zeros(NT);
for i=2:NT
    for j=2:NZ-1
        W(i,j)=0.;
        m=0;
        tol=1.;
        while (tol>10e-7)
            Wold(i,j)=W(i,j);
            n=2*m+1;
            Wm(i,j)= (4/pi/n)*(sin(n*pi*Z(j)/2))*exp(-(n*n*pi*pi/4)*T(i));
            W(i,j)=Wold(i,j)+Wm(i,j);
            tol=abs(W(i,j)-Wold(i,j));
            m=m+1;
        end
    end
end

for i=2:NT
    UU(i)=0.;
    m=0;
    tol=1.;
    while (tol>10e-7)
        UUold(i)=UU(i);
        n=2*m+1;
        UUm(i)= (8/pi/pi/n/n)*exp(-(n*n*pi*pi/4)*T(i));
        UU(i)=UUold(i)+UUm(i);
        tol=abs(UU(i)-UUold(i));
        m=m+1;
    end
    U(i)=1.-UU(i);
end
subplot(211)
plot(T,(1.-U))
title('Average Degree of Consolidation')
xlabel('T')
ylabel('U')
subplot(212)
plot(W,Z)
title('Pore Pressure Ratio')
xlabel('U/U_0')
ylabel('Depth, Z')
```

Main Program "FDM_III.m"

```
% "FDM_III.m"
% Finite Difference Method used to solve 2-D steady state seepage of soil
% _____
clear
disp('Specify the Number of Layers');
layer=input('Number of Layers ');
lx=180;
```

```
ly=40;
x=linspace(0,180,180);
a=length(x);
y=linspace(0,40,40);
k=500;
b=length(y);
h(1:b,1:a)=50.;
%_____
% Specify Permeability Value for Each Layer
%_____
k=zeros(4,1);   % includes permeability value
k(1)=0.0025;
k(2)=0.0005;
k(3)=0.0025;
k(4)=0.0005;

dx=1;
dy=1;
%_____
% Assigning "k" value
%_____
R=zeros(b,a);
L=zeros(b,a);
A=zeros(b,a);
B=zeros(b,a);

if layer==1
    disp('Specify if the Soil is Isotropic(1) or Anisotropic(2)?');
    T=input('Type of Soil ');
    if T==1
        for i=1:180
            for j=1:40
                R(j,i)=2*k(2)*k(2)/(k(2)+k(2))/dx^2;
                L(j,i)=2*k(2)*k(2)/(k(2)+k(2))/dx^2;
                A(j,i)=2*k(2)*k(2)/(k(2)+k(2))/dy^2;
                B(j,i)=2*k(2)*k(2)/(k(2)+k(2))/dy^2;
            end
        end
    else
        for i=1:180
            for j=1:40
                R(j,i)=2*k(1)*k(1)/(k(1)+k(1))/dx^2;
                L(j,i)=2*k(1)*k(1)/(k(1)+k(1))/dx^2;
                A(j,i)=2*k(2)*k(2)/(k(2)+k(2))/dy^2;
                B(j,i)=2*k(2)*k(2)/(k(2)+k(2))/dy^2;
            end
        end
    end
end
%_____
% If There Are Two Isotropic Layers with: kx1=ky1=0.0025 & kx2=ky2=0.0005
%_____
if layer==2
    for j=1:40
        if j<21
            for i=1:180
                R(j,i)=2*k(3)*k(3)/(k(3)+k(3))/dx^2;
                L(j,i)=2*k(3)*k(3)/(k(3)+k(3))/dx^2;
```

```
                A(j,i)=2*k(3)*k(3)/(k(3)+k(3))/dy^2;
                B(j,i)=2*k(3)*k(3)/(k(3)+k(3))/dy^2;
            end
        else
            for i=1:180
                R(j,i)=2*k(4)*k(4)/(k(4)+k(4))/dx^2;
                L(j,i)=2*k(4)*k(4)/(k(4)+k(4))/dx^2;
                A(j,i)=2*k(4)*k(4)/(k(4)+k(4))/dy^2;
                B(j,i)=2*k(4)*k(4)/(k(4)+k(4))/dy^2;
            end
        end
    end
end
%_____
% Specifying the Initial Conditions
%_____
k=1;
h(1,1:60)=60;
h(1,121:180)=40;
for k=1:5000
    for j=1:40
        if j==1
            for i=1:60
                head(j,i)=60;
            end
            for i=61:120
                head(j,i)=0;
            end
            for i=121:180
                head(j,i)=40.;
            end
        elseif j>1 & j<6
            for i=1:180
                if i==1
                    head(j,i)=(R(j,i)*h(j,i+1)+B(j,i)*h(j+1,i)+A(j,i)*h(j-
                    1,i))/(R(j,i)+B(j,i)+A(j,i));
                elseif i==180
                    head(j,i)=(L(j,i)*h(j,i-1)+B(j,i)*h(j+1,i)+A(j,i)*h(j-
                    1,i))/(L(j,i)+B(j,i)+A(j,i));
                elseif i==60
                    head(j,i)=(L(j,i)*h(j,i-1)+B(j,i)*h(j+1,i)+A(j,i)*h(j-
                    1,i))/(L(j,i)+B(j,i)+A(j,i));
                elseif i==121
                    head(j,i)=(R(j,i)*h(j,i+1)+B(j,i)*h(j+1,i)+A(j,i)*h(j-
                    1,i))/(R(j,i)+B(j,i)+A(j,i));
                elseif i>60 & i<121
                    if j<5
                    head(j,i)=0;
                    else
                    head(j,i)=(R(j,i)*h(j,i+1)+L(j,i)*h(j,i-
                    1)+B(j,i)*h(j+1,i))/(R(j,i)+L(j,i)+B(j,i));
                    end
                else
                    head(j,i)=(R(j,i)*h(j,i+1)+L(j,i)*h(j
                    ,i-1)+B(j,i)*h(j+1,i)+A(j,i)*h(j-1,i))/
                    (R(j,i)+L(j,i)+B(j,i)+A(j,i));
                end
```

```
                end
            elseif j==40
                for i=1
                    head(j,i)=(R(j,i)*h(j,i+1)+A(j,i)*h(j-1,i))/(R(j,i)+A(j,i));
                end
                for i=2:179
                    head(j,i)=(R(j,i)*h(j,i+1)+L(j,i)*h(j,i-1)+A(j,i)*h(j-1,i))/
                    (R(j,i)+L(j,i)+A(j,i));
                end
                for i=180
                    head(j,i)=(L(j,i)*h(j,i-1)+A(j,i)*h(j-1,i))/(L(j,i)+A(j,i));
                end
            else
                for i=1:180
                    if i==1
                        head(j,i)=(R(j,i)*h(j,i+1)+B(j,i)*h(j+1,i)+A(j,i)*h(j-
                        1,i))/(R(j,i)+B(j,i)+A(j,i));
                    elseif i==180
                        head(j,i)=(L(j,i)*h(j,i-1)+B(j,i)*h(j+1,i)+A(j,i)*h(j-
                        1,i))/(L(j,i)+B(j,i)+A(j,i));
                    else
                        head(j,i)=(R(j,i)*h(j,i+1)+L(j,i)*h(j
                        ,i-1)+B(j,i)*h(j+1,i)+A(j,i)*h(j-1,i))/
                        (R(j,i)+L(j,i)+B(j,i)+A(j,i));
                    end
                end
            end
        end
        h=head;
    end
contourf(h,100);
hold on
vx=zeros(b,a);
vy=zeros(b,a);
[vx,vy]=gradient(h);
for j=1:5
    for i=60:121
        vx(j,i)=0;
        vy(j,i)=0;
    end
end
set(gca,'YDir','rev');
streamslice(-vx,-vy,'cubic');
xlabel('Distance')
ylabel('Elevation')
hold off;
```

10

Finite Element Method

10.1 Introduction

In the main introduction to this book, we have identified the essential elements of "modeling and computing". For an engineering problem at hand, we begin with the idealization of the problem and identify essential physics underlying it. Thereafter invoking the available knowledge of the basic laws and principles, a mathematical model is formulated. This model is to be in terms of a set of differential equations governing all the relevant aspects of the problem and the conditions existing along the boundaries of the solution domain. Finally, a complete solution to these equations is sought. For simple problems, it may be possible to find analytical solutions to the governing equations but often that is not the case for real world problems. This is exactly why approximate solutions to such problems are developed through 'numerical methods'. We have looked at one such numerical method earlier in the previous chapter. Similar to FDM, in the 'Finite Element Method' (FEM), the governing differential equations are transformed into a set of algebraic equations using a different approach. The essential difference between these two methods is in the way one carries out the process of discretization.

In this chapter, we provide a simple introduction to the FEM. Figure 10.1-1 presents an outline of all the essential elements of the FEM for solving a boundary (or an initial) value problem. The general formulation is illustrated in the context of the general problems of flow of pore water and deformation of soil skeleton encountered frequently in geotechnical engineering. Within the chapter, discussion ranges from the Direct Stiffness Method (DSM) and the Galerkin's Method of Weighted Residual (GMWR) to the basics of more elaborate FEM for both 1-D and 2-D problems. In Figure 10.1-1, necessary steps of obtaining a general numerical solution of a physical problem are presented. As one expects, every numerical approximation of a problem will surely result in certain amount of error in the solution due to number of sources.

10.2 Direct Stiffness Method

In order to get initiated, let us consider a simple method of the *direct stiffness* approach. In fact, FEM had its initial development rooted in this direct engineering approach. Considering a 1-D problem of deformation, a bar shown below is taken as an assemblage of several small 'elements' of uniform cross section. In this method, first assuming a simple form of variation of the displacement over an element, a load-displacement relationship (element stiffness equation) is formulated directly. Then by assembly of these element equations, a global stiffness equation is obtained for the entire bar. Finally, this equation is solved by incorporating the boundary conditions. In the following, this method is illustrated with the help of an example problem (Figure 10.2-1).

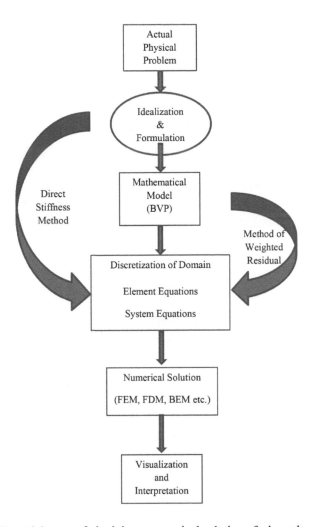

Figure 10.1-1 Essential steps of obtaining a numerical solution of a boundary value problem

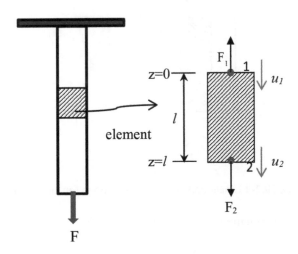

Figure 10.2-1 A hanging bar problem: 1-D deformation

Let us first represent the displacements at the top and bottom of the element as,

$$u(z = 0) = u_1 \tag{10.2.1a}$$

$$u(z = l) = u_2 \tag{10.2.1b}$$

We then assume the displacement within the element to vary as,

$$u(z) = N_1(z)u_1 + N_2(z)u_2 \tag{10.2.2}$$

where $N_i(z)$ are called the "approximation functions", with subscript 'i' being the DOF number and are obtained assuming a linear variation between the two adjacent points (also called nodal points) as,

$$N_1(z) = a_0 + a_1 z \tag{10.2.3a}$$

$$N_2(z) = b_0 + b_1 z \tag{10.2.3b}$$

The coefficients of 10.2.3 can be evaluated as: $a_0 = 1$, $a_1 = -1/l$, $b_0 = 0$ and $b_1 = 1/l$ using the approximation functions shown in Figure 10.2-2. Thus, the functions are obtained as,

$$N_1(z) = 1 - z/l \tag{10.2.4a}$$

$$N_2(z) = z/l \tag{10.2.4b}$$

It is now more appropriate to write the relation (10.2.2) in a vector form,

$$u(z) = \begin{bmatrix} N_1(z) & N_2(z) \end{bmatrix} \begin{Bmatrix} u_1 \\ u_2 \end{Bmatrix} \tag{10.2.5}$$

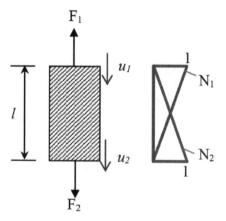

Figure 10.2-2 Linear approximation functions for the bar member

For a linear elastic bar element, the stiffness relation is,

$$F = \frac{EA}{L}\delta \tag{10.2.6}$$

where $\dfrac{EA}{L}$ is the stiffness of a single bar and E is the elasticity modulus, A is cross sectional area and L is the length of the bar. Equation (10.2.6) is called the "force-displacement relation". For the bar element in Figure 10.2-3, there need to be written two similar relations considering the number of DOF that go,

$$F_1 = -\frac{EA}{l}\left(u_2 - u_1\right) \tag{10.2.7a}$$

$$F_2 = \frac{EA}{l}\left(u_2 - u_1\right) \tag{10.2.7b}$$

and in matrix form as,

$$\left\{\begin{matrix} F_1 \\ F_2 \end{matrix}\right\} = \frac{EA}{L}\begin{bmatrix} 1 & -1 \\ -1 & 1 \end{bmatrix}\left\{\begin{matrix} u_1 \\ u_2 \end{matrix}\right\} \tag{10.2.8}$$

Equation (10.2.8) for a multi-dimensional or multi-DOF system becomes,

$$\left\{F^e\right\} = \left[K^e\right]\left\{u^e\right\} \tag{10.2.9}$$

where the superscript 'e' stands for the 'element' and K^e is the global stiffness matrix. Also, $F^e = \left\{\begin{matrix} F_1 \\ F_2 \end{matrix}\right\}$ and $U^e = \left\{\begin{matrix} U_1 \\ U_2 \end{matrix}\right\}$ are the nodal forces and displacements and $K^e = \begin{bmatrix} 1 & -1 \\ -1 & 1 \end{bmatrix}$ is the element stiffness matrix.

Example 10.2-1: Let us consider a bar with varying cross sectional area A, elasticity modulus E and which is under a tensile force, T as seen in the figure below. Our objective is to determine variation of the displacement, $u(z)$ in the bar.

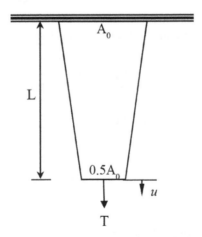

Figure 10.2-3 Non-uniform solid column

One Element Solution

Using a single uniform bar element (Figure 10.2-4), one can easily obtain the stiffness relation below with the boundary conditions: $u(z = 0) = u_1 = 0$, $F(z = l) = F_2 = T$:

$$0.75\frac{EA_0}{L}\begin{bmatrix} 1 & -1 \\ -1 & 1 \end{bmatrix}\begin{Bmatrix} \overset{0}{\cancel{u_1}} \\ u_2 \end{Bmatrix} = \begin{Bmatrix} F_1 \\ \underset{T}{\cancel{F_2}} \end{Bmatrix} \tag{10.2.10}$$

The above can readily be solved to yield $u_2 = 1.33 \ TL/EA_0$.

Two-Element Solution

Using now two uniform bar elements (Figure 10.2-5), we derive one stiffness relation for each element. Thus,

$$\frac{7}{4}\frac{EA_0}{L}\begin{bmatrix} 1 & -1 \\ -1 & 1 \end{bmatrix}\begin{Bmatrix} u_1 \\ u_2 \end{Bmatrix} = \begin{Bmatrix} F_1 \\ F_2 \end{Bmatrix} \tag{10.2.11a}$$

$$\frac{5}{4}\frac{EA_0}{L}\begin{bmatrix} 1 & -1 \\ -1 & 1 \end{bmatrix}\begin{Bmatrix} u_2 \\ u_3 \end{Bmatrix} = \begin{Bmatrix} F_2 \\ F_3 \end{Bmatrix} \tag{10.2.11b}$$

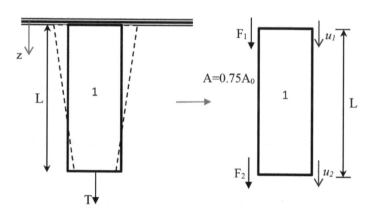

Figure 10.2-4 One element model

with $k_1 = \frac{7}{4}\frac{EA_0}{L}\begin{bmatrix} 1 & -1 \\ -1 & 1 \end{bmatrix}$ and $k_2 = \frac{5}{4}\frac{EA_0}{L}\begin{bmatrix} 1 & -1 \\ -1 & 1 \end{bmatrix}$. In order to obtain a single numerical

solution, these relations are then assembled to give:

$$\begin{bmatrix} k_1 & -k_1 & 0 \\ -k_1 & k_1+k_2 & -k_2 \\ 0 & -k_2 & k_2 \end{bmatrix}\begin{Bmatrix} \overset{0}{\cancel{u_1}} \\ u_2 \\ u_3 \end{Bmatrix} = \begin{Bmatrix} F_1 \\ \underset{0}{\cancel{F_2}} \\ \underset{0}{\cancel{F_3}} \end{Bmatrix}, \text{ with the conditions: } u_1 = 0; \ F_2 = 0; \ F_3 = T \tag{10.2.12}$$

If we apply the same boundary conditions and consider that there is no force in the middle and solve the linear system for the tip displacement using the reduced linear form of,

$$\begin{bmatrix} k_1+k_2 & -k_2 \\ -k_2 & k_2 \end{bmatrix}\begin{Bmatrix} u_2 \\ u_3 \end{Bmatrix} = \begin{Bmatrix} 0 \\ T \end{Bmatrix} \tag{10.2.13}$$

we get $u_3 = 1.371\dfrac{TL}{EA_0}$.

Analytical Solution

Exact solution to this problem is possible once the variation of A is mathematically determined and the variation of strain is integrated over the length. Therefore, we have,

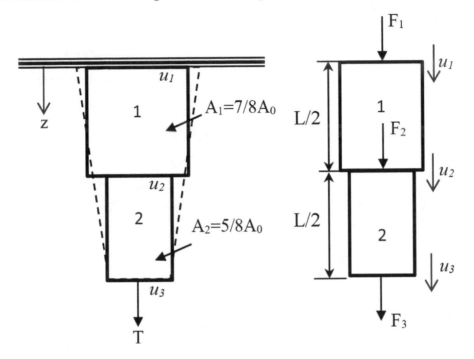

Figure 10.2-5 Two-element model

$$A(z) = A_0\left(1 - \frac{z}{2L}\right)$$

(10.2.14)

Thus, the strain is calculated as,

$$\varepsilon(z) = \frac{\sigma(z)}{E} = \frac{T}{EA_0\left(1 - \dfrac{z}{2L}\right)}$$

(10.2.15)

which we integrate and get the displacement variation as,

$$u_z = \int_0^z \varepsilon(z) = \int_0^z \frac{T}{EA_0\left(1 - \dfrac{z}{2L}\right)} dz = \frac{2TL}{EA_0} \ln\frac{2L}{2L - z}$$

(10.2.16)

Finally, the exact value of the tip displacement is:

$$u(z = L) = \frac{2TL}{EA_0} \ln(2) = 1.3863\frac{TL}{EA_0}$$

Comparison of the results can be seen below in Figure 10.2-6.

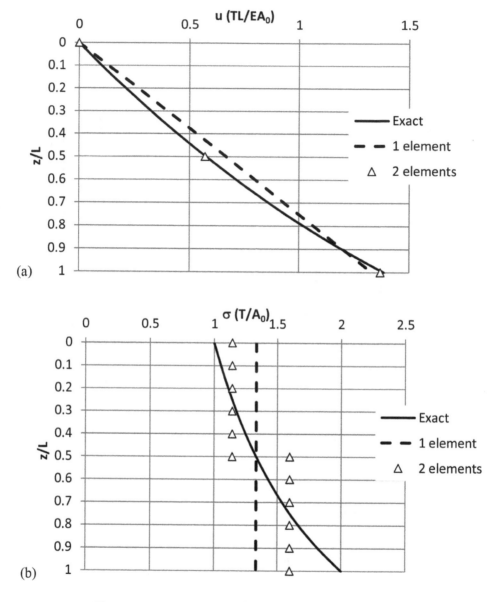

Figure 10.2-6 (a) Displacement variation, (b) Stress variation

10.3 Galerkin Method of Weighted Residual

In this section we present the classical Galerkin Method of Weighted Residual (abbreviated as GMWR or MWR) to develop approximate solution of a differential equation. The central idea is to seek an approximate solution by minimizing a weighted error (resulting from the approximation) over the solution domain. Later the same idea will be used to develop a general method of finite element formulation.

Let us consider a simple differential equation of the form below:

$$D\big(y(x),x\big)=0 \tag{10.3.1}$$

with prescribed boundary conditions, $y(a) = y(b) = 0$ for $a < x < b$. Here D is the differential operator and x is the independent variable. Let $y^*(x) = \sum_{i=1}^{n}\{c_i N_i(x)\}$ be an approximate solution where c_i is the constant coefficient and $N_i(x)$ are the trial functions. Since this is an approximate solution, the governing equation will not be satisfied and thus will result in a residual error $R(x)$. That is,

$$D(y^*(x), x) \neq 0 \tag{10.3.2}$$

$$D(y^*(x), x) = R(x) \tag{10.3.3}$$

In MWR, we seek an approximate solution by reducing a weighted average of this residual error to zero, as represented below,

$$\int_a^b w_i(x) R(x) dx = 0 \tag{10.3.4}$$

where $w_i(x)$ are the weighting functions. If $w_i(x) = N_i(x)$, this is called Galerkin MWR and so,

$$\int_a^b N_i(x) R(x) dx = 0, \ i = 1, 2, \ldots, n \tag{10.3.5}$$

This method is illustrated through few simple examples below.

Example 10.3-1: Let us consider the governing differential equation to be:

$$D(y(x), x) = \frac{d^2 y}{dx^2} - 10x^2 - 5 = 0 \text{; with } y(0) = y(1) = 0$$

Let $y^*(x) = \sum_{i=1}^{n}\{c_i N_i(x)\}$ and choose $N_1(x) = x(x - 1)$. Thus, $y^*(x) = c_1 x(x - 1)$, $\frac{dy^*}{dx} = c_1(2x - 1)$ and $\frac{d^2 y^*}{dx^2} = 2c_1$ yielding the residual, $R(x) = 2c_1 - 10x^2 - 5 \neq 0$. Then using (10.3.4), we find

$$\int_0^1 x(x - 1)\left[2c_1 - 10x^2 - 5\right] dx = 0$$

$$\rightarrow c_1 = 4$$

Therefore the approximate solution is obtained as, $y^*(x) = 4x(x - 1)$. The exact solution of $D(y, x)$ is derived also,

$$y(x) = 0.833x^4 + 2.5x^2 - 3.33x \tag{10.3.6}$$

Let us now consider a two term approximation with $N_1(x) = x(x - 1)$ and $N_2(x) = x^2(x - 1)$, $y^*(x) = c_1 x(x - 1) + c_2 x^2(x - 1)$ with the second derivative being,

$$\frac{d^2 y^*}{dx^2} = 2c_1 + 2c_2(3x - 1) \tag{10.3.7}$$

Hence the residual is obtained as,

$$R(x) = 2c_1 + 2c_2(3x-1) - 10x^2 - 5 \neq 0 \qquad (10.3.8)$$

Again using the GMWR, for $i = 1$,

$$\int_0^1 x(x-1)\left[2c_1 + 2c_2(3x-1) - 10x^2 - 5\right]dx = 0 \qquad (10.3.9)$$

and for $i = 2$,

$$\int_0^1 x^2(x-1)\left[2c_1 + 2c_2(3x-1) - 10x^2 - 5\right]dx = 0$$

$$\rightarrow c_1 = 3.1667, \ c_2 = 1.6667 \qquad (10.3.10)$$

The final solution is: $y^*(x) = 1.1875x(x - 1) + 1.6667x^2(x - 1)$. Solutions are plotted in Figure 10.3-1.

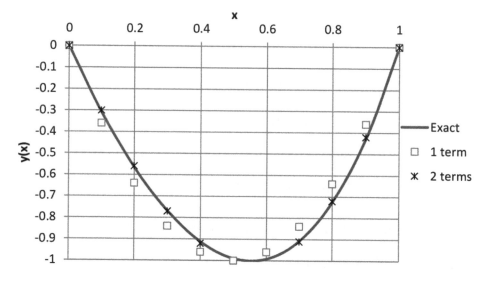

Figure 10.3-1 Comparison of approximations

Example 10.3-2: We now consider the displacement field in a bar under tension (Figure 10.3-2).

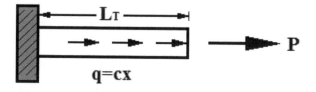

Figure 10.3-2 Bar under tension

For a bar under tension experiencing elongation of $u(x)$, the stress-strain relationship can be easily written using linear elasticity as,

$$\sigma(x) = E\varepsilon(x) = E\frac{du(x)}{dx} \tag{10.3.11}$$

From the force equilibrium,

$$F_{int} - F_{ext} = 0 \tag{10.3.12}$$

$$A\frac{\sigma(x)}{dx} + q = 0 \tag{10.3.13}$$

Substituting (10.3.11) into (10.3.13) yields,

$$AE\frac{d^2u(x)}{dx^2} + q = 0 \tag{10.3.14}$$

and finally,

$$\frac{d^2u(x)}{dx^2} + \frac{cx}{AE} = 0 \tag{10.3.15}$$

The boundary conditions below complete the problem,

$$AE\frac{du(x)}{dx} = P\ @\ x = L_T,\ u = 0\ @\ x = 0 \tag{10.3.16}$$

The exact solution of the problem in terms of displacement and stress can be easily obtained as,

$$u(x) = \left(\frac{2P + cL_T^2}{2AE}\right)x - \frac{c}{6AE}x^3 \tag{10.3.17a}$$

$$\sigma(x) = \left(\frac{2P + cL_T^2}{2A}\right) - \frac{c}{2A}x^2 \tag{10.3.17b}$$

The Galerkin problem is now expressed through (10.3.4) as,

$$\int_0^{L_T} w_i(x)\left(\frac{d^2u(x)}{dx^2} + \frac{cx}{AE}\right)dx = 0,\ i = 1,2...n \tag{10.3.18}$$

Before we proceed by assuming a weighting function for $w_i(x)$, it is necessary to do integration by parts to the terms of this equation which, after minor manipulation goes like this:

$$\left[w_i\left(\frac{du}{dx}\right)\right]_0^{L_T} + \int_0^{L_T}\left(-\left(\frac{dw_i}{dx}\right)\frac{du(x)}{dx} + w_i\frac{cx}{AE}\right)dx = 0,\ i = 1,2...n \tag{10.3.19}$$

Using now $u = c_1N_1(x) + c_2N_2(x) = c_1x + c_2x^2$ as the trial functions and taking the partial derivatives of the functions with respect to the variable x, (10.3.19) can be rewritten as,

$$\frac{PL_T}{AE} + \int_0^{L_T}\left(-(c_1 + 2c_2x) + x\frac{cx}{AE}\right)dx = 0 \tag{10.3.20a}$$

$$\frac{PL_T}{AE} + \int_0^{L_T}\left((-2x)(c_1 + 2c_2 x) + x^2\frac{cx}{AE}\right)dx = 0 \tag{10.3.20b}$$

which results in the following solution of coefficients:

$$c_1 = \frac{P}{AE} + \frac{7cL_T^2}{12AE} \tag{10.3.21a}$$

$$c_2 = -\frac{cL_T}{4AE} \tag{10.3.21b}$$

Solution for the displacement and stress fields take the final form of,

$$u(x) = \left(\frac{12P + 7cL_T^2}{12AE}\right)x - \frac{cL_T}{4AE}x^2 \tag{10.3.22a}$$

$$\sigma(x) = \left(\frac{12P + 7cL_T^2}{12A}\right) - \frac{cL_T}{2A}x \tag{10.3.22b}$$

For unit values of material properties and loads, Figure 10.3-3 presents a comparison of the approximate results (for displacements and stresses) from GMWR with the exact solution.

Example 10.3-3: Another common problem in engineering is the flow of a fluid under steady state condition with a source term (Figure 10.3-4). The governing equation is derived considering the mass balance of an element of a porous medium following the steps as:

$$q_z A - q_{z+\delta z}A + S(z)\delta z = 0 \tag{10.3.23}$$

$$q_z - \left(q_z + \frac{dq_z}{dz}\delta z\right) + S(z)\delta z = 0 \tag{10.3.24}$$

$$-\frac{d}{dz}\left(-k\frac{d\varphi}{dz}\right)\delta z + S(z)\delta z = 0 \tag{10.3.25}$$

$$\frac{d}{dz}\left(k\frac{d\varphi}{dz}\right)\delta z + S(z)\delta z = 0 \tag{10.3.26}$$

where in the relations, q_z is the rate of flow in the z-direction, $S(z)$ is the source term supplying the fluid, φ is the total head and k is the coefficient of hydraulic conductivity or permeability as is more common in classical soil mechanics terminology. Therefore,

$$D(\varphi) = -\frac{d}{dz}\left(k\frac{d\varphi}{dz}\right) - S = 0, \; 0 \le z < H \tag{10.3.27}$$

along with the boundary conditions,

$$-k\frac{d\varphi}{dz} - q_0 = 0 \; @ \; z=0 \tag{10.3.28a}$$

$$\varphi = \varphi_H \; @ \; z=H \tag{10.3.28b}$$

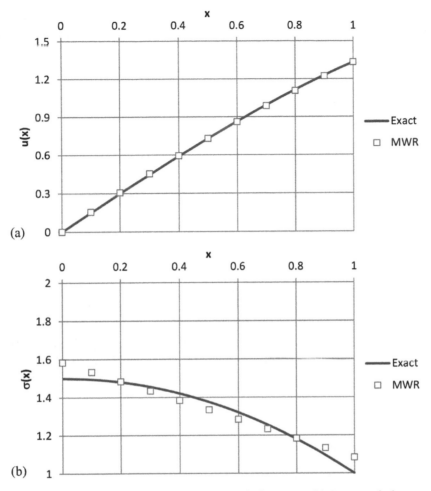

Figure 10.3-3 Comparison of results for (a) Displacement, (b) Stress variations

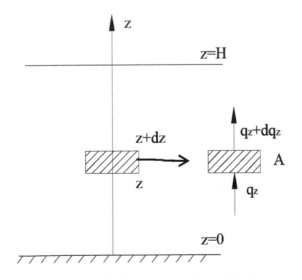

Figure 10.3-4 Steady-state flow of fluid

It should be noted here that the soil skeleton of the porous medium (i.e. soil) is assumed rigid in this problem and the variation of total head and the flow is of our interest. It is again possible to derive an analytical solution for this problem. So, for constant k and S, the variation of the total head is readily obtained as,

$$\varphi(z) = \frac{SH^2}{2k}\left[1-\left(\frac{z}{H}\right)^2\right] + \frac{q_0 H}{k}\left(1-\frac{z}{H}\right) + \varphi_H \tag{10.3.29}$$

Approximate Solution Using GMWR

If we use the general GMWR to solve a differential equation, let

$$\varphi^* = \sum_{j=1}^{N} c_j \phi_j(z) \tag{10.3.30}$$

Now if we substitute this approximation for φ in the governing equation (10.3.27), the equation is not satisfied and we are left with a residual error of,

$$R(\varphi^*) = -\frac{d}{dz}\left(k\frac{d\varphi^*}{dz}\right) - S \neq 0, \quad 0 \leq z < H \tag{10.3.31}$$

Again carrying out (10.3.4) in a similar manner to previous problems to minimize the residual error in an average sense using the function 'ϕ_i' as weighting functions, we have,

$$\int_0^H \phi_i\left[-\frac{d}{dz}(k\frac{d\varphi^*}{dz}) - S\right]dz = 0, \quad i=1,2,\cdots,n \tag{10.3.32}$$

Performing integration by parts yields,

$$\int_0^H \frac{d\phi_i}{dz}k\frac{d\varphi^*}{dz}dz - \int_0^H S\phi_i dz + \left[-\phi_i k\frac{d\varphi^*}{dz}\right]_0^H = 0 \tag{10.3.33}$$

Using now,

$$\varphi^* = c_1\left(1-\frac{z}{H}\right) + c_2\left(1-\left(\frac{z}{H}\right)^2\right) + \varphi_H \tag{10.3.34}$$

as the trial function which satisfies the boundary conditions as prescribed, these equations are obtained:

$$\int_0^H \left(\frac{-1}{H}\right)k\left(\frac{-c_1}{H} + \frac{-2c_2 z}{H}\right)dz - \int_0^H \left(S\left(1-\frac{z}{H}\right)dz\right) - q_0 = 0 \tag{10.3.35a}$$

$$\int_0^H \left(\frac{-2}{H}\right)z \, k\left(\frac{-c_1}{H} + \frac{-2c_2 z}{H}\right)dz - \int_0^H \left(S\left(1-\left(\frac{z}{H}\right)^2\right)dz\right) - q_0 = 0 \tag{10.3.35b}$$

leading to,

$$\frac{k}{H^2}\left(c_1 H + c_2 H^2\right) - S\left(\frac{H}{2}\right) - q_0 = 0 \tag{10.3.36a}$$

$$\frac{k}{H^2}\left(c_1 H^2 + \frac{4}{3}c_2 H^3\right) - S\left(H - \frac{H^2}{3}\right) - q_0 = 0 \qquad (10.3.36b)$$

Solving (10.3.35) simultaneously gives the two coefficients,

$$c_1 = \frac{S\dfrac{H^2}{2} + q_0 H - 3\left[SH\left(1 - \dfrac{5H}{6}\right) - q_0(H-1)\right]}{k} \qquad (10.3.37a)$$

$$c_2 = \frac{3}{k}\left[S\left(1 - \frac{5H}{6}\right) - q_0\frac{(H-1)}{H}\right] \qquad (10.3.37b)$$

Total head is calculated by substituting (10.3.37) into (10.3.34) and plotted for both exact and approximate variation as below:

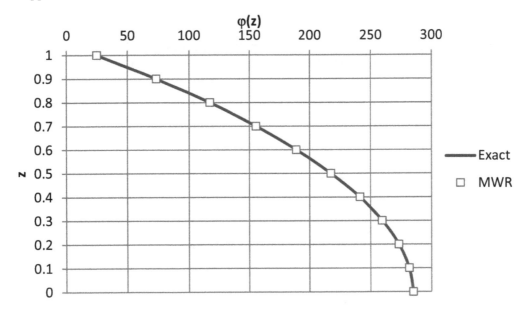

Figure 10.3-5 Total head variation for $q_0 = 0.1$, $H = 1$ m, $S = 5$, $k = 0.01$ m/s, $\varphi_h = 25$

10.4 FEM: 1-D Problems

In this section, we present the FEM formulation using again the basic idea of Galerkin Method of Weighted Residual (GMWR) in the context of simple one dimensional (1-D) problems. The essential idea here is to subdivide the solution domain into smaller 'elements' and the approximation functions are defined over these elements but not over the entire domain as was done in the previous section. For that, FE formulations of a number of problems that are common in geomechanics, are presented.

10.4.1 Steady State Flow

Let us reconsider the problem of 1-D steady state flow of water through soil as shown in Figure 10.3-4. Here, for the equations governing the flow discussed in the previous section, we apply the classical GMWR. The solution domain is divided into small elements and the *approximating functions* are defined in piecewise fashion over the elements. For the ease of illustration, only three elements are considered as can be seen in Figure 10.4-1. For the field variable of total head, it can be written that,

$$\varphi = a_j \phi_j = a_1 \phi_1 + a_2 \phi_2 + a_3 \phi_3 + a_4 \phi_4 \tag{10.4.1}$$

Considering now that, $a_j = \varphi_j^e$ and $\phi_j = N_j^e$ over an element, we can write,

$$\varphi = \sum_{j=1}^{3} \varphi_j^e N_j^e = \left(\varphi_1^1 \right) N_1^1 + \left(\varphi_2^1 \right) N_2^1$$

$$\varphi_1^1 = \psi_1 \qquad + \left(\varphi_1^2 \right) N_1^2 + \left(\varphi_2^2 \right) N_2^2 \qquad \varphi_2^3 = \psi_4$$

$$\varphi_2^1 = \varphi_1^2 = \psi_2 \qquad + \left(\varphi_1^3 \right) N_1^3 + \left(\varphi_2^3 \right) N_2^3$$

$$\varphi_2^2 = \varphi_1^3 = \psi_3 \tag{10.4.2}$$

for the given figure below. Here φ_j^e are the nodal total heads in terms of local node numbers and Ψ_j ($j = 1..4$) are the nodal total heads in terms of global node numbers.

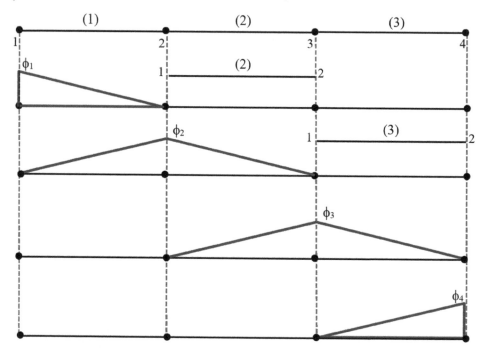

Figure 10.4-1 Three-element mesh from two-noded bar elements and element shape functions

Combining the shape functions lead to the following Figure 10.4-2. Note that we are considering a linear variation of $\varphi(z)$ over an element such that,

$$\varphi(z) = a_1 z + a_2 \tag{10.4.3}$$

$$\varphi(z_i) = a_1 z_i + a_2 = \varphi_1^e \tag{10.4.4a}$$

$$\varphi\left(z_{i+1}\right)=a_1 z_{i+1}+a_2=\varphi_2^e \tag{10.4.4b}$$

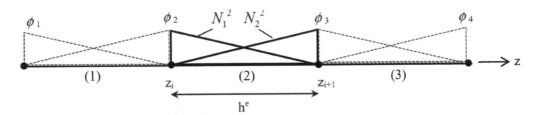

Figure 10.4-2 Linear bar element shape functions

Solving (10.4.4) for the coefficients yields,

$$a_1=\frac{\varphi_2^e-\varphi_1^e}{z_{i+1}-z_i} \tag{10.4.5a}$$

$$a_2=\frac{\varphi_2^e z_{i+1}-\varphi_1^e z_i}{z_{i+1}-z_i} \tag{10.4.5b}$$

and substituting these in $\varphi(z)$ gives,

$$\varphi\left(z\right)=N_1^e\varphi_1^e+N_2^e\varphi_2^e \tag{10.4.6}$$

where

$$N_1^e=\frac{z_2-z}{h^e} \tag{10.4.7a}$$

$$N_2^e=\frac{z-z_1}{h^e} \tag{10.4.7b}$$

Later in the formulation of FEM, first derivatives of shape functions with respect to spatial coordinates will be needed; therefore,

$$\frac{dN_1}{dz}=\frac{-1}{h^e} \tag{10.4.8a}$$

$$\frac{dN_2}{dz}=\frac{1}{h^e} \tag{10.4.8b}$$

As we note in the above, φ is first written in terms of the basis ('approximating', 'trial', later 'shape functions' will all be used interchangeably) functions as,

$$\varphi=\sum_{j=1}^{n}a_j\phi_j\left(z\right) \tag{10.4.9}$$

Then, we express φ in terms of the shape functions defined for individual elements,

$$\varphi=\sum_{e=1}^{n}\varphi_j^e N_j^e \tag{10.4.10}$$

that are used in the weak form of the governing equation written as,

$$\int_0^H \phi_i \left[\frac{d}{dz}(K \frac{d\varphi}{dz}) + S \right] dz = 0, \ i = 1, 2, \cdots n \tag{10.4.11}$$

$$\int_0^H \frac{d\phi_i}{dz} K \frac{d\varphi}{dz} dz - \int_0^H S\phi_i dz - \left[\phi_i K \frac{d\varphi}{dz} \right]_0^H = 0 \tag{10.4.12}$$

Finally, substituting $\varphi = \sum_{e=1}^n \varphi_j^e N_j^e$ in (10.4.12), we get the strong form as,

$$\sum_{e=1}^n \left(\int_{\Omega^e} \frac{dN_i}{dz} K^e \frac{dN_j}{dz} dz)\varphi_j^e - \int_{\Omega^e} s^e N_i dz - N_i K^e \frac{d\varphi}{dz} \Big|_{z_i}^{z_{i+1}} \right) = 0 \tag{10.4.13}$$

Element matrix and vectors then become:

$$\sum_{e=1}^M K_{ij}^e \varphi_j^e = \sum_{e=1}^M F_i^e \tag{10.4.14}$$

$$K_{ij}^e = K^e \int_{z_i}^{z_{i+1}} \frac{dN_i}{dz} \frac{dN_j}{dz} dz, \ i = 1, 2, j = 1, 2 \tag{10.4.15}$$

$$K_{11}^e = K^e \int_{z_i}^{z_{i+1}} \frac{dN_1}{dz} \frac{dN_1}{dz} dz = K^e \int_{z_i}^{z_{i+1}} (-\frac{1}{h^e})(-\frac{1}{h^e}) dz = \frac{K^e}{h^e} \tag{10.4.16}$$

$$K_{12}^e = K^e \int_{z_i}^{z_{i+1}} \frac{dN_1}{dz} \frac{dN_2}{dz} dz = K^e \int_{z_i}^{z_{i+1}} (-\frac{1}{h^e}) \frac{1}{h^e} dz = -\frac{K^e}{h^e} \tag{10.4.17}$$

$$K_{21}^e = K^e \int_{z_i}^{z_{i+1}} \frac{dN_2}{dz} \frac{dN_1}{dz} dz = K^e \int_{z_i}^{z_{i+1}} \frac{1}{h^e} (-\frac{1}{h^e}) dz = -\frac{K^e}{h^e} \tag{10.4.18}$$

$$K_{22}^e = K^e \int_{z_i}^{z_{i+1}} \frac{dN_2}{dz} \frac{dN_2}{dz} dz = K^e \int_{z_i}^{z_{i+1}} \frac{1}{h^e} \frac{1}{h^e} dz = \frac{K^e}{h^e} \tag{10.4.19}$$

Element flow matrix is:

$$\left[K^e \right] = \frac{K^e}{h^e} \begin{bmatrix} 1 & -1 \\ -1 & 1 \end{bmatrix} \tag{10.4.20}$$

with the element DOF,

$$\varphi^e = \begin{Bmatrix} \varphi_1^e \\ \varphi_2^e \end{Bmatrix} \tag{10.4.21}$$

and the right hand side (RHS) force vectors are,

$$F_i^e = S^e \int_{z_i}^{z_{i+1}} N_i dz + q_i^e, \ i = 1, 2 \tag{10.4.22}$$

$$F_1^e = S^e \int_{z_1}^{z_2} N_1 dz + q_1^e = \frac{S^e h^e}{2} + q_1^e \tag{10.4.23}$$

$$F_2^e = S^e \int_{z_2}^{z_3} N_2 dz + q_2^e = \frac{S^e h^e}{2} + q_2^e \tag{10.4.24}$$

$$\{F^e\} = \frac{S^e h^e}{2} \begin{Bmatrix} 1 \\ 1 \end{Bmatrix} + \begin{Bmatrix} q_1^e \\ q_2^e \end{Bmatrix} \tag{10.4.25}$$

Final stiffness equation is then:

$$\left[K^e\right]\{\varphi^e\} = \{F^e\} \tag{10.4.26}$$

Example 10.4-1: Develop the FE formulation for the 1-D flow problem.

Solution:

We will go about the solution of this exercise in a manner that will summarize the steps of developing a complete FE solution to a given 1-D problem.

Steps in FEM:

(1) Formulate the governing equation and boundary conditions for the given problem. See Figure 10.4-3 below.

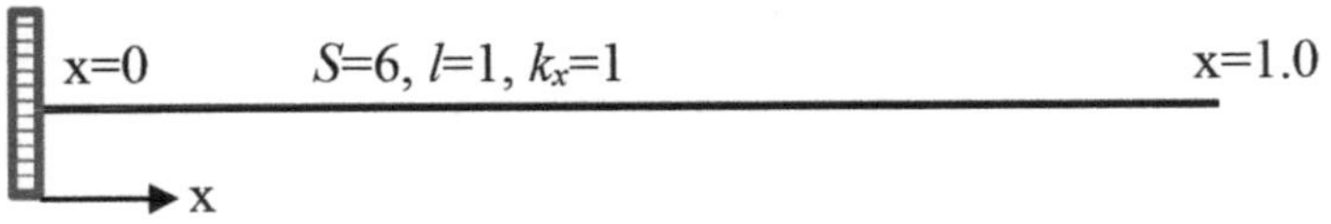

x=0   S=6, *l*=1, k_x=1   x=1.0

➔ x

Figure 10.4-3 Bar through which flow takes place

Governing differential equation and the boundary conditions in terms of DOF (*h*) is given below, notice the change in the head variable:

$$-\frac{d}{dx}\left(k_x \frac{dh}{dx}\right) - S = 0 \qquad 0 \le x \le 1$$

$$k_x \frac{dh}{dx} = q_o = 0 \qquad @\, x = 0$$

$$h = h_o = 0 \qquad @\, x = 1$$

(2) Choose the element type and discretize the domain into a number of those elements as in Figure 10.4-4 below.

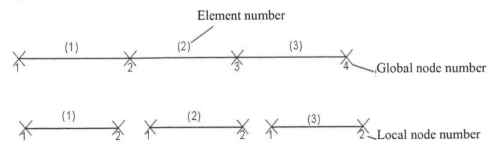

Figure 10.4-4 Two-noded bar elements

where H is the global DOF and h is the local DOF. Connecting DOFs are then,

$$h_1^1 = H_1;$$

$$h_2^1 = h_1^2 = H_2;$$

$$h_2^2 = h_1^3 = H_3; \; h_2^3 = H_4$$

(3) Use the GMWR develop the local element stiffness equation (omitting the internal steps),

$$\int_\Omega \phi_i R_\Omega(h^*) = 0 \rightarrow k^e = \frac{k_x}{l^e}\begin{bmatrix} 1 & -1 \\ -1 & 1 \end{bmatrix} = \frac{1}{\frac{1}{3}}\begin{bmatrix} 1 & -1 \\ -1 & 1 \end{bmatrix}$$

$$h^e = \begin{Bmatrix} h_1^e \\ h_2^e \end{Bmatrix}$$

$$f^e = \frac{S^e l^e}{2}\begin{Bmatrix} 1 \\ 1 \end{Bmatrix} + \begin{Bmatrix} q_1^e \\ q_2^e \end{Bmatrix} = \frac{6\frac{1}{3}}{2}\begin{Bmatrix} 1 \\ 1 \end{Bmatrix} + \begin{Bmatrix} q_1^e \\ q_2^e \end{Bmatrix} = \begin{Bmatrix} 1 \\ 1 \end{Bmatrix} + \begin{Bmatrix} q_1^e \\ q_2^e \end{Bmatrix}$$

(4) Assemble the element matrices and vectors to get the global system equations

$$k^e = \frac{k_x}{l^e}\begin{bmatrix} 1 & -1 \\ -1 & 1 \end{bmatrix} = \frac{1}{1/3}\begin{bmatrix} 1 & -1 \\ -1 & 1 \end{bmatrix} = 3\begin{bmatrix} 1 & -1 \\ -1 & 1 \end{bmatrix}$$

$$\sum_{e=1}^{3} k^e h^e = f^e$$

$$[K]\{H\} = \{F\}$$

Connectivity matrix: $b_{ij} = \begin{bmatrix} 1 & 2 \\ 2 & 3 \\ 3 & 4 \end{bmatrix}$

$$\begin{bmatrix} k_{11}^1 & k_{12}^1 & 0 & 0 \\ k_{21}^1 & k_{22}^1 + k_{11}^2 & k_{12}^2 & 0 \\ 0 & k_{21}^2 & k_{22}^2 + k_{11}^3 & k_{12}^3 \\ 0 & 0 & k_{21}^3 & k_{22}^3 \end{bmatrix}\begin{Bmatrix} H_1 \\ H_2 \\ H_3 \\ H_4 \end{Bmatrix} = \begin{Bmatrix} F_1 \\ F_2 \\ F_3 \\ F_4 \end{Bmatrix}$$

$$\begin{bmatrix} 3 & -3 & 0 & 0 \\ -3 & 6 & -3 & 0 \\ 0 & -3 & 6 & -3 \\ 0 & 0 & -3 & 3 \end{bmatrix}\begin{Bmatrix} H_1 \\ H_2 \\ H_3 \\ H_4 \end{Bmatrix} = \begin{Bmatrix} 1 \\ 2 \\ 2 \\ 1 \end{Bmatrix}$$

(5) Modify the global system equation (obtained in Step 4) applying the boundary conditions in terms of eliminating the necessary rows and columns,

Natural boundary condition @$x = 0$, $k\dfrac{dh}{dx} = q_o = 0$

Essential boundary condition @$x = 1$, $h = h_o = 0$

$$\begin{bmatrix} 3 & -3 & 0 \\ -3 & 6 & -3 \\ 0 & -3 & 6 \end{bmatrix} \begin{Bmatrix} H_1 \\ H_2 \\ H_3 \end{Bmatrix} = \begin{Bmatrix} 1 \\ 2 \\ 2 \end{Bmatrix}$$

(6) Solve the final reduced system of equations considering a certain accuracy

$$\{H\} = [K]^{-1}\{F\} = \{3.0000 \quad 2.6667 \quad 1.6667 \quad 0\}$$

Computer Implementation

The formulation presented in the previous section is implemented into a simple MATLAB® program, "**MYFEM_1D.m**". All of the steps are included in the same program. This program is then used to analyze the example problem presented above. All the data (both input and the results) are clearly identified in the program. See Figure 10.4-5 below.

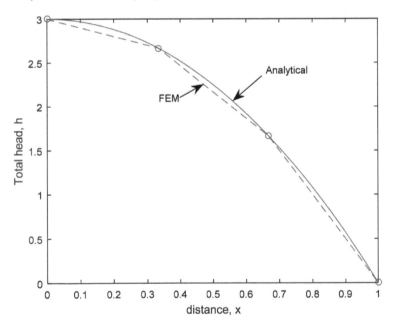

Figure 10.4-5 Solution to 1-D steady state seepage equation

10.4.2 General Second Order Differential Equation

The FE formulation for the equation governing the 1-D flow problem can be readily extended to a general second order differential equation in the form of,

$$a\frac{d^2u}{dx^2} + b\frac{du}{dx} + cu = f(x), \ 0 \le x \le L \tag{10.4.27}$$

Given that the boundary conditions are defined,

$$u(0) = u_0 \tag{10.4.28a}$$

$$u(L) = u_L \tag{10.4.28b}$$

with proper definition of parameters a, b, c and f in the above, this equation becomes the governing equation for several other physical problems as will be illustrated subsequently. Applying the weighted residual method to (10.4.27) gives,

$$\int_0^L \phi_i \left(a\frac{d^2u^*}{dx^2} + b\frac{du^*}{dx} + cu^* - f \right) dx = 0, \ i = 1, 2, \cdots n \tag{10.4.29}$$

Substituting for an approximate $u^* = \sum_{e=1}^{M} u_j^e N_j^e$, e being the element number in (10.4.29) and

taking $\phi_i \approx N_i^e$, we get

$$\sum_{e=1}^{M} K_{ij}^e u_j^e = \sum_{e=1}^{M} F_i^e \tag{10.4.30}$$

where

$$K_{ij}^e = -a\int_{x_i}^{x_{i+1}} \frac{dN_i}{dz}\frac{dN_j}{dz} dx + b\int_{x_i}^{x_{i+1}} N_i \frac{dN_j}{dz} dx + c\int_{x_i}^{x_{i+1}} N_i N_j dx \tag{10.4.31a}$$

$$F_i^e = \int_{x_i}^{x_{i+1}} f \cdot N_i dx + q_i^e \tag{10.4.31b}$$

In the above form, if we use the linear shape functions,

$$N_1 = \frac{x_{i+1} - x}{h_e} \tag{10.4.32a}$$

$$N_2 = \frac{x - x_i}{h_e} \tag{10.4.32b}$$

The element matrices in (10.4.29) can be written in the following matrix form:

$$\left[K^e\right] = \int_{x_i}^{x_{i+1}} \left(-a\begin{bmatrix} N_1' \\ N_2' \end{bmatrix}\begin{bmatrix} N_1' & N_2' \end{bmatrix} + b\begin{bmatrix} N_1 \\ N_2 \end{bmatrix}\begin{bmatrix} N_1' & N_2' \end{bmatrix} + c\begin{bmatrix} N_1 \\ N_2 \end{bmatrix}\begin{bmatrix} N_1 & N_2 \end{bmatrix} \right) dx \tag{10.4.33a}$$

$$\left\{F^e\right\} = \int_{x_i}^{x_{i+1}} f \cdot \begin{bmatrix} N_1 \\ N_2 \end{bmatrix} dx + \left\{q_i^e\right\} \tag{10.4.33b}$$

If we use linear shape functions in (10.4.33), the element matrices can be readily evaluated by integration,

$$\left[K^e\right] = -\frac{a}{h^e}\begin{bmatrix} 1 & -1 \\ -1 & 1 \end{bmatrix} + \frac{b^e}{2}\begin{bmatrix} 1 & -1 \\ -1 & 1 \end{bmatrix} + \frac{c^e h^e}{6}\begin{bmatrix} 1 & -1 \\ -1 & 1 \end{bmatrix} \tag{10.4.34}$$

and for $f(x) = f^e$ the force vector takes the below form,

$$\{F^e\} = \frac{f^e h^e}{2}\begin{Bmatrix}1\\1\end{Bmatrix} + \begin{Bmatrix}q_1^e\\q_2^e\end{Bmatrix}$$

(10.4.35)

Computer Implementation

Figure 10.4-6 shows a comparison of solution of a typical boundary value problem defined in (10.4.27) using ten two-noded bar elements for a linearly varying source term, *f(x)*. See MATLAB program "**FEM_1D_SS.m**".

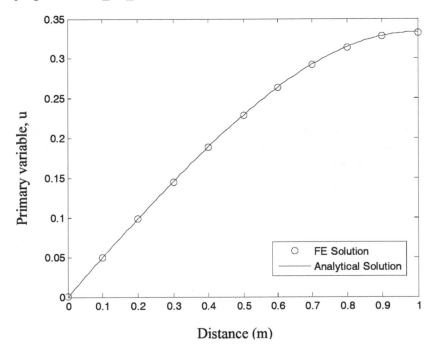

Figure 10.4-6 Solution of a general 2nd order differential equation using finite elements

10.4.3 General 1-D Time Dependent Problem

A general differential equation governing time dependent variation of a field variable can be written in the form given as:

$$\frac{d}{dx}\left(a\frac{du}{dx}\right) + b\frac{du}{dx} + cu = d\frac{df}{dt} + e\frac{du}{dt}$$

(10.4.36)

The prescribed boundary conditions are:

$$u(0) = u_0$$ (10.4.37a)

$$u(L) = u_L$$ (10.4.37b)

$$a\left(\frac{du}{dx}\right)_{x=L} = Q_L$$ (10.4.37c)

Following the same steps of the formulation presented in the previous section, general FE stiffness equation can be readily obtained as:

$$\sum_{e=1}^{M}\left(C^e\dot{u}^e + K^e u^e\right) = \sum_{e=1}^{M}\left(\dot{f}^e + Q^e\right) \tag{10.4.38}$$

where

$$C_{ij}^e = -e^e \int_{x_A}^{x_B} N_i^e N_j^e dx \tag{10.4.39}$$

$$K_{ij}^e = -a^e \int_{x_A}^{x_B} \frac{dN_i^e}{dx}\frac{dN_j^e}{dx} dx + b^e \int_{x_A}^{x_B} N_i^e \frac{dN_j^e}{dx} dx + c^e \int_{x_A}^{x_B} N_i^e N_j^e dx \tag{10.4.40}$$

$$\dot{f}_i^e = d^e f^e \int_{x_A}^{x_B} N_i^e dx \tag{10.4.41}$$

$$Q_i^e = \sum_{i=1}^{n} N_i(x_j^e)Q_j^e \tag{10.4.42}$$

It is strongly recommended to the readers to derive the above form of (10.4.38) by themselves as a matter of exercise. Using now a two-noded linear element, the element matrices and vectors can be obtained as,

$$\left[K^e\right] = -\frac{a^e}{h_e}\begin{bmatrix} 1 & -1 \\ -1 & 1 \end{bmatrix} + \frac{b^e}{2}\begin{bmatrix} -1 & 1 \\ -1 & 1 \end{bmatrix} + \frac{c^e h_e}{6}\begin{bmatrix} 2 & 1 \\ 1 & 2 \end{bmatrix} \tag{10.4.43}$$

$$\left[C^e\right] = -\frac{c^e h_e}{6}\begin{bmatrix} 2 & 1 \\ 1 & 2 \end{bmatrix} \tag{10.4.44}$$

$$\{F^e\} = \frac{d^e h_e}{2}\begin{Bmatrix} 1 \\ 1 \end{Bmatrix} \tag{10.4.45}$$

$$\{Q^e\} = \begin{Bmatrix} Q_1^e \\ Q_2^e \end{Bmatrix} \tag{10.4.46}$$

After assembling the element equations, the global system of equations for the entire system written for $t = t^*$ is,

$$[C]\{\dot{u}^*\} + [K]\{u^*\} = \dot{f}^*\{F\} + \{Q\} \tag{10.4.47}$$

Let,

$$u^* = \theta\{u\}_{n+1} + (1-\theta)\{u\}_n \tag{10.4.48}$$

$$\dot{u}^* = \frac{\{u\}_{n+1} - \{u\}_n}{\Delta t_{n+1}} \tag{10.4.49}$$

$$\dot{f}^* = \frac{f_{n+1} - f_n}{\Delta t_{n+1}} = \theta\dot{f}_{n+1} + (1-\theta)\dot{f}_n \tag{10.4.50}$$

An example variation of displacement and force with time can be seen in Figure 10.4-7. Here θ is the controlling parameter of time stepping. For $\theta = 0$, the numerical scheme is called the forward difference scheme, for $\theta = \frac{1}{2}$ it is called "Crank-Nicholson", for $\theta = \frac{2}{3}$ it is the "Galerkin Method" and $\theta = 1$ is called the "Backward Difference" scheme. The reader will surely recall the Finite Difference Method that was presented in Chapter 9.

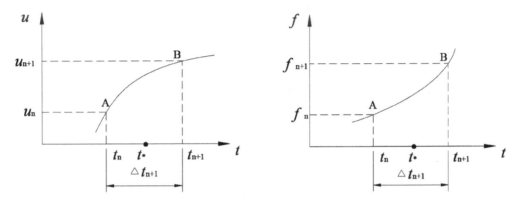

Figure 10.4-7 Variation of unknown values in time

Combining equations (10.4.47)-(10.4.50), we have:

$$[C]\frac{\{u\}_{n+1} - \{u\}_n}{\Delta t_{n+1}} + [K]\left(\theta\{u\}_{n+1} + (1-\theta)\{u\}_n\right) = \frac{f_{n+1} - f_n}{\Delta t_{n+1}}\{F\} + \{Q\} \qquad (10.4.51)$$

leading to,

$$\{[C] + \theta[K](\Delta t_{n+1})\}\{u\}_{n+1} = \{[C] - (1-\theta)[K](\Delta t_{n+1})\}\{u\}_n + f_{n+1}\{F\} - f_n\{F\} + (\Delta t_{n+1})\{Q\}$$
$$(10.4.52)$$

and thus,

$$\left[\hat{K}\right]\{u\}_{n+1} = \{\hat{R}\} \qquad (10.4.53)$$

where

$$\left[\hat{K}\right] = [C] + \theta\Delta t_{n+1}[K] \qquad (10.4.54)$$

$$\{\hat{R}\} = \{\hat{R}_n\} + \{\hat{R}_{n+1f}\} + \{\hat{R}_{nf}\} \qquad (10.4.55)$$

with

$$\{\hat{R}_n\} = \{[C] - (1-\theta)[K]\Delta t_{n+1}\}\{u\}_n \qquad (10.4.56a)$$

$$\{\hat{R}_{n+1f}\} = f_{n+1}\{F\} \qquad (10.4.56b)$$

$$\{\hat{R}_{nf}\} = -f_n\{F\} + \Delta t_{n+1}\{Q\} \qquad (10.4.56c)$$

The above equation can be solved for $\{u\}_{n+1}$ (the unknown primary variable at time t_{n+1}) if $\{u\}_n$ (the values at previous time, t_n) are known. The stability with the limited conditions, the solution in time domain can be obtained by a successive solution of (10.4.54) for each time step.

Computer Implementation

Figure 10.4-8 shows the absolute value variation of a time-dependent variable with time and depth considering Eq. (10.4.36). The formulation discussed in this section is implemented in the MATLAB program "**FEM_1D_TR.m**". The main program and the associated subprograms are provided at the end of this chapter.

10.4.4 1-D Contaminant Migration

Figure 10.4-9 presents the problem of a simple 1-D contaminant migration in a homogeneous layer of soil with a constant source of concentration, c_o at the surface. We recall the governing equation for contaminant transport from Chapter 7,

$$nD\frac{\partial^2 c}{\partial z^2} - nv\frac{\partial c}{\partial z} = (n + \rho K_d)\frac{\partial c}{\partial t} + n\lambda c \qquad (10.4.57)$$

If we re-write the general 1-D equation for which the FE formulation has already been developed along with the related MATLAB program which has been written to solve the following general second order differential equation, we have,

$$a\frac{\partial^2 u}{\partial z^2} + b\frac{\partial u}{\partial z} + cu = d\frac{\partial f}{\partial t} + e\frac{\partial u}{\partial t} \qquad (10.4.58)$$

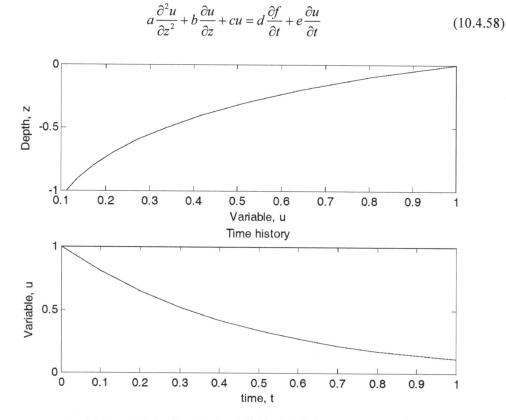

Figure 10.4-8 Variation of time-dependent variable in time and depth

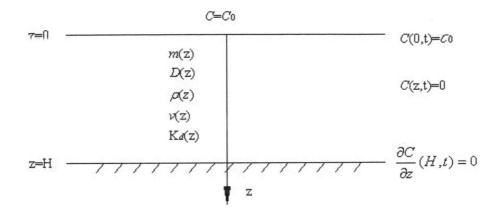

$$C=C_0$$

$z=0$ ——————————————————— $C(0,t)=C_0$

$m(z)$
$D(z)$
$\rho(z)$
$v(z)$
$K_d(z)$

$C(z,t)=0$

$z=H$ —/—/—/—/—/—/—/—/—/—/— $\dfrac{\partial C}{\partial z}(H,t)=0$

z

Figure 10.4-9 1-D transport of contaminants

Now all we need to do is to make necessary substitutions for the coefficients of all the terms which reduces the equation into 1-D form of contaminant migration. Thus, we should do the following substitutions: $a = nD$, $u = c$, $x = z$, $b = -nv$, $c = -n\lambda$, $d = 0$, $e = n + \rho K_d$.

Computer Implementation

See the FE program called "**FEM_1D_TR.m**" for the above-described formulation of the 1-D contaminant migration. Figure 10.4-10 is obtained using this program.

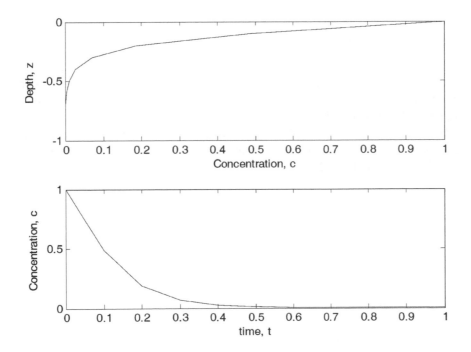

Figure 10.4-10 Variation of concentration of contaminants in time and depth (per Figure 10.4-9), $n = 0.4$, $K_d = 1.2$, $v = 0.002$, $D = 0.02$, $\lambda = 100$

10.4.5 1-D Consolidation

Figure 10.4-8 presents Terzaghi's 1-D consolidation problem of a homogenous layer of fine grained soil subjected to a surface loading, $f(t)$. The general governing equation of 1-D consolidation is recalled:

$$-\frac{k_z}{\rho_w g}\frac{\partial^2 p_w}{\partial z^2} = m_v\left[\frac{\partial f}{\partial t}-\frac{\partial p_w}{\partial t}\right] \qquad (10.4.59)$$

where p_w is the only field variable, pore water pressure. As the consolidation is a quasi-static process requiring the field variable to be a function of time, there has to be an initial condition to begin the solution in time. This is defined as:

$$p_w(z,0)=0 \qquad (10.4.60)$$

Boundary conditions associated with the problem are:

$$p_w(z,t)=0 \qquad (10.4.61a)$$

$$\frac{\partial p_w}{\partial t}(H,L)=0 \qquad (10.4.61b)$$

Eq. (10.4.59) can be modeled by a general 1-D second order differential equation if we make the following substitution for the coefficients a, b, c and d into (10.4.36) such that,

$$b=c=0,\ x=z,\ a=-\frac{k_z}{\rho_w g},\ d=m_v,\ e=-m_v \qquad (10.4.62)$$

Computer Implementation

Figure 10.4-12 presents the pore pressure variation with time and depth as obtained for a column of soil under constant step load seen in Figure 10.4-11 for various locations and times. The finite element program written in MATLAB, "**FEM_Consol1D.m**" is used to solve 1-D consolidation and compared with the analytical solution as derived in Chapter 7. Surely, other more sophisticated numerical formulations can also be implemented for multi-dimensional consolidation problem (see Zienkiewicz et al. 1980; Zienkiewicz et al. 1999; Ulker and Rahman, 2009; Ulker et al. 2012).

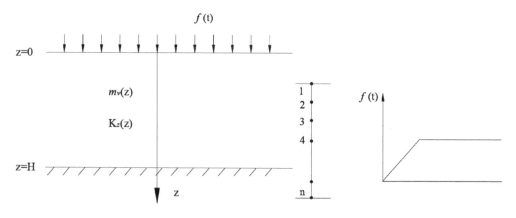

Figure 10.4-11 1-D consolidation of a soil column under step load and its FE model

10.4.6 Principle of Virtual Work

We have seen the FE formulation using the GMWR. For a stress-deformation problem, we often use an alternative formulation using the "principle of virtual work" (PVW). This principle simply states that *"for any kinematically admissible infinitesimal virtual displacement (δx), at equilibrium, the work done by internal forces (Internal Virtual Work, IVW) equals the work done by external forces (External Virtual Work, EVW) on the same virtual displacement such that"*:

$$IVW=EVW \qquad\qquad (10.4.63)$$

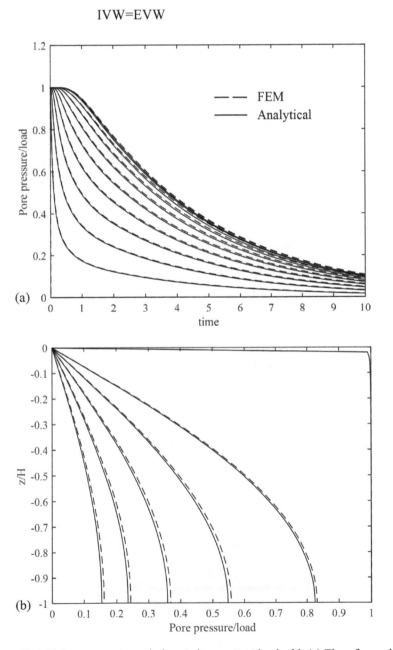

Figure 10.4-12 Pore pressure variation under constant load with (a) Time for various depths, (b) Depth for various times

Here kinematic admissibility requires that: (i) the elements must not violate compatibility, and (ii) boundary conditions have to be satisfied. For the virtual displacement to be kinematically admissible, $x + \delta x$ should be a kinematically admissible displacement as well. While a rigid body only translates and rotates, a deformable body experiences shape and dimension changes. If we choose virtual displacements that are not just rigid body displacements, the virtual work done will not be zero. For example, if we consider the spring of length, L, given below (Figure 10.4-13), and apply a small elongation, dx, to the spring, the force F will do work,

$$dW = Fdx \qquad\qquad (10.\,4.64)$$

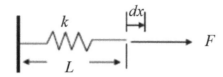

Figure 10.4-13 Linear spring under stretch

This work is not zero since by stretching the spring, its internal energy, E_p, (also called the strain energy or the potential energy of the spring) is increased. In this case the IVW done on the δx, the virtual displacement of the spring, will be,

$$IVW = T(\delta x) \qquad\qquad (10.4.65)$$

where T is the internal force that develops in the spring. EVW is also written as,

$$EVW = F(\delta x) \qquad\qquad (10.4.66)$$

Equalizing the two relations and adding the force-displacement relation of,

$$T = k(dx) \qquad\qquad (10.4.67)$$

yields,

$$(\delta x)F = (\delta x)k(dx) \qquad\qquad (10.4.68)$$

where k is the spring constant. Equation (10.4.68) yields the equilibrium relation of,

$$F = k(dx) \qquad\qquad (10.4.69)$$

For any small change in δx in the FE configuration, we need to find dx that satisfies the PVW. For multi-dimensions, terms in (10.4.68) become tensors,

$$\{\delta x\}^T \{F\} = \{\delta x\}^T [K]\{dx\} \qquad\qquad (10.4.70)$$

Bar Element

Example 10.4-2: Write the PVW for the bar seen in Figure 10.4-14 below.

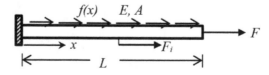

Figure 10.4-14 Bar under body and surface forces

Solution:

$$\text{EVW} = F|_{(x=L)}\left(\delta x\right)|_{(x=L)} + F_i\left(\delta x\right)|_{(x=x_i)} + \int_0^L f(x)(\delta x)\,dx \tag{10.4.71}$$

$$\text{IVW} = \int_0^L \left[\sigma(x)A(x)\right]\left(\delta\varepsilon(x)\right)dx \tag{10.4.72}$$

where $\delta\varepsilon(x)$ is the virtual strain. Since IVW = EVW we have,

$$\int_0^L \delta\varepsilon^T \left[\sigma(x)A(x)\right]dx = \int_0^L \delta x^T f(x)\,dx + \delta x^T|_{(x=x_i)} F_i + \delta x^T|_{(x=L)} F_L \tag{10.4.73}$$

Again for multi-dimensions, this relation can be written as,

$$\int_\Omega \{\delta\varepsilon\}^T \{\sigma\}\,d\Omega = \int_\Gamma \{\delta\phi\}^T \{\Phi\}\,d\Gamma + \int_\Omega \{\delta\phi\}^T \{\Upsilon\}\,d\Omega \tag{10.4.74}$$

where the left hand side (LHS) represents the internal response of the system as the IVW and the first and the second terms of the RHS are the surface tractions and the body forces, respectively. Now recalling the FE discretization, the virtual displacements and strains are discretized the same way as the admissible displacements and strains are. Thus, using the Bubnov-Galerkin approach we can write,

$$\{\delta\phi\} = [N]\{\delta\phi_e\} \tag{10.4.75a}$$

$$\{\delta\varepsilon\} = [B]\{\delta\phi_e\} \tag{10.4.75b}$$

where again $\underset{\sim}{B} = \underset{\sim}{\partial}\underset{\sim}{N}$ is the strain-displacement matrix with $\underset{\sim}{\partial}$ being the gradient matrix. Substituting (10.4.75a) and (10.4.75b) into (10.4.74) results in,

$$\int_\Omega \{\delta\phi_e\}^T [B]^T \{\sigma\}\,d\Omega = \int_\Gamma \{\delta\phi_e\}^T [N]^T \{\Phi\}\,d\Gamma + \int_\Omega \{\delta\phi_e\}^T [N]^T \{\Upsilon\}\,d\Omega \tag{10.4.76}$$

We can also approximate the stresses using the Hooke's law as,

$$\underset{\sim}{\sigma} = \underset{\sim}{E}\underset{\sim}{B}\underset{\sim}{\phi}_e \tag{10.4.77}$$

where $\underset{\sim}{E}$ is the material constitutive matrix, and use it in (10.4.76) getting,

$$\{\delta\phi_e\}^T \left(\int_\Omega [B]^T [E][B]\,d\Omega\right)\{\phi_e\} = \{\delta\phi_e\}^T \int_\Gamma [N]^T \{\Phi\}\,d\Gamma + \{\delta\phi_e\}^T \int_\Omega [N]^T \{\Upsilon\}\,d\Omega \tag{10.4.78}$$

which is another form of (10.4.68). Here $\{\delta\varphi_e\}^T$ and $\{\varphi_e\}$ are kept outside the integral since they are independent of coordinate axes. In (10.4.78), the integral term on the right is called the "stiffness matrix",

$$[K] = \int_\Omega [B]^T [E][B]\,d\Omega \tag{10.4.79}$$

and is a function of material behavior. It is simply possible to derive the "stiffness relation" from (10.4.78) as,

$$\underset{\sim}{K}\underset{\sim}{D} = \underset{\sim}{F} \tag{10.4.80}$$

where $\underset{\sim}{K}$ is the global stiffness matrix obtained from the assembly of $\underset{\sim}{k}$, $\underset{\sim}{D}$ is the global DOF vector and $\underset{\sim}{F}$ is the global force vector including both the surface and the body forces. Equation (10.4.80) is the stiffness relation of the system under consideration and applicable to the approximate solution of many engineering systems in many fields ranging from solid, soil and fluid mechanics to biomechanics.

E and A being the material properties, the stiffness relation for the bar element seen in Figure (10.4-11) is derived considering the equilibrium of forces in the x-direction and writing from the Hooke's law that $\sigma = E\varepsilon$ with $\varepsilon = \dfrac{u_2 - u_1}{L_e}$. Therefore,

$$\underset{\sim}{k}\underset{\sim}{u} = \underset{\sim}{f} = k \begin{bmatrix} 1 & -1 \\ -1 & 1 \end{bmatrix} = \begin{Bmatrix} f_1 \\ f_2 \end{Bmatrix} \tag{10.4.81}$$

where $\underset{\sim}{f}$ is the force vector and $k = \dfrac{EA}{L_e}$. In order to derive the stiffness matrix $\underset{\sim}{k}$, the shape function matrix, $\underset{\sim}{N}$ is first obtained for the given geometry of a single element in the local coordinates for,

$$\underset{\sim}{X} = \begin{bmatrix} 1 & 0 \\ 1 & L_e \end{bmatrix} \tag{10.4.82a}$$

as,

$$\underset{\sim}{N} = \begin{bmatrix} \dfrac{L_e - x}{L_e} & \dfrac{x}{L_e} \end{bmatrix} \tag{10.4.82b}$$

The stiffness matrix is then calculated using the following relation, the derivation of which is through the PVW given earlier. Therefore,

$$\underset{\sim}{k} = \int_{\Omega} \underset{\sim}{B}^T \underset{\sim}{E} \underset{\sim}{B} d\Omega = \int \begin{pmatrix} \dfrac{-1}{L_e} \\ \dfrac{1}{L_e} \end{pmatrix} E \left(\dfrac{-1}{L_e} \ \dfrac{1}{L_e} \right) A dx = \dfrac{EA}{L_e} \begin{bmatrix} 1 & -1 \\ -1 & 1 \end{bmatrix} \tag{10.4.83}$$

10.4.7 Coupled Flow and Deformation: 1-D Quasi-Static Solution

Terzaghi's 1-D consolidation theory is the basis for understanding general settlement behavior of soil under applied loads. While the drainage is allowed in the soil, under external load, pore water leaves the medium causing settlement in the soil as a result of deformation of the soil skeleton (composed of solid grains). This is already discussed earlier. However, Terzaghi's theory is one dimensional and evaluates only the pore pressures. Biot (1941) developed a more general mathematical formulation of this process which accounts for both the displacement and the pore pressures in multi-dimensions. He then extended the theory to include dynamics (Biot,

1955, 1962). In this section, the FE formulation of the quasi-static (QS) form of Biot's theory is presented in its 1-D form using the PVW. Thus, this section will be a practical example of the use of the PVW within the context of soil mechanics.

The FE QS formulation of the governing equations where the inertial terms associated with the relative fluid acceleration as well as the solid part motion are neglected, is presented in this section. The final two equations which are resultant, are given as,

$$\frac{\partial \sigma}{\partial z} + \rho g = 0 \tag{10.4.84}$$

$$\frac{\partial \dot{u}}{\partial z} + \frac{\partial}{\partial z}\left[\frac{k}{\rho_f g}\left(-\frac{\partial p_w}{\partial z} + \rho_f g\right)\right] + \frac{n}{K_f}\dot{p}_w = 0 \tag{10.4.85}$$

Weak Form

The weak form of the above equations is presented. If we pre-multiply the equations by virtual displacement δ_u and pressure δ_{pw}, respectively, we get,

$$\int_\Omega \delta u^T \nabla \sigma d\Omega + \int_\Omega \delta u^T \rho g d\Omega = 0 \tag{10.4.86}$$

$$\int_\Omega \delta p_w^{\ T} \nabla \dot{u} d\Omega + \int_\Omega \delta p_w^{\ T} \nabla^T \left[\frac{k}{\rho_f g}\left(-\nabla p_w + \rho_f g\right)\right] d\Omega + \int_\Omega \delta p_w^{\ T} \frac{n}{K_f}\dot{p}_w d\Omega = 0 \tag{10.4.87}$$

By doing integration by parts and substituting the constitutive relation for effective stress to (10.4.86),

$$\int_\Omega (\nabla \delta u)^T \left(\underset{\approx}{D}\nabla u\right) d\Omega - \int_\Omega (\nabla \delta u)^T (p_w) d\Omega = \int_\Gamma \delta u^T \sigma d\Gamma + \int_\Omega \delta u^T \rho g d\Omega \tag{10.4.88}$$

$$\int_\Omega \delta p_w^{\ T} \nabla \dot{u} d\Omega + \int_\Omega (\nabla \delta p_w)^T \frac{[k]}{\rho_f g}(\nabla p_w) d\Omega + \int_\Omega \delta p_w^{\ T} \frac{n}{K_f}\dot{p}_w d\Omega$$

$$= \int_\Gamma \delta p_w^{\ T} \frac{[k]}{\rho_f g}\nabla p_w d\Gamma + \int_\Omega (\nabla \delta p_w)^T \frac{[k]}{\rho_f g}(\rho_f g) d\Omega \tag{10.4.89}$$

Finite Element Approximation

Approximating the field variables u and p_w in the regular FE sense yields,

$$\left(\int_\Omega (\nabla N_u)^T [D](\nabla N_u) d\Omega\right) U - \left(\int_\Omega (\nabla N_u)^T \{m\} N_p d\Omega\right) P_w = \int_\Gamma N_u^{\ T} \sigma d\Gamma + \int_\Omega N_u^{\ T} \rho g d\Omega \tag{10.4.90}$$

$$\left(\int_\Omega Np^T \{m\}^T \nabla Nu d\Omega\right) \dot{U} + \left(\int_\Omega (\nabla Np)^T \frac{[k]}{\rho_f g}\nabla Np d\Omega\right) P_w + \left(\int_\Omega Np^T \frac{n}{K_f}Np d\Omega\right) \dot{P}_w$$

$$= \int_\Gamma (Np)^T \frac{[k]}{\rho_f g}(\nabla p) d\Gamma + \int_\Omega (Np)^T \frac{[k]}{\rho_f g}(\rho_f g) d\Omega \tag{10.4.91}$$

These equations can also be written as,

$$\mathbf{K}_S\{U\} - \mathbf{C}\{P_w\} = \{F_s\} \tag{10.4.92a}$$

$$\mathbf{C}^T\{\dot{U}\} + \mathbf{K}_f\{P_w\} + \mathbf{C}_f\{\dot{P}_w\} = \{F_f\} \tag{10.4.92b}$$

or in matrix form as,

$$\begin{bmatrix} 0 & 0 \\ \mathbf{C}^\mathbf{T} & \mathbf{C}_\mathbf{f} \end{bmatrix}\begin{Bmatrix} \dot{\mathbf{U}} \\ \dot{\mathbf{P}}_\mathbf{w} \end{Bmatrix} + \begin{bmatrix} \mathbf{K}_\mathbf{s} & -\mathbf{C} \\ 0 & \mathbf{K}_\mathbf{f} \end{bmatrix}\begin{Bmatrix} \mathbf{U} \\ \mathbf{P}_\mathbf{w} \end{Bmatrix} = \begin{Bmatrix} F_s \\ F_f \end{Bmatrix} \tag{10.4.93}$$

where the sub-matrices are evaluated as,

$$\mathbf{K}_\mathbf{s} = \int_\Omega [B_u]^T \underset{\approx}{D}[B_u]d\Omega \tag{10.4.94a}$$

$$\mathbf{K}_\mathbf{f} = \int_\Omega [B_p]^T \frac{[k]}{\rho_f g}[B_p]d\Omega \tag{10.4.94b}$$

$$\mathbf{C} = \int_\Omega [B_u]^T \{m\}[N_p]d\Omega \tag{10.4.94c}$$

$$\mathbf{C}_\mathbf{f} = \int_\Omega [N_p]^T \frac{n}{K_f}[N_p]d\Omega \tag{10.4.94d}$$

$$\mathbf{M}_\mathbf{s} = \int_\Omega [N_u]^T \rho[N_u]d\Omega \tag{10.4.94e}$$

$$\mathbf{M}_\mathbf{sf} = \int_\Omega [B_p]^T \frac{[k]}{g}[N_u]d\Omega \tag{10.4.94f}$$

$$F_s = \int_\Gamma [N_u]^T \{\sigma\}d\Gamma + \int_\Omega N_u^T \rho g d\Omega \tag{10.4.94g}$$

$$F_f = \int_\Gamma [N_p]^T \left(\frac{[k]}{\rho_f g}\right)(n^T p)d\Gamma + \int_\Omega [N_p]^T \frac{[k]}{\rho_f g}(\rho_f g)d\Omega \tag{10.4.94h}$$

Eight node quadrilateral shape functions (Q8) are used for quadratic approximation of the displacement DOF and four node bilinear shape functions (Q4) are used for pressure DOF. Gaussian 2×2 integration rule is chosen for the numerical integration of both fields. More information regarding the element types and their formulations are given subsequently in this chapter.

This formulation is reduced to the numerical solution of consolidation problem when the compressibility matrix is taken as zero under a very large K_f modulus. In the undrained case, $[\mathbf{K}_f]$ matrix becomes zero. If we also neglect the compressibility of water then $[\mathbf{C}_f]$ also becomes zero and for steady state, the resulting equations are,

$$\begin{bmatrix} \mathbf{K}_\mathbf{s} & -\mathbf{C} \\ \mathbf{C}^\mathbf{T} & 0 \end{bmatrix}\begin{Bmatrix} \mathbf{U} \\ \mathbf{P}_\mathbf{w} \end{Bmatrix} = \begin{Bmatrix} \mathbf{F}_\mathbf{s} \\ \mathbf{F}_\mathbf{f} \end{Bmatrix} \tag{10.4.95}$$

The above FE formulation has only a unique solution when the number of the terms defining u is greater than the number of p_w variables which is generally used in *incompressible elasticity* problems. For instance, If we use bilinear shape functions in discretizing p_w, then we need to use quadratic shape functions in discretizing u as is also the case in this chapter. It should also be noted here that the solution of Eq. (10.4.93) requires one of the readily available time-integration methods (i.e. Newmark method) as there are several time derivative terms in the equation. However, such solution techniques are outside the scope of this book and are left for another discussion on dynamic analysis.

Computer Implementation

The above formulation is implemented into a MATLAB program called "**FE_Coupled_1D_QS.m**". Results are compared with the analytical solutions given in Ch. 7 (Figure 10.4-15).

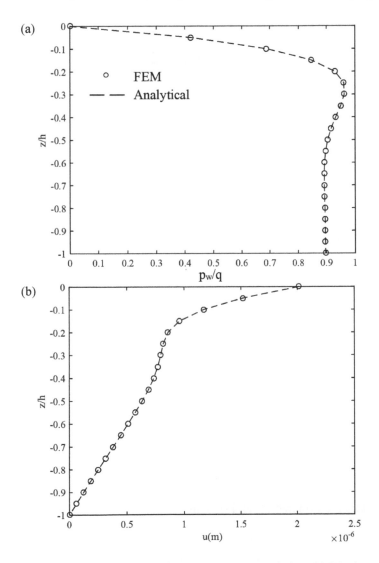

Figure 10.4-15 Absolute value of (a) Pore pressure variation, (b) Displacement distribution with depth under periodic load

10.4.8 Approximation of Field Variables: General Considerations

Recall the shape functions, $N_i(x)$ defining the approximate variation of the field over an element presented earlier as:

$$u = \sum_{e=1}^{n_{el}} u_j^e N_j^e \tag{10.4.96}$$

where n_{el} is the number of members which we shall, from now on, call "elements, e" through which the approximation is performed for the field variables (or the DOF) using the "nodal values", u_j^e, defined for each element. Thus, $u_j^e = \{u_i \; u_{i+1}\}$ is the DOF vector, 'i' being the nodal points. The inter-element approximation is made possible by using the *shape functions*. Besides, the choice of shape functions can now be done more freely using a variety of higher order polynomial functions. Therefore, for the simplest bar element (Figure 10.4-16) if we redefine the linear functions (see Figure 10.4-17) we get,

$$N_1 = \frac{x_{i+1} - x}{h_e}, \quad N_2 = \frac{x - x_i}{h_e} \tag{10.4.97}$$

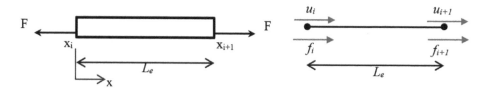

Figure 10.4-16 Bar element and its FE model

For a two-noded bar element we can write,

$$u = a_0 + a_1 x = \begin{bmatrix} 1 & x \end{bmatrix} \begin{pmatrix} a_0 \\ a_1 \end{pmatrix} = \underset{\sim}{P} \underset{\sim}{a} \tag{10.4.98}$$

as the generalized DOF linear relation. For the coordinates, $x = x_1$ and $x = x_2$, the nodal DOF vector takes the form,

$$\underbrace{\begin{Bmatrix} u_1 \\ u_2 \end{Bmatrix}}_{\underset{\sim}{u}} = \underbrace{\begin{bmatrix} 1 & x_1 \\ 1 & x_2 \end{bmatrix}}_{\underset{\sim}{X}} \underbrace{\begin{pmatrix} a_0 \\ a_1 \end{pmatrix}}_{\underset{\sim}{a}} \tag{10.4.99}$$

where $\underset{\sim}{X}$ is the coordinates matrix. Then taking the inverse of (10.4.99) results in,

$$\begin{pmatrix} a_0 \\ a_1 \end{pmatrix} = \frac{1}{x_2 - x_1} \begin{bmatrix} x_2 & -x_1 \\ -1 & 1 \end{bmatrix} \begin{Bmatrix} u_1 \\ u_2 \end{Bmatrix} \tag{10.4.100}$$

Substituting this relation into (10.4.98) yields,

$$u = \begin{bmatrix} \dfrac{x_2 - x}{x_2 - x_1} & \dfrac{x - x_1}{x_2 - x_1} \end{bmatrix} \begin{Bmatrix} u_1 \\ u_2 \end{Bmatrix} \tag{10.4.101}$$

where

$$\underset{\sim}{N} = \begin{bmatrix} N_1 & N_2 \end{bmatrix} = \begin{bmatrix} \dfrac{x_2 - x}{x_2 - x_1} & \dfrac{x - x_1}{x_2 - x_1} \end{bmatrix} \tag{10.4.102}$$

is the shape function matrix which is plotted in Figure 10.4-17. For some situations we might like to use a higher order approximation for the variation of the field variable. For a 1-D problem if we use a three-noded bar element, the shape functions should now be quadratic in nature and the generalized DOF is obtained accordingly,

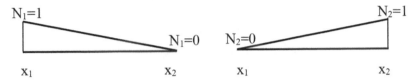

$N_1=1$ $N_1=0$ $N_2=0$ $N_2=1$

x_1 x_2 x_1 x_2

Figure 10.4-17 Linear shape functions of the bar element

$$u = a_0 + a_1 x + a_2 x^2 = \begin{bmatrix} 1 & x & x^2 \end{bmatrix} \begin{pmatrix} a_0 \\ a_1 \\ a_2 \end{pmatrix} \tag{10.4.103}$$

When the nodal coordinates are used in this relation, it becomes,

$$\underbrace{\begin{Bmatrix} u_1 \\ u_2 \\ u_3 \end{Bmatrix}}_{u} = \underbrace{\begin{bmatrix} 1 & x_1 & x_1^2 \\ 1 & x_2 & x_2^2 \\ 1 & x_3 & x_3^2 \end{bmatrix}}_{\underset{\sim}{X}} \underbrace{\begin{Bmatrix} a_0 \\ a_1 \\ a_2 \end{Bmatrix}}_{\underset{\sim}{a}} \tag{10.4.104}$$

Following the same steps, we can readily obtain the shape functions for the three-noded bar,

$$\underset{\sim}{N} = \begin{bmatrix} N_1 & N_2 & N_3 \end{bmatrix} = \begin{bmatrix} \dfrac{(x_2 - x)(x_3 - x)}{(x_2 - x_1)(x_3 - x_1)} & \dfrac{(x_1 - x)(x_3 - x)}{(x_1 - x_2)(x_3 - x_1)} & \dfrac{(x_1 - x)(x_2 - x)}{(x_1 - x_3)(x_2 - x_3)} \end{bmatrix} \tag{10.4.105}$$

Figure 10.4-18 shows these quadratic shape functions. Hence generally speaking, the following relation holds for all types of elements,

$$\underset{\sim}{\phi} = \underset{\sim}{N} \underset{\sim}{\phi}_e \tag{10.4.106}$$

where ϕ is the approximation of the field variables and ϕ_e is the vector of DOF at the nodal points. The formulations of other basic finite elements are presented in the subsequent sections.

10.4.9 Assembly of Elements

The assembly process has been mentioned earlier without any details. This important step of FEM is handled by combining the stiffness relations of individual elements in terms of using the DOF and the nodal forces of the connecting nodes. Such a process, within the context of 1-D bar elements is presented. For three bar elements (Figure 10.4-19), the assembly process is made possible only by intra-element continuity that requires,

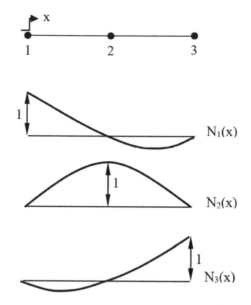

Figure 10.4-18 Quadratic shape functions of the bar element

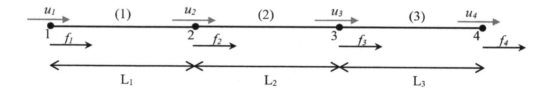

Figure 10.4-19 Three-element mesh from two noded bar elements

$$u_1^{(1)} = u_1; \; u_2^{(2)} = u_2^{(1)}; \; u_2^{(2)} = u_1^{(3)}; \; u_2^{(3)} = u_4 \tag{10.4.107}$$

where the superscripts show the element number and the subscripts show the DOF number. If we write the element stiffness matrix for each element we get,

$$k^i = \int_\Omega \underset{\sim}{B}^T \underset{\sim}{E} \underset{\sim}{B} d\Omega = \int \begin{pmatrix} \dfrac{-1}{L_e} \\ \dfrac{1}{L_e} \end{pmatrix} E \left(\dfrac{-1}{L_e} \; \dfrac{1}{L_e} \right) A dx = \left(\dfrac{EA}{L} \right)^i \begin{bmatrix} 1 & -1 \\ -1 & 1 \end{bmatrix} = \begin{bmatrix} k_{11}^i & k_{12}^i \\ k_{21}^i & k_{22}^i \end{bmatrix}, \; i = 1,2,3 \tag{10.4.108}$$

which are assembled to yield the "global stiffness matrix",

$$[K]_{4\times 4} = \begin{bmatrix} k_{11}^1 & k_{12}^1 & 0 & 0 \\ k_{21}^1 & k_{22}^1 + k_{11}^2 & k_{12}^2 & 0 \\ 0 & k_{21}^2 & k_{22}^2 + k_{11}^3 & k_{12}^3 \\ 0 & 0 & k_{21}^3 & k_{22}^3 \end{bmatrix} \tag{10.4.109}$$

where the diagonal '*jj*' elements have contributions from both elements that are sharing the node. One thing that is noticed from the global stiffness matrix is that it is sparse and symmetric. It should also satisfy the following relation to be positive and definite as well,

$$\underset{\sim}{x}^T \underset{\sim}{K} \underset{\sim}{x} > 0 \tag{10.4.110}$$

A positive definite stiffness matrix is a symmetric matrix $[K]_{n \times n}$ for which all eigenvalues are positive. One way to tell if a matrix is definitely positive is to check that all its pivots are positive. Since $[K]_{n \times n}$ will be a symmetric $n \times n$ matrix with n being the total number of DOF and the LHS of (10.4.110) is its associated quadratic form, we want to make sure (10.4.110) is satisfied for every $x \neq 0$. In order to have a unique solution to,

$$\underset{\sim}{K} \underset{\sim}{D} = \underset{\sim}{F} \tag{10.4.111}$$

We need to also make sure that $[K]_{n \times n}$ is invertible (or non-singular), so that zero is not an eigenvalue. Similar to K, RHS force vector is also assembled similarly,

$$[F]_{4 \times 1} = \begin{Bmatrix} f_1^{(1)} \\ f_2^{(1)} + f_1^{(2)} \\ f_2^{(2)} + f_1^{(3)} \\ f_2^{(3)} \end{Bmatrix} \tag{10.4.112}$$

10.4.10 Conditions on Shape Functions

An important question that comes into one's mind regarding the FE analysis is: How does the accuracy of the approximate solution depends on our choice of approximation functions? Accuracy is satisfied by convergence, which in turn, is satisfied by using more terms in the polynomial series used to represent the shape functions. In a mesh, we simply add more nodes and elements into the mesh that replaces the original structure. This way the sequence of trial solutions must approach the exact solution in terms of displacements, stresses and strain energies. Simply put, in order to achieve convergence of an FE analysis, the choice of shape functions is an important task. One needs to know how to select the most appropriate interpolation functions based upon certain criteria. The necessary ones are:

- Continuity (intra-element)
- Compatibility (inter-element)
- Completeness

which can be called "*the 3C's for convergence*" as a requirement to satisfy convergence. The deformations in the structural system are gathered from the values of the displacements at the nodes and the shape functions within elements. Within each element, *compatibility of deformation* is assured by the choice of simple polynomial functions for interpolation allowing *continuous representation of the displacement* field. However, this does not ensure that the displacements are compatible between element edges. In order to avoid gaps or overlaps in the process of mesh generation, special care is required. The specification of continuity depends on how strains are defined in terms of derivatives of the displacement fields. A physical problem can be described by the static condition $\delta f = 0$, where $f = f(\varphi)$ is the functional. If f contains derivatives of the field variable φ to the order m, then it requires that within each element, φ, which is the approximate field chosen as a trial function, must contain a complete polynomial of degree m so that φ is continuous within elements and the completeness requirements are also met. The continuity of φ must be maintained across element boundaries which requires that the trial function φ and its derivatives through order m-1 must be continuous across element edges. In most natural formulations in solid mechanics and soil mechanics, strains are defined by first derivatives of the displacement fields. In such cases, a simple continuity of the displacement field across element edges suffices. This is called "C^0-continuity". Compatibility between adjacent elements for problems which require only C^0-continuity can be easily assured if the displacements along the side of an element depend only on the displacement field specified at the nodes placed on that edge. As for *completeness*, the FE mesh needs to represent constant virtual displacements (δu) (or real displacement, u) as well as constant virtual and real strains, $\delta \varepsilon$ and ε, respectively within an element.

10.5 FEM: 2-D Problems

In this section, the FE formulation for two common 2-D problems are presented. These problems consist of a steady state seepage problem and a stress analysis problem presented in the subsequent sections. The FE formulations presented in this section are mainly the extensions of the GMWR formulation used for the 1-D problems presented earlier.

10.5.1 Steady State Seepage: Scalar Field Variable

Physical Problem and Governing Equations

As for the application of 2-D FEM, let us first consider a general 2-D PDE of the following form:

$$\frac{\partial}{\partial x}\left(a_{11} \frac{\partial u}{\partial x} + a_{12} \frac{\partial u}{\partial y} \right) + \frac{\partial}{\partial y}\left(a_{21} \frac{\partial u}{\partial x} + a_{22} \frac{\partial u}{\partial y} \right) + a_{00}u + f = 0 \tag{10.5.1}$$

With proper identification and definition of various parameters and the field variables, the above equation governs various problems ranging from heat transfer to groundwater flow, from electrostatics to magneto-statics etc. Here, a steady state seepage problem is considered (Figure 10.5-1).

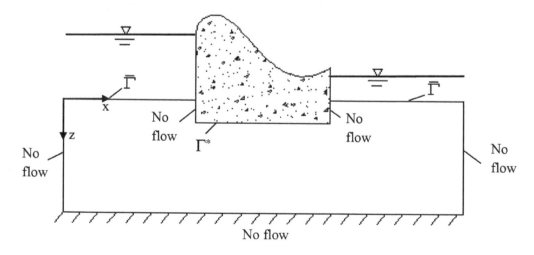

Figure 10.5-1 Seepage through a soil layer under a concrete dam

For this problem we rewrite the equation in Ch. 9 (9.5.1) in a slightly modified fashion as:

$$-\frac{\partial}{\partial x}\left(-k_x\frac{\partial\varphi}{\partial x}\right) - \frac{\partial}{\partial z}\left(k_z\frac{\partial\varphi}{\partial z}\right) + l\varphi + f - Q \tag{10.5.2}$$

with the boundary conditions:

$$\varphi = \bar{\varphi} \text{ on } \bar{\Gamma} \tag{10.5.3a}$$

$$-K\frac{\partial\varphi}{\partial n} = \bar{q} \text{ on } \bar{\Gamma} \tag{10.5.3b}$$

where φ is the total head, $\bar{\varphi}$ is the specified total head, k_x and k_z are the coefficients of permeability in the x and z directions, respectively, $\bar{q}$ is the specified total flux, l is the source term coefficient and f is the source or the sink term. Equations (10.5.2) and (10.5.3) govern the steady state seepage of groundwater through the foundation soils beneath a concrete dam. The flow of water is governed by the D'Arcy's law in terms of total flux in both directions as,

$$q_x = -k_x\frac{\partial\varphi}{\partial x} \tag{10.5.4a}$$

$$q_z = -k_z\frac{\partial\varphi}{\partial z} \tag{10.5.4b}$$

Application of Method of Weighted Residual

Using (10.3.4) again by incorporating the 2-D steady state flow gives,

$$\int_{\Omega^e} w\left(-\frac{\partial q_x}{\partial x} - \frac{\partial q_z}{\partial z} + l\varphi + f\right)dxdz = 0 \tag{10.5.5}$$

while noting that,

$$-w\frac{\partial}{\partial x}q_x = -\frac{\partial}{\partial x}(wq_x) + \frac{\partial w}{\partial x}q_x \tag{10.5.6a}$$

$$-w\frac{\partial}{\partial z}q_z = -\frac{\partial}{\partial z}(wq_z) + \frac{\partial w}{\partial z}q_z \tag{10.5.6b}$$

If we apply the divergence theorem, we get,

$$\int_{\Omega^e}\frac{\partial}{\partial x}(wq_x)dxdz = \oint_{\Gamma^e} wq_x n_x ds \tag{10.5.7a}$$

$$\int_{\Omega^e}\frac{\partial}{\partial z}(wq_z)dxdz = \oint_{\Gamma^e} wq_z n_z ds \tag{10.5.7b}$$

Using (10.5.6) and (10.5.7) in (10.5.5) gives,

$$\int_{\Omega^e}\left[\frac{\partial w}{\partial x}\left(k_x\frac{\partial\varphi}{\partial x}\right) + \frac{\partial w}{\partial z}\left(k_z\frac{\partial\varphi}{\partial z}\right) + lw\varphi + wf\right]dxdz - \oint_{\Gamma^e} w\left(k_x\frac{\partial\varphi}{\partial x}n_x + k_z\frac{\partial\varphi}{\partial z}n_z\right)ds = 0$$

$$\tag{10.5.8}$$

Finite Element Formulation

Following the same approach used in section 10.4, the approximation for the field variable φ is:

$$\varphi(x,z) \approx \varphi^e(x,z) = \sum_{j=1}^{n} \phi_j^e(x,z)\varphi_j^e \tag{10.5.9}$$

Substituting (10.5.9) into (10.5.8) we write,

$$\int_{\Omega^e}\left[\frac{\partial w}{\partial x}\left(k_x\sum_{j=1}^{n}\frac{\partial \phi_j^e}{\partial x}\varphi_j^e\right) + \frac{\partial w}{\partial z}\left(k_z\sum_{j=1}^{n}\frac{\partial \phi_j^e}{\partial z}\varphi_j^e\right) - lw\sum_{j=1}^{n}\phi_j^e\varphi_j^e - wf\right]dxdz - \oint_{\Gamma^e} wq_n ds = 0 \tag{10.5.10}$$

For the weighting function, we choose $w = \varphi_1, \varphi_2, \dots, \varphi_n$. Thus, Eq. (10.5.10) becomes,

$$\sum_{j=1}^{n}\left\{\int_{\Omega^e}\left[\frac{\partial \phi_j^e}{\partial x}\left(k_x\frac{\partial \phi_j^e}{\partial x}\right) + \frac{\partial \phi_j^e}{\partial x}\left(k_x\frac{\partial \phi_j^e}{\partial x}\right) + l\phi_i^e\phi_j^e\right]dxdz\right\}\varphi_j^e - \int_{\Omega^e}f\phi_i^e dxdz - \oint_{\Gamma^e}\phi_i^e q_n ds = 0 \tag{10.5.11}$$

for $i = 1, 2,\dots, n$. The above equation comes down to,

$$\sum_{j=1}^{n}K_{ij}^e\varphi_j^e = f_i^e + Q_i^e \tag{10.5.12}$$

where

$$K_{ij}^e = \int_{\Omega^e}\left[\frac{\partial \phi_j^e}{\partial x}\left(k_x\frac{\partial \phi_j^e}{\partial x}\right) + \frac{\partial \phi_j^e}{\partial z}\left(k_z\frac{\partial \phi_j^e}{\partial z}\right) + l\phi_i \phi_j\right]dxdz \tag{10.5.13a}$$

$$f_i^e = \int_{\Omega^e}f\psi_i^e dxdz \tag{10.5.13b}$$

$$Q_i^e = \oint_{\Gamma^e}\psi_i^e q_n ds \tag{10.5.13c}$$

The above equation in matrix notation can be written as:

$$\left[K_{ij}^e\right]\{\varphi^e\} = \{f^e\} + \{Q^e\} \tag{10.5.14}$$

As was done in getting Eq. (10.4.30) for the global stiffness matrix, taking $\phi_i \approx N_i$ through the GMWR,

$$K_{ij}^e = \int_{\Omega^e}\left(\frac{\partial N_i}{\partial x}k_x\frac{\partial N_j}{\partial x} + \frac{\partial N_i}{\partial z}k_z\frac{\partial N_j}{\partial z} + lN_iN_j\right)dxdz \tag{10.5.15}$$

Considering k_x, k_s and l to remain constant within an element,

$$K_{ij}^e = k_x\left[S_{ij}^{xx}\right] + k_z\left[S_{ij}^{zz}\right] + l\left[S_{ij}^0\right] \tag{10.5.16}$$

where

$$\left[S_{ij}^{xx}\right] = \int_{\Omega^e}\frac{\partial N_i}{\partial x}\frac{\partial N_j}{\partial x}dxdz \tag{10.5.17a}$$

$$\left[S_{ij}^{zz}\right] = \int_{\Omega^e}\frac{\partial N_i}{\partial z}\frac{\partial N_j}{\partial z}dxdz \tag{10.5.17b}$$

$$\left[S_{ij}^0 \right] = \int_{\Omega^e} l N_i N_j \, dx dz \tag{10.5.17c}$$

Let us now evaluate the element matrices. Eq. (10.5.17a) is evaluated as,

$$\left[S_{ij}^{xx} \right] = \int_0^a \int_0^b \frac{\partial N_i}{\partial x} \frac{\partial N_j}{\partial x} \, dx dz \tag{10.5.18}$$

by using,

$$N_1 = \left(1 - \frac{x}{a} \right) \left(1 - \frac{z}{b} \right) \tag{10.5.19}$$

$$\frac{\partial N_1}{\partial x} = -\frac{1}{a} \left(1 - \frac{z}{b} \right) \tag{10.5.20}$$

See the other bilinear shape functions used here at Eq. 10.6.26. So, the first entry of the matrix becomes,

$$S_{11}^{xx} = \int_0^a dx \int_0^b \frac{1}{a^2} \left(1 - \frac{z}{b} \right)^2 dz = \frac{b}{3a} \tag{10.5.21}$$

Similarly other elements of $\left[S_{ij}^{xx} \right]$ can be evaluated along with other element matrices and the vectors f_i^e. Finally, we can write these element matrices and vectors in the following form,

$$\left[S_{ij}^{xx} \right] = \frac{b}{6a} \begin{bmatrix} 2 & -2 & -1 & 1 \\ -2 & 2 & 1 & -1 \\ -1 & 1 & 2 & -2 \\ 1 & -1 & -2 & 2 \end{bmatrix} \tag{10.5.22}$$

$$\left[S_{ij}^{zz} \right] = \frac{a}{6b} \begin{bmatrix} 2 & 1 & -1 & -2 \\ 1 & 2 & -2 & -1 \\ -1 & -2 & 2 & 1 \\ -2 & -1 & 1 & 2 \end{bmatrix}. \tag{10.5.23}$$

$$\left[S_{ij}^0 \right] = \frac{ab}{36} \begin{bmatrix} 4 & 2 & 1 & 2 \\ 2 & 4 & 2 & 1 \\ 1 & 2 & 4 & 2 \\ 2 & 1 & 2 & 4 \end{bmatrix} \tag{10.5.24}$$

$$\{f\} = \frac{1}{4} fab \{1 \ \ 1 \ \ 1 \ \ 1\}^T \tag{10.5.25}$$

Computer Implementation

The above FE formulation is implemented in the MATLAB® program "**FEM_2D_SEEP.m**" which is used to analyze the steady state seepage under a concrete dam storing a 20 m high reservoir of water. All the input data is clearly defined in the listing of the program. Figure 10.5-2 shows the FE results in terms of equi-potential lines and flow lines being orthogonal to one another. As the FE mesh is refined, the lines get smoother.

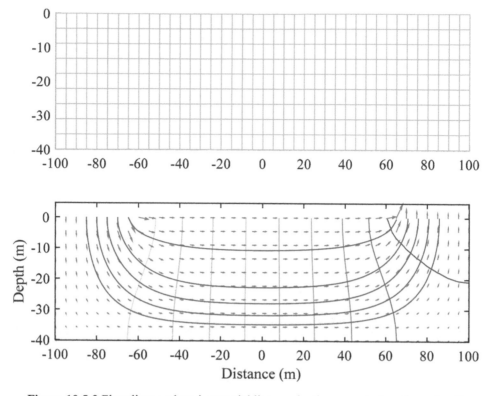

Figure 10.5-2 Flow lines and equi-potential lines under the concrete dam through soil along with the mesh

10.5.2 Stress Analysis: Vector Field Variable

In this section, FE formulation of the 2-D stress-analysis problem is presented. For the situation shown in Figure 10.5-3 and in many other geotechnical problems, we need to evaluate stress and strain variations in the soil so that optimum design procedures can be developed for structures and their foundations.

Physical Problem and Governing Equations

In many geotechnical problems, loading in one direction (i.e. z-direction) to be extended over a long distance is considered. For such situations, assuming a *plane strain* stress state allowing us to work only with the deformation and stress states in a plane (i.e. x-y plane) is plausible. In that case, the following will hold:

$$\varepsilon_{zz} = \gamma_{xz} = \gamma_{yz} = 0 \tag{10.5.26}$$

Constitutive relations are necessary to set a valid material behavior under applied loading. Linear elasticity is one of the simplest cases of such relations. Let us recall the stress-strain relationship,

$$\begin{Bmatrix} \sigma_x \\ \sigma_x \\ \sigma_{xy} \end{Bmatrix} = \begin{bmatrix} c_{11} & c_{12} & 0 \\ c_{21} & c_{22} & 0 \\ 0 & 0 & c_{33} \end{bmatrix} \begin{Bmatrix} \varepsilon_x \\ \varepsilon_y \\ \gamma_{xy} \end{Bmatrix} \tag{10.5.27}$$

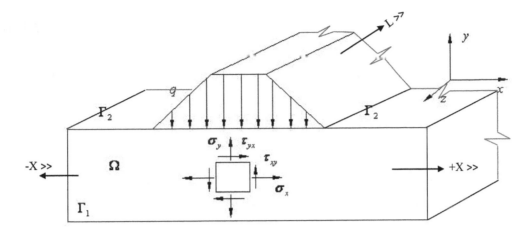

Figure 10.5-3 2-D soil embankment

where

$$c_{11} = c_{22} = \frac{E(1-\mu)}{(1+\mu)(1-2\mu)} \tag{10.5.28a}$$

$$c_{33} = \frac{E}{2(1+\mu)} = G \tag{10.5.28b}$$

$$c_{12} = c_{21} = \frac{E\mu}{(1+\mu)(1-2\mu)} \tag{10.5.28c}$$

Stress-strain relationship helps associate stress state to the deformation state through a linear elastic material behavior. Equations of equilibrium are necessary to define the governing equations of the static response of a solid body under loading. These are given as:

$$\frac{\partial \sigma_x}{\partial x} + \frac{\partial \tau_{yx}}{\partial y} + f_x = 0 \tag{10.5.29a}$$

$$\frac{\partial \sigma_y}{\partial y} + \frac{\partial \tau_{xy}}{\partial x} + f_y = 0 \tag{10.5.29b}$$

The kinematics in terms of "strain-displacement" relations complete the equation set for the response of a deformable soil system. Therefore,

$$\{\varepsilon\} = [\partial]\{\delta\} = \begin{Bmatrix} \varepsilon_x \\ \varepsilon_y \\ \gamma_{xy} \end{Bmatrix} = \begin{bmatrix} \dfrac{\partial}{\partial x} & 0 \\ 0 & \dfrac{\partial}{\partial y} \\ \dfrac{\partial}{\partial y} & \dfrac{\partial}{\partial x} \end{bmatrix} \begin{Bmatrix} u \\ v \end{Bmatrix} \tag{10.5.30}$$

The boundary conditions prescribed over the defined surfaces must satisfy the following:

$$t_x = \sigma_x n_x + \tau_{yx} n_y = \hat{t}_x \tag{10.5.31a}$$

$$t_y = \tau_{xy} n_x + \sigma_y n_y = \hat{t}_y \tag{10.5.31b}$$

as the natural boundary conditions and,

$$u = \hat{u} \tag{10.5.32a}$$

$$v = \hat{v} \tag{10.5.32b}$$

are prescribed as the essential boundary conditions.

Application of Galerkin's Method

Governing PDEs in terms of displacement variables are first written as:

$$\frac{\partial}{\partial x}\left(c_{11}\frac{\partial u}{\partial x} + c_{12}\frac{\partial v}{\partial y}\right) + \frac{\partial}{\partial y}\left(c_{33}\frac{\partial u}{\partial y} + c_{33}\frac{\partial v}{\partial x}\right) + f_x = 0 \tag{10.5.33a}$$

$$\frac{\partial}{\partial x}\left(c_{33}\frac{\partial u}{\partial y} + c_{33}\frac{\partial v}{\partial x}\right) + \frac{\partial}{\partial y}\left(c_{12}\frac{\partial u}{\partial x} + c_{22}\frac{\partial v}{\partial y}\right) + f_y = 0 \tag{10.5.33b}$$

Similarly the boundary conditions can also be expressed in terms of (u, v),

$$\left(c_{11}\frac{\partial u}{\partial x} + c_{12}\frac{\partial u}{\partial y}\right)n_x + \left(c_{33}\frac{\partial u}{\partial y} + c_{33}\frac{\partial v}{\partial x}\right)n_y = \hat{t}_x \tag{10.5.34a}$$

$$\left(c_{33}\frac{\partial u}{\partial y} + c_{33}\frac{\partial v}{\partial x}\right)n_x + \left(c_{12}\frac{\partial u}{\partial x} + c_{22}\frac{\partial v}{\partial y}\right)n_y = \hat{t}_y \tag{10.5.34b}$$

Weak Form

Following the same approach (as we have been using), the weak form of the governing equation can be written as:

$$\int_{\Omega_e}\left[\frac{\partial w_1}{\partial x}\left(c_{11}\frac{\partial u}{\partial x} + c_{12}\frac{\partial v}{\partial y}\right) + \frac{\partial w_1}{\partial y}\left(c_{33}\frac{\partial u}{\partial y} + c_{33}\frac{\partial v}{\partial x}\right) + w_1 f_x\right]dxdy$$

$$-\oint_\Gamma w_1\left[\left(c_{11}\frac{\partial u}{\partial x} + c_{12}\frac{\partial u}{\partial y}\right)n_x + \left(c_{33}\frac{\partial u}{\partial y} + c_{33}\frac{\partial v}{\partial x}\right)n_y\right] = 0 \tag{10.5.35a}$$

$$\int_{\Omega_e}\left[\frac{\partial w_1}{\partial x}\left(c_{33}\frac{\partial u}{\partial y} + c_{33}\frac{\partial v}{\partial x}\right) + \frac{\partial w_1}{\partial y}\left(c_{12}\frac{\partial u}{\partial x} + c_{22}\frac{\partial v}{\partial y}\right) - w_1 f_y\right]dxdy$$

$$-\oint_\Gamma w_1\left[\left(c_{33}\frac{\partial u}{\partial y} + c_{33}\frac{\partial v}{\partial x}\right)n_x + \left(c_{12}\frac{\partial u}{\partial x} + c_{22}\frac{\partial v}{\partial y}\right)n_y\right] = 0 \tag{10.5.35b}$$

FE Formulation

The whole domain Ω is discretized into a set of elements Ω^e, and the primary field variables u and v are approximated as:

$$u = \sum_{j=1}^n u_j^e N_j^e(x, y) \tag{10.5.36}$$

$$v = \sum_{j=1}^{n} v_j^e N_j^e(x, y) \tag{10.5.37}$$

which can be also written as,

$$
\left\{ \begin{matrix} u \\ v \end{matrix} \right\} = \begin{bmatrix} N_1 & N_2 & \cdots & N_n & 0 & 0 & \cdots & 0 \\ 0 & 0 & \cdots & 0 & N_1 & N_2 & \cdots & N_n \end{bmatrix} \begin{Bmatrix} u_1 \\ u_2 \\ \vdots \\ u_n \\ v_1 \\ v_2 \\ \vdots \\ v_n \end{Bmatrix} \tag{10.5.38}
$$

where n will depend on the number of nodes in Ω^e. Equation (10.5.38) can also be written as:

$$\left\{ \begin{matrix} u \\ v \end{matrix} \right\} = [N]\{\delta\} \tag{10.5.39}$$

Using the Galerkin's approach, we can choose:

$$w_1 = N_i^e \tag{10.5.40a}$$

$$w_2 = N_i^e \tag{10.5.40b}$$

with $i = 1, 2, \ldots, n$. Using (10.5.35)-(10.5.40), the following FE equations are formulated:

$$\begin{bmatrix} K^{11} & K^{12} \\ K^{21} & K^{22} \end{bmatrix} \left\{ \begin{matrix} u \\ v \end{matrix} \right\} = \left\{ \begin{matrix} F^1 \\ F^2 \end{matrix} \right\} \tag{10.5.41}$$

where

$$K_{ij}^{11} = \int_{\Omega^e} \left(c_{11} \frac{\partial N_i}{\partial x} \frac{\partial N_j}{\partial x} + c_{33} \frac{\partial N_i}{\partial y} \frac{\partial N_j}{\partial y} \right) dxdy \tag{10.5.42a}$$

$$K_{ij}^{12} = K_{ij}^{21} = \int_{\Omega^e} \left(c_{12} \frac{\partial N_i}{\partial x} \frac{\partial N_j}{\partial y} + c_{33} \frac{\partial N_i}{\partial y} \frac{\partial N_j}{\partial x} \right) dxdy \tag{10.5.42b}$$

$$K_{ij}^{22} = \int_{\Omega^e} \left(c_{33} \frac{\partial N_i}{\partial x} \frac{\partial N_j}{\partial x} + c_{22} \frac{\partial N_i}{\partial y} \frac{\partial N_j}{\partial y} \right) dxdy \tag{10.5.42c}$$

$$F_i^1 = \int_{\Omega^e} N_i f_x dxdy + \oint_{\Gamma^e} N_i t_x ds \tag{10.5.42d}$$

$$F_i^2 = \int_{\Omega^e} N_i f_y dxdy + \oint_{\Gamma^e} N_i t_y ds \tag{10.5.42e}$$

The previous equations can also be presented in the following form:

$$\left[K^e \right] \{\delta^e\} = \{f^e\} + \{Q^e\} \tag{10.5.43}$$

$$[K^e] = \int_{\Omega^e} [B]^T [D][B] \, dx dy \tag{10.5.44}$$

$$\{f^e\} = \int_{\Omega^e} [N] \begin{Bmatrix} f_x \\ f_y \end{Bmatrix} dx dy \tag{10.5.45}$$

$$\{Q^e\} = \oint_{\Gamma^e} [N] \begin{Bmatrix} t_x \\ t_y \end{Bmatrix} ds \tag{10.5.46}$$

$$\{\delta\} = \{u_1 \ u_2 \ldots u_n \ v_1 \ v_2 \ldots v_n\}^T \tag{10.5.47}$$

Computer Implementation

The above FE formulation is also implemented in MATLAB with "**FEM_2D_STRESS.m**". Following problem in Figure 10.5-4 has been solved and the results are given below. Goal is to evaluate the displacements and stresses in the two CST elements the formulation of which is given in the next section for F = 800 N, H = 120 mm, L = 160 mm, E = 3 MPa, v = 0.25. The main program with the functions are presented at the end of the chapter.

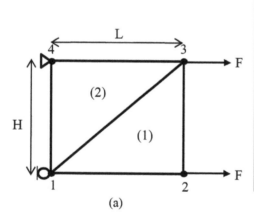

DOF #	u,v (m)	Stress #	σ(Pa)
1	0	σ_{11}	10.2916
2	0	σ_{12}	0.5184
3	4.065×10^{-4}	σ_{13}	0.3888
4	7.069×10^{-5}	σ_{21}	9.7084
5	3.641×10^{-4}	σ_{22}	2.4271
6	-3.888×10^{-5}	σ_{23}	-0.3888
7	0		
8	0		

(a)   (b)

Figure 10.5-4 2-D plane stress problem, σ_{11} = horizontal stress of element 1, σ_{22} = vertical stress of element 2, σ_{13} = shear stress of element 1

10.6 Basic Element Formulations

In order to be able to solve the problems presented in the previous two sections, we need to formulate basic 1-D and 2-D finite elements using approximations of DOF in various degrees such as in sections 10.4 and 10.5. Then they can be used in a mesh generated to discretize the domain of interest provided the elements are well-behaved and satisfy the convergence requirements as well as they pass the "patch test". The patch test takes a patch of elements and makes sure that the elements exhibit constant stress condition across the patch under statically determinate boundary conditions. In this section, we present a number of common 2-D elements that pass the patch test available in classical FE literature and their mathematical formulations.

10.6.1 Various 2-D Elements

In classical FEM, there are many 2-D elements used to discretize the domain of interest. Their master element formulations are presented in this section. Among those elements, the constant strain triangle (CST), linear strain triangle (LST), the 4-noded quadrilateral (Q4) and the eight-noded quadrilateral (Q8) are presented here.

Constant Strain Triangle (CST)

As it can be seen in Figure 10.6-1, CST is the simplest of the 2-D elements. It has three nodes with two DOF at each node in two directions, hence six DOF per element. The DOF approximation is as follows,

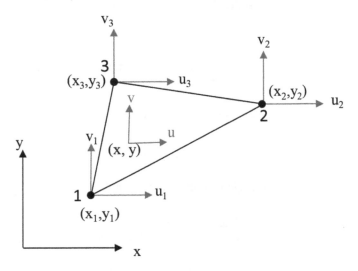

Figure 10.6-1 CST element

$$\phi = a_0 + a_1 x + a_2 y \tag{10.6.1a}$$

or

$$\phi = \begin{Bmatrix} u \\ v \end{Bmatrix} = \begin{Bmatrix} a_0 + a_1 x + a_2 y \\ a_3 + a_4 x + a_5 y \end{Bmatrix} \tag{10.6.1b}$$

leading to,

$$\begin{Bmatrix} \phi_1 \\ \phi_2 \\ \phi_3 \end{Bmatrix} = \begin{bmatrix} 1 & x_1 & y_1 \\ 1 & x_2 & y_2 \\ 1 & x_3 & y_3 \end{bmatrix} \begin{Bmatrix} a_0 \\ a_1 \\ a_2 \end{Bmatrix} \tag{10.6.2}$$

Then the $\underset{\sim}{N}$ becomes,

$$\underset{\sim}{N} = \underset{\sim}{P} \underset{\sim e}{P}^{-1} = \begin{bmatrix} 1 & x & y \end{bmatrix} \begin{bmatrix} 1 & x_1 & y_1 \\ 1 & x_2 & y_2 \\ 1 & x_3 & y_3 \end{bmatrix}^{-1} \tag{10.6.3}$$

which in the actual formulation is written as,

$$\underset{\sim}{N} = \begin{bmatrix} N_1 & 0 & N_2 & 0 & N_3 & 0 \\ 0 & N_1 & 0 & N_2 & 0 & N_3 \end{bmatrix} \tag{10.6.4}$$

The general formulae for the shape functions are then,

$$N_1 = \frac{a_1 + b_1 x + c_1 y}{2A} \tag{10.6.5a}$$

$$N_2 = \frac{a_2 + b_2 x + c_2 y}{2A} \tag{10.6.5b}$$

$$N_3 = \frac{a_3 + b_3 x + c_3 y}{2A} \tag{10.6.5c}$$

where A is the area of the triangle, $A = \dfrac{1}{2} \det [\, P_e \,] = \dfrac{1}{2} \det \begin{bmatrix} 1 & x_1 & y_1 \\ 1 & x_2 & y_2 \\ 1 & x_3 & y_3 \end{bmatrix}$ and

$$\begin{bmatrix} a_1 = x_2 y_3 - x_3 y_2 & b_1 = y_2 - y_3 & c_1 = x_3 - x_2 \\ a_2 = x_3 y_1 - x_1 y_3 & b_2 = y_3 - y_1 & c_2 = x_1 - x_3 \\ a_3 = x_1 y_2 - x_2 y_1 & b_3 = y_1 - y_2 & c_3 = x_2 - x_1 \end{bmatrix} \tag{10.6.6}$$

which for a special case of the triangle resting on the ground (i.e. $x_1 = y_1 = y_2 = 0$) yields,

$$\underset{\sim}{N} = \begin{bmatrix} N_1 & N_2 & N_3 \end{bmatrix} = \left[\left(1 - \frac{x}{x_1} + y\frac{(x_3 - x_2)}{x_2 y_3} \right) \quad \left(\frac{x}{x_2} - \frac{y x_3}{x_2 y_3} \right) \quad \left(\frac{y}{y_3} \right) \right] \tag{10.6.7}$$

As it can be seen the shape functions of the CST are linear functions of x and y (Figure 10.6-2). The approximation of the strains is such that,

$$\{\phi\}_{2\times 1} = [N]_{2\times 6} \{\phi_e\}_{6\times 1} = \begin{Bmatrix} u(x,y) \\ v(x,y) \end{Bmatrix} = \begin{bmatrix} N_1 & 0 & N_2 & 0 & N_3 & 0 \\ 0 & N_1 & 0 & N_2 & 0 & N_3 \end{bmatrix} \begin{Bmatrix} u_1 \\ v_1 \\ u_2 \\ v_2 \\ u_3 \\ v_3 \end{Bmatrix} \tag{10.6.8}$$

which is used to approximate the strains defined as,

$$\underset{\sim}{\varepsilon} = \begin{Bmatrix} \varepsilon_x \\ \varepsilon_y \\ \gamma_{xy} \end{Bmatrix} = \begin{Bmatrix} \dfrac{\partial u}{\partial x} \\ \dfrac{\partial v}{\partial y} \\ \dfrac{\partial u}{\partial y} + \dfrac{\partial v}{\partial x} \end{Bmatrix} = \begin{bmatrix} \partial/\partial x & 0 \\ 0 & \partial/\partial y \\ \partial/\partial y & \partial/\partial x \end{bmatrix} \begin{Bmatrix} u \\ v \end{Bmatrix} \tag{10.6.9}$$

such that,

$$
\left\{ \begin{array}{c} \varepsilon_x \\ \varepsilon_y \\ \gamma_{xy} \end{array} \right\} = \left[\partial \right]_{3\times2} \left[N \right]_{2\times6} \left\{ \phi_e \right\}_{6\times1} = \left[\begin{array}{cc} \partial/\partial x & 0 \\ 0 & \partial/\partial y \\ \partial/\partial y & \partial/\partial x \end{array} \right] \left[\begin{array}{cccccc} N_1 & 0 & N_2 & 0 & N_3 & 0 \\ 0 & N_1 & 0 & N_2 & 0 & N_3 \end{array} \right] \left\{ \begin{array}{c} u_1 \\ v_1 \\ u_2 \\ v_2 \\ u_3 \\ v_3 \end{array} \right\} \qquad (10.6.10)
$$

where we define the strain-displacement matrix as,

$$
\underset{\sim}{B} = \left[\begin{array}{cc} \partial/\partial x & 0 \\ 0 & \partial/\partial y \\ \partial/\partial y & \partial/\partial x \end{array} \right] \left[\begin{array}{cccccc} N_1 & 0 & N_2 & 0 & N_3 & 0 \\ 0 & N_1 & 0 & N_2 & 0 & N_3 \end{array} \right] \qquad (10.6.11)
$$

or

$$
\underset{\sim}{B} = \left[\begin{array}{cccccc} \dfrac{\partial N_1(x,y)}{\partial x} & 0 & \dfrac{\partial N_2(x,y)}{\partial x} & 0 & \dfrac{\partial N_3(x,y)}{\partial x} & 0 \\[2ex] 0 & \dfrac{\partial N_1(x,y)}{\partial y} & 0 & \dfrac{\partial N_2(x,y)}{\partial y} & 0 & \dfrac{\partial N_3(x,y)}{\partial y} \\[2ex] \dfrac{\partial N_1(x,y)}{\partial y} & \dfrac{\partial N_1(x,y)}{\partial x} & \dfrac{\partial N_2(x,y)}{\partial y} & \dfrac{\partial N_2(x,y)}{\partial x} & \dfrac{\partial N_3(x,y)}{\partial y} & \dfrac{\partial N_3(x,y)}{\partial x} \end{array} \right]
$$

$$
= \dfrac{1}{2A} \left[\begin{array}{cccccc} b_1 & 0 & b_2 & 0 & b_3 & 0 \\ 0 & c_1 & 0 & c_2 & 0 & c_3 \\ c_1 & b_1 & c_2 & b_2 & c_3 & b_3 \end{array} \right] \qquad (10.6.12)
$$

which clearly indicates that inside each element, all components of strain are constant, hence the name *Constant Strain Triangle*. It should be noted here that though the displacement field is continuous across element boundaries, the strains and stresses are not. It is suggested to use CST where strain gradients are small and also where there is transition from coarse to finer mesh and vice versa. It is typically advised to avoid using it in critical areas where there are high stress concentrations, corners etc. or where there will be high element bending. At those locations use of "Linear Strain Triangle" which is presented next, is preferable.

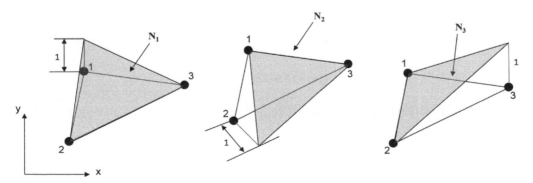

Figure 10.6-2 Shape functions of the CST element

Linear Strain Triangle (LST)

Linear Strain Triangle is a modified form of CST in terms of the nodes that exist on the edges and of the degree the DOF are approximated within the element. The variation of DOF along the edges between each nodal point is quadratic like a three-noded bar element. So roughly, the edges of LST behave as if each side is a three-noded second degree-polynomial. Figure 10.6-3 shows the DOF vectors and nodes of an LST. LST has six nodes with additional nodes being in the middle of the edges. Each node has two DOF with a total of 12 DOF in the element. The horizontal and vertical displacements are written in terms of generalized DOF as,

$$u = a_0 + a_1 x + a_2 y + a_3 x^2 + a_4 xy + a_5 y^2 \tag{10.6.13a}$$

$$v = a_6 + a_7 x + a_8 y + a_9 x^2 + a_{10} xy + a_{11} y^2 \tag{10.6.13b}$$

where a_i for $i = 1, \ldots 11$ are constants obtained from,

$$\underset{\sim}{a} = \underset{\approx}{X}^{-1} \underset{\sim}{\phi}_e \tag{10.6.14}$$

where X is the coordinate matrix. (10.6.2) can be rewritten as,

$$\underset{\sim}{\phi} = \underset{\approx}{P}\underset{\sim}{a} = \begin{bmatrix} 1 & x & y & x^2 & xy & y^2 & 0 & 0 & 0 & 0 & 0 & 0 \\ 0 & 0 & 0 & 0 & 0 & 0 & 1 & x & y & x^2 & xy & y^2 \end{bmatrix} \begin{Bmatrix} a_0 \\ a_1 \\ \vdots \\ a_{11} \end{Bmatrix} \tag{10.6.15}$$

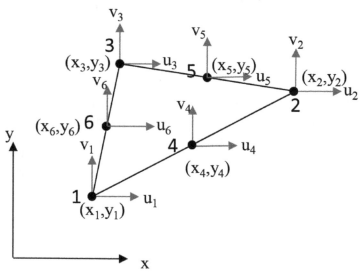

Figure 10.6-3 LST element

The shape function matrix is obtained similarly by substituting (10.6.14) into (10.6.15) as,

$$\underset{\sim}{\phi} = \underset{\approx}{P}\underset{\approx}{X}^{-1}\underset{\sim}{\phi}_e = \underset{\approx}{N}\underset{\sim}{\phi}_e \tag{10.6.16}$$

which is nothing but,

$$\{\phi\}_{2\times1} = [N]_{2\times12}\{\phi_e\}_{12\times1} = \begin{Bmatrix} u\,(x,y) \\ v\,(x,y) \end{Bmatrix} = \begin{bmatrix} N_1 & 0 & N_2 & 0 & N_3 & 0 & N_4 & 0 & N_5 & 0 & N_6 & 0 \\ 0 & N_1 & 0 & N_2 & 0 & N_3 & 0 & N_4 & 0 & N_5 & 0 & N_6 \end{bmatrix} \begin{Bmatrix} u_1 \\ v_1 \\ u_2 \\ v_2 \\ u_3 \\ v_3 \\ u_4 \\ v_4 \\ u_5 \\ v_5 \\ u_6 \\ v_6 \end{Bmatrix} \quad (10.6.17)$$

For a special case of a right-angled LST with width, b and height, h, the shape functions can be evaluated as,

$$N_1 = 1 - \frac{3x}{b} - \frac{3y}{h} + \frac{2x^2}{b^2} + \frac{4xy}{bh} + \frac{2y^2}{h^2} \qquad (10.6.18\text{a})$$

$$N_2 = -\frac{x}{b} + \frac{2x^2}{b^2} \qquad (10.6.18\text{b})$$

$$N_3 = -\frac{y}{h} + \frac{2y^2}{h^2} \qquad (10.6.18\text{c})$$

$$N_4 = \frac{4xy}{bh} \qquad (10.6.18\text{d})$$

$$N_5 = \frac{4y}{h} - \frac{4xy}{bh} - \frac{4y^2}{h^2} \qquad (10.6.18\text{e})$$

$$N_6 = \frac{4x}{b} - \frac{4xy}{bh} - \frac{4x^2}{b^2} \qquad (10.6.18\text{f})$$

Two of the shape functions of the LST element can be seen in Figure 10.6-4.

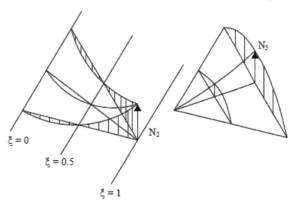

Figure 10.6-4 Two basic variations of shape functions of the LST element

The following relations can be derived for the strain field,

$$\varepsilon_x = a_1 + 2a_3 x + a_4 y \tag{10.6.19a}$$

$$\varepsilon_y = a_8 + a_{10} x + 2a_{12} y \tag{10.6.19b}$$

$$\gamma_{xy} = a_2 + a_7 + \left(a_4 + 2a_9\right)x + \left(2a_5 + a_{10}\right)y \tag{10.6.19c}$$

suggesting the variation is linear with respect to x and y. This helps provide better results than the CST in a mesh if bending or high stress concentrations are going to be issues in the problem. Element strains are approximated the same way using (10.6.19).

Four Noded Bilinear Rectangle (Q4)

Four noded bilinear rectangle is the simplest of the 2-D rectangular elements. There are four nodes at the corners with two DOF at each (total of eight DOF), see Figure 10.6-5. Its displacement field is,

$$\varphi = \begin{Bmatrix} u \\ v \end{Bmatrix} = \begin{Bmatrix} a_0 + a_1 x + a_2 y + a_3 xy \\ a_4 + a_5 x + a_6 y + a_7 xy \end{Bmatrix} \tag{10.6.20}$$

containing the additional 'bilinear' term as compared to triangular elements. Following the same steps of deriving shape functions we start off by writing a very general case,

$$\varphi(x,z) = a_1 + a_2 x + a_3 y + a_4 xy \tag{10.6.21}$$

Substituting respective coordinates one by one accordingly yields,

$$\varphi_1 = \varphi(-l_1,-l_2) = a_1 - a_2 l_1 - a_3 l_2 + a_4 l_1 l_2 \tag{10.6.22a}$$

$$\varphi_2 = \varphi(l_1,-l_2) = a_1 + a_2 l_1 - a_3 l_2 - a_4 l_1 l_2 \tag{10.6.22b}$$

$$\varphi_3 = \varphi(l_1,l_2) = a_1 + a_2 l_1 + a_3 l_2 + a_4 l_1 l_2 \tag{10.6.22c}$$

$$\varphi_4 = \varphi(-l_1,l_2) = a_1 - a_2 l_1 + a_3 l_2 - a_4 l_1 l_2 \tag{10.6.22d}$$

We can write this in a matrix form as,

$$\begin{Bmatrix} a_1 \\ a_2 \\ a_3 \\ a_4 \end{Bmatrix} = \begin{bmatrix} 1 & -l_1 & -l_2 & l_1 l_2 \\ 1 & l_1 & -l_2 & -l_1 l_2 \\ 1 & l_1 & l_2 & l_1 l_2 \\ 1 & -l_1 & l_2 & -l_1 l_2 \end{bmatrix}^{-1} \begin{Bmatrix} \varphi_1 \\ \varphi_2 \\ \varphi_3 \\ \varphi_4 \end{Bmatrix} \tag{10.6.23}$$

or

$$\begin{Bmatrix} a_1 \\ a_2 \\ a_3 \\ a_4 \end{Bmatrix} = \begin{bmatrix} 1/4 & 1/4 & 1/4 & 1/4 \\ -1/4l_1 & 1/4l_1 & 1/4l_1 & -1/4l_1 \\ -1/4l_2 & -1/4l_2 & 1/4l_2 & 1/4l_2 \\ 1/4l_1 l_2 & -1/4l_1 l_2 & 1/4l_1 l_2 & -1/4l_1 l_2 \end{bmatrix} \begin{Bmatrix} \varphi_1 \\ \varphi_2 \\ \varphi_3 \\ \varphi_4 \end{Bmatrix} \tag{10.6.24}$$

Substituting a_1, a_2, a_3, a_4 from Eq. (10.6.24) in Eq. (10.6.21), we get:

$$[\varphi]=[N_1 \quad N_2 \quad N_3 \quad N_4]\begin{Bmatrix}\varphi_1\\\varphi_2\\\varphi_3\\\varphi_4\end{Bmatrix} \tag{10.6.25}$$

Following the same steps again, below are obtained:

$$N_1 = \frac{(l_1-x)(l_2-y)}{4l_1l_2} \tag{10.6.26a}$$

$$N_2 = \frac{(l_1+x)(l_2-y)}{4l_1l_2} \tag{10.6.26b}$$

$$N_3 = \frac{(l_1+x)(l_2+y)}{4l_1l_2} \tag{10.6.26c}$$

$$N_4 = \frac{(l_1-x)(l_2+y)}{4l_1l_2} \tag{10.6.26d}$$

and are placed in the approximation relation as,

$$\{\varphi\}_{2\times1}=[N]_{2\times8}\{\varphi_e\}_{8\times1}=\begin{Bmatrix}u(x,y)\\v(x,y)\end{Bmatrix}=\begin{bmatrix}N_1 & 0 & N_2 & 0 & N_3 & 0 & N_4 & 0\\0 & N_1 & 0 & N_2 & 0 & N_3 & 0 & N_4\end{bmatrix}\begin{Bmatrix}u_1\\v_1\\u_2\\v_2\\u_3\\v_3\\u_4\\v_4\end{Bmatrix} \tag{10.6.27}$$

Figure 10.6-6 shows a typical variation of shape functions for the Q4 element. Strains and stresses are evaluated the same way using the strain-displacement relation and the Hooke's law. For strains we write,

$$\varepsilon_x = a_1 + a_3 y \tag{10.6.28a}$$

$$\varepsilon_y = a_6 + a_7 x \tag{10.6.28b}$$

$$\gamma_{xy} = a_2 + a_5 + a_3 x + a_7 y \tag{10.6.28c}$$

or in a more general fashion,

$$\begin{Bmatrix}\varepsilon_x\\\varepsilon_y\\\gamma_{xy}\end{Bmatrix}=\frac{1}{4l_1l_2}\begin{bmatrix}-(l_2-y) & 0 & (l_2-y) & 0 & \cdots\\0 & -(l_1-x) & 0 & -(l_1+x) & \cdots\\-(l_1-x) & -(l_2-y) & -(l_1+x) & (l_2-y) & \cdots\end{bmatrix}\begin{Bmatrix}u_1\\v_1\\u_2\\v_2\\u_3\\\cdots\end{Bmatrix} \tag{10.6.29}$$

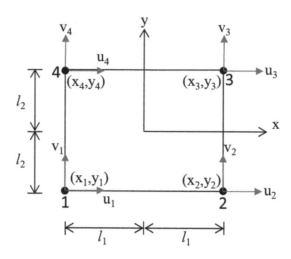

Figure 10.6-5 Q4 element

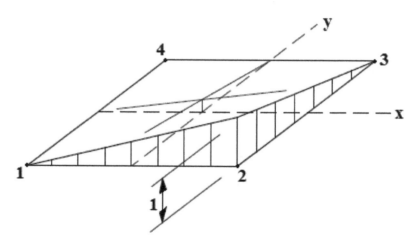

Figure 10.6-6 Q4 shape function for node number two

Eight Noded Quadratic Rectangle (Q8)

Eight noded quadratic rectangle (Q8) which is also called "the serendipity rectangle" has higher order terms in the shape functions. There are four nodes at the corners and four nodes at the mid-points of the sides with two DOF at each (total of 16 DOF), see Figure 10.6-7. Its displacement field is,

$$\phi = \left\{ \begin{matrix} u \\ v \end{matrix} \right\} = \left\{ \begin{matrix} a_0 + a_1 x + a_2 y + a_3 x^2 + a_4 xy + a_5 y^2 + a_6 x^2 y + a_7 xy^2 \\ a_8 + a_9 x + a_{10} y + a_{11} x^2 + a_{12} xy + a_{13} y^2 + a_{14} x^2 y + a_{15} xy^2 \end{matrix} \right\} \qquad (10.6.30)$$

containing the additional 'multiplied quadratic' two terms as compared to LST and Q4 combined. Strains are evaluated the same way as,

$$\varepsilon_x = a_1 + 2a_3 x + a_4 y + 2a_6 xy + a_7 y^2 \qquad (10.6.31a)$$

$$\varepsilon_y = a_{10} + a_{12} x + 2a_{13} y + a_{14} x^2 + 2a_{15} xy \qquad (10.6.31b)$$

$$\gamma_{xy} = a_2 + a_9 + (a_4 + 2a_{11})x + (2a_5 + a_{12})y + a_6x^2 + 2(a_7 + a_{14})xy + a_{15}y^2 \qquad (10.6.31c)$$

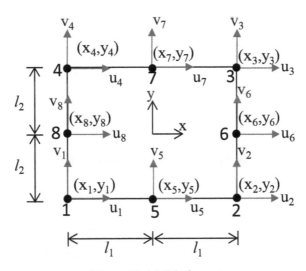

Figure 10.6-7 Q8 element

Shape functions are obtained similarly as:

$$N_1 = \frac{\{(l_2 - y)l_2x^2 + l_1l_2xy + (l_1 - x)l_1y^2 - l_1^2l_2^2\}}{4l_1^2l_2^2} \qquad (10.6.32a)$$

$$N_2 = \frac{\{(l_2 - y)l_2x^2 - l_1l_2xy + (l_1 + x)l_1y^2 - l_1^2l_2^2\}}{4l_1^2l_2^2} \qquad (10.6.32b)$$

$$N_3 = \frac{\{(l_2 + y)l_2x^2 + l_1l_2xy + (l_1 + x)l_1y^2 - l_1^2l_2^2\}}{4l_1^2l_2^2} \qquad (10.6.32c)$$

$$N_4 = \frac{\{(l_2 + y)l_2x^2 - l_1l_2xy + (l_1 - x)l_1y^2 - l_1^2l_2^2\}}{4l_1^2l_2^2} \qquad (10.6.32d)$$

$$N_5 = \frac{\{(-l_2 + y)l_2x^2 - l_1^2l_2y + l_1^2l_2^2\}}{2l_1^2l_2^2} \qquad (10.6.32e)$$

$$N_6 = \frac{\{(-x - l_1)l_2x^2 - l_1l_2^2y + l_1^2l_2^2\}}{2l_1^2l_2^2} \qquad (10.6.32f)$$

$$N_7 = \frac{\{(-l_2 - y)l_2x^2 + l_1^2l_2y + l_1^2l_2^2\}}{2l_1^2l_2^2} \qquad (10.6.32g)$$

$$N_8 = \frac{\{(x - l_1)l_1x^2 - l_1^2l_2x + l_1^2l_2^2\}}{2l_1^2l_2^2} \qquad (10.6.32h)$$

Figure 10.6-8 shows the two typical variations of shape functions for the quadratic quadrilateral, Q8.

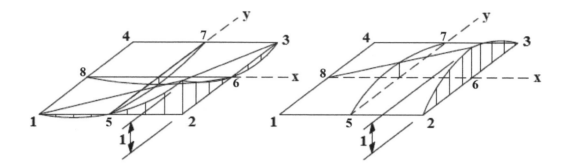

Figure 10.6-8 Q8 shape functions (N_2 and N_6)

The 3-D elements and the special type of elements such as the plate bending, infinite, crack etc. are out of the scope of this book and are thus not discussed here. Interested readers can refer to Reddy (1993), Hughes (1987), Cook et al. (2001) or Zienkiewicz and Taylor (2000).

Example 10.6-1: For a four-noded rectangular element (Q4) seen in Figure 10.6-9, obtain the shape function matrix, N, the strain-displacement matrix, B and the stiffness matrix, K. Consider the element behavior linear elastic with $E = 33.75$ units for elasticity modulus and $v = 0.25$ Poisson's ratio, $a = 1$ unit, $b = 1$ unit and $t = 1$ unit, obtain the strain ε, and stress σ, vectors in terms of the DOFs. The element will be in plane stress condition.

Solution:
The coordinate transformation can be done through,

$$r = \frac{x - \bar{x}}{a} \tag{10.6.33a}$$

$$s = \frac{y - \bar{y}}{b} \tag{10.6.33b}$$

Displacement DOF (u, v) can be interpolated using the shape function matrix as:

$$\begin{Bmatrix} u \\ v \end{Bmatrix} = \begin{bmatrix} N_1 & N_2 & N_3 & N_4 & 0 & 0 & 0 & 0 \\ 0 & 0 & 0 & 0 & N_1 & N_2 & N_3 & N_4 \end{bmatrix} \begin{Bmatrix} u_1 \\ u_2 \\ u_3 \\ u_4 \\ v_1 \\ v_2 \\ v_3 \\ v_4 \end{Bmatrix} \tag{10.6.34}$$

where

$$N_1 = \frac{1}{4}(1 - r)(1 - s) \tag{10.6.35a}$$

$$N_2 = \frac{1}{4}(1 + r)(1 - s) \tag{10.6.35b}$$

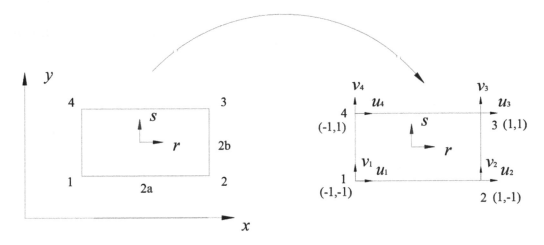

Figure 10.6-9 Quadrilateral as mapped from a master element in *s-r* coordinates

$$N_3 = \frac{1}{4}(1+r)(1+s) \qquad (10.6.35c)$$

$$N_4 = \frac{1}{4}(1-r)(1+s) \qquad (10.6.35d)$$

The strain vector is now obtained the usual way using the $[\partial]$ operator and the DOF vector $\{\delta^e\}$ as:

$$\{\varepsilon\} = [\partial][N]\{\delta^e\} = [B]\{\delta^e\} \qquad (10.6.36)$$

resulting in

$$\{\varepsilon\} = \begin{bmatrix} \dfrac{s-1}{4a} & \dfrac{1-s}{4a} & \dfrac{1+s}{4a} & -\dfrac{1+s}{4a} & 0 & 0 & 0 & 0 \\[2mm] 0 & 0 & 0 & 0 & \dfrac{r-1}{4b} & -\dfrac{1+r}{4b} & \dfrac{1+r}{4b} & \dfrac{1-r}{4b} \\[2mm] \dfrac{r-1}{4b} & -\dfrac{1+r}{4b} & \dfrac{1+r}{4b} & \dfrac{1-r}{4b} & \dfrac{s-1}{4a} & \dfrac{1-s}{4a} & \dfrac{1+s}{4a} & -\dfrac{1+s}{4a} \end{bmatrix} \qquad (10.6.37)$$

Note that:

$$\frac{\partial}{\partial x} = \frac{\partial}{\partial r}\frac{\partial r}{\partial x} = \frac{1}{a}\frac{\partial}{\partial r} \qquad (10.6.38)$$

$$\frac{\partial}{\partial y} = \frac{\partial}{\partial s}\frac{\partial s}{\partial x} = \frac{1}{b}\frac{\partial}{\partial s} \qquad (10.6.39)$$

Now we can evaluate the element stiffness matrix using the following equation which will yield an 8×8 symmetric matrix containing 36 independent terms. All these terms are evaluated through integration. Note that in case of a CST element since $[B]$ and $[D]$ were constants, no integration is required to evaluate the stiffness matrix. The integration in this case is straightforward but tedious. The stiffness matrix now is,

$$[K^e] = \frac{Etab}{(1+\mu)(1-2\mu)} \int_{-1}^{1} \int_{-1}^{1} B^T DBdrds \qquad (10.6.40)$$

where

$$B = \begin{bmatrix} \dfrac{s-1}{4a} & \dfrac{1-s}{4a} & \dfrac{1+s}{4a} & -\dfrac{1+s}{4a} & 0 & 0 & 0 & 0 \\[2mm] 0 & 0 & 0 & 0 & \dfrac{r-1}{4b} & -\dfrac{1+r}{4b} & \dfrac{1+r}{4b} & \dfrac{1-r}{4b} \\[2mm] \dfrac{r-1}{4b} & -\dfrac{1+r}{4b} & \dfrac{1+r}{4b} & \dfrac{1-r}{4b} & \dfrac{s-1}{4a} & \dfrac{1-s}{4a} & \dfrac{1+s}{4a} & -\dfrac{1+s}{4a} \end{bmatrix} \qquad (10.6.41a)$$

$$D = \begin{bmatrix} 1-\mu & \mu & 0 \\ \mu & 1-\mu & 0 \\ 0 & 0 & 1-2\mu\Big/\mu \end{bmatrix} \qquad (10.6.41b)$$

Let us, for the purpose of illustration, evaluate only the first term of the stiffness matrix;

$$K_{11}^e = \frac{Etb}{16a(1+2\mu)} \int_{-1}^{1} \int_{-1}^{1} (s-1)^2 \, drds + \frac{Eta}{32b(1+\mu)} \int_{-1}^{1} \int_{-1}^{1} (r-1)^2 \, drds \qquad (10.6.42)$$

which becomes,

$$K_{11}^e = \frac{Etb}{16a(1+2v)} \frac{2(s-1)^3}{3}\Big|_{-1}^{1} + \frac{Eta}{32b(1+v)} \frac{2(r-1)^3}{3}\Big|_{-1}^{1} = \frac{Etb}{16a(1+2v)} \frac{16}{3} + \frac{Eta}{32b(1+v)} \frac{16}{3} = 12$$

$$(10.6.43)$$

Note that in the above integration the integrands are quadratic terms in r and s. Therefore, we may perform numerical integration using Gauss-Quadrature method. In this case, we may use the following formula,

$$[K^e] = ab \sum_{i=1}^{2} \sum_{j=1}^{2} w_i w_j [B(r_i, s_j)]^T [D][B(r_i, s_j)] \qquad (10.6.44)$$

and the following Gauss points of integration,

$$r_i = \pm\frac{\sqrt{3}}{3}, \quad s_j = \pm\frac{\sqrt{3}}{3} \qquad (10.6.45)$$

along with the weight factors, $w_i = w_j = 1$. Gauss-Quadrature integration method can be elaborated elsewhere for interested readers.

10.6.2 Consistent Nodal Loads

In FEM, loads are evaluated to be applied on a mesh either on element edges or at the nodes. Either way, they need to be integrated within the surface (for tractions) or in the volume (for body forces) of the domain. The concentrated forces, (F_c) are applied at nodes such that the mesh is refined around those locations to be able to avoid singularities. Thus we simply write,

$$R_c = N^T F_c \tag{10.6.46}$$

for evaluating the consistent nodal load for concentrated loads, R_g. The distributed loads are evaluated along the element edges using the following equations,

$$R_d = \int_S N^T F_d dS \tag{10.6.47}$$

where F_d is the distributed traction vector at the nodes of the edge of concern. Total consistent load vector is then evaluated as,

$$R = R_c + R_d \tag{10.6.48}$$

Example 10.6-2: Evaluate the consistent nodal load vector for the top surface in the given Q4 configuration in Figure 10.6-10.

Solution:

Load vector $R = \begin{Bmatrix} F_4 \\ F_3 \end{Bmatrix}$ is to be evaluated using the given distributed load, q using (10.6.47),

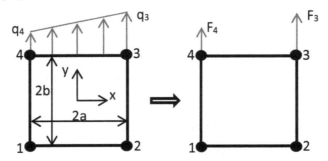

Figure 10.6-10 Consistent nodal loads for Q4

where we need only N_3 and N_4 which can be evaluated by plugging in $y = b$ into (10.6.26) giving the forms,

$$N_3 = \frac{a+x}{2a} \tag{10.6.49a}$$

$$N_4 = \frac{a-x}{2a} \tag{10.6.49b}$$

So we have $N = \frac{1}{2a}[a-x \quad a+x]$. F_d is the interpolated values of the applied distributed load at the nodes. Thus for an element thickness of t,

$$F_d = \frac{N}{t} \begin{Bmatrix} q_4 \\ q_3 \end{Bmatrix} \tag{10.6.50}$$

$$R_d = \int_S N^T \frac{N}{t} \begin{Bmatrix} q_4 \\ q_3 \end{Bmatrix} t dx = \int_{-a}^{a} \frac{1}{2a} \begin{bmatrix} a-x \\ a+x \end{bmatrix} \frac{1}{2a}[a-x \quad a+x] \begin{Bmatrix} q_4 \\ q_3 \end{Bmatrix} dx \tag{10.6.51}$$

where we have $dS = tdx$. Carrying out the integration in equation (10.6.51) results in,

$$R = \begin{Bmatrix} F_4 \\ F_3 \end{Bmatrix} = \frac{a}{3} \begin{bmatrix} 2 & 1 \\ 1 & 2 \end{bmatrix} \begin{Bmatrix} q_4 \\ q_3 \end{Bmatrix} \qquad (10.6.52)$$

Example 10.6-3: Redo the previous example for the given Q8 element. See Figure 10.6-11 below.

Solution:

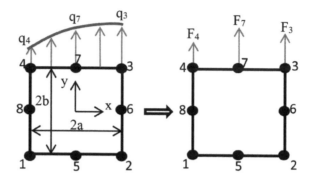

Figure 10.6-11 Consistent nodal load for Q8

Now the load vector is $R = \begin{Bmatrix} F_4 \\ F_7 \\ F_3 \end{Bmatrix}$ to be evaluated using N_7 as well in addition to N_3 and N_4

which are obtained through (10.6.46) by simply putting $y = b$ as,

$$N_3 = \frac{x(x+a)}{2a^2} \qquad (10.6.53a)$$

$$N_4 = \frac{x(x-a)}{2a^2} \qquad (10.6.53b)$$

$$N_7 = \frac{(a^2 - x^2)}{a^2} \qquad (10.6.53c)$$

So we have $N = \frac{1}{2a^2}[x(x-a) \quad 2(a^2 - x^2) \quad x(x+a)]$. Now F_d becomes,

$$F_d = \frac{N}{t} \begin{Bmatrix} q_4 \\ q_7 \\ q_3 \end{Bmatrix} \qquad (10.6.54)$$

and the consistent nodal load vector is evaluated,

$$R_d = \int_S N^T \frac{N}{t} \begin{Bmatrix} q_4 \\ q_7 \\ q_3 \end{Bmatrix} t\,dx = \int_{-a}^{a} \frac{1}{2a^2} \begin{bmatrix} x(x-a) \\ 2(a^2 - x^2) \\ x(x+a) \end{bmatrix} \frac{1}{2a^2}[x(x-a) \quad 2(a^2 - x^2) \quad x(x+a)] \begin{Bmatrix} q_4 \\ q_7 \\ q_3 \end{Bmatrix} dx$$

$$(10.6.55)$$

yielding,

$$R = \begin{Bmatrix} F_4 \\ F_7 \\ F_3 \end{Bmatrix} = \frac{a}{15} \begin{bmatrix} 4 & 2 & -1 \\ 2 & 16 & 2 \\ -1 & 2 & 4 \end{bmatrix} \begin{Bmatrix} q_4 \\ q_7 \\ q_3 \end{Bmatrix} \tag{10.6.56}$$

For the special case of uniformly distributed load ($q_3 = q_4 = q_7 = q$), the load vector takes the form,

$$R = 2aq \begin{Bmatrix} 1/6 \\ 2/3 \\ 1/6 \end{Bmatrix} \tag{10.6.57}$$

10.7 The Principle of Minimum Potential Energy

For the problems in solid mechanics, the formulation of FE equations is sometimes easier when we use *the principle of minimum potential energy*. A mechanical system has potential energy that can be expressed in terms of its initial and final configurations without reference to whatever deformation history or path takes the system from initial to final configuration (Langhaar, 1962). Of all displacement states of a body or a structure subjected to external loading that satisfies the boundary conditions (i.e. imposed displacements), the displacement state that also satisfies the equilibrium equations is such that the total potential energy is a minimum for the stable equilibrium. So we can write the total potential energy as,

$$\Pi(U_1, U_2, \cdots, U_n) \tag{10.7.1}$$

Given the strain energy, U_e, the potential energy with forces, $-W$, the total potential energy is evaluated as,

$$\Pi = U_e - W \tag{10.7.2}$$

Then the following must be satisfied:

$$\frac{\partial \Pi}{\partial U_i} = 0 \tag{10.7.3}$$

The principle is valid whether or not the load vs. deformation relation is linear.

Example 10.7-1: For the linear spring of Figure 10.4-10, calculate the total potential energy.

Solution:

This simple system's potential energy is written as,

$$\Pi = U_e + W$$
$$U_e = \frac{1}{2}k(dx)^2$$
$$W = -Fdx$$
$$\Pi = \frac{1}{2}k(dx)^2 - Fdx$$

which can be regarded as the total internal and external work done in changing the configuration from the reference state of $dx = 0$ to the displaced state $dx \neq 0$. Using the law of conservation of energy, we could 'bypass' the stationary potential energy principle and equate the work done by the load to strain energy stored in the linear spring. Thus,

$$\frac{1}{2}F(dx) = \frac{1}{2}k(dx)^2 \rightarrow dx = \frac{F}{k}$$

Elastic Strain Energy

For a plane strain state, the strain energy per unit volume is:

$$U_e = \frac{1}{2}\left(\sigma_x \varepsilon_x + \sigma_y \varepsilon_y + \tau_{xy}\gamma_{xy}\right) \tag{10.7.4}$$

or also stated as:

$$U_e = \frac{1}{2}\{\sigma\}^T \{\varepsilon\} = \frac{1}{2}\{\varepsilon\}^T [D]\{\varepsilon\} \tag{10.7.5}$$

The total strain energy for an element Ω_e should now be written,

$$U_e = \frac{1}{2}\{\varepsilon\}^T [D]\{\varepsilon\}\, dxdy \tag{10.7.6}$$

Recalling,

$$\{\varepsilon\} = [\partial]\{\delta\} \tag{10.7.7}$$

along with

$$\{\varepsilon\} = [\partial][\psi]\{\delta^e\} = [B]\{\delta^e\} \tag{10.7.8}$$

$$[\psi] = \begin{bmatrix} \psi_1 & \psi_2 & \cdots & \psi_n & 0 & 0 & \cdots & 0 \\ 0 & 0 & \cdots & 0 & \psi_1 & \psi_2 & \cdots & \psi_n \end{bmatrix} \tag{10.7.9}$$

$$[B] = \begin{bmatrix} \dfrac{\partial\psi_1}{\partial x} & \dfrac{\partial\psi_2}{\partial x} & \cdots & \dfrac{\partial\psi_n}{\partial x} & 0 & 0 & \cdots & 0 \\[2mm] 0 & 0 & \cdots & 0 & \dfrac{\partial\psi_1}{\partial y} & \dfrac{\partial\psi_2}{\partial y} & \cdots & \dfrac{\partial\psi_n}{\partial y} \\[2mm] \dfrac{\partial\psi_1}{\partial y} & \dfrac{\partial\psi_2}{\partial y} & \cdots & \dfrac{\partial\psi_n}{\partial y} & \dfrac{\partial\psi_1}{\partial x} & \dfrac{\partial\psi_2}{\partial x} & \cdots & \dfrac{\partial\psi_n}{\partial x} \end{bmatrix} \tag{10.7.10}$$

$$\{\varepsilon\} = [B]\{\delta^e\} \tag{10.7.11}$$

$$\{\varepsilon\}^T = \{\delta^e\}^T [B]^T \tag{10.7.12}$$

we therefore have,

$$U_e = \frac{1}{2}\int_{\Omega_e} \{\delta\}^T [B]^T [D][B]\{\delta\}\, dxdy \tag{10.7.13}$$

giving us the stiffness matrix,

$$[K^e] = \int_{\Omega_e} [B]^T [D][B]\, dxdy \tag{10.7.14}$$

yielding

$$U_e - \frac{1}{2}\{\delta\}^T \left[K^e\right]\{\delta\}$$ (10.7.15)

Work done by the nodal forces is:

$$W = f_{1x}u_1 + f_{2x}u_2 + \cdots + f_{nx}u_n + f_{1y}v_1 + f_{2y}v_2 + \cdots + f_{ny}v_n$$ (10.7.16)

$$W = \{\delta\}^T \{f\}$$ (10.7.17)

Now using (10.7.2), we get:

$$\Pi = U_e - W = \frac{1}{2}\{\delta\}^T \left[K^e\right]\{\delta\} - \{\delta\}^T \{f\}$$ (10.7.18)

and (10.7.3),

$$\frac{\partial \Pi}{\partial \delta_i} = 0, \ i = 1, 2, \cdots, 2n$$ (10.7.19)

The stiffness relation is then evaluated as:

$$\left[K^e\right]\{\delta^e\} = \{f^e\}$$ (10.7.20)

finalizing the formulation through the potential energy principle.

10.8 Isoparametric Element Formulation

10.8.1 Introduction

The formulations of elements presented so far pertain to elements with typical linear edges and uniform spacing of nodal points. We should regard them as special cases of a more general formulation for which the coordinate system maintains its position regardless of the physical size and shape of the element. This is important in the sense that this way the element edges can deform in various ways in a prescribed mesh to be able to represent the distortion more efficiently and accurately. Therefore, we need somewhat "distorted elements" and a technique for their FE formulation. Figure 10.8-1 illustrates this need in a rather simple example. Here, as the undistorted rectangular elements are used in a refined mesh, there will be a problem in the elements around the circular hole where the element formulation is not described due to distortion.

The general idea behind isoparametric formulation of a multi-dimensional element is that it is possible to use "curved" edges in a mesh which allows modeling complex geometries. Except for triangular elements, elements are restricted to rectangular shapes. Although it is easy to formulate rectangles, most complicated geometries are extremely hard to discretize using rectangular elements. Thus, there is a need for a special formulation in which complete transformation of coordinate axes will be possible such that they are approximated using the same type of shape functions as the deformations of nodal points are. Rectangular elements are called the "master elements" while their deformed or curved counterparts are called the "physical elements". The generation of different elements from a single master element which facilitates the evaluation of element matrices in a mesh is the basis for isoparametric formulation. In this section, we will look into the formulation of a plane quadrilateral element. A number of examples showing the effectiveness of mapping is given first.

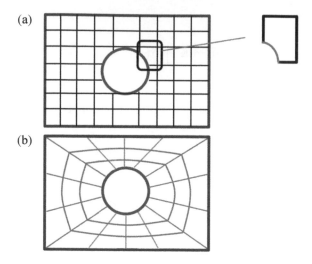

Figure 10.8-1 (a) Rectangular elements and the problem in the mesh, (b) Distorted elements

Example 10.8-1: Determine the following mapping relations for the given elements between the coordinate systems.

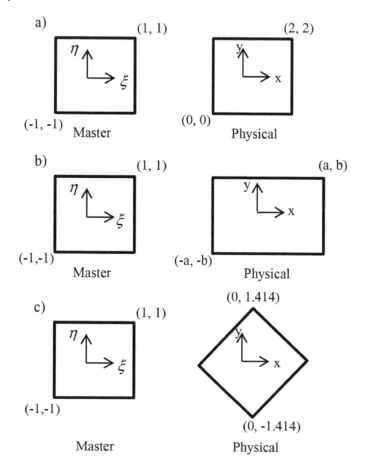

Solution:

(a) Translation: $x = 1 + \xi,\ y = 1 + \eta$

(b) Expansion: $x = a\xi,\ y = b\eta$

(c) Rotation: $x = \dfrac{\xi - \eta}{\sqrt{2}},\quad y = \dfrac{\xi + \eta}{\sqrt{2}}$

The goal of isoparametric formulation is to derive the element stiffness matrix and the stiffness relation for distorted elements. This is done by "mapping" from the described master element coordinates called the "natural coordinates" where the shape functions can be obtained in the "natural coordinate system", (ξ, η) onto the existing physical element as shown in Figure 10.8-2. For a general form of mapping, we want to obtain, $\underset{\sim}{x} = \begin{Bmatrix} x \\ y \end{Bmatrix} = f\left(\begin{Bmatrix} \xi \\ \eta \end{Bmatrix} \right)$, hence the name "isoparametric" which pertains to using the same interpolation for approximating the field variables also used for calculating the coordinates, as in (10.4.104). Therefore, we can safely write,

$$\underset{\sim}{x} = \underset{\sim}{N}\underset{\sim}{c} \tag{10.8.1}$$

where $\underset{\sim}{c} = \begin{Bmatrix} x_1 \\ y_1 \\ \vdots \end{Bmatrix}$ is the coordinates vector for the physical element in consideration. Firstly,

we will derive the stiffness relation for the simplest 1-D bar element and expand it further subsequently for multi-dimensions.

10.8.2 Isoparametric Formulation for the Bar Element

For the given master and physical configurations seen in Figure 10.8-3, for which $\underset{\sim}{c} = \begin{Bmatrix} x_1 \\ x_2 \\ x_3 \end{Bmatrix}$, shape functions are defined on the master element as $\underset{\sim}{N} = \underset{\sim}{N}(\xi)$.

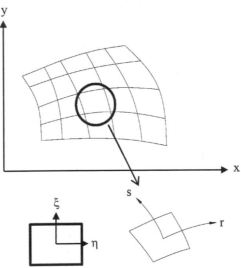

Figure 10.8-2 Isoparametric distorted elements

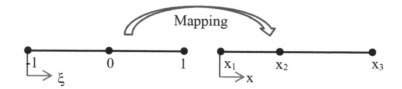

Figure 10.8-3 Mapping for the bar element

We need to use the chain rule of differentiation here and write,

$$\frac{\partial \underset{\sim}{N}(\xi)}{\partial \xi} = \frac{\partial x}{\partial \xi} \frac{\partial \underset{\sim}{N}(\xi)}{\partial x} \tag{10.8.2}$$

where

$$J = \frac{\partial x}{\partial \xi} \tag{10.8.3}$$

is called the "*Jacobian*". Since we define the physical coordinates in terms of the natural coordinates we will have the Jacobian computed readily. If we define the strain-displacement matrix,

$$\underset{\sim}{B} = \frac{\partial \underset{\sim}{N}(\xi)}{\partial x} = \frac{\partial \underset{\sim}{N}(\xi)}{\partial \xi} \frac{\partial \xi}{\partial x} \tag{10.8.4}$$

where $\dfrac{\partial \xi}{\partial x}$ is clearly the inverse of Jacobian. Thus, rewriting the B matrix using (10.8.2) gives,

$$\underset{\sim}{B} = \frac{\partial \underset{\sim}{N}(\xi)}{\partial x} = \frac{\partial \underset{\sim}{N}(\xi)}{\partial \xi} J^{-1} \tag{10.8.5}$$

Now the stiffness matrix can be derived on the natural coordinates as,

$$[K] = \int_{x_1}^{x_3} B(\xi)^T \, EAB(\xi) \, J d\xi \tag{10.8.6}$$

with $dx = Jd\xi$ transformation. To get the J, we need to substitute (10.8.1) into (10.8.3) giving,

$$J = \frac{\partial \left(\underset{\sim}{N} \right)}{\partial \xi} \underset{\sim}{c} \tag{10.8.7}$$

where the shape function matrix is defined in the master coordinates as,

$$\underset{\sim}{N}(\xi) = \left[\frac{1-\xi}{2} \quad \frac{1+\xi}{2} \right] \tag{10.8.8}$$

Then for a bar with length L, we have,

$$J = \frac{\partial \left(\underset{\sim}{N} \right)}{\partial \xi} \underset{\sim}{c} = \left[-\frac{1}{2} \quad \frac{1}{2} \right] \begin{Bmatrix} 0 \\ L \end{Bmatrix} = \frac{L}{2} \tag{10.8.9}$$

which yields the stiffness matrix obtained previously in (10.4.108) when substituted in (10.8.6).

Example 10.8-2: Obtain the stiffness relation of the three-noded bar below using isoparametric formulation. See Figure 10.8-4.

Figure 10.8-4 Three-noded bar

Solution:

The shape functions of this bar defined in master element are,

$$\underset{\sim}{N}(\xi) = \left[\frac{\xi^2 - \xi}{2} \quad 1 - \xi^2 \quad \frac{\xi^2 + \xi}{2} \right] \tag{10.8.10}$$

and the Jacobian is,

$$J = \frac{\partial \left(\underset{\sim}{N} \right)}{\partial \xi} \underset{\sim}{c} = \left[\frac{2\xi - 1}{2} - 2\xi \quad \frac{2\xi + 1}{2} \right] \left\{ \begin{matrix} 0 \\ \alpha L \\ L \end{matrix} \right\} = \xi L (1 - 2\alpha) + \frac{L}{2} \tag{10.8.11}$$

leading to,

$$\underset{\sim}{B} = \frac{\partial \underset{\sim}{N}(\xi)}{\partial x} = \frac{\partial \underset{\sim}{N}(\xi)}{\partial \xi} J^{-1} \tag{10.8.12}$$

For the special case of $\alpha = 0.25$, Jacobian value corresponds to stretching with $J = \frac{L}{2}(1 + \xi)$ where, at the left node with $\xi = -1$, we get $J = 0$ yielding infinite strain which is not realistic. Also for $\alpha = 0.125$, Jacobian becomes negative which means there will be folding in the element containing such a node. This leads to a non-converged solution. Thus, the $J > 0$ condition must be satisfied at all times. We can now finally evaluate the stiffness relation of this bar element,

$$\underset{\sim}{K} = \int_{-1}^{1} \underset{\sim}{B}^T \underset{\sim}{E} \underset{\sim}{B} J d\xi \tag{10.8.13}$$

which is a rational function of linear Jacobian. The nature of (10.8.13) is that, it is linear in the first derivative of N matrix yielding a rational function. Thus, computation of (10.8.13) requires "numerical integration". The standard "Gauss-Quadrature" method as mentioned previously can be used to carry out the numerical integration in the spatial domain. It should be noted here that the integration rule used to evaluate the stiffness matrix that would be exact if there were no distortion is called the "full integration rule" and that the stiffness matrix is then calculated at the integration points.

10.8.3 Isoparametric Formulation for 2-D Elements

2-D isoparametric element formulation is presented in this section. For the Q4 element, the master element is given in Figure 10.8-5 where a distorted Q4 is also shown. Recalling equation (10.6.26), the shape function matrix is now written in terms of $\underset{\sim}{N} = \underset{\sim}{N}(\xi, \eta)$, similar to (10.8.8) for 1-D. Then we write the geometric mapping functions as,

$$\underset{\sim}{x} = \underset{\sim}{N}(\xi, \eta) \underset{\sim}{c} \tag{10.8.14a}$$

$$y = \underset{\sim}{N}(\xi,\eta)\underset{\sim}{c} \tag{10.8.14b}$$

and the interpolation functions as,

$$\underset{\sim}{u} = \underset{\sim}{N}(\xi,\eta)\underset{\sim}{u}_e \tag{10.8.15a}$$

$$\underset{\sim}{v} = \underset{\sim}{N}(\xi,\eta)\underset{\sim}{v}_e \tag{10.8.15b}$$

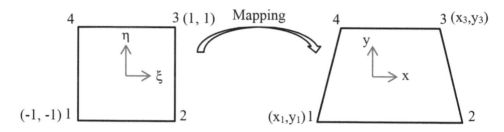

Figure 10.8-5 Mapping for the Q4 element

Now recalling the general form of the B matrix in (10.6.12) and strains in (10.6.29), we write the derivatives with respect to new coordinates as:

$$\begin{Bmatrix} \dfrac{\partial u}{\partial \xi} \\[2ex] \dfrac{\partial u}{\partial \eta} \end{Bmatrix} = \begin{bmatrix} \dfrac{\partial x}{\partial \xi} & \dfrac{\partial y}{\partial \xi} \\[2ex] \dfrac{\partial x}{\partial \eta} & \dfrac{\partial y}{\partial \eta} \end{bmatrix} \begin{Bmatrix} \dfrac{\partial u}{\partial x} \\[2ex] \dfrac{\partial u}{\partial y} \end{Bmatrix} \tag{10.8.16}$$

where Jacobian is now a matrix in the form of,

$$\underset{\sim}{J} = \begin{bmatrix} \dfrac{\partial x}{\partial \xi} & \dfrac{\partial y}{\partial \xi} \\[2ex] \dfrac{\partial x}{\partial \eta} & \dfrac{\partial y}{\partial \eta} \end{bmatrix} \tag{10.8.17}$$

After some algebra, it can be shown that:

$$\begin{Bmatrix} \dfrac{\partial u}{\partial x} \\[2ex] \dfrac{\partial u}{\partial y} \end{Bmatrix} = \dfrac{1}{|J|} \begin{bmatrix} J_{11} & -J_{12} \\[1ex] -J_{21} & J_{22} \end{bmatrix} \begin{Bmatrix} \dfrac{\partial u}{\partial \xi} \\[2ex] \dfrac{\partial u}{\partial \eta} \end{Bmatrix} \tag{10.8.18a}$$

$$\begin{Bmatrix} \dfrac{\partial v}{\partial x} \\[2ex] \dfrac{\partial v}{\partial y} \end{Bmatrix} = \dfrac{1}{|J|} \begin{bmatrix} J_{11} & -J_{12} \\[1ex] -J_{21} & J_{22} \end{bmatrix} \begin{Bmatrix} \dfrac{\partial v}{\partial \xi} \\[2ex] \dfrac{\partial v}{\partial \eta} \end{Bmatrix} \tag{10.8.18b}$$

Combination of the two previous relations gives,

$$\begin{Bmatrix} \dfrac{\partial u}{\partial x} \\[2mm] \dfrac{\partial u}{\partial y} \\[2mm] \dfrac{\partial v}{\partial x} \\[2mm] \dfrac{\partial v}{\partial y} \end{Bmatrix} = \begin{bmatrix} \underset{\sim}{J}^{-1} & 0 \\[2mm] 0 & \underset{\sim}{J}^{-1} \end{bmatrix} \begin{Bmatrix} \dfrac{\partial u}{\partial \xi} \\[2mm] \dfrac{\partial u}{\partial \eta} \\[2mm] \dfrac{\partial v}{\partial \xi} \\[2mm] \dfrac{\partial v}{\partial \eta} \end{Bmatrix}$$

(10.8.19)

It can readily be proven that:

$$\begin{Bmatrix} \dfrac{\partial u}{\partial \xi} \\[2mm] \dfrac{\partial u}{\partial \eta} \\[2mm] \dfrac{\partial v}{\partial \xi} \\[2mm] \dfrac{\partial v}{\partial \eta} \end{Bmatrix} = \begin{bmatrix} \dfrac{\partial N_1}{\partial \xi} & 0 & \dfrac{\partial N_2}{\partial \xi} & 0 & \cdots \\[2mm] \dfrac{\partial N_1}{\partial \eta} & 0 & \dfrac{\partial N_2}{\partial \eta} & 0 & \cdots \\[2mm] 0 & \dfrac{\partial N_1}{\partial \xi} & 0 & \dfrac{\partial N_2}{\partial \xi} & \cdots \\[2mm] 0 & \dfrac{\partial N_1}{\partial \eta} & 0 & \dfrac{\partial N_2}{\partial \eta} & \cdots \end{bmatrix} \begin{Bmatrix} \underset{\sim}{u}_e \\[2mm] \underset{\sim}{v}_e \end{Bmatrix}$$

(10.8.20)

The B matrix can now be derived as,

$$\underset{\sim}{B}(\xi,\eta) = \begin{bmatrix} 1 & 0 & 0 & 0 \\ 0 & 0 & 0 & 1 \\ 0 & 1 & 1 & 0 \end{bmatrix} \begin{bmatrix} \underset{\sim}{J}^{-1} & 0 \\ 0 & \underset{\sim}{J}^{-1} \end{bmatrix} \begin{bmatrix} \dfrac{\partial N_1}{\partial \xi} & 0 & \dfrac{\partial N_2}{\partial \xi} & 0 & \cdots \\[2mm] \dfrac{\partial N_1}{\partial \eta} & 0 & \dfrac{\partial N_2}{\partial \eta} & 0 & \cdots \\[2mm] 0 & \dfrac{\partial N_1}{\partial \xi} & 0 & \dfrac{\partial N_2}{\partial \xi} & \cdots \\[2mm] 0 & \dfrac{\partial N_1}{\partial \eta} & 0 & \dfrac{\partial N_2}{\partial \eta} & \cdots \end{bmatrix}$$

(10.8.21)

Finally, the stiffness matrix is now obtained as,

$$\underset{\sim}{K} = \int_a \underset{\sim}{B}(x,y)^T \, \underset{\sim}{E}\underset{\sim}{B}(x,y)\,dxdy = \int_{-1}^{1}\int_{-1}^{1} \underset{\sim}{B}(\xi,\eta)^T \, \underset{\sim}{E}\underset{\sim}{B}(\xi,\eta)\det(J)\,d\xi d\eta$$

(10.8.22)

Evaluating the stiffness matrix in physical coordinates completes our discussion on the isoparametric formulation. Other element formulations and discussion on 3-D isoparametric elements which is nothing but a mere extension of the current form to another dimension, can be found in large amount of respective sources on FE formulation and hence is not discussed here.

Shape Functions in Natural Coordinates

The shape functions of the main 2-D elements whose formulations are presented, defined previously in natural coordinates, are given here. For the Q4 element they are,

$$N_1 = \frac{1}{4}(1-\xi)(1-\eta) \qquad (10.8.23a)$$

$$N_2 = \frac{1}{4}(1+\xi)(1-\eta) \qquad (10.8.23b)$$

$$N_3 = \frac{1}{4}(1+\xi)(1+\eta) \qquad (10.8.23c)$$

$$N_4 = \frac{1}{4}(1-\xi)(1+\eta) \qquad (10.8.23d)$$

For the LST,

$$N_1 = \xi(2\xi - 1) \qquad (10.8.24a)$$

$$N_2 = \eta(2\eta - 1) \qquad (10.8.24b)$$

$$N_3 = \zeta(2\zeta - 1) \qquad (10.8.24c)$$

$$N_4 = 4\xi\eta \qquad (10.8.24d)$$

$$N_5 = 4\zeta\eta \qquad (10.8.24e)$$

$$N_6 = 4\xi\zeta \qquad (10.8.24f)$$

where $\zeta = 1 - \xi - \eta$. Finally for the Q8 they are,

$$N_1 = \frac{(1-\xi)(\eta-1)(\xi+\eta+1)}{4} \qquad (10.8.25a)$$

$$N_2 = \frac{(1+\xi)(\eta-1)(-\xi+\eta+1)}{4} \qquad (10.8.25b)$$

$$N_3 = \frac{(1+\xi)(\eta+1)(\xi+\eta-1)}{4} \qquad (10.8.25c)$$

$$N_4 = \frac{(-1+\xi)(\eta+1)(\xi-\eta+1)}{4} \qquad (10.8.25d)$$

$$N_5 = \frac{(-\eta+1)(1-\xi^2)}{2} \qquad (10.8.25e)$$

$$N_6 = \frac{(\xi+1)(1-\eta^2)}{2} \qquad (10.8.25f)$$

$$N_7 = \frac{(\eta+1)(1-\xi^2)}{2} \qquad (10.8.25g)$$

$$N_8 = \frac{(1-\xi)(1-\eta^2)}{2} \qquad (10.8.25h)$$

In this chapter, the basics of the classical FEM are presented. It was intended to make a brief introduction to how to discretize a given differential equation with one or more field variables using various formulations. The discussions on the Direct Stiffness Method and Galerkin's

Method of Weighted Residual are also placed in relation to the finite element analysis. The Principle of Virtual Work is also described and its use is illustrated for a 1-D coupled problem whose decoupled form was formulated through the GMWR earlier.

A number of problems in 1-D and 2-D are defined and solved numerically. The reader is strongly advised to go through the example problems and check their calculations with the ones provided in this chapter and also with the analytical solutions given in the respective chapter. MATLAB codes provided at the end of the chapter should be run for various scenarios of material parameters and boundary conditions. Thus, the difference in results in terms of both field and state variables (i.e. DOF and stresses) should be observed. Possible interpretations should be made for better understanding the concept. The idea here is to present the chapter in a more practical manner but keeping the theoretical basis as much and as involved as possible at the same time for the readers. We hope that graduate students, researchers and engineers in their early stages of career will find it particularly beneficial.

10.9. Exercise Problems

10.9.1 Refering to example 10.4-2 and the Figure 10.4-14, write the PVW for the bar by constraining the right end also and placing the tip force into the mid-point of the bar.

10.9.2 Consider a column of soil under its own weight acting as a distributed load, w, (with unit weight γ, cross sectional area A, and elasticity modulus, E) as seen below. *Under no additional external forces*, do the following:

- Derive the governing equation for the displacement of the column, u, using the soil element on the left and its static equilibrium. Here F_b is the *body force* of the element shown.
- Solve the governing equation you obtained in the first part considering the boundary conditions shown in the figure analytically for the variation of the axial displacement $u(z)$ and also get the axial stress $\sigma(z)$. What is the displacement at the top and stress at the bottom?

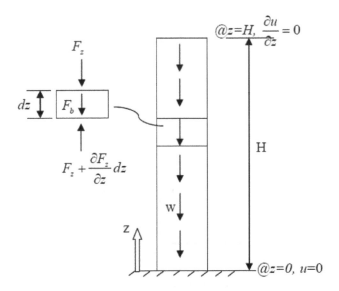

Figure 10.9-1 Column of soil under own weight

10.9.3 Redo the previous problem of column of soil under its own weight with the Method of Weighted Residuals (MWR) using,
- One term approximation (single term of a generalized polynomial)
- Two-term approximation (an additional higher order term)
  Hint: Your approximate solution, $u^*(z)$ could be in the form of $u^*(z) = c_i N_i(z)$ where c_i are constant coefficients and $N_i(z)$ are trial functions.

10.9.4 Obtain an approximate solution to the given differential equation using MWR considering the following:
- Two-term solution (your trial solution must have two trial functions)
- Another higher order term added where the exact solution is captured better.

$$D(y(x), x) = \frac{d^2y}{dx^2} + 2\frac{dy}{dx} - 5x^2 + 3 = 0, \ y(0) = y(-1) = 0 \text{ for } -1 \le x \le 0$$

10.9.5 Plot the variation of $u(z)$ and $\sigma(z)$ with the depth z of 10.9.2 above. Although you can use any software, MATLAB is recommended.

10.9.6 Plot the one term solution of the MWR in problem (10.9.3) along with the analytical solution you had developed. Verify your approximation.

10.9.7 For the given constant strain triangle (CST):
- (a) Obtain the shape function matrix, N. Here, you should get started by obtaining the generalized DOF for the given coordinate values, defined as, $u(x, y) = a_0 + a_1x + a_2y$, $v(x, y) = a_3 + a_4x + a_5y$ and obtain the N matrix of the CST for the given configuration. Hint: X = coordinates matrix containing the given (x_i, y_i) couples.
- (b) Get the strain-displacement matrix B using the N. Also calculate the stiffness matrix K. Consider the element behavior linear elastic with $E = 25$ MPa as the elasticity modulus and $n = 0.2$ as Poisson's ratio and obtain the strain ε, and stress σ vectors in terms of the DOFs.

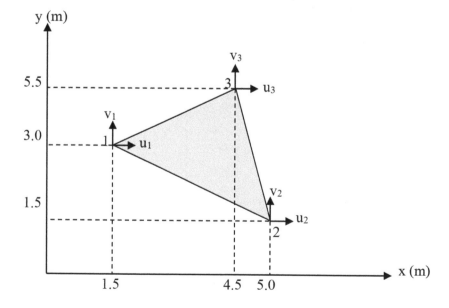

Figure 10.9-2 CST element

10.9.8 For the given situations, calculate the "consistent nodal load vectors" for the loaded edges.

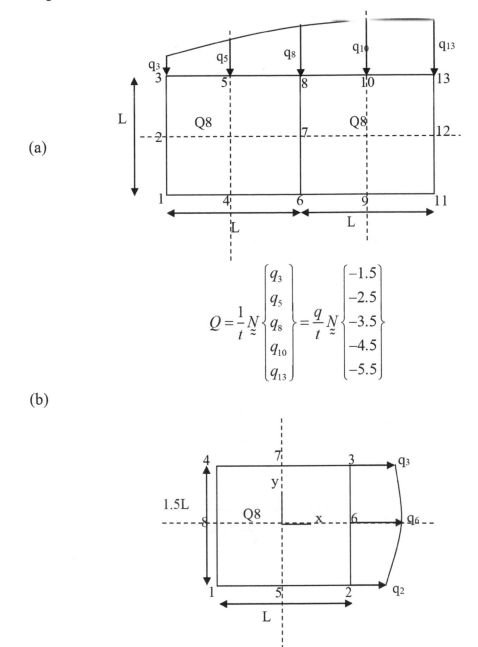

Figure 10.9-3 (a) Two Q8 elements loaded at the top, (b) One Q8 element loaded along the right edge

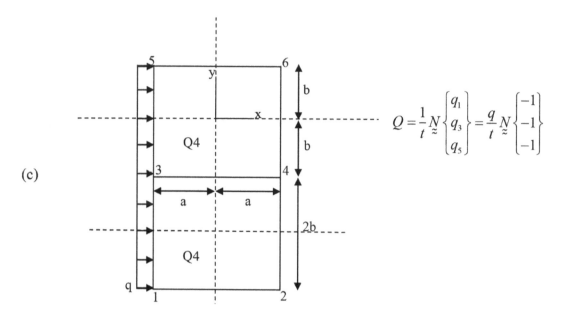

$$Q = \frac{1}{t}\underset{\sim}{N}\left\{\begin{matrix}q_1\\q_3\\q_5\end{matrix}\right\} = \frac{q}{t}\underset{\sim}{N}\left\{\begin{matrix}-1\\-1\\-1\end{matrix}\right\}$$

(c)

Figure 10.9-3 (c) Two Q4 elements loaded uniformly on the left edges

10.9.9 Consider the quasi-static consolidation of a foundation soil under constant loading as shown in the figure below. At the layer surface, effective vertical stress and the shear stress vanish and the pore pressure equals the applied load from the effective stress principle. The displacements as well as the gradient of pore pressure are specified as zero at the bottom impermeable boundary. Evaluate the pore pressure variation in time and depth using the FE formulation presented. Verify your results with the analytical solution of Schiffman et al. (1969) which can be seen below in Figure 10.9-2 as an example solution using parabolic elements with a time factor $T = 0.1$.

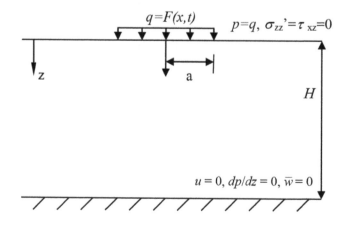

Figure 10.9-4 Consolidation of a single layer under constant step load

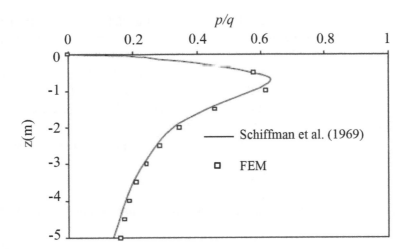

Figure 10.9-5 Pore pressure variation in depth as compared to Schiffman et al. (1969)

10.9.10. Use the MATLAB program called "**MYFEM_1D.m**" to do the following.
  (a) Compute the total head, ϕ, variation along the length of the 1-D horizontal soil layer using 10 elements. Modify the necessary parts of the code to get the FE solution. Plot the variation with the length as well as the analytical solution given in the following form: $\varphi_{exact} = 3(1 - x^2)$ for $S = 6$ m^3/hr, $k_x = 1$m/hr and $L = 1$ m.

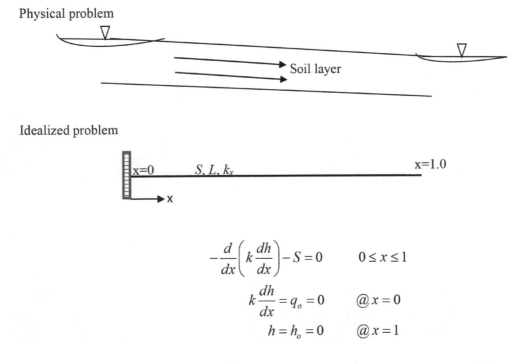

$$-\frac{d}{dx}\left(k\frac{dh}{dx}\right) - S = 0 \qquad 0 \le x \le 1$$

$$k\frac{dh}{dx} = q_o = 0 \qquad @\, x = 0$$

$$h = h_o = 0 \qquad @\, x = 1$$

Figure 10.9-6 Physical problem of 1-D seepage, its idealized problem and boundary conditions

(b) Exercise your new code (for 10 elements) to determine the effect of the soil permeability on the total head variation by changing it each time you run the code. Try $k_x = 0.01$ m/sn, $k_x = 0.001$ m/sn, $k_x = 0.0001$ m/sn, $k_x = 0.00001$ m/s. Show the variation of all on the same plot. Note that the entire soil will have to have the same permeability each time to be able to model its effect.

(c) Determine the effect of the source term, S, (fluid flux pumped into the system) on the total head through a parametric study.

(d) Change the boundary conditions in the way below and redo the problem using 10 elements. How do the results change? $k\dfrac{dh}{dx} = q_0 = 0$ @$x = 1$; $h = h_0 = 0.5$ @ $x = 0.2$; $h = h_0 = 0$ @ $x = 0$

10.9.11 Consider the embankment and its foundation soil below. Idealize the profile in 1-D and solve for the displacements, u_i (i = layer number) of each layer using the DSM ($A = 1$ m²). Calculate the stresses caused by elastic displacements. Neglect the stress distribution due to the embankment load in the soil and take the embankment load as a single force applied at the surface.

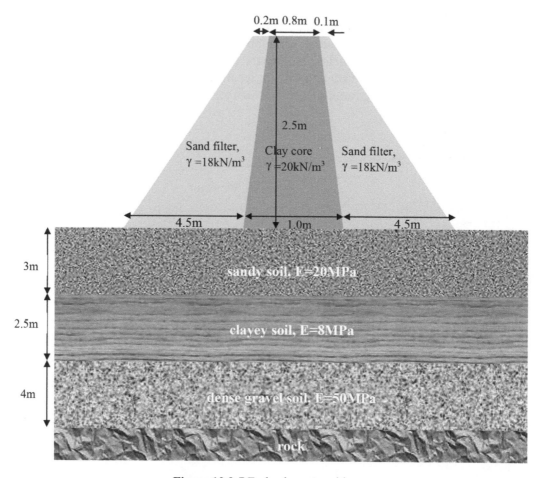

Figure 10.9-7 Embankment problem

10.9.12 Use the MATLAB program called "**FEM_2D_SEEP.m**" to do the following.

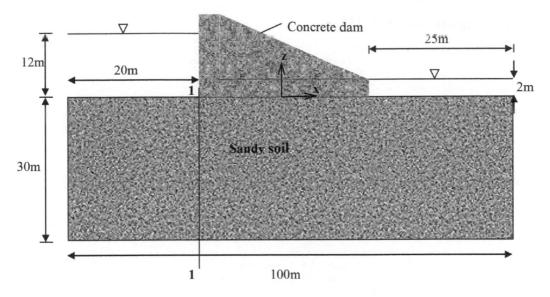

Figure 10.9-8 2-D seepage under a concrete dam

Provide your mesh everytime it changes. No need to plot the same mesh for each section if it does not change! Show eight flow lines and at least 12 equipotential lines in each part.

(a) Compute the 2-D equipotential and flow lines for the above soil layer considering the steady state seepage problem. Take $k_x = k_z = 0.00005$ m/s. Plot the total head variation with depth (φ-z) at the front corner of the dam (along section 1-1).

(b) Refine the mesh by a factor of two in both directions. That is, there will have to be twice as many elements in both directions as your previous analysis. Comment on the differences in the results and replot the (φ-z) variation at section 1-1. Then refine the mesh again by twice as much and comment on the new results. What can you say about the mesh refinement process in your analyses? Do we need to keep refining the mesh and get better results or is there a limit beyond which the results do not get considerably better? If so, what is that limit in this problem?

(c) Determine the effect of the soil anisotropic permeability on the total head variation by changing it each time you run the code. Try $k_x/k_z = 5$, $k_x/k_z = 10$, $k_x/k_z = 100$. Plot the (φ-z) variations at section 1-1 also.

(d) Determine the effect of the total head difference between the front and back of the concrete dam on the results by doing a parametric study. Choose the ratios $\phi_{left}/\phi_{right} = 2, 3, 4$ and 6 (current value).

(e) Specify a thin layer of clay with 1 m thickness and a very low permeability of $k_x = k_z = 1$ E-7 m/s at the surface. Reanalyze the problem and comment on the results. Feel free to choose any dimensions, total head differences and other material parameters etc. in addition to the ones given in the figure. Choose a moderate mesh of Q4 elements. Make sure that your FE mesh captures the thickness of the clay layer.

MATLAB® Scripts

Main program "MYFEM_1D.m"

```
%-------------------------------------------------------------------
% To solve the ordinary differential equation given as
% -a u''    = s,  0 < x < H
% u(1) = ubval  and  u'(0) = fbval
%
% Variable descriptions
% k = element matrix f = element vector kk = system matrix ff = system
% vector index = a vector containing system dofs associated with each
% element ubdof = a vector containing dofs associated with boundary
% conditions ubval = a vector containing boundary condition values
% associated with the dofs in 'bcdof'
clear all
clc
%-------------------------------------------------------------------
%   input data
%-----------------------------------
nel=3;                  % number of elements
nnel=2;                 % number of nodes per element
L=1;                    % length
k=1;                    % material constant i.e. permeability
s=6;                    % source term
dx=L/nel;               % coordinate increments
nnode=nel+1;            % total number of nodes in system
sdof=nnode;             % total system dofs
%-------------------------------------
%   input data for nodal coordinate values
%-------------------------------------
x(1)=0.0;
for iel=2:nnode
    x(iel)=x(iel-1)+dx;
end
%---------------------------------------------------
%   input data for nodal connectivity for each element
%---------------------------------------------------
for i=1:nnode
    b(i,1)=i;
    b(i,2)=i+1;
end
%-------------------------------------------
% input data for coefficients of the ODE [may be different for each
% element]
%-------------------------------------------
a=k*ones(nel,1);        % permeability vector
s=s*ones(nel,1);        % source term vector
%-----------------------------------
%   input data for boundary conditions
%-----------------------------------
nub=1;                  % number of nodes with specified values of 'u'
ubdof= [4 ];            % nodal numbers of the nodes with specified 'u' values
ubval= [0.];            % the corresponding values of 'u'
nfb=1;                  % number of nodes with specified values of flux
fbdof= [1];             % nodal numbers of the nodes with specified flux
fbval=[0.0];            % the corresponding values of specified flux
%-------------------------------------------
% initialization of matrices and vectors
%-------------------------------------------
```

```
ff=zeros(sdof,1);                    % initialization of system force vector
kk=zeros(sdof,sdof);                 % initialization of system matrix
index=zeros(nnel,1);                 % initialization of index vector
%----------------------------------------------------------------
%   computation of element matrices and vectors and their assembly
%----------------------------------------------------------------
for iel=1:nel                        % loop for the total number of elements
    ae=a(iel);
    se=s(iel);
    nl=b(iel,1); nr=b(iel,2);        % extract nodes for (iel)-th element
    xl=x(nl); xr=x(nr);              % extract nodal coord values for the
                                       element
    eleng=xr-xl;                      % element length
% extract system dofs associated with element
    edof = nnel;
    start = (iel-1)*(nnel-1);
    for i=1:edof
        index(i)=start+i;
    end
% compute element matrix
k=(ae/eleng).*[1 -1;-1 1];
% compute element vector
    f=(se*eleng/2.).*[1;1];
%-----------------------------------------------------------------------
% assemble element matrices and vectors to get the system matrix and
% vector
%-----------------------------------------------------------------------
    edof = length(index);
    for i=1:edof
        ii=index(i);
        ff(ii)=ff(ii)+f(i);
        for j=1:edof
            jj=index(j);
            kk(ii,jj)=kk(ii,jj)+k(i,j);
        end
    end
end
%----------------------------------------------------------
%   apply the natural boundary condition
%----------------------------------------------------------
for ifb=1:nfb
    ff(fbdof(ifb))=ff(fbdof(ifb))-fbval(ifb);
end
%-----------------------------------------------
%   apply the essential boundary conditions
%-----------------------------------------------
n=length(ubdof);
sdof=size(kk);
for i=1:n
    c=ubdof(i);
    for j=1:sdof
        kk(c,j)=0;
    end
    kk(c,c)=1;
    ff(c)=ubval(i);
end
%-----------------------------------------------------------
%   solve the matrix equation
```

```
%------------------------------------------------------------
uu=kk\ff;
%------------------------------------------------------------
%   print and plot the results
% ----------------------------------------------------------
plot(x,uu','--ro')
title(' distribution of the primary variable')
xlabel('distance, x')
ylabel('primary variable, u')
hold on
x = 0.0:0.01:1.0;
u = 3*(1.-x.^2);
plot(x,u);num=1:sdof;store=[num' uu];
```

Main program "FEM_1D_SS.m"

```
% ***********************************************************************
*
% To solve the ordinary differential equation given as
% a u'' + b u' + c u = d,  0 < x < H
% u(0) = bcval  and  u'(1) = fbval
% Variable descriptions
% k = element matrix
% f = element vector
% kk = system matrix
% ff = system vector
% index = a vector containing system dofs associated with each element
% bcdof = a vector containing dofs associated with boundary conditions
% bcval = a vector containing boundary condition values associated with
% the dofs in 'bcdof'
%***********************************************************************
*
% clear all
% clc
%----------------------------------------
%  input data for control parameters
%----------------------------------------
H=1;                    % length of domain
nel=10;                 % number of elements
nnel=2;                 % number of nodes per element
ndof=1;                 % number of dofs per node
d0=0.05;                % initial source term
del_de=0.1;             % increment of the source term
nnode=nel+1;            % total number of nodes in system
sdof=nnode*ndof;        % total system dofs
eleng=H/nel;            % element length
%----------------------------------------
%  input data for nodal coordinate values
%----------------------------------------
for i=1:nnode
    gcoord(i)=(i-1)*eleng;
end
%-----------------------------------------------------
%  input data for nodal connectivity for each element
%-----------------------------------------------------
for i=1:nel
    nodes(i,1)=i;
    nodes(i,2)=i+1;
```

```
end
%--------------------------------------------
%   input data for coefficients of the ODE [may be different for each element]
%--------------------------------------------
ae=-ones(1,nel);                  % coefficient 'a' of the diff eqn
be=zeros(1,nel);                  % coefficient 'b' of the diff eqn
ce=zeros(1,nel);                  % coefficient 'c' of the diff eqn
de(1)=d0;
for i=2:nel
    de(i)=de(i-1)+del_de;         % coeffecient 'd' [ value of the source
                                  term]
end
%----------------------------------------
%   input data for boundary conditions
%----------------------------------------
ncb=1;                % number of nodes with specified values of 'u'
bcdof=[1];            % nodal numbers of the nodes with specified 'u' values
bcval=[0.];           % the corresponding values of 'u'

nfb=1;                % number of nodes with specified values of flux
fbdof=[ nnode];       % nodal numbers of the nodes with specified flux
fbval=[0.0];          % the corresponding values of specified flux
%----------------------------------------
%   initialization of matrices and vectors
%----------------------------------------
ff=zeros(sdof,1);                 % initialization of system force vector
kk=zeros(sdof,sdof);              % initialization of system matrix
mm=zeros(sdof,sdof);              % initialization of system mass matrix/not
                                  to be used
index=zeros(nnel*ndof,1);         % initialization of index vector
%-----------------------------------------------------------------
%   computation of element matrices and vectors and their assembly
%-----------------------------------------------------------------
for iel=1:nel                     % loop for the total number of elements
    acoef=ae(iel);
    bcoef=be(iel);
    ccoef=ce(iel);
    dcoef=de(iel);

    nl=nodes(iel,1); nr=nodes(iel,2);     % extract nodes for (iel)-th
                                          element
    xl=gcoord(nl); xr=gcoord(nr);         % extract nodal coord values for
                                          the element
    eleng=xr-xl;                          % element length
    index=feeldof1(iel,nnel,ndof);        % extract system dofs associated
                                          with element
    [k]=feodel1(acoef,bcoef,ccoef,eleng)  % compute element matrix
    sm=size(k);
    [m]=zeros(sm(1),sm(2));               % only to be defined for the
                                          assembly function
    f=fefl1(dcoef,xl,xr)                  % compute element vector
    [kk,mm,ff]=feasmbl1(kk,mm,ff,k,m,f,index)   % assemble matrices and
                                                vectors
end
%-----------------------------------------------------------------
%   apply the natural boundary condition at the last node
%-----------------------------------------------------------------
```

```
for ifb=1:nfb
    ff(fbdof(ifb))=ff(fbdof(ifb))-fbval(ifb);
end
%----------------------------
%   apply boundary conditions
%----------------------------
[kk,ff]=feapplybc(kk,ff,bcdof,bcval);
%----------------------------
%  solve the matrix equation
%----------------------------
uu=kk\ff;
%----------------------------------------------------------
%   print and plot the results
% ----------------------------------------------------------
plot(gcoord,uu','ro')
title(' distribution of the primary variable')
xlabel('distance, x')
ylabel('primary variable, u')
hold on
%u = dsolve('D2u = -x', 'u(0) = 0', '-Du(1) = 0', 'x');
% Analytical solution
syms x
x = 0.0:0.01:1.0;
u = -1/6*x.^3+1/2*x;
plot(x,u);
num=1:1:sdof;
store=[num' uu]
%----------------------------------------------------------
```

Subprogram "fef1l.m"

```
function [f]=fef1l(dcoef,xl,xr)
%----------------------------------------------------------------
% Purpose:
% element vector for f(x)=1
% using linear element
% Synopsis:
% [f]=fef1l(dcoef,xl,xr)
% Variable Description:
% f - element vector (size of 2x1)
% xl - coordinate value of the left node
% xr - coordinate value of the right node
%----------------------------------------------------------------
% element vector
eleng=xr-xl;            % element length
fu=[1;1];
f=(dcoef*eleng/2.)*fu;
end
%----------------------------------------------------------------
```

Subprogram "feode1l.m"

```
function [k]=feode1l(acoef,bcoef,ccoef,eleng)
%----------------------------------------------------------------
% Purpose:
% element matrix for (a u'' + b u' + c u)
% using linear element
% Synopsis:
% [k]=feode1l(acoef,bcoef,ccoef,eleng)
```

```
% Variable Description:
% k - element matrix (size of 2x2)
% acoef - coefficient of the second order derivative term
% bcoef - coefficient of the first order derivative term
% ccoef - coefficient of the zero-th order derivative term
% eleng - element length
%-------------------------------------------------------------------
% element matrix
a1=-(acoef/eleng);  a2=bcoef/2; a3=ccoef*eleng/6;
k=[ a1-a2+2*a3  -a1+a2+a3;...
    -a1+a2+a3   a1+a2+2*a3];
end
%-------------------------------------------------------------------
```

Subprogram "feeldof1.m"

```
function [index]=feeldof1(iel,nnel,ndof)
%-----------------------------------------------------------
% Purpose:
% Compute system dofs associated with each element in one-
% dimensional problem
% Synopsis:
% [index]=feeldof1(iel,nnel,ndof)
% Variable Description:
% index - system dof vector associated with element "iel"
% iel - element number whose system dofs are to be determined
% nnel - number of nodes per element
% ndof - number of dofs per node
%-----------------------------------------------------------
edof = nnel*ndof;
start = (iel-1)*(nnel-1)*ndof;
for i=1:edof
    index(i)=start+i;
end
end
%-----------------------------------------------------------
```

Subprogram "feasmbl1.m"

```
function [kk,mm,ff]=feasmbl1(kk,mm,ff,k,m,f,index)
%-----------------------------------------------------------
% Purpose:
% Assembly of element matrices into the system matrix &
% Assembly of element vectors into the system vector
% Synopsis:
% [kk,mm,ff]=feasmbl1(kk,mm,ff,k,f,index)
% Variable Description:
% kk - system flow  matrix
% mm - system vol. change matrix
% ff - system vector
% k  - element matrix
% f  - element vector
% index - d.o.f. vector associated with an element
%-----------------------------------------------------------
edof = length(index);
for i=1:edof
    ii=index(i);
    ff(ii)=ff(ii)+f(i);
```

```
        for j=1:edof
            jj=index(j);
            kk(ii,jj)=kk(ii,jj)+k(i,j);
            mm(ii,jj)=mm(ii,jj)+m(i,j);
        end
end
end
%------------------------------------------------------------
```

Subprogram "feapplybc.m"

```
function [kk,ff]=feapplybc(kk,ff,bcdof,bcval)
%------------------------------------------------------------
% Purpose:
% Apply constraints to matrix equation [kk]{x}={ff}
% Synopsis:
% [kk,ff]=feapplybc(kk,ff,bcdof,bcval)
% Variable Description:
% kk - system matrix before applying constraints
% ff - system vector before applying constraints
% bcdof - a vector containging constrained d.o.f
% bcval - a vector containing contained value
% If there are constraints at d.o.f=2 and 10 and their values are 0.0 and
% 2.5, then, bcdof(1)=2 and bcdof(2)=10; and bcval(1)=1.0 and bcval(2)=2.5.
%------------------------------------------------------------
n=length(bcdof);
sdof=size(kk);
for i=1:n
    c=bcdof(i);
    for j=1:sdof
        kk(c,j)=0;
    end
    kk(c,c)=1;
    ff(c)=bcval(i);
end
end
%------------------------------------------------------------
```

Main program "FEM_1D_TR.m"

```
% To solve the ordinary differential equation given as
% a u'' +b u' + c u = d (df/dt)+e(du/dt),  0 < x < H
% u(0) = bcval  and  u'(1) = fbval
%
% Variable descriptions
% k = element matrix
% f = element vector
% kk = system matrix
% ff = system vector
% index = a vector containing system dofs associated with each element
% bcdof = a vector containing dofs associated with boundary conditions
% bcval = a vector containing boundary condition values associated with
%           the dofs in 'bcdof'
%-------------------------------------------------------------------------
clear all
clf
%-----------------------------------
%  input data for control parameters
%-----------------------------------
```

```
nel=10;                    % number of elements
nnel=2;                    % number of nodes per element
ndof=1;                    % number of dofs per node
nnode=11;                  % total number of nodes in system
sdof=nnode*ndof;           % total system dofs
%-------------------------------------------
%  input data for nodal coordinate values
%-------------------------------------------
z=0:.1:1.;
gcoord=z;
%-----------------------------------------------------------
%  input data for nodal connectivity for each element
%-----------------------------------------------------------
for iel=1:nel
    nodes(iel,1)=iel;
    nodes(iel,2)=iel+1;
end
%---------------------------------------------
%  input data for coefficients of the ODE [may be different for each element]
%---------------------------------------------
rhog= 1.;                             % unit eight of water
for iel=1:nel
    perm(iel)=1.;                     % coefficient of permeability
    mv(iel)= 1.;                      % coefficient of volume change
    ae(iel)= -(perm(iel)/rhog);       % coefficient 'a' of the diff eqn
    be(iel)=0.;                       % coefficient 'b' of the diff eqn
    ce(iel)=0.;;                      % coefficient 'c' of the diff eqn
    de(iel)= mv(iel);                 % coeffecient 'd' [ value of the source
                                           term]
    ee(iel)= -mv(iel);                % coefficient 'e' [ time dependent term]
end
%---------------------------------------------------------------------------
%  input data for initial condition
for i=1:nnode
    uu(i)=1.;
end
uu=uu';
%---------------------------------------------------------------------------
%  input data for boundary conditions
%-------------------------------------------
ncb=1;                     % number of nodes with specified values of 'u'
bcdof= [1];                % nodal numbers of the nodes with specified 'u' values
bcval= [0];                % the corresponding values of 'u'
nfb=1;                     % number of nodes with specified values of flux
fbdof= [11];               % nodal numbers of the nodes with specified flux
fbval=[0];                 % the corresponding values of specified flux
%-----------------------------------------------------------
%  input data for the loading history and time integration
%-----------------------------------------------------------
t0=0.;
dt=0.1;
nt=11;
tf=t0+(nt-1)*dt;
t=t0:dt:tf;
for i=1:nt
    load(i)=1.;
end
```

```
theta= 0.5;
%----------------------------------------------------------------
%   initialization of matrices and vectors
%----------------------------------------------------------------
ff=zeros(sdof,1);            % initialization of system force vector
kk=zeros(sdof,sdof);         % initialization of system flow matrix
mm=zeros(sdof,sdof);         % initialization of  system vol. change matrix
index=zeros(nnel*ndof,1);    % initialization of index vector
%----------------------------------------------------------------
%   computation of element matrices and vectors and their assembly
%----------------------------------------------------------------
for iel=1:nel               % loop for the total number of elements
    acoef=ae(iel);
    bcoef=be(iel);
    ccoef=ce(iel);
    dcoef=de(iel);
    ecoef=ee(iel);
    nl=nodes(iel,1); nr=nodes(iel,2);   % extract nodes for (iel)-th element
    xl=gcoord(nl); xr=gcoord(nr);       % extract nodal coord values for the
                                           element
    eleng=xr-xl;                        % element length
    index=feeldof1(iel,nnel,ndof);      % extract system dofs associated with
                                           element
    [k,m]=feode21(acoef,bcoef,ccoef,ecoef,eleng);  % element matrices
    f=fef11(dcoef,xl,xr);                          % element vector
    [kk,mm,ff]=feasmbl1(kk,mm,ff,k,m,f,index);     % assembly
end
%   solution in the time domain: solve the final matrix equation
%----------------------------------------------------------------
kk=mm+theta*dt*kk;
mm=mm-(1.-theta)*dt*kk;
%----------------------------------------------------------------
ut(:,1)=uu;
for it=2:nt
    ff=mm*ut(:,it-1)+(load(it)-load(it-1))*ff;
%----------------------------------------------------------------
%   apply the natural boundary condition at the last node
%----------------------------------------------------------------
    for ifb=1:nfb
        ff(fbdof(ifb))=ff(fbdof(ifb))+dt*fbval(ifb);
    end
%----------------------------------------
%   apply boundary conditions
%----------------------------------------
    [kk,ff]=feapplybc(kk,ff,bcdof,bcval);
%----------------------------------------------------------------
%   Solve the final Equation
%----------------------------------------------------------------
    ut(:,it)=kk\ff;
end
%----------------------------------------------------------------
%   print and plot the results
%----------------------------------------------------------------
subplot(211)
plot(ut,-gcoord)
title('pore pressure distribution')
ylabel('depth, z')
xlabel('pore pressure, u)')
```

```
subplot(212)
plot(t,ut')
title('pore pressure time history')
ylabel('pore pressure, u')
xlabel('time, t')
%----------------------------------------------------------------
```

Subprogram "feode2l.m"

```
function [k,m]=feode2l(acoef,bcoef,ccoef,ecoef,eleng)
%----------------------------------------------------------------
% Purpose:
% element matrix for (a u'' + b u' + c u)
% using linear elements
% Synopsis:
% [k,m]=feode2l(acoef,bcoef,ccoef,ecoef,eleng)
% Variable Description:
% k - element matrix (size of 2x2)
% m - element volume change matrix
% acoef - coefficient of the second order derivative term
% bcoef - coefficient of the first order derivative term
% ccoef - coefficient of the zero-th order derivative term
% ecoef - coeffecient of volume change
% eleng - element length
%----------------------------------------------------------------
% element matrix
a1=-(acoef/eleng);  a2=bcoef/2;  a3=ccoef*eleng/6;
k=[ a1-a2+2*a3      -a1+a2+a3;...
    -a1+a2+a3        a1+a2+2*a3];
% element volume change matrix
cterm= -ecoef*eleng/6;
m= [2*cterm cterm; cterm 2*cterm];
%----------------------------------------------------------------
end
```

Main program "FEM_Consol1D.m"

```
%----------------------------------------
%  FE Solution of 1-D Flow-Deformation Coupled Problem (Consolidation)
% (For Constant (q=q0) Loading, Including the Compressibility of Water)
%----------------------------------------
% disp ('Enter the # of elements')
% nel=input ('prompt')
% disp('How many plots would you like to see between t1=1sec and tlast=nt')
% np=input ('prompt')
clear all
close all
N=6;
%----------------------------------------
%  input data for FEM
%----------------------------------------
nel=50;                 % number of elements
nnel=2;                 % number of nodes per element
ndof=1;                 % number of dofs per node for one half of the system
                          assuming same shape func. for u and p
nnode=nel+1;            % total number of nodes in system
sdof=nnode*ndof;        % total dofs for u part
%----------------------------------------
```

```
%   input data for nodal coordinate values
%----------------------------------------
H=10;                       % Depth of layer
dz=H/nel;                   % Depth increment
z=0:dz:H;
gcoord=z;
zz=z/H;
%-------------------------------------------------------
%   input data for nodal connectivity for each element
%-------------------------------------------------------
for iel=1:nel;
    nodes(iel,1)=iel;
    nodes(iel,2)=iel+1;
end
%-----------------------------------------
%   input data for material
%-----------------------------------------
E=500000;                          % Elasticity Modulus (kPa)
mu=0.4;                            % Poisson's Ratio
kz=0.0001;                         % Permeability (m/s)
kf=20000000;                       % Compressibility of fluid
n=0.4;                             % Porosity
rho=1.9;                           % Density of soil+water (t/m^3)
rhof=1;                            % Density of water
g=10;                              % Acc. of Gravity
D=E*(1-mu)/((1+mu)*(1-2*mu));      % Constrained Modulus of Soil Skeleton
%-----------------------------------------------------------------
%   input data for the loading history and time integration
%-----------------------------------------------------------------
q0=1;                              % Load amplitude (kN)
t0=0.;
dt=0.01;
nt=1000;
Tcc=nt*dt;
tf=t0+(nt-1)*dt;
t=t0:dt:tf;
for i=1:nt
    load(i)=q0;
end
bf=rhof*g;
theta=1;
%-------------------------------------------
%   input data for boundary conditions
%-------------------------------------------
ncb=2;                     % number of nodes with specified values of 'u and p'
bcdof= [nnode nnode+1]; % nodal numbers of the nodes with specified 'u and p'
                           values
bcval= [0 0];              % the corresponding values of 'u and p'
%-------------------------------------------
%   initialization of global matrices and vectors
%-------------------------------------------
Cf=zeros(sdof,sdof);    % initialization of system comp. matrix
Ks=zeros(sdof,sdof);    % initialization of system stiffness matrix
Kf=zeros(sdof,sdof);    % initialization of system flow matrix
C=zeros(sdof,sdof);     % initialization of system coupling matrix
F=zeros(sdof,1);        % initialization of system force vector
P=zeros(sdof,1);        % initialization of pressure vector coming from u
                           eqn.
```

```
Z=zeros(sdof,1);          % initialization of system force vector coming from p
                            eqn.
%------------------------------------------------------------------
%  computation of system matrices and vectors
%------------------------------------------------------------------
na=nodes(iel,1);    nb=nodes(iel,2);   % extract nodes for (iel)-th element
za=gcoord(na);      zb=gcoord(nb);     % extract nodal coord values for the
                                         element
eleng=zb-za;                           % element length
% Construction of Coupling Matrix
C(1,1)=-1/2;
for j=1:nel
    C(j,j+1)=-1/2;
    C(j+1,j)=1/2;
    C(j+1,j+1)=0;
end
C(sdof,sdof)=1/2;
% Construction of Stiffness Matrix
Ks(1,1)=D/eleng;
for j=1:nel
    Ks(j,j+1)=-D/eleng;
    Ks(j+1,j)=-D/eleng;
    Ks(j+1,j+1)=2*D/eleng;
end
Ks(sdof,sdof)=(D/eleng);
% Construction of Flow Matrix
Kf(1,1)=kz/(bf*eleng);
for j=1:nel
    Kf(j,j+1)=-kz/(bf*eleng);
    Kf(j+1,j)=-kz/(bf*eleng);
    Kf(j+1,j+1)=2*kz/(bf*eleng);
end
Kf(sdof,sdof)=kz/(bf*eleng);
% Construction of Fluid Comp. Matrix
Cf(1,1)=(n*eleng)/(3*kf);
for j=1:nel
    Cf(j,j+1)=(n*eleng)/(6*kf);
    Cf(j+1,j)=(n*eleng)/(6*kf);
    Cf(j+1,j+1)=(2*n*eleng)/(3*kf);
end
Cf(sdof,sdof)=(n*eleng)/(3*kf);
%----------------------------------------------------------
%  Construction of the final matrices
%  A(Xn+1)=B(Xn)+F
%----------------------------------------------------------
A=[theta*Ks -theta*C;theta*C' (theta^2*dt*Kf+theta*Cf)];
B=[(theta-1)*Ks -(theta-1)*C;theta*C' -((theta-theta^2)*dt*Kf-theta*Cf)];
F(1)=load(1);
S=[F;Z];
%----------------------------------------------------------
%  Initial Values at the end of first time step!!
%----------------------------------------------------------
X(1:sdof,1)=zeros(sdof,1);          % Initial Displacement
X(sdof+1:2*sdof,1)=load(1);         % Initial Pore Pressure
%----------------------------------------------------------
%  Solution of the final matrix equation in the time domain
%----------------------------------------------------------
for it=2:nt
```

```
      S(1)=theta*load(it)+(1-theta)*load(it-1);
%------------------------------------------------------------
%    Apply essential boundary conditions
%------------------------------------------------------------
      size=length(S);
      for i=1:ncb
          y=bcdof(i);
          for j=1:size
              A(y,j)=0;          % make the yth row zero
          end
          A(y,y)=1;              % make the yth diagonal element one
          S(y)=bcval(i);         % adjust the RHS force vector accordingly
      end
      R=B*X(:,it-1)+S;
%------------------------------------------------------------
%    Solve the Final Equation System
%------------------------------------------------------------
      X(:,it)=A\R;
end
%------------------------------------------------------------
%    Separate u and p
%------------------------------------------------------------
m=size;
uu=X(1:sdof,1:nt);
uur=real(uu);
uu=uu';
pp=X(sdof+1:m,1:nt);
ppr=real(pp);
for i=1:nt
    PPR(:,i)=ppr(:,nt-i+1);
end
pp_FEM=abs(pp');
%------------------------------------------------------------
%    Analytical Solution For 1-D Consolidation
%------------------------------------------------------------
Hd=2*H;
mv=1/D;
cv=kz/(mv*g);
Pp=0;
for n=1:2000
    f1=exp(-cv*(n*pi/Hd)^2*t);
    f2=(2-2*cos(n*pi))/(n*pi)*sin(n*pi*z/Hd);
    Pp=Pp+q0*f1'*f2;
end
pp_exact=abs(Pp);
dt=nt/N;
%------------------------------------------------------------
%    Plot the results
%------------------------------------------------------------
% Pore Pressure
figure
plot(t,pp_exact(:,1:5:sdof)/q0,'k')
hold on
plot(t,pp_FEM(:,1:5:sdof)/q0,'--k')
title('absolute value of pore pressure distribution in time')
ylabel('Pore pressure/load')
xlabel('time')
figure
```

```
plot(pp_exact(1:dt:nt,:)/q0,-zz,'k')
hold on
plot(pp_FEM(1:dt:nt,:)/q0,-zz,'--k')
xlabel('Pore pressure/load')
ylabel('z/H')
title('absolute value of pore pressure distribution in depth')
%-----------------------------------------------------------
% Displacement
%-----------------------------------------------------------
figure
subplot(211)
plot(uu(1:100:nt,:),-gcoord,'k')
title('Vertical Displacement Distribution')
ylabel('depth, z')
xlabel('displacement, u')
% Time History
subplot(212)
plot(t,uu(:,1:5:sdof)','k')
title('Displacement Time History')
ylabel('displacement, u')
xlabel('time, t')
```

Main program "FE_Coupled_1D_QS.m"

```
% 1-D Quasi-Static Coupled Flow-Deformation Solution
% For Periodic (q=q0*e^iwt) Loading
% Including the Compressibility of Water
% Body forces and all inertial terms are ignored
%--------------------------------------------------
%  QS Formulation
%--------------------------------------------------
clear all
close all
clc
%-----------------------------------------
%  input data for FEM
%-----------------------------------------
nel=100;              % number of elements
np=4;                 % number of plots to present
Np=5;                 % plot number to present
nnel=2;               % number of nodes per element
ndof=1;               % number of dofs per node per field
nnode=nel+1;          % total number of nodes in system
sdof=nnode*ndof;      % total system dofs per field
%-----------------------------------------
%  input data for nodal coordinate values
%-----------------------------------------
H=10;                 % Depth of layer
dz=H/nel;             % Depth increment
z=0:dz:H;
gcoord=z;
zz=z/H;
%---------------------------------------------------------
%  input data for nodal connectivity for each element
%---------------------------------------------------------
for iel=1:nel
    nodes(iel,1)=iel;
    nodes(iel,2)=iel+1;
```

```
end
%-------------------------------------------
%  input data for material
%-------------------------------------------
E=750000;                              % Elasticity Modulus (kPa)
mu=0.2;                                % Poisson's Ratio
kz=0.001;                              % Permeability (m/s)
kf=3330000;                            % Compressibility of water
n=0.3333;                              % Porosity
rho=2;                                 % Density of soil (t/m^3)
rhof=1;                                % Fluid Unit Weight
g=9.81;                                % Acc. of Gravity
D=E*(1-mu)/((1+mu)*(1-2*mu));          % Constrained Modulus
bf=rhof*g;
%-----------------------------------------------------------------
%  input data for the loading history and time integration
%-----------------------------------------------------------------
q0=1;                    % Load amplitude (kN)
w=3.379; %1;             % Frequency
T=2*pi/w;                % Period of load
t0=0.;
dt=0.01;
nt=5000;
tf=t0+(nt-1)*dt;
t=t0:dt:tf;
for i=1:nt
    load(i)=q0*exp(sqrt(-1)*w*t(i));
end
% Newmark Parameters
gama=0.5;     % must be > 0.5 for numerical damping
beta=0.25*(gama+0.5)^2;
%-----------------------------------------
%  input data for boundary conditions
%-----------------------------------------
ncb=2;                   % number of nodes with specified values of 'u and p'
bcdof= [nnode nnode+1]; % dof numbers of the nodes with specified 'u and p'
                         values
bcval= [0 0];            % the corresponding values of 'u and p'
%-----------------------------------------
%  initialization of global matrices and vectors
%-----------------------------------------
Cf=zeros(sdof,sdof);     % initialization of system comp. matrix
Ks=zeros(sdof,sdof);     % initialization of system stiffness matrix
Kf=zeros(sdof,sdof);     % initialization of system flow matrix
C=zeros(sdof,sdof);      % initialization of system coupling matrix
F=zeros(sdof,1);         % initialization of system force vector
P=zeros(sdof,1);         % initialization of pressure vector coming from u
                           eqn.
Z=zeros(sdof,1);         % initialization of system force vector coming from p
                           eqn.
Vert_Stress=zeros(nel*2,1);
f_int=[zeros(2*sdof,1)];
%-----------------------------------------------------------------
%  computation of system matrices and vectors
%-----------------------------------------------------------------
na=nodes(iel,1); nb=nodes(iel,2);   % extract nodes for (iel)-th element
za=gcoord(na); zb=gcoord(nb);       % extract nodal coord values for the
                                      element
```

```
eleng=zb-za;                           % element length
% Construction of Coupling Matrix
C(1,1)=-1/2;
for j=1:nel
    C(j,j+1)=-1/2;
    C(j+1,j)=1/2;
    C(j+1,j+1)=0;
end
C(sdof,sdof)=1/2;
% Construction of Stiffness Matrix
Ks(1,1)=D/eleng;
for j=1:nel
    Ks(j,j+1)=-D/eleng;
    Ks(j+1,j)=-D/eleng;
    Ks(j+1,j+1)=2*D/eleng;
end
Ks(sdof,sdof)=(D/eleng);
% Construction of Flow Matrix
Kf(1,1)=kz/(bf*eleng);
for j=1:nel
    Kf(j,j+1)=-kz/(bf*eleng);
    Kf(j+1,j)=-kz/(bf*eleng);
    Kf(j+1,j+1)=2*kz/(bf*eleng);
end
Kf(sdof,sdof)=kz/(bf*eleng);
% Construction of Fluid Comp. Matrix
Cf(1,1)=(n*eleng)/(3*kf);
for j=1:nel
    Cf(j,j+1)=(n*eleng)/(6*kf);
    Cf(j+1,j)=(n*eleng)/(6*kf);
    Cf(j+1,j+1)=(2*n*eleng)/(3*kf);
end
Cf(sdof,sdof)=(n*eleng)/(3*kf);
%------------------------------------------------------------
%  Construction of the final matrices
%------------------------------------------------------------
cc=[zeros(sdof,sdof) zeros(sdof,sdof);C' Cf];
kk=[Ks -C;zeros(sdof,sdof) Kf];
LHS=(gama/(beta*dt))*cc+kk;
F(1)=load(1);
S=[F;Z];
%------------------------------------------------------------
%  Initial Conditions
%------------------------------------------------------------
X(:,1)=zeros(2*sdof,1);        % Initial DOF u and p
V(:,1)=zeros(2*sdof,1);        % Initial velocity
A(:,1)=zeros(2*sdof,1);        % Initial acceleration
%------------------------------------------------------------
%  Solution of the final matrix equation in the time domain
%------------------------------------------------------------
for it=1:nt
    S(1)=load(it);
%------------------------------------------------------------
%  Apply essential boundary conditions
%------------------------------------------------------------
    R1=cc*((((gama/(beta*2))-1)*dt*A(:,it)+((gama/beta)-1)*V(:,it)+(gama/
    (beta*dt))*X(:,it)));
    RHS=R1+S;
```

```
    size=length(LHS);
    for i=1:ncb
        y=bcdof(i);
        for j=1:size
            LHS(y,j)=0;                  % make the yth row zero
        end
        LHS(y,y)=1;                      % make the yth diagonal element one
        RHS(y)=bcval(i);                 % adjust the RHS force vector accordingly
    end
%------------------------------------------------------------
%   Solve the Final Equation System
%------------------------------------------------------------
    % U-P
    X(:,it+1)=LHS\RHS;
    % U'-P'
    V(:,it+1)=(gama/(beta*dt))*(X(:,it+1)-X(:,it))-(((gama/beta)-1)*V(:,it))-
    ((gama/(beta*2))-1)*dt*A(:,it);
    % U"-P"
    A(:,it+1)=(1/(beta*dt^2))*(X(:,it+1)-X(:,it)-dt*V(:,it))-((1/(beta*2))-
    1)*A(:,it);
end
%------------------------------------------------------------
%   Separate u and p
%------------------------------------------------------------
m=size;
uu=X(1:sdof,1:nt);
uur=real(uu);
pp=X(sdof+1:m,1:nt);
ppr=real(pp);
for i=1:nt
    PPR(:,i)=ppr(:,nt-i+1);
end
ppabs=abs(pp);
uuabs=abs(uu);
%------------------------------------------------------------
%   Plot the results
%------------------------------------------------------------
Feur=uur(1:sdof,1:nt/np:nt);
Fepr=ppr(1:sdof,1:nt/np:nt)/q0;
Feu=uuabs(1:sdof,nt);
Fep=ppabs(1:sdof,nt)/q0;
%% Time History
% % Displacement
figure
subplot(211)
plot(t,uur(Np,:),'--')
title('displacement time history')
ylabel('u')
xlabel('t')
subplot(212)
% % Pore pressure
plot(t,ppr(Np,:))
title('pore pressure time history')
ylabel('p')
xlabel('t')
%------------------------------------------------------------
% Variation with Depth
%------------------------------------------------------------
```

```
figure
% Displacement
subplot(211)
plot(Feur,-zz)
title('displacement distribution')
ylabel('z/H')
xlabel('u/q0')
% Pore pressure
subplot(212)
plot(Fepr,-zz)
title('pore pressure distribution')
ylabel('z/H')
xlabel('p/q0')
axis ([-1.4 1.4 -1 0])
%-------------------------------------------------------------
% Absolute Values
%-------------------------------------------------------------
figure
subplot(211)
plot(Fep,-zz,'--ko')
title('absolute value of pore pressure distribution')
ylabel('z/h')
xlabel('p/q0')
subplot(212)
plot(Feu,-zz,'--ko')
title('absolute value of displacement distribution')
ylabel('z/h')
xlabel('u(m)')
```

Main program "FEM_2D_SEEP.m"

```
%-----------------------------------------------------------------------------
% FEM_2D_SEEP
% A Finite Element Program for a 2-D steady state seepage problem
% Variable descriptions
% k = element matrix
% f = element vector
% kk = system matrix
% ff = system vector
% gcoord = coordinate values of each node
% nodes = nodal connectivity of each element
% index = a vector containing system dofs associated with each element
% bcdof = a vector containing dofs associated with boundary conditions
% bcval = a vector containing boundary condition values associated with
%           the dofs in 'bcdof'
%-----------------------------------------------------------------------------
clear all
clf
%----------------------------------
%  input data for control parameters
%----------------------------------
hy=40;              % depth
lx=200;             % length
k_x=0.1;            % horizontal permeability
k_y=0.1;            % vertical permeability
```

```
nx=41;                 % number of divisions in x-dir
ny=10;                 % number of divisions in y-dir
ncb_left=8;            % number of nodes on the left with specified head
ncb_right=8;           % number of nodes on the right with specified head
head_left=20;          % left head value
head_right=0;          % right head value
nnel=4;                % number of nodes per element
ndof=1;                % number of dofs per node
N_pot=10;              % number of equi-potential lines to display
N_flow=5;              % number of flow lines to display
node_start=4;          % starting node number for flow line display on the left
dx=lx/(nx-1);          % length increment
dy=hy/(ny-1);          % depth increment
mx=nx-1;
my=ny-1;
nel=mx*my;             % number of elements
nnode=nx*ny;           % total number of nodes in system
sdof=nnode*ndof;       % total system dofs


%----------------------------------------------
%  input data for nodal coordinate values
%  gcoord(i,j) where i->node no. and j->x or y
%----------------------------------------------
x=-lx/2:dx:lx/2;
y=0:-dy:-hy;
[X,Y]=meshgrid(x,y);
Z=zeros(ny,nx);
subplot(2,1,1)
mesh(X,Y,Z)
view(0,90)

for ix=1:nx
    for iy=1:ny
        in=iy+(ix-1)*ny;
        gcoord(in,1)=X(iy,ix);
        gcoord(in,2)=Y(iy,ix);
    end
end
%----------------------------------------------------------
%  input data for nodal connectivity for each element
%  nodes(i,j) where i-> element no. and j-> connected nodes
%----------------------------------------------------------
nodes=zeros(nel,4);
for ix=1:mx
    for iy=1:my
        iel=iy+(ix-1)*my;
        nodes(iel,1)=iel+(ix);
        nodes(iel,2)=nodes(iel,1)+ny;
        nodes(iel,3)=nodes(iel,2)-1;
        nodes(iel,4)=nodes(iel,3)-ny;
    end
end
%---------------------------------------------------------------------------
%  input data for coefficients of the ODE [may be different for each element]
%---------------------------------------------------------------------------
perm=zeros(nel,2);
kx=k_x*ones(1,ny-1);
ky=k_y*ones(1,ny-1);
```

```
for ix=1:mx
    ii=1+(ix-1)*my;
    jj=my+(ix-1)*my;
    perm(ii:jj,1)=kx';
    perm(ii:jj,2)=ky';
end
for iel=1:nel
    ce(iel)=0.;
end
%-----------------------------------
%  input data for boundary conditions
%-----------------------------------
ncb=ncb_left+ncb_right;
for i=1:ncb_left
    bcval(i)=head_left;
    bcdof(i)=(i-1)*ny+1;
end
for i=ncb_left+1:ncb
    bcval(i)=head_right;
    bcdof(i)=((i+nx-ncb)-1)*ny+1;
end
nfb=0;
%----------------------------------------
%  initialization of matrices and vectors
%----------------------------------------
ff=zeros(sdof,1);          % initialization of system force vector
kk=zeros(sdof,sdof);       % initialization of system matrix
index=zeros(nnel*ndof,1);  % initialization of index vector
%----------------------------------------------------------------
%  computation of element matrices and vectors and their assembly
%----------------------------------------------------------------
for iel=1:nel               % loop for the total number of elements
    kyel=perm(iel,2);
    kxel=perm(iel,1);
    el=ce(iel);
    for i=1:nnel
        nd(i)=nodes(iel,i);     % extract connected node for (iel)th
                                  element
        x(i)=gcoord(nd(i),1);   % extract x value of the node
        y(i)=gcoord(nd(i),2);   % extract y value of the node
    end
    xleng = (x(2)-x(1));        % length of the element in x-axis
    yleng = (y(4)-y(1));        % length of the element in y-axis
    index=feeldof(nd,nnel,ndof);  % extract system dofs associated with
                                    element
    k=felp2dr4(kxel,kyel,el,xleng,yleng);    % compute element matrix
    [kk]=feasmbl2(kk,k,index);
end
%----------------------------
%  apply boundary conditions
%----------------------------
[kk,ff]=feapplybc(kk,ff,bcdof,bcval);
%----------------------------
%  solve the matrix equation
%----------------------------
fsol=kk\ff;
```

```
head=zeros(nx,ny);
for ix=1:nx
    ii=1+(ix-1)*ny;
    jj=ny+(ix-1)*ny;
    head(ix,1:ny)=(fsol(ii:jj))';
end
head=head';
%----------------------------
%  plot the variable
%----------------------------
subplot(212)
contour(X,Y,head,N_pot)
hold on
[Dx,Dy]=gradient(head);
quiver(X,Y,-Dx,Dy)
hold on
XL1=-(100-(node_start-1)*dx);
XL2=XL1+(N_flow-1)*dx;
SX=XL1:dx:XL2;
lsy=length(SX);
SY=zeros(1,lsy);
streamline(X,Y,-Dx,Dy,SX,SY)
hold off
%----------------------------------------------------------------
```

Subprogram "feeldof.m"

```
function [index]=feeldof(nd,nnel,ndof)
%----------------------------------------------------------------
% Purpose:
% Compute system dofs associated with each element
% Synopsis:
% [index]=feeldof(nd,nnel,ndof)
% Variable Description:
% index - system dof vector associated with element "iel"
% nd - connected node for (iel)th element
% nnel - number of nodes per element
% ndof - number of dofs per node
%----------------------------------------------------------------
edof = nnel*ndof;
k=0;
for i=1:nnel
    start = (nd(i)-1)*ndof;
    for j=1:ndof
        k=k+1;
        index(k)=start+j;
    end
end
end
%----------------------------------------------------------------
```

Subprogram "felp2dr4.m"

```
function [k]=felp2dr4(kxel,kyel,el,xleng,yleng)
%----------------------------------------------------------------
%  Purpose:
% element matrix for two-dimensional Laplace's equation
% using four-node bilinear rectangular element
% Synopsis:
```

```
% [k]=felp2dr4(kxel,kyel,el,xleng,yleng)
% Variable Description:
% k - element stiffness matrix (size of 4x4)
% xleng - element size in the x-axis
% yleng - element size in the y-axis
%----------------------------------------------------------------
k1=(kxel*xleng/(6*yleng))*[ 2 -2 -1 1;-2  2  1 -1;-1  1  2 -2;1 -1 -2  2];
k2=(kyel*yleng/(6*xleng))*[ 2  1 -1 -2;1  2 -2 -1;-1 -2  2  1;-2 -1  1  2];
k3=(el*xleng*yleng/36)* [ 4  2  1  2;2  4  2  1;1  2  4  2;2  1  2  4;];
k=k1+k2+k3;
end
%----------------------------------------------------------------
```

Subprogram "feasmbl2.m"

```
function [kk]=feasmbl2(kk,k,index)
%----------------------------------------------------------------
% Purpose:
% Assembly of element matrices into the system matrix
%
% Synopsis:
% [kk]=feasmbl2(kk,k,index)
%
% Variable Description:
% kk - system matrix
% k  - element matrix
% index - d.o.f. vector associated with an element
%----------------------------------------------------------------
edof = length(index);
for i=1:edof
    ii=index(i);
    for j=1:edof
        jj=index(j);
        kk(ii,jj)=kk(ii,jj)+k(i,j);
    end
end

end
%----------------------------------------------------------------
```

Main program "FEM_2D_STRESS.m"

```
% *********************************************************************
% FEM_2D_STRESS
% A plane stress analysis of a solid using linear triangular elements
%
% *********************************************************************
% Variable descriptions
% k = element matrix
% f = element vector
% kk = system matrix
% ff = system vector
% disp = system nodal displacement vector
% eldisp = element nodal displacement vector
% stress = matrix containing stresses
% strain = matrix containing strains
% gcoord = coordinate values of each node
% nodes = nodal connectivity of each element
% index = a vector containing system dofs associated with each element
```

```
% bcdof = a vector containing dofs associated with boundary conditions
% bcval = a vector containing boundary condition values associated with
%            the dofs in 'bcdof'
%-----------------------------------------------------------------------------
clear all
%-----------------------------------
%  input data for control parameters
%-----------------------------------
nel=2;                      % number of elements
nnel=3;                     % number of nodes per element
ndof=2;                     % number of dofs per node
nnode=4;                    % total number of nodes in system
emodule=3000000.;           % elastic modulus
poisson=0.25;               % Poisson's ratio
f_input_1=800;              % force applied at node 9 in x-axis
f_input_2=800;              % force applied at node 10 in x-axis
sdof=nnode*ndof;            % total system dofs
edof=nnel*ndof;             % degrees of freedom per element
%--------------------------------------------
% input data for nodal coordinate values
% gcoord(i,j) where i->node no. and j->x or y
%--------------------------------------------
gcoord=[0.0  0.0; 120.0  0.0; 120.0  160.0; 0.0  160.0];
%------------------------------------------------------------
% input data for nodal connectivity for each element
% nodes(i,j) where i-> element no. and j-> connected nodes
%------------------------------------------------------------
nodes=[1 2 3; 1 3 4];
%-------------------------------------
%  input data for boundary conditions
%-------------------------------------
bcdof=[1 2 7 8];            % these dofs are constrained
bcval=[0 0 0 0];            % whose described values are 0
%-----------------------------------------
%  initialization of matrices and vectors
%-----------------------------------------
ff=zeros(sdof,1);           % system force vector
kk=zeros(sdof,sdof);        % system stiffness matrix
disp=zeros(sdof,1);         % system displacement vector
eldisp=zeros(edof,1);       % element displacement vector
stress=zeros(nel,3);        % matrix containing stress components
strain=zeros(nel,3);        % matrix containing strain components
index=zeros(edof,1);        % index vector
kinmtx2=zeros(3,edof);      % kinematic matrix
matmtx=zeros(3,3);          % constitutive matrix
%---------------------------
%  force vector
%---------------------------
ff(3)=f_input_1;            % force applied at node 9 in x-axis
ff(5)=f_input_2;            % force applied at node 10 in x-axis
%----------------------------------------------------------------
%  computation of element matrices and vectors and their assembly
%----------------------------------------------------------------
matmtx=fematiso(1,emodule,poisson);     % compute constitutive matrix

for iel=1:nel               % loop for the total number of elements
    nd(1)=nodes(iel,1);     % 1st connected node for (iel)-th element
    nd(2)=nodes(iel,2);     % 2nd connected node for (iel)-th element
```

```
        nd(3)=nodes(iel,3);      % 3rd connected node for (iel)-th element
        x1=gcoord(nd(1),1); y1=gcoord(nd(1),2);% coord values of 1st node
        x2=gcoord(nd(2),1); y2=gcoord(nd(2),2);% coord values of 2nd node
        x3=gcoord(nd(3),1); y3=gcoord(nd(3),2);% coord values of 3rd node
        index=feeldof(nd,nnel,ndof);% extract system dofs associated with element
%------------------------------------------------------------
%  find the derivatives of shape functions
%------------------------------------------------------------
        area=0.5*(x1*y2+x2*y3+x3*y1-x1*y3-x2*y1-x3*y2);  % area of triangule
        area2=area*2;
        dhdx=(1/area2)*[(y2-y3) (y3-y1) (y1-y2)];  % derivatives w.r.t. x-axis
        dhdy=(1/area2)*[(x3-x2) (x1-x3) (x2-x1)];  % derivatives w.r.t. y-axis
        kinmtx2=fekine2d(nnel,dhdx,dhdy);          % compute kinematic matrix
        k=kinmtx2'*matmtx*kinmtx2*area;            % element stiffnes matrix
        kk=feasmbl2(kk,k,index);                   % assemble element matrices
end
%-----------------------------------
%   apply boundary conditions
%-----------------------------------
[kk,ff]=feapplybc(kk,ff,bcdof,bcval)
%-----------------------------------
% solve the matrix equation
%-----------------------------------
disp=kk\ff;
%-------------------------------------------
% element stress computation
%-------------------------------------------

for ielp=1:nel          % loop for the total number of elements
    nd(1)=nodes(ielp,1); % 1st connected node for (iel)-th element
    nd(2)=nodes(ielp,2); % 2nd connected node for (iel)-th element
    nd(3)=nodes(ielp,3); % 3rd connected node for (iel)-th element
    x1=gcoord(nd(1),1); y1=gcoord(nd(1),2);% coord values of 1st node
    x2=gcoord(nd(2),1); y2=gcoord(nd(2),2);% coord values of 2nd node
    x3=gcoord(nd(3),1); y3=gcoord(nd(3),2);% coord values of 3rd node
    index=feeldof(nd,nnel,ndof);% extract system dofs associated with element
%-------------------------------------------------------------
%   extract element displacement vector
%-------------------------------------------------------------
    for i=1:edof
        eldisp(i)=disp(index(i));
    end
    area=0.5*(x1*y2+x2*y3+x3*y1-x1*y3-x2*y1-x3*y2);  % area of triangule
    area2=area*2;
    dhdx=(1/area2)*[(y2-y3) (y3-y1) (y1-y2)];  % derivatives w.r.t. x-axis
    dhdy=(1/area2)*[(x3-x2) (x1-x3) (x2-x1)];  % derivatives w.r.t. y-axis
    kinmtx2=fekine2d(nnel,dhdx,dhdy);          % compute kinematic matrix
    estrain=kinmtx2*eldisp;                    % compute strains
    estress=matmtx*estrain;                    % compute stresses
    for i=1:3
        strain(ielp,i)=estrain(i);             % store for each element
        stress(ielp,i)=estress(i);             % store for each element
    end
end
%-------------------------------------------
% print fem solutions
%-------------------------------------------
num=1:sdof;
```

```
displace=[num' disp]                    % print nodal displacements

for i=1:nel
    stresses=[i stress(i,:)]            % print stresses
end
%-------------------------------------------------------------------
```

Subprogram "fematiso.m"

```
function [matmtrx]=fematiso(iopt,elastic,poisson)

%-------------------------------------------------------------------
% Purpose:
% Get the constitutive equation for isotropic material
%
% Synopsis:
% [matmtrx]=fematiso(iopt,elastic,poisson)
%
% Variable Description:
% elastic - elastic modulus
% poisson - Poisson's ratio
% iopt=1 - plane stress analysis
% iopt=2 - plane strain analysis
% iopt=3 - axisymmetric analysis
% iopt=4 - three dimensional analysis
%-------------------------------------------------------------------
if iopt==1          % plane stress
    matmtrx= elastic/(1-poisson*poisson)* ...
    [1  poisson 0; ...
    poisson  1  0; ...
    0  0  (1-poisson)/2];

elseif iopt==2      % plane strain
    matmtrx= elastic/((1+poisson)*(1-2*poisson))* ...
    [(1-poisson)  poisson 0;
    poisson (1-poisson)  0;
    0  0  (1-2*poisson)/2];

elseif iopt==3      % axisymmetry
    matmtrx= elastic/((1+poisson)*(1-2*poisson))* ...
    [(1-poisson)  poisson  poisson  0;
    poisson  (1-poisson)   poisson  0;
    poisson  poisson  (1-poisson)   0;
    0   0    0   (1-2*poisson)/2];

else      % three-dimension
    matmtrx= elastic/((1+poisson)*(1-2*poisson))* ...
    [(1-poisson)  poisson  poisson  0   0    0;
    poisson  (1-poisson)   poisson  0   0    0;
    poisson  poisson  (1-poisson)   0   0    0;
    0   0    0    (1-2*poisson)/2   0    0;
    0   0    0    0    (1-2*poisson)/2   0;
    0   0    0    0    0   (1-2*poisson)/2];

end

end

%-------------------------------------------------------------------
```

Subprogram "fekine2d.m"

```
function [kinmtx2]=fekine2d(nnel,dhdx,dhdy)
%-------------------------------------------------------------------------
% Purpose:
% Determine the kinematic equation between strains and displacements
% for 2-D solids
% Synopsis:
% [kinmtx2]=fekine2d(nnel,dhdx,dhdy)
% Variable Description:
% nnel - number of nodes per element
% dhdx - derivatives of shape functions with respect to x
% dhdy - derivatives of shape functions with respect to y
%-------------------------------------------------------------------------
for i=1:nnel
    i1=(i-1)*2+1;
    i2=i1+1;
    kinmtx2(1,i1)=dhdx(i);
    kinmtx2(2,i2)=dhdy(i);
    kinmtx2(3,i1)=dhdy(i);
    kinmtx2(3,i2)=dhdx(i);
end

end
```

Appendix

A.1 Fourier Series and Fourier Transform

In this section, the fundamentals of Fourier series and Fourier transform representations of general mathematical functions are recalled.

A.1.1 Sinusoidal Function

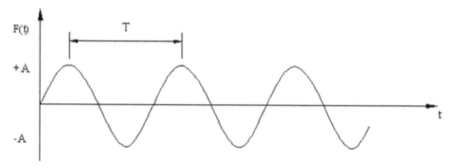

$$F(t) = A \sin (\omega t)$$

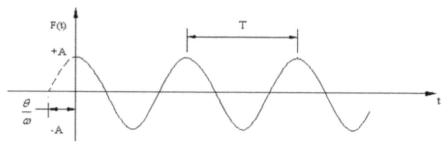

$$F(t) = A \sin (\omega t + \theta)$$

Figure A.1-1 Sinusoidal functions

- Period: $T = \dfrac{2\pi}{\omega}$

- Frequency: $f = \dfrac{1}{T} = \dfrac{\omega}{2\pi}$

- Angular frequency: $\omega = \dfrac{2\pi}{T} = 2\pi f$

- Phase angle; θ

A.1.2 Frequency Spectrum

$$F(t) = A \sin(\omega t + \theta)$$

$$= A \cos\left(\omega t + \theta - \frac{\pi}{2}\right) \tag{A.1.1}$$

$$F(t) = A \cos(\omega t + \theta') \tag{A.1.2}$$

A.1.3 Exponential Form

Euler's relation:

$$e^{i\alpha} = \cos\alpha + i\sin\alpha, \quad \cos\alpha = \mathrm{Re}(e^{i\alpha}) \tag{A.1.3}$$

and

$$\sin\alpha = \mathrm{Im}(e^{i\alpha})$$

$$F(t) = A\,\mathrm{Im}(e^{i(\omega t + \theta)}) \tag{A.1.4a}$$

$$F(t) = A\,\mathrm{Re}(e^{i(\omega t + \theta)}) \tag{A.1.4b}$$

$$\cos\alpha = \frac{e^{i\alpha} + e^{-i\alpha}}{2} \tag{A.1.5a}$$

$$\sin\alpha = \frac{e^{i\alpha} - e^{-i\alpha}}{2} \tag{A.1.5b}$$

A.1.4 General Periodic Function: Fourier Series Representation

A general periodic function shown below is defined as:

$$F(t) = F(t + nT)$$

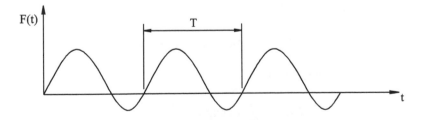

Figure A.1-2 An arbitrary periodic function

The above function (in which the same form is repeated over a period, T) can be represented by a Fourier series as defined below,

$$F(t) = a_0 + \sum_{n=1}^{\infty}(a_n \cos n\omega_1 + b_n \sin n\omega_1) \qquad\qquad \text{(A.1.6)}$$

$$a_0 = \int_0^T F(t)dt \qquad\qquad \text{(A.1.7a)}$$

$$a_n = \frac{2}{T}\int_0^T F(t)\cos n\omega_1 dt \qquad\qquad \text{(A.1.7b)}$$

$$b_n = \frac{2}{T}\int_0^T F(t)\sin n\omega_1 dt \qquad\qquad \text{(A.1.7c)}$$

$$f_1 = \frac{1}{T}, \quad \omega_1 = \frac{2\pi}{T} \qquad\qquad \text{(A.1.7d)}$$

$$f_n = \frac{n}{T}, \quad \omega_n = \frac{2\pi n}{T} \qquad\qquad \text{(A.1.7e)}$$

Alternatively, the above series may also be written in the following form:

$$F(t) = A_0 + \sum_{i=1}^{\infty} A_n \cos(n\omega_1 - \theta_n) \qquad\qquad \text{(A.1.8a)}$$

$$A_n = \sqrt{a_n^2 + b_n^2} \qquad\qquad \text{(A.1.8b)}$$

$$\theta_n = \tan^{-1}\left(\frac{b_n}{a_n}\right) \qquad\qquad \text{(A.1.8c)}$$

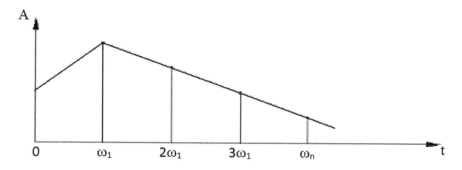

Figure A.1-3 Fourier magnitude

Exponential form of Fourier series:

$$F(t) = \sum_{n=-\infty}^{\infty} c_n \exp(in\omega_1 t) \qquad\qquad \text{(A.1.9)}$$

$$c_n = \frac{1}{T}\int_0^T F(t)\exp(in\omega_1 t)dt \qquad\qquad \text{(A.1.10)}$$

Example:

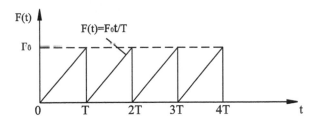

Figure A.1-4 Variation of $F(t)$

$$a_0 = \frac{1}{T}\int_0^T \frac{F}{T}t\,dt = \frac{F_0}{2} \tag{A.1.11}$$

$$a_n = \frac{2}{T}\int_0^T \frac{F_0}{T}t\cos n\omega_1 t\,dt$$

$$= \frac{2F_0}{T^2}\int_0^T t\cos n\omega_1 t\,dt$$

$$= \frac{2F_0}{T^2}\left[t\frac{\sin n\omega_1 t}{n\omega_1}\Big|_0^T - \int_0^T \frac{\sin n\omega_1 t}{n\omega_1}dt\right]$$

$$= \frac{2F_0}{T}\left(T\frac{\sin 2n\pi}{n\omega_1} + \frac{\cos 2n\pi}{n\omega_1} - \frac{1}{(n\omega_1)^2}\right)$$

$$= 0 \tag{A.1.12}$$

$$b_n = \frac{2}{T}\int_0^T \frac{F_0}{T}t\sin 2n\omega_1 dt = -\frac{F_0}{n\pi} \tag{A.1.13}$$

$$F(t) = \frac{F_0}{2} - \sum_{n=1}^{\infty}\left(\frac{F_0}{n\pi}\sin 2n\pi f_1 t\right) \tag{A.1.14}$$

A.1.5 Non-Periodic Function: Fourier Transform Representation

Fourier Series $\longrightarrow$ Fourier Transform

Only a periodic function can be represented by a Fourier series. For a non-periodic function, we use a Fourier transform. The basic idea is to treat a non-periodic function as a periodic function with an infinite period. Here, we briefly recall the development of a Fourier transform representation.

Recall the Fourier series representation of a periodic function:

$$F(t) = \sum_{n=-\infty}^{\infty} c_n \exp(in\omega_1 t) \tag{A.1.15a}$$

$$c_n = \frac{1}{T} \int_0^T F(t) \exp(-in\omega_1 t)\, dt \tag{A.1.15b}$$

where,

$$\omega_1 = \frac{2\pi}{T}, \quad \frac{1}{T} = \frac{\omega_1}{2\pi} = \frac{\Delta\omega}{2\pi} \tag{A.1.15c}$$

Note: $T\dfrac{\omega_1}{2\pi} = 1$ and $n\omega_1 = n\Delta\omega = \omega_n$. If T is finite, ω_n are discrete frequencies. Since, $\dfrac{T}{2\pi}\Delta\omega = 1$, we can rewrite Eq. (A.1.15) as:

$$F(t) = \sum_{n=-\infty}^{\infty} \frac{T}{2n}\Delta\omega c_n \exp(in\omega_1 t) \tag{A.1.16}$$

where,

$$\boxed{\begin{aligned} F(t) &= \frac{\Delta\omega}{2n} \sum_{n=-\infty}^{\infty} Tc_n \exp(in\omega_1 t) \\ c(\omega_n) &= \int_0^T F(t)\exp(-in\omega_1 t)\, dt \end{aligned}} \tag{A.1.17}$$

As $T \to \infty$, $\Delta\omega \to d\omega$, $\omega_n \to \omega$

$$\boxed{\begin{aligned} F(t) &= \frac{1}{2\pi} \int_{-\infty}^{\infty} c(\omega_n)\exp(in\omega t)\, d\omega \\ c(\omega_n) &= \int_{-\infty}^{\infty} F(t)\exp(-i\omega t)\, dt \end{aligned}} \tag{A.1.18}$$

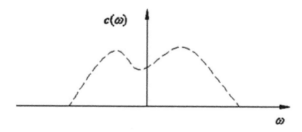

Figure A.1-5 Variation of $c(\omega)$

A.1.6 Discrete Fourier Transform (DFT), Fast Fourier Transform (FFT) and Inverse Fast Fourier Transform (IFFT)

Evaluation of Fourier transform is possible only for very simple functions. For a more general function, a discrete form of Fourier transform (DFT) is used. For the numerical evaluation of the discrete Fourier transform, a fast algorithm [Discrete Fast Fourier Transform (DFFT)] is used. For the numerical evaluation of the inverse of FT, Inverse Fast Fourier Transform (IFFT) is used. Here a brief sketch of the process is presented. These operations can be easily carried out by simple MATLAB® commands.

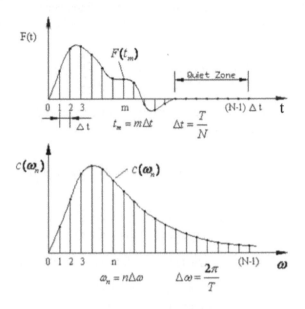

Figure A.1-6 Variation of given functions

$$F(t) = \frac{\Delta\omega}{\pi} \sum_{n=-\infty}^{\infty} c(\omega_n) \exp(i\omega_n t)$$

$$c(\omega_n) = \int_0^T F(t) \exp(-i\omega_n t) dt$$

Let
$$t_m = m\Delta t, \ \omega_n = n\Delta\omega = n\frac{2\pi}{T}$$

Discrete Fourier transform (DFT) pair

$$F(t_m) = \frac{\Delta\omega}{2n} \sum_{n=0}^{N-1} c(\omega_n) \exp\left(\frac{i2\pi nm}{N}\right)$$

$$c(\omega_n) = \Delta t \sum_{m=0}^{N-1} F(t_m) \exp\left(-\frac{i2\pi nm}{N}\right)$$

A.2 Laplace Transform

In mathematics, the Laplace transform is one of the best known and most widely used integral transforms. It is commonly used for solving differential and integral equations where an algebraic equation can be easily obtained from an ordinary differential equation. In physics, it finds some areas of application in the analyses of linear time-invariant systems. In such cases,

Laplace transform is often interpreted as the transformation from the time-domain, in which inputs and outputs are functions of time, to the frequency-domain, where the same input and output functions are of complex angular frequency, or radians per unit time.

The Laplace transform of a function $f(t)$, defined for all real numbers of $t \geq 0$, is the function $\hat{f}(s)$ defined as,

$$L\{f(t)\} = \hat{f}(s) = \lim_{\alpha \to 0^-} \longrightarrow \int_{\alpha}^{\infty} f(t)e^{-st}dt \qquad (A2.1)$$

Noting that the limit assures the inclusion of the entire Dirac delta function $\delta(t)$ at 0 if there is an impulse at 0. The Laplace variable 's' is in complex form as $s = a+ib$. Inverse Laplace transform is given by,

$$L^{-1}\{\hat{f}(s)\} = f(t) = \frac{1}{2\pi i} \int_{\gamma - i\infty}^{\gamma + i\infty} f(s)e^{st}ds \qquad (A2.2)$$

where γ is a real number depending on the region of convergence of $\hat{f}(s)$. Linearity of operator is a property of Laplace transform used frequently in solving linear dynamic systems. That is,

$$L\{\alpha f(t) + \beta g(t)\} = \alpha L\{f(t)\} + \beta L\{g(t)\} \qquad (A2.3)$$

where differentials and integrals can be treated as products and divisions, respectively. The Laplace transform of the derivative of a function is written as,

$$L\{\dot{f}(t)\} = s\hat{f}(t) - \lim_{\alpha \to 0^-} \longrightarrow f(\alpha) \qquad (A2.4)$$

where the second term on the RHS denotes an initial condition that is assumed to be equal to zero in the models without an initial condition.

A.3 MATLAB® Commands: FFT, IFFT, FFTSHIFT

A.3.1 FFT Discrete Fourier Transform

FFT(X) is the discrete Fourier transform (DFT) of vector X. For matrices, the FFT operation is applied to each column. For N-D arrays, the FFT operation operates on the first non-singleton dimension.

FFT(X,N) is the N-point FFT, padded with zeros if X has less than N points and truncated if it has more.

FFT(X, [], DIM) or FFT(X,N,DIM) applies the FFT operation across the dimension DIM. For length N input vector x, the DFT is a length N vector X, with elements:

X(k) = sum x(n)*exp(-j*2*pi*(k – 1)*(n – 1)/N), $1 \leq k \leq N$.
The inverse DFT (computed by IFFT) is given by:
x(n) = (1/N) sum X(k)*exp(j*2*pi*(k – 1)*(n – 1)/N), $1 \leq n \leq N$.

A.3.2 IFFT Inverse Discrete Fourier Transform

IFFT(X) is the inverse discrete Fourier transform of X. IFFT(X,N) is the N-point inverse transform.

IFFT(X,[],DIM) or IFFT(X,N,DIM) is the inverse discrete Fourier transform of X across the dimension DIM.

A.3.3 FFTSHIFT Shift Zero-Frequency Component to Center of Spectrum

For vectors, FFTSHIFT(X) swaps the left and right halves of X. For matrices, FFTSHIFT(X) swaps the first and third quadrants and the second and fourth quadrants. For N-D arrays, FFTSHIFT(X) swaps "half-spaces" of X along each dimension.

FFTSHIFT(X,DIM) applies the FFTSHIFT operation along the dimension DIM.

FFTSHIFT is useful for visualizing the Fourier transform with the zero-frequency component in the middle of the spectrum.

A.4 Solution Flow Chart for the Analysis of a Viscoelastic Material

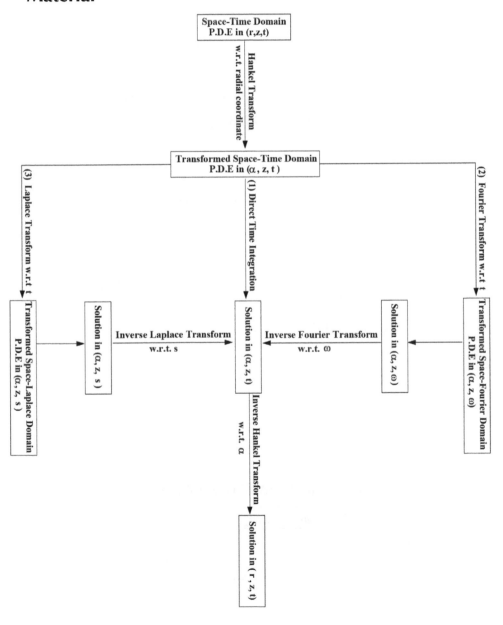

A.5 Analytical Solution of Wave-Induced Porous Soil Layer Response

The matrix system given in (7.8.9) for the quasi-static formulation can be written as,

$$
\begin{bmatrix}
A & B & A_{13} & B \\
B & C & B & A_{24} \\
A_{31} & D & A & B \\
D & A_{42} & B & C
\end{bmatrix}
\begin{bmatrix}
U_x \\
U_z \\
\bar{W}_x \\
\bar{W}_z
\end{bmatrix}
= [M]\{X\} = 0
\tag{A.5.1}
$$

where

$$
A = -m^2\kappa,\ B = im\kappa DD,\ A_{13} = \frac{i}{\Pi_{1x}} - m^2\kappa,\ C = DD^2\kappa,\ A_{24} = \frac{i}{\Pi_{1z}} + DD^2\kappa,\ A_{31} = -m^2 + \kappa_2 DD^2,
$$

$$
D = im\,(\kappa + \kappa_1 + \kappa_2)\,DD,\ A_{42} = -m^2\kappa_2 + DD^2.
$$

Then the solution can be obtained by writing the characteristic equation from det[M] = 0 as,

$$
\alpha_1 DD^6 + \alpha_2 DD^4 + \alpha_3 DD^2 + \alpha_4 = 0
\tag{A.5.2}
$$

Here, the coefficients are:

$$
\alpha_1 = A_{51}A_{84}
\tag{A.5.3a}
$$

$$
\alpha_2 = A_{52}\{A_{52}[A_{76}{}^2 + \kappa - A_{77}] + 2A_{51}A_{52} + 2A_{13}\,(A_{51} - \kappa A_{76}) + A_{83} + \kappa A_{84}\}
$$
$$
- A_{51}A_{52}{}^2 + A_{84}A_{13}
\tag{A.5.3b}
$$

$$
\alpha_3 = A_{52}{}^2\left\{A_{51}A_{53}\frac{\kappa_2}{\kappa^2} - A_{75}\,(1 - 2A_{61})\right\} + A_{75}A_{52}\left\{\frac{A_{83}}{\kappa} - A_{52}\right\}
\tag{A.5.3c}
$$

$$
\alpha_4 = m^2 A_{84}\{m^2 A_{13} + A_{52}{}^2\}
\tag{A.5.3d}
$$

where

$$
A_{51} = \kappa(1-\kappa),\ A_{52} = m^2\,\kappa,\ A_{53} = A_{13} + A_{52},\ A_{61} = (\kappa + \kappa_1 + \kappa_2),\ A_{62} = \left(\frac{\beta\Pi_2}{n} + \frac{i}{\Pi_1}\right),\ A_{64} = \beta\Pi_2,
$$

$$
A_{75} = \left(A_{62} - \frac{A_{64}}{n}\right),\ A_{76} = (A_{61} - \kappa),\ A_{77} = (1 + \kappa_2^2),\ A_{83} = A_{13}(A_{61}^2 - A_{77}),\ A_{84} = \kappa_2 A_{75}
$$

The roots of the characteristic equation (A.5.2) are evaluated as:

$$
\eta_{1,2} = \pm\sqrt{\frac{\varphi^2 + 2\alpha_2(2\alpha_2 - \varphi) - 12\alpha_1\alpha_3}{6\alpha_1\varphi}}
\tag{A.5.4a}
$$

$$
\eta_{3,4} = \pm\sqrt{\frac{(-1 + i\sqrt{3})\varphi^2 - 4\alpha_2\varphi + 4(1 + i\sqrt{3})(3\alpha_1\alpha_3 - \alpha_2{}^2)}{12\alpha_1\varphi}}
\tag{A.5.4b}
$$

$$\eta_{5,6} = \pm\sqrt{\frac{(-1 - i\sqrt{3})\varphi^2 - 4\alpha_2\varphi + 4(1 - i\sqrt{3})(3\alpha_1\alpha_3 - \alpha_2{}^2)}{12\alpha_1\varphi}} \qquad (A.5.4c)$$

where,

$$\begin{aligned} \varphi &= (\xi + \alpha_1 12\sqrt{3}\zeta)^{1/3} \\ \xi &= 36\alpha_1\alpha_3\alpha_2 - 108\alpha_1{}^2\alpha_4 - 8\alpha_2{}^3 \\ \zeta &= 4\alpha_1\alpha_3{}^3 - \alpha_2{}^2\alpha_3{}^2 - 18\alpha_1\alpha_2\alpha_3\alpha_4 + 27\alpha_1{}^2\alpha_4{}^2 + 4\alpha_4\alpha_2{}^3 \end{aligned} \qquad (A.5.5)$$

Then the desired displacements are found as:

$$U_x = a_1 e^{\eta_1 \bar{z}} + a_2 e^{\eta_2 \bar{z}} + a_3 e^{\eta_3 \bar{z}} + a_4 e^{\eta_4 \bar{z}} + a_5 e^{\eta_5 \bar{z}} + a_6 e^{\eta_6 \bar{z}} \qquad (A.5.6a)$$

$$U_z = a_1 b_1 e^{\eta_1 \bar{z}} + a_2 b_2 e^{\eta_2 \bar{z}} + a_3 b_3 e^{\eta_3 \bar{z}} + a_4 b_4 e^{\eta_4 \bar{z}} + a_5 b_5 e^{\eta_5 \bar{z}} + a_6 b_6 e^{\eta_6 \bar{z}} \qquad (A.5.6b)$$

$$\bar{W}_x = a_1 c_1 e^{\eta_1 \bar{z}} + a_2 c_2 e^{\eta_2 \bar{z}} + a_3 c_3 e^{\eta_3 \bar{z}} + a_4 c_4 e^{\eta_4 \bar{z}} + a_5 c_5 e^{\eta_5 \bar{z}} + a_6 c_6 e^{\eta_6 \bar{z}} \qquad (A.5.6c)$$

$$\bar{W}_z = a_1 d_1 e^{\eta_1 \bar{z}} + a_2 d_2 e^{\eta_2 \bar{z}} + a_3 d_3 e^{\eta_3 \bar{z}} + a_4 d_4 e^{\eta_4 \bar{z}} + a_5 d_5 e^{\eta_5 \bar{z}} + a_6 d_6 e^{\eta_6 \bar{z}} \qquad (A.5.6d)$$

where b_i, c_i, d_i (i = 1, 2, 3, 4, 5, 6) form the elements of V_i which are the eigenvectors corresponding to eigenvalues (roots) of the characteristic equation and written as

$$V_i = \begin{Bmatrix} 1 \\ b_i \\ c_i \\ d_i \end{Bmatrix} \qquad (A.5.7)$$

Using (A.1), one can obtain the b_i, c_i, d_i with $[M]\{V_i\} = 0$ as follows: For FD, we have,

$$\begin{bmatrix} A & B & A_{13} & B \\ B & C & B & A_{24} \\ A_{31} & D & A & B \\ D & A_{42} & B & C \end{bmatrix} \begin{bmatrix} 1 \\ b_i \\ c_i \\ d_i \end{bmatrix} = 0 \qquad (A.5.8)$$

$$\begin{bmatrix} C & B & A_{24} \\ D & A & B \\ A_{42} & B & C \end{bmatrix}_{DD=\lambda_i} \begin{bmatrix} b_i \\ c_i \\ d_i \end{bmatrix} = -\begin{bmatrix} B \\ A_{31} \\ D \end{bmatrix} \qquad (A.5.9)$$

$$\begin{bmatrix} b_i \\ c_i \\ d_i \end{bmatrix} = \begin{bmatrix} C & B & A_{24} \\ D & A & B \\ A_{42} & B & C \end{bmatrix}^{-1} \begin{bmatrix} -B \\ -A_{31} \\ -D \end{bmatrix} \qquad (A.5.10)$$

where in the 3×3 matrix, *DD*s are replaced by λ_i, implicitly. From here,

$$
\begin{bmatrix} b_i \\ c_i \\ d_i \end{bmatrix} = \frac{1}{\begin{vmatrix} C & B & A_{24} \\ D & A & B \\ A_{42} & B & C \end{vmatrix}} \begin{bmatrix} \begin{vmatrix} A & B \\ B & C \end{vmatrix} & \begin{vmatrix} A_{24} & B \\ C & B \end{vmatrix} & \begin{vmatrix} B & A_{24} \\ A & B \end{vmatrix} \\ \begin{vmatrix} B & D \\ C & A_{42} \end{vmatrix} & \begin{vmatrix} C & A_{24} \\ A_{42} & C \end{vmatrix} & \begin{vmatrix} A_{24} & C \\ B & D \end{vmatrix} \\ \begin{vmatrix} D & A \\ A_{42} & B \end{vmatrix} & \begin{vmatrix} B & C \\ B & A_{42} \end{vmatrix} & \begin{vmatrix} C & B \\ D & A \end{vmatrix} \end{bmatrix} \begin{bmatrix} -B \\ -A_{31} \\ -D \end{bmatrix} \tag{A.5.11}
$$

and b_i, c_i, d_i are obtained as;

$$
\begin{bmatrix} b_i \\ c_i \\ d_i \end{bmatrix} = \frac{\begin{bmatrix} (AC - B^2) & (A_{24}B - BC) & (B^2 - A_{24}A) \\ (BA_{42} - CD) & (C^2 - A_{24}A_{42}) & (DA_{24} - CB) \\ (DB - A_{42}A) & (BA_{42} - CB) & (CA - BD) \end{bmatrix} \begin{bmatrix} -B \\ -A_{31} \\ -D \end{bmatrix}}{\left[C(AC - B^2) - B(DC - BA_{42}) + A_{24}(DB - A_{42}A) \right]} \tag{A.5.12}
$$

$$
\begin{bmatrix} b_i \\ c_i \\ d_i \end{bmatrix} = \frac{\begin{bmatrix} -B(AC - B^2) - A_{31}(A_{24}B - BC) - D(B^2 - A_{24}A) \\ -B(BA_{42} - CD) - A_{31}(C^2 - A_{24}A_{42}) - D(DA_{24} - CB) \\ -B(DB - A_{42}A) - A_{31}(BA_{42} - CB) - D(CA - BD) \end{bmatrix}}{\left[C(AC - B^2) - B(DC - BA_{42}) + A_{24}(DB - A_{42}A) \right]} \tag{A.5.13}
$$

$$
\begin{bmatrix} b_i \\ c_i \\ d_i \end{bmatrix} = \frac{\begin{bmatrix} B^3 - BAC - A_{31}A_{24}B + A_{31}BC - DB^2 + DA_{24}A \\ -B(BA_{42} - CD) - A_{31}(C^2 - A_{24}A_{42}) - D(DA_{24} - CB) \\ BD(D - B) + BA_{42}(A - A_{31}) + C(BA_{31} - AD) \end{bmatrix}}{\left[C(AC - B^2) - B(DC - BA_{42}) + A_{24}(DB - A_{42}A) \right]} \tag{A.5.14}
$$

If we write these separately, we have

$$
b_i = \left(\frac{B^3 + B\left[C(A_{31} - A) - DB - A_{31}A_{24} \right] + DA_{24}A}{B^2(A_{42} - C) - BD(C - A_{24}) + A(C^2 - A_{24}A_{42})} \right)_{DD = \lambda_i} \tag{A.5.15a}
$$

$$
c_i = \left(\frac{-B^2 A_{42} + 2BCD - D^2 A_{24} - A_{31}(C^2 - A_{24}A_{42})}{B^2(A_{42} - C) - BD(C - A_{24}) + A(C^2 - A_{24}A_{42})} \right)_{DD = \lambda_i} \tag{A.5.15b}
$$

$$
d_i = \left(\frac{BD(D - B) + BA_{42}(A - A_{31}) + C(A_{31}B - AD)}{B^2(A_{42} - C) - BD(C - A_{24}) + A(C^2 - A_{24}A_{42})} \right)_{DD = \lambda_i} \tag{A.5.15c}
$$

Using the boundary conditions stated before,

At $z = -H$,

$$\bar{U}_x = a_1 e^{-\lambda_1} + a_2 e^{\lambda_1} + a_3 e^{-\lambda_2} + a_4 e^{\lambda_2} + u_5 e^{-\lambda_3} + a_6 e^{\lambda_3} = 0 \tag{A.5.16a}$$

$$\bar{U}_z = a_1 b_1 e^{-\lambda_1} + a_2 b_2 e^{\lambda_1} + a_3 b_3 e^{-\lambda_2} + a_4 b_4 e^{\lambda_2} + a_5 b_5 e^{-\lambda_3} + a_6 b_6 e^{\lambda_3} = 0 \tag{A.5.16b}$$

$$\bar{W}_z = a_1 d_1 e^{-\lambda_1} + a_2 d_2 e^{\lambda_1} + a_3 d_3 e^{-\lambda_2} + a_4 d_4 e^{\lambda_2} + a_5 d_5 e^{-\lambda_3} + a_6 d_6 e^{\lambda_3} = 0 \tag{A.5.16c}$$

At $z = 0$,

$$\sigma'_{zz} = D\frac{\partial U_z}{\partial z} + \lambda \frac{\partial U_x}{\partial x} = \frac{D}{H}(a_1 b_1 \lambda_1 - a_2 b_2 \lambda_1 + a_3 b_3 \lambda_2 - a_4 b_4 \lambda_2 + a_5 b_5 \lambda_3 - a_6 b_6 \lambda_3)$$
$$+ ik\lambda(a_1 + a_2 + a_3 + a_4 + a_5 + a_6) = 0 \tag{A.5.17}$$

where $\;D = \lambda + 2G = \dfrac{E(1-v)}{(1+v)(1-2v)} \cdot \cdot$

$$\tau_{xz} = G\left(\frac{\partial U_x}{\partial z} + \frac{\partial U_z}{\partial x}\right) = \frac{G}{H}(a_1 \lambda_1 - a_2 \lambda_1 + a_3 \lambda_2 - a_4 \lambda_2 + a_5 \lambda_3 - a_6 \lambda_3)$$
$$+ ikG(a_1 b_1 + a_2 b_2 + a_3 b_3 + a_4 b_4 + a_5 b_5 + a_6 b_6) = 0 \tag{A.5.18}$$

$$P = \left(ik(1+c_1) + \frac{\lambda_1}{H}(b_1 + d_1)\right)a_1 + \left(ik(1+c_2) - \frac{\lambda_1}{H}(b_2 + d_2)\right)a_2$$
$$+ \left(ik(1+c_3) + \frac{\lambda_2}{H}(b_3 + d_3)\right)a_3 + \left(ik(1+c_4) - \frac{\lambda_2}{H}(b_4 + d_4)\right)a_4 \tag{A.5.19}$$
$$+ \left(ik(1+c_5) + \frac{\lambda_3}{H}(b_5 + d_5)\right)a_5 + \left(ik(1+c_6) - \frac{\lambda_3}{H}(b_6 + d_6)\right)a_6 = -\frac{\gamma_w hn}{2K_f \cosh(kd)}$$

These equations can be written in matrix form to find the coefficients a_i as,

$$\begin{bmatrix} \xi_{11}e^{-\lambda_1} & \xi_{12}e^{\lambda_1} & \xi_{13}e^{-\lambda_2} & \xi_{14}e^{\lambda_2} & \xi_{15}e^{-\lambda_3} & \xi_{16}e^{\lambda_3} \\ \xi_{21}e^{-\lambda_1} & \xi_{22}e^{\lambda_1} & \xi_{23}e^{-\lambda_2} & \xi_{24}e^{\lambda_2} & \xi_{25}e^{-\lambda_3} & \xi_{26}e^{\lambda_3} \\ \xi_{31}e^{-\lambda_1} & \xi_{32}e^{\lambda_1} & \xi_{33}e^{-\lambda_2} & \xi_{34}e^{\lambda_2} & \xi_{35}e^{-\lambda_3} & \xi_{36}e^{\lambda_3} \\ \xi_{41} & \xi_{42} & \xi_{43} & \xi_{44} & \xi_{45} & \xi_{46} \\ \xi_{51} & \xi_{52} & \xi_{53} & \xi_{54} & \xi_{55} & \xi_{56} \\ \xi_{61} & \xi_{62} & \xi_{63} & \xi_{64} & \xi_{65} & \xi_{66} \end{bmatrix} \begin{Bmatrix} a_1 \\ a_2 \\ a_3 \\ a_4 \\ a_5 \\ a_6 \end{Bmatrix} = \begin{Bmatrix} 0 \\ 0 \\ 0 \\ 0 \\ 0 \\ -q_0/Q \end{Bmatrix} \tag{A.5.20}$$

$$\xi_{1j} = 1, \xi_{2j} = b_j, \xi_{3j} = d_j, \xi_{4j} = \frac{K}{h}b_j \eta_j + ik\lambda, \xi_{5j} = G\left(\frac{\eta_j}{h} + ikb_j\right),$$

$$\xi_{6j} = ik(1 + c_j) + \frac{\eta_j}{h}(b_j + d_j).$$

A.6 Semi-Analytical Solution of Wave-Induced Multi-Layer Porous Soil Response

The continuity conditions (CC) are written in open forms as,

CC-1:

$$(a_{11}e^{-\eta_1 l_j} + a_{21}e^{-\eta_2 l_j} + a_{31}e^{-\eta_3 l_j} + a_{41}e^{-\eta_4 l_j} + a_{51}e^{-\eta_5 l_j} + a_{61}e^{-\eta_6 l_j})_j$$

$$= (a_{12}e^{-\eta_1 l_j} + a_{22}e^{-\eta_2 l_j} + a_{32}e^{-\eta_3 l_j} + a_{42}e^{-\eta_4 l_j} + a_{52}e^{-\eta_5 l_j} + a_{62}e^{-\eta_6 l_j})_{j+1} \qquad (A.6.1)$$

CC-2:

$$(a_1 b_1 e^{-\eta_1 l_j} + a_2 b_2 e^{-\eta_2 l_j} + a_3 b_3 e^{-\eta_3 l_j} + a_4 b_4 e^{-\eta_4 l_j} + a_5 b_5 e^{-\eta_5 l_j} + a_6 b_6 e^{-\eta_6 l_j})_j$$

$$= (a_1 b_1 e^{-\eta_1 l_j} + a_2 b_2 e^{-\eta_2 l_j} + a_3 b_3 e^{-\eta_3 l_j} + a_4 b_4 e^{-\eta_4 l_j} + a_5 b_5 e^{-\eta_5 l_j} + a_6 b_6 e^{-\eta_6 l_j})_{j+1} \qquad (A.6.2)$$

C.C-3:

$$\left(\frac{k_z}{\rho_w g}\frac{K_f}{n}\right)_j \left\{ \begin{array}{l} e^{-\eta_1^j l_j}\frac{\eta_1}{h}\left[i\psi(1+c_1)+(b_1+d_1)\frac{\eta_1}{h}\right]_j a_1^j + e^{-\eta_2^j l_j}\frac{\eta_2}{h}\left[i\psi(1+c_2)+(b_2+d_2)\frac{\eta_2}{h}\right]_j a_2^j + \\[2mm] e^{-\eta_3^j l_j}\frac{\eta_3}{h}\left[i\psi(1+c_3)+(b_3+d_3)\frac{\eta_3}{h}\right]_j a_3^j + e^{-\eta_4^j l_j}\frac{\eta_4}{h}\left[i\psi(1+c_4)+(b_4+d_4)\frac{\eta_4}{h}\right]_j a_4^j + \\[2mm] e^{-\eta_5^j l_j}\frac{\eta_5}{h}\left[i\psi(1+c_5)+(b_5+d_5)\frac{\eta_5}{h}\right]_j a_5^j + e^{-\eta_6^j l_j}\frac{\eta_6}{h}\left[i\psi(1+c_6)+(b_6+d_6)\frac{\eta_6}{h}\right]_j a_6^j \end{array} \right\}$$

$$= \left(\frac{k_z}{\rho_w g}\frac{K_f}{n}\right)_{j+1} \left\{ \begin{array}{l} e^{-\eta_1^{j+1} l_j}\frac{\eta_1}{h}\left[i\psi(1+c_1)+(b_1+d_1)\frac{\eta_1}{h}\right]_{j+1} a_1^{j+1} + e^{-\eta_2^{j+1} l_j}\frac{\eta_2}{h}\left[\begin{array}{l} i\psi(1+c_2) \\ +(b_2+d_2)\frac{\eta_2}{h}\end{array}\right]_{j+1} a_2^{j+1} + \\[3mm] e^{-\eta_3^{j+1} l_j}\frac{\eta_3}{h}\left[i\psi(1+c_3)+(b_3+d_3)\frac{\eta_3}{h}\right]_{j+1} a_3^{j+1} + e^{-\eta_4^{j+1} l_j}\frac{\eta_4}{h}\left[\begin{array}{l} i\psi(1+c_4) \\ +(b_4+d_4)\frac{\eta_4}{h}\end{array}\right]_{j+1} a_4^{j+1} + \\[3mm] e^{-\eta_5^{j+1} l_j}\frac{\eta_5}{h}\left[i\psi(1+c_5)+(b_5+d_5)\frac{\eta_5}{h}\right]_{j+1} a_5^{j+1} + e^{-\eta_6^{j+1} l_j}\frac{\eta_6}{h}\left[\begin{array}{l} i\psi(1+c_6) \\ +(b_6+d_6)\frac{\eta_6}{h}\end{array}\right]_{j+1} a_6^{j+1} \end{array} \right\}$$

$$(A.6.3)$$

CC-4:

$$
\left(i\psi\lambda_j + D_j b_1^j \frac{\eta_1^j}{h}\right) a_1^j e^{-\eta_1^j l_j} + \left(i\psi\lambda_j + D_j b_2^j \frac{\eta_2^j}{h}\right) a_2^j e^{-\eta_2^j l_j} + \left(i\psi\lambda_j + D_j b_3^j \frac{\eta_3^j}{h}\right) a_3^j e^{-\eta_3^j l_j}
$$

$$
+ \left(i\psi\lambda_j + D_j b_4^j \frac{\eta_1^j}{h}\right) a_4^j e^{-\eta_4^j l_j} + \left(i\psi\lambda_j + D_j b_5^j \frac{\eta_5^j}{h}\right) a_5^j e^{-\eta_5^j l_j} + \left(i\psi\lambda_j + D_j b_6^j \frac{\eta_6^j}{h}\right) a_6^j e^{-\eta_6^j l_j}
$$

$$
- \left(i\psi\lambda_{j+1} + D_{j+1} b_1^{j+1} \frac{\eta_1^{j+1}}{h}\right) a_1^{j+1} e^{-\eta_1^{j+1} l_j} - \left(i\psi\lambda_{j+1} + D_{j+1} b_2^{j+1} \frac{\eta_2^{j+1}}{h}\right) a_2^{j+1} e^{-\eta_2^{j+1} l_j}
$$

$$
- \left(i\psi\lambda_{j+1} + D_{j+1} b_3^{j+1} \frac{\eta_3^{j+1}}{h}\right) a_3^{j+1} e^{-\eta_3^{j+1} l_j} - \left(i\psi\lambda_{j+1} + D_{j+1} b_4^{j+1} \frac{\eta_1^{j+1}}{h}\right) a_4^{j+1} e^{-\eta_4^{j+1} l_j}
$$

$$
- \left(i\psi\lambda_{j+1} + D_{j+1} b_5^{j+1} \frac{\eta_5^{j+1}}{h}\right) a_5^{j+1} e^{-\eta_5^{j+1} l_j} - \left(i\psi\lambda_{j+1} + D_{j+1} b_6^{j+1} \frac{\eta_6^{j+1}}{h}\right) a_6^{j+1} e^{-\eta_6^{j+1} l_j} = 0
$$

$$(A.6.4)$$

CC-5:

$$
G_j \begin{bmatrix} \left(i\psi b_1^j + \frac{\eta_1^j}{h}\right) a_1^j e^{-\eta_1^j l_j} + \left(i\psi b_2^j + \frac{\eta_2^j}{h}\right) a_2^j e^{-\eta_2^j l_j} + \left(i\psi b_3^j + \frac{\eta_3^j}{h}\right) a_3^j e^{-\eta_3^j l_j} + \\ \left(i\psi b_4^j + \frac{\eta_1^j}{h}\right) a_4^j e^{-\eta_4^j l_j} + \left(i\psi b_5^j + \frac{\eta_5^j}{h}\right) a_5^j e^{-\eta_5^j l_j} + \left(i\psi b_6^j + \frac{\eta_6^j}{h}\right) a_6^j e^{-\eta_6^j l_j} \end{bmatrix} -
$$

$$
G_{j+1} \begin{bmatrix} \left(i\psi b_1^{j+1} + \frac{\eta_1^{j+1}}{h}\right) a_1^{j+1} e^{-\eta_1^{j+1} l_j} + \left(i\psi b_2^{j+1} + \frac{\eta_2^{j+1}}{h}\right) a_2^{j+1} e^{-\eta_2^{j+1} l_j} + \left(i\psi b_3^{j+1} + \frac{\eta_3^{j+1}}{h}\right) a_3^{j+1} e^{-\eta_3^{j+1} l_j} + \\ \left(i\psi b_4^{j+1} + \frac{\eta_1^{j+1}}{h}\right) a_4^{j+1} e^{-\eta_4^{j+1} l_j} + \left(i\psi b_5^{j+1} + \frac{\eta_5^{j+1}}{h}\right) a_5^{j+1} e^{-\eta_5^{j+1} l_j} + \left(i\psi b_6^{j+1} + \frac{\eta_6^{j+1}}{h}\right) a_6^{j+1} e^{-\eta_6^{j+1} l_j} \end{bmatrix} = 0
$$

$$(A.6.5)$$

CC-6:

$$
\left(\frac{K_f}{n}\right)_j \begin{Bmatrix} e^{-\eta_1^j l_j}\left[i\psi(1+c_1) + (b_1+d_1)\frac{\eta_1}{h}\right]_j a_1^j + e^{-\eta_2^j l_j}\left[i\psi(1+c_2) + (b_2+d_2)\frac{\eta_2}{h}\right]_j a_2^j + \\ e^{-\eta_3^j l_j}\left[i\psi(1+c_3) + (b_3+d_3)\frac{\eta_3}{h}\right]_j a_3^j + e^{-\eta_4^j l_j}\left[i\psi(1+c_4) + (b_4+d_4)\frac{\eta_4}{h}\right]_j a_4^j + \\ e^{-\eta_5^j l_j}\left[i\psi(1+c_5) + (b_5+d_5)\frac{\eta_5}{h}\right]_j a_5^j + e^{-\eta_6^j l_j}\left[i\psi(1+c_6) + (b_6+d_6)\frac{\eta_6}{h}\right]_j a_6^j \end{Bmatrix} -
$$

$$\left(\frac{K_f}{n}\right)_{j+1} \left\{ \begin{array}{l} e^{-\eta_1^{j+1}l_j}\left[i\psi\left(1+c_1\right)+\left(b_1+d_1\right)\dfrac{\eta_1}{h}\right]_{j+1} a_1^{j+1} + e^{-\eta_2^{j+1}l_j}\left[i\psi\left(1+c_2\right)+\left(b_2+d_2\right)\dfrac{\eta_2}{h}\right]_{j+1} a_2^{j+1} + \\[3mm] e^{-\eta_3^{j+1}l_j}\left[i\psi\left(1+c_3\right)+\left(b_3+d_3\right)\dfrac{\eta_3}{h}\right]_{j+1} a_3^{j+1} + e^{-\eta_4^{j+1}l_j}\left[i\psi\left(1+c_4\right)+\left(b_4+d_4\right)\dfrac{\eta_4}{h}\right]_{j+1} a_4^{j+1} + \\[3mm] e^{-\eta_5^{j+1}l_j}\left[i\psi\left(1+c_5\right)+\left(b_5+d_5\right)\dfrac{\eta_5}{h}\right]_{j+1} a_5^{j+1} + e^{-\eta_6^{j+1}l_j}\left[i\psi\left(1+c_6\right)+\left(b_6+d_6\right)\dfrac{\eta_6}{h}\right]_{j+1} a_6^{j+1} \end{array} \right\} = 0$$

(A.6.6)

These equations including the ones coming from the boundary conditions can be written in a matrix form to find the coefficients $a_i^{\,j}$ and $a_i^{\,j+1}$ for the layers in a general form as,

$$
\begin{bmatrix}
\begin{bmatrix} \xi_{11} & \cdots & \xi_{16} \\ \vdots & \ddots & \vdots \\ \xi_{31} & \cdots & \xi_{36} \end{bmatrix} & & \left[\underset{\sim}{0}_{3x(M-6)}\right] & & \\
\begin{bmatrix} \xi_{41} & \cdots & \xi_{412} \\ \vdots & \ddots & \vdots \\ \xi_{111} & \cdots & \xi_{1112} \end{bmatrix} & & \left[\underset{\sim}{0}_{6x(M-12)}\right] & & \\
\left[\underset{\sim}{0}_{6x6}\right] & \begin{bmatrix} \xi_{127} & \cdots & \xi_{1218} \\ \vdots & \ddots & \vdots \\ \xi_{197} & \cdots & \xi_{1918} \end{bmatrix} & \left[\underset{\sim}{0}_{6x(M-18)}\right] & & \\
& & \ddots & & \\
\left[\underset{\sim}{0}_{3x(M-6)}\right] & & & \begin{bmatrix} \xi_{(M-2)(M-5)} & \cdots & \xi_{(M-2)M} \\ \vdots & \ddots & \vdots \\ \xi_{M(M-5)} & \cdots & \xi_{MM} \end{bmatrix}
\end{bmatrix}_{MxM}
\begin{Bmatrix} a_1^1 \\ \vdots \\ a_6^1 \\ a_1^2 \\ \vdots \\ a_6^2 \\ \vdots \\ \vdots \\ a_1^N \\ \vdots \\ a_6^N \end{Bmatrix}_{Mx1}
$$

$$
= \begin{Bmatrix} 0 \\ 0 \\ F \\ 0 \\ \vdots \\ \vdots \\ 0 \end{Bmatrix}_{Mx1}
$$

(A.6.7)

where $M = 6N$ and for $s = 1\ldots 6$,

$$\xi_{1s} = i\psi\lambda^1 + D^1 b_s^1 \frac{\eta_s^1}{h},$$

$$\xi_{2s} = G_1\left(i\psi b_s^1 + \frac{\eta_s^1}{h}\right),$$

$$\xi_{3s} = i\psi(1 + c_s^1) + \frac{\eta_s^1}{h}(b_s^1 + d_s^1).$$

For $j = 1,...N-1;\ s = 1 + 6(j-1)....6j,$

$$\xi_{[4+6(j-1)],s} = e^{-\eta_{s-6(j-1)}^j l_j},$$

$$\xi_{[5+6(j-1)],s} = b_{s-6(j-1)}^j e^{-\eta_{s-6(j-1)}^j l_j},$$

$$\xi_{[6+6(j-1)],s} = c_{s-6(j-1)}^j e^{-\eta_{s-6(j-1)}^j l_j}$$

$$\xi_{[7+6(j-1)],s} = d_{s-6(j-1)}^j e^{-\eta_{s-6(j-1)}^j l_j}$$

$$\xi_{[8+6(j-1)],s} = e^{-\eta_{s-6(j-1)}^j l_j}\left(i\psi D^j + \lambda^j b_{s-6(j-1)}^j \frac{\eta_{s-6(j-1)}^j}{h}\right),$$

$$\xi_{[9+6(j-1)],s} = e^{-\eta_{s-6(j-1)}^j l_j}\left(i\psi\lambda^j + D^j b_{s-6(j-1)}^j \frac{\eta_{s-6(j-1)}^j}{h}\right),$$

$$\xi_{[10+6(j-1)],s} = G_j e^{-\eta_{s-6(j-1)}^j l_j}\left(i\psi b_{s-6(j-1)}^j + \frac{\eta_{s-6(j-1)}^j}{h}\right)$$

$$\xi_{[11+6(j-1)],s} = \left(\frac{K_f}{n}\right)^j e^{-\eta_{s-6(j-1)}^j l_j}\left[i\psi(1 + c_{s-6(j-1)}^j) + (b_{s-6(j-1)}^j + d_{s-6(j-1)}^j)\frac{\eta_{s-6(j-1)}^j}{h}\right]$$

For $j = 2,...N;\ s = 1 + 6(j-1)...6j,$

$$\xi_{[4+6(j-2)],s} = -e^{-\eta_{s-6(j-1)}^j l_{j-1}},$$

$$\xi_{[5+6(j-2)],s} = -b_{s-6(j-1)}^j e^{-\eta_{s-6(j-1)}^j l_{j-1}},$$

$$\xi_{[6+6(j-2)],s} = -c_{s-6(j-1)}^j e^{-\eta_{s-6(j-1)}^j l_{j-1}},$$

$$\xi_{[7+6(j-2)],\,s} = -d^j_{s-6(j-1)} e^{-\eta^j_{s-6(j-1)}l_{j-1}},$$

$$\xi_{[8+6(j-2)],\,s} = -e^{-\eta^j_{s-6(j-1)}l_{j-1}} \left(i\psi\, D^j + \lambda^j b^j_{s-6(j-1)} \frac{\eta^j_{s-6(j-1)}}{h} \right),$$

$$\xi_{[9+6(j-2)],\,s} = -e^{-\eta^j_{s-6(j-1)}l_{j-1}} \left(i\psi\lambda^j + D^j b^j_{s-6(j-1)} \frac{\eta^j_{s-6(j-1)}}{h} \right),$$

$$\xi_{[10+6(j-2)],\,s} = -G_j e^{-\eta^j_{s-6(j-1)}l_{j-1}} \left(i\psi\, b^j_{s-6(j-1)} + \frac{\eta^j_{s-6(j-1)}}{h} \right)$$

$$\xi_{[11+6(j-2)],\,s} = -\left(\frac{K_f}{n}\right)^j e^{-\eta^j_{s-6(j-1)}l_{j-1}} \left[i\psi\,(1 + c^j_{s-6(j-1)}) + (b^j_{s-6(j-1)} + d^j_{s-6(j-1)}) \frac{\eta^j_{s-6(j-1)}}{h} \right]$$

and finally for $s = M-5,\dots M$;

$$\xi_{[M-2],s} = e^{-\eta_{(s-(M-6))}},$$

$$\xi_{[M-1],s} = b_{(s-(M-6))} e^{-\eta_{(s-(M-6))}},$$

$$\xi_{M,s} = d_{(s-(M-6))} e^{-\eta_{(s-(M-6))}}$$

References

Abbo, A.J. and Sloan, S.W. (1995), "A smooth hyperbolic approximation to the Mohr-Coulomb yield criterion", *Computers & Structures*, 54(3): 427-441.

Anandarajah, A. (2010), Computational Methods in Elasticity and Plasticity: Solids and Porous Media, Springer Science and Business Media, LLC.

Biot, M.A. (1941), "General theory of three dimensional consolidation", *Journal of Applied Physics*, 12: 155-164.

Biot, M.A. (1955), "Theory of elasticity and consolidation for a porous anisotropic solid", *Journal of Applied Physics*, 26: 182-185.

Biot, M.A. (1962), "Mechanics of deformation and acoustic propagation in porous media", *Journal of Applied Physics*, 33: 1482-1498.

Booker, J.R. (1989), Lecture Notes, University of Sydney, Sydney NSW.

Booker, J.R. and Small, J.C. (1985), "Finite layer analysis of layered viscoelastic materials under three-dimensional loading conditions", *Int. J. Numer. Methods in Engg.*, 21: 1709-1727.

Chen, W.F. and Mizuno, E. (1990), Nonlinear Analysis in Soil Mechanics: Theory and Implementation, Elsevier Science Publishers, Amsterdam.

Christensen, R.M. (1982), Theory of Viscoelasticity: An Introduction, Academic Press, New York.

Cook, R.D., Malkus, D.S., Plesha, M.E. and Witt, R.J. (2001), Concepts and Applications of Finite Element Analysis. Fourth Edition, John Wiley and Sons, NY.

De Souza Neto, E.A., Peric, D. and Owen, D.R.J. (2008), Computational Methods for Plasticity: Theory and Applications, John Wiley and Sons, UK.

Desai, C.S. and Christian, J. (1977). Numerical Methods in Geotechnical Engineering, McGraw Hill, New York.

Drucker, D.C. (1949), "Some implications of work hardening and ideal plasticity", *Quart. Appl. Math.*, 7(2): 411-418.

Drucker, D.C. (1966), "Concept of path independence and material stability for soils", Kravtchenko and Sirieys (Eds), Rheol. Mech. Soils Proc. IUTAM Sym., Grenoble, Springer, Berlin, pp. 23-43.

Flügge, W. (1975), Viscoelasticity, Springer-Verlag: Berlin, New York.

Gutierrez-Lemini, D. (2014), Engineering Viscoelasticity, Springer Science, New York.

Hill, R. (1950), The Mathematical Theory of Plasticity, Clarendon Press, Oxford.

Honig, G. and Hirdes, U. (1984), "A method for the numerical inversion of Laplace transforms", *J. Comp. and Applied Math.*, 10: 113-132.

Hughes, T.J.R. (1987), The Finite Element Method: Linear Static and Dynamic Finite Element Analysis, Prentice-Hall, Inc., Englewood Cliffs, New Jersey.

Hwon, Y.W. and Bang, H. (1996), The Finite Element Method using MATLAB, CRC Press.

Lai, J.S. (1976), "Predicting permanent deformation of asphalt concrete from creep tests", *Transportation Research Board*, Transportation Research Record 616, pp. 41-43.

Lambe, T.W. and Whitman, V.W. (1969), Soil Mechanics, John Wiley and Sons, NY.

Langhaar, H. (1962), Energy Methods in Applied Mechanics, John Wiley and Sons, NY.

Madsen O.S. (1978), "Wave-induced pore pressure and effective stresses in porous bed", *Geotechnique*, 28(4): 377-393.

McWhorter, D. and Sunada, D.K. (1981), Groundwater Hydrology and Hydraulics. Water Resources Publications, Littleton, CO. 492 pp.

Mroz, Z. and Zienkiewicz, O.C. (1984), "'Uniform formulation of constitutive laws for clays", *Mechanics of Engineering Materials*. C.S. Desai and R.H. Gallagher (Eds), pp. 415-449, Wiley.

Ogata, A. and Banks, R.B. (1961), "A solution of the differential equation of longitudinal dispersion in porous media." Profl. Paper No. 411-A, *U.S. Geological Survey*, Washington, D.C.

Okusa, S. (1985), "Wave induced stresses in unsaturated submarine sediments", *Geotechnique*, 35(4): 517-532.

Osterberg, J.O. (1957), "Influence values for vertical stresses in semi-infinite mass due to embankment loading", *Proc. Fourth Int. Conf. Soil Mechanics and Foundation Engineering*, 1: 393-396.

Pastor, M., Zienkiewicz, O.C. and Leung, K.H. (1985), "Simple model for transient soil loading in earthquake analysis: II. Non-associative models for sands", *Int. J. Numer. Anal. Mthd Geomech.*, 9: 477-498.

Pastor, M., Zienkiewicz, O.C. and Chan, A.C. (1990), "Theme/feature paper: Generalized plasticity and the modeling of soil behavior", *Int. J. Numer. Anal. Mthds Geomech.*, 14: 151-190.

Peric, D. (1993), "On a class of constitutive equations in viscoplasticity: Formulation and computational issues", *Int. J. Numer. Mthds. Eng.*, 36(8): 1365-1393.

Prevost, J.H. (1980), "Mechanics of continuous porous media", *Int. Journal of Engineering Science*, 18: 787-800.

Prevost, J.H. (1982), "Nonlinear transient phenomena in saturated porous media", *Computer Methods in Applied Mechanics and Engineering*, 20: 3-8.

Puzrin, A.M. (2012), Constitutive Modeling in Geomechanics, Springer-Verlag, Berlin.

Rahman, M.S., El-Zahaby, K. and Booker, J.R. (1994), "A semi-analytical method for the wave induced seabed response", *Int. J. Numer. Analy. Methods in Geomech*, 18: 213-236.

Reddy, J.N. (1993), An Introduction to the Finite Element Method, McGraw Hill.

Schapery, R.A. (1999), "Nonlinear viscoelastic and viscoplastic constitutive equations with growing damage", *International Journal of Fracture*, 97: 33-36.

Schiffman, R.L., Chen, A.T. and Jordan, J.C. (1969), "An analysis of consolidation theories", *Proceedings ASCE*, 95(SM1).

Simo, J.C. and Hughes, T.J.R. (1998), Computational Inelasticity, Springer-Verlag, New York, Inc.

Smith, G.D. (1965), Numerical Solution of Partial Differential Equations. Oxford University Press.

Terzaghi, K. (1923), "Die Berechnung der Durchlassigkeitsziffer des Tones aus dem Verlauf der hydrodynamische Spannungserscheinungen", *Sitzber. Akad. Wiss. Wien*, Abt. 132:2a, 125-138.

Terzaghi, K. (1923), Erdbaumechanik auf Bodenphysikalischer Grundlage, Deuticke, Wien.

Toth, J. (1962) "A theory of groundwater motion in small drainage basins in central Alberta, Canada", *Journal of Geophysical Research*, 67(11): 4375-4388.

Truesdell, C. (1957), "Sulle basi della thermomeccanica", *Rendiconti Lincei*, Series 8, 22(33-38): 158-166.

Truesdell, C. (1962), "Mechanical basis of diffusion", *J. of Chemical Physics*, 37: 23-36.

Ulker, M.B.C. (2009), "Dynamic response of saturated porous media: Wave induced response and instability of seabed", *Doctoral Dissertation*, North Carolina State University, Department of Civil, Construction and Environmental Engineering, Raleigh, NC.

Ulker, M.B.C. and Rahman, M.S. (2009), "Response of saturated porous media: Different formulations and their applicability", *Int. J. Analy. Numer. Mthds in Geomec.*, 33(5): 633-664.

Ulker, M.B.C. (2012), "Pore pressure, stress distributions and instantaneous liquefaction of two-layer soil under waves", *ASCE J. Waterway, Port, Coastal, Ocean, Engg.*, 138(6): 435-450.

Ulker, M.B.C. (2015), "Semianalytical solution to the wave-induced dynamic response of saturated layered porous media", *ASCE J. Waterway, Port, Coastal, Ocean, Engg.*, DOI. 10.1061/(ASCE) WW.1943-5460.0000272, 141(1), 06014001.

Uzan, J. (1996), "Asphalt concrete characterization for pavement performance prediction", *Asphalt Paving Technology*, 65: 573-607.

Wood, D.M. (2004), Geotechnical Modeling, Taylor and Francis, New York.

Xu, Q. (2004), "Modeling and Computing for Layered Pavements under Vehicle Loading", *Doctoral Dissertation*, North Carolina State University, Department of Civil, Construction and Environmental Engineering, Raleigh, NC, U.S.A.

Xu, Q. and Rahman, M.S. (2008), "Finite element analyses of layered viscoelastic system under vertical circular loading", *Int. J. Numer. Analy. Methods in Geomech.*, 32: 897-913.

Ziegler, H. (1959), "A modification of Prager's hardening rule", *Appl. Math.*, 17: 55-65.

Zienkiewicz, O.C., Chang, C.T. and Bettess, P. (1980), "Drained, undrained, consolidating and dynamic behavior assumptions in soils", *Geotechnique*, 30(4): 385-395.

Zienkiewicz, O.C. and Mroz, Z. (1984), "Generalized plasticity formulation and applications to geomechanics", *Mechanics of Engineering Materials*. C.S. Desai and R.H. Gallagher (Eds). pp. 655-679, Wiley.

Zienkiewicz, O.C., Chan, A.H.C., Pastor, M., Paul, D.K. and Shiomi, T. (1990), "Static and dynamic behaviour of soils: A rational approach to quantitative solutions. I. Fully saturated problems", Proceedings of the Royal Society of London. Series A, *Mathematical and Physical Sciences*, 429(1877): 285-309.

Zienkiewicz, O.C., Chang, C.H., Pastor, M., Schrefler, B.A. and Shiomi, T. (1999), Computational Geomechanics with Special Reference to Earthquake Engineering, Wiley, New York.

Zienkiewicz, O.C. and Taylor, R.L. (2000), The Finite Element Method, 5th Ed. Butterworth-Heinemann, Oxford.

Index

Milton Keynes UK
Ingram Content Group UK Ltd.
UKHW050457071024
449327UK00015B/410